Evolution of Antimicrobial Peptides

Piyush Baindara • Santi Mohan Mandal
Editors

Evolution of Antimicrobial Peptides

From Self-Defense to Therapeutic Applications

Springer

Editors
Piyush Baindara
Animal Sciences Research Center, Division of Animal Sciences
University of Missouri
Columbia, MO, USA

Santi Mohan Mandal
Department of Bioscience and Biotechnology
Indian Institute of Technology Kharagpur
Kharagpur, WB, India

ISBN 978-3-031-67514-0 ISBN 978-3-031-67515-7 (eBook)
https://doi.org/10.1007/978-3-031-67515-7

This Springer imprint is published by the registered company Springer Nature Switzerland AG
The registered company address is: Gewerbestrasse 11, 6330 Cham, Switzerland

This book is dedicated to my parents, Mother Nature, and my family who make me able to evolve every single day to dedicate more and achieve perfection in my work toward scientific contributions.

Preface

Antibiotics have been used for more than five decades to treat infectious diseases and for improvement of health. However, indiscriminate use of antibiotics caused many health problems and led to a rapid emergence of drug-resistant pathogens, which become a major threat to global health care. The increasing appearance of multidrug-resistant pathogens and infective agents alarmed an urgent and constant demand for the investigation of new alternatives to conventional antibiotics.

In the recent past, antimicrobial peptides (AMPs) received great attention as a promising alternative to conventional antibiotics due to their unique properties over antibiotics. AMPs are naturally occurring integral components of the host defense system and are functional throughout all orders of life from prokaryotes to eukaryotes. AMPs are potent molecules with a broad spectrum of biological activities against a group of infectious or pathogenic agents, such as bacteria, fungi, viruses, and parasites. AMPs also have demonstrated therapeutic potential against non-infectious diseases like chronic inflammation and cancer. In addition, AMPs were found to have potential activities against slow-growing or even non-growing bacteria, which are usually associated with biofilms in pathological conditions such as cystic fibrosis, dental plaque formation involved in caries and periodontitis, and colonization of indwelling medical devices. In recent years, studies on AMPs and their applications have become one of the hot spots in the areas of agricultural science, physiology, medicine, and thus therapeutic applications.

The scope of this book is to summarize the knowledge and enrich the understanding of the discovery, screening, and characterization of natural AMPs for therapeutic applications using molecular biology, proteomics, genomics, and bioinformatics approaches. This book compiles the recent studies and knowledge about the therapeutic applications of AMPs in different areas such as oral health, skin infections, gut health, respiratory infections, reproductive health, innate immunity, and overall host defense.

Columbia, MO, USA — Piyush Baindara
Kharagpur, WB, India — Santi Mohan Mandal

Acknowledgments

I acknowledge the hard work of all the contributors and reviewers to complete this book project.

Contents

About the Editors

Piyush Baindara is currently working as a Senior Research Associate in the Somatic Cell Genome Editing program at the Animal Sciences Research Center, University of Missouri, Columbia, USA. He received his PhD in microbiology from the CSIR-Institute of Microbial Technology, Chandigarh, India. His PhD research focus was on bacterial antimicrobial peptides. Subsequently, he pursued his postdoctoral research at the University of Arkansas for Medical Sciences, Little Rock, and the University of Missouri, Columbia, USA. He has published more than 40 research articles in reputed peer-reviewed journals. His research interests include the discovery and characterization of novel antimicrobial peptides including bacteriocins, lanthipeptides, and human defensins, antimicrobial drug resistance, peptide-based vaccine development, infectious agents and disease pathogenesis, anticancer therapeutics, and NETs (Neutrophil Extracellular Traps) in infectious and autoimmune diseases.

Santi Mohan Mandal obtained his PhD in the field of molecular microbiology and continued research with a major focus on antimicrobial chemotherapy. He visited University of Texas Medical Branch-USA and National University of Singapore for his postdoctoral training. He worked as an Assistant Professor of Microbiology at Vidyasagar University, India. He also served as a Technical Officer at the Department of Bioscience and Biotechnology, Indian Institute of Technology Kharagpur, Kharagpur, India. He has published more than 150 research papers in reputed journals and conferred several prestigious awards for his research contribution. Currently, he is engaged as an Associate Scientist at the Department of Chemistry and Biochemistry, University of California San Diego, USA.

Chapter 1
Antimicrobial Peptides as Future Therapeutics: Challenges and Possibilities

Mitali Mishra

Abbreviations

ACE1	Angiotensin-converting enzyme 1
AIDS	Acquired Immunodeficiency Syndrome
AMPs	Antimicrobial peptides
AMR	Antimicrobial resistance
ARIs	Antibiotic-resistant infections
CAGR	Compound annual growth rate
CDC	Centers for Disease Control and Prevention
DRAMP	Data repository of AMPs
EMA	European Medicines Agency
FDA	Food and Drug Administration
HIV	Human Immunodeficiency Virus
LPPS	Liquid phase peptide synthesis
LTP	Lipid transfer protein
MRSA	Methicillin-resistant *Staphylococcus aureus*
MDR	Multidrug resistance
PAMPs	Plant antimicrobial peptides
SPPS	Solid phase peptide synthesis
WHO	World Health Organization

1 Introduction

Multidrug resistance in pathogenic bacteria is on the rise globally, recording multiple morbidity and mortality in hospital-acquired and community infections. Undoubtedly, antibiotic-resistant infections (ARIs) have successfully emerged as one of the leading global threats to public health, including both developed and developing nations. According to a report released by the Centers for Disease

M. Mishra (✉)
Department of Microbiology, Icahn School of Medicine at Mount Sinai, New York, NY, USA

P. Baindara, S. M. Mandal (eds.), *Evolution of Antimicrobial Peptides*,
https://doi.org/10.1007/978-3-031-67515-7_1

Control and Prevention (CDC) in 2019, around 2.8 million people in the USA were affected with drug-resistant bacterial infections, and annual deaths due to such infections were reported to be over 35,000 people. Even the World Health Organization (WHO) has been alarmed about the escalating threat posed by multidrug-resistant (MDR) microorganisms and estimated deaths of up to ten million people by 2050 (GLASS 2021). Various factors can contribute to the development of ARIs, including primarily overuse or misuse of antibiotics. In addition, antimicrobial resistance (AMR) was also reported to be associated with socioeconomic problems like overpopulation, prolonged hospital admission, poor hygiene and sanitation, use of antibiotics in the agricultural sector, livestock industry, and inclusion in household products (da Cunha et al. 2017; Aslam et al. 2018). The bacterial pathogens that have been shown to co-evolve simultaneously with the development of antibiotics, are known to have either intrinsic (like mutations in target protein, selective permeability and efflux of antimicrobials) or acquired (presence of mobile genetic elements, and resistant genes) and phenotypic resistance (persister formation). Very recently, in a report shared by Global Antimicrobial Resistance and Use Surveillance System (GLASS) Report: Early Implementation 2020) (GLASS 2021), the WHO issued an emergency alert on the need to develop new alternative options that can overcome the therapeutic failures of conventional antibiotics, with novel mode of action and minimal side effects. In the quest for new antibiotics, many researchers have highlighted the importance of antimicrobial peptides (AMPs).

Antimicrobial peptides (AMPs) represent a diverse class of bioactive molecules that exhibit a broad spectrum of antimicrobial activity against pathogens like bacteria, fungi, viruses, and also parasites (Almaaytah 2022; Chen and Lu 2020). AMPs are a group of small peptides with less than 100 amino acid residues with an overall positive charge imparted by the significant portion of hydrophobic and positively charged residues, which makes them an ideal candidate for treating bacterial pathogens (Baltzer and Brown 2011). In general, AMPs have a lower likelihood of resistance development, emphasizing their potential to be used in therapeutic applications, as evident from many such clinical trials.

These peptides are often produced by a myriad of organisms, from bacteria to plants, insects to mammals, including humans, which is why they have also been termed "nature's antibiotics." Table 1.1 summarizes a few such examples of AMPs from various sources along with their availability and known mode of action.

Unlike antibiotics, which are mostly synthetic or derived from microorganisms, these AMPs are naturally available and have evolved as an integral part of an innate immune defense mechanism, protecting against invading pathogens and maintaining microbial homeostasis (Mahlapuu et al. 2020). Contrary to conventional antibiotics, which usually require target-specific binding to trigger activity, AMPs exert their antimicrobial effects rapidly through multifaceted mechanisms, including disruption of microbial membranes, interference with essential intracellular processes, and induce cell death within a short period, often making it difficult for microbes to develop resistance. Besides modulating the immune system, AMPs were also reported to neutralize endotoxins; activate immune cells, such as macrophages and neutrophils; and prevent pathogens invasion and abrupt biofilm formation (Baltzer

Table 1.1 List of some important AMPs derived from different sources with examples

Sources	Group of AMPs	Examples	Presence/distribution	Mechanism of action	References
Humans	Cathelicidins	LL-37 and human cationic antimicrobial protein (hCAP-18)	Secreted by leukocytes (like neutrophils, monocytes, B cells, NK cells), epithelial cells of skin, lung, colon, squamous epithelia, and body fluids like nasal lavage and plasma	Antimicrobial and anti-inflammatory properties, wound repair, and angiogenesis	Brandenburg et al. (2012), Kosciuczuk et al. (2012)
	Defensins	α-Defensin (human neutrophil peptides [HNPs] 1–4)	Secreted by granulocytes and lymphocytes	Antimicrobial activity against bacteria (both gram-negative and gram-positive), fungi, and viruses (HSV-1, cytomegalovirus, HIV-1, and IAV); immunomodulatory properties like chemotactic activity for monocytes, T cells, and dendritic cells; and recruitment of the phagocytes, neutrophil, and monocytes to the site of inflammation	Holly et al. (2017), Ganz (2003)
		β-Defensin (human β-defensins [HBDs] 1–4)	Secreted by leukocytes and epithelial cells (skin and salivary glands)		
		Human enteric defensins (HDs) 5 and 6	Secreted by Paneth cells at the small intestinal crypt		
	Histatins	HIS-1, HIS-3, or HIS-5	Produced in saliva, submandibular, and parotid glands	Maintenance of oral homeostasis and high antifungal activity against pathogens like *Candida albicans* and *Cryptococcus* spp.	Kavanagh and Dowd (2004), Khurshid et al. (2017)

(continued)

Table 1.1 (continued)

Sources	Group of AMPs	Examples	Presence/distribution	Mechanism of action	References
Plants	Thionins	Purothionin, viscotoxins, and crambin	Mainly found in seeds, monocots (grains), dicotyledonous plants (like mistletoes), and cereals	Toxicity affects the cell membrane of bacteria and fungi	Apel (1991), Stec (2006)
	Lipid transfer proteins (LTPs)	Major LTPs such as LTP1, LTP2, LTPc, LTPd, and LTPg	Universally found in land plants and flowering plants	Disrupt bacterial cell membrane with lipid binding and transfer properties and protect phytopathogens.	Edqvist et al. (2018), Wong et al. (2019)
	Defensins	PsD1, Rs-AFP1, Rs-AFP2, NaD1, DmAMP1, and Sd5	Found in stomatal and peripheral cells and ubiquitously present in the plant kingdom but first isolated from seeds of monocot and dicot plant species	Exhibit broad antifungal activity (toward phytopathogenic fungi like *Fusarium graminearum* and human pathogenic fungi such as *Candida albicans*) as well as lesser-known antibacterial properties; provides abiotic stress adaptation	Sher Khan et al. (2019), Hegedus and Marx (2013)
Fungi	PAF	*Penicillium chrysogenum* antifungal protein (PAF)	Secreted as a small, highly-basic and cysteine-rich antifungal protein from *Penicillium chrysogenum*	Have greater antifungal activity against filamentous fungi by hyperpolarization of the plasma membrane, increase in K^+ efflux and Ca^{2+} influx, generation of oxidative stress, and apoptosis	Binder et al. (2010), Marx et al. (2008)
	AFP	*Aspergillus giganteus* antifungal protein (AFP)	Isolated from culture supernatants of *Aspergillus giganteus*	Disrupts plasma membrane integrity and membrane permeabilization as a means for their antifungal spectrum and activity against *Fusarium* spp.	Meyer and Jung (2018), Hegedus and Marx (2013)
	Peptaibols	Alamethicin-F30, trichokonin VI (TK VI), and trichorzianine	Produced majorly in *Trichoderma* spp. as a secondary metabolite	Antibacterial, antifungal, and antitumor activities and neuroactive properties by membrane permeabilizing and pore formation	Hou et al. (2022)

Insects	Cecropins	Cecropin A, B, C, D, and P1	From the pupae of the giant silk moth *Hyalophora cecropia*	Antimicrobial activity against both gram-positive and gram-negative bacteria by lysing bacterial cellular membranes and, in some instances, cercopins exhibit antifungal activity and low toxicity against normal mammalian cells	Manniello et al. (2021), Wu et al. (2018)
	Attacins	Attacins A–F	Purified from the hemolymph of *Hyalophora cecropia*	Mostly active against gram-negative bacteria (particularly *Escherichia coli*) by blocking the synthesis of major outer membrane proteins (OMPs)	Manniello et al. (2021)
	Ponericins	Ponericins G, W, and L	Found in the venom of the predatory ant *Pachycondyla goeldii*	Antibacterial activities against both gram-positive and gram-negative bacteria, insecticidal activity against cricket larvae, and hemolytic activities	Orivel et al. (2001)
	Diptericins	Diptericins A–C	Found in immunized larvae of the dipterian *Phormia terraenovae* and also in *Drosophila melanogaster* and *Sarcophaga peregrina*	Antibacterial activity against gram-negative bacteria (*E. coli* K12 and *Erwinia* Spp.) by acting on the cytoplasmic membrane	Wu et al. (2018)

(continued)

Table 1.1 (continued)

Sources	Group of AMPs	Examples	Presence/distribution	Mechanism of action	References
Fish	Hepcidins	HAMP1 and HAMP2	Widely found in many fish species like teleosts, sea bass, rainbow trouts, etc. It is constitutively expressed in the liver, kidney (head and posterior), spleen, skin, gills, gut, heart, muscle, and brain	Perform multiple functions of regulation of iron metabolism, immunomodulatory roles, and antimicrobial activity by disrupting membrane integrity of both gram-positive and gram-negative bacteria as well as yeasts (*Saccharomyces cerevisiae*)	Chaturvedi et al. (2018), Masso-Silva and Diamond (2014), Yulema Valero et al. (2013)
	Piscidins	Piscidin-1, piscidin-2, pleurocidins, misgurin, and epinecidin	Found in tilapia, Atlantic cod, and striped bass and detected in gill, skin, and intestine mainly, but their expression is dependent on the isoforms	Exhibit broader antibacterial (both gram-positive and gram-negative pathogens) and antifungal activities by disrupting membrane integrity and cellular efflux with the formation of pores and immunomodulatory roles and antitumor activities	Yulema Valero et al. (2013)
	Cathelicidins	CATH-1, CATH-2, aCATH, and acCATH	Found in bony fishes like rainbow trout and Atlantic salmon and expressed in the gill, skin, spleen, head, and kidney	Potent antibacterial activity toward both gram-negative and gram-positive pathogens and antifungal activity to *C. albicans*	Kosciuczuk et al. (2012)

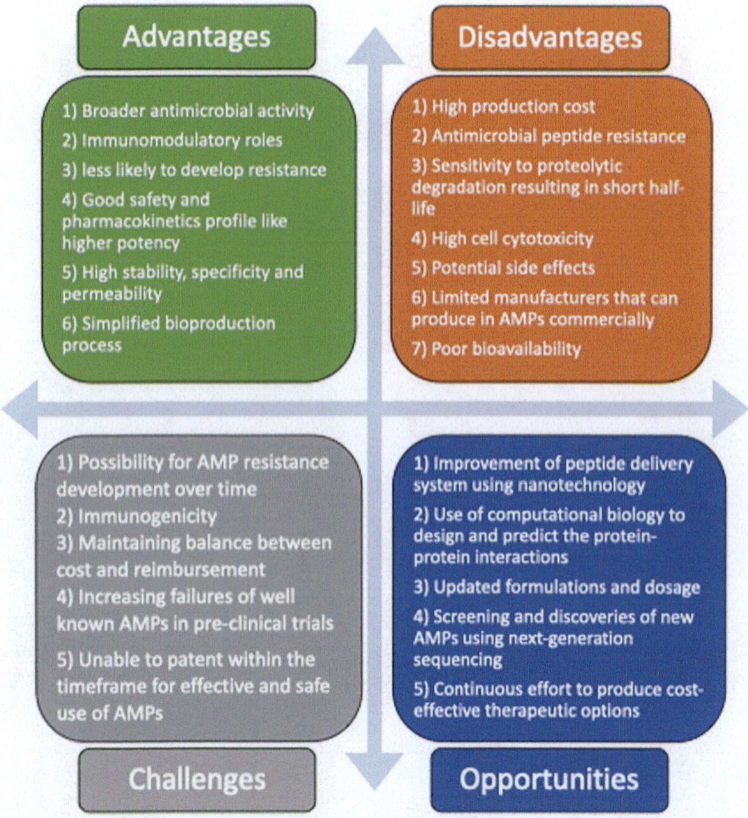

Fig. 1.1 Graphical overview of pros and cons of antimicrobial peptide development including challenges and future perspectives

and Brown 2011; Mishra et al. 2017). In addition, it is possible to find AMPs as anticancer agents (Mishra et al. 2017; de la Fuente-Nunez et al. 2017). Additionally, AMPs can also be used for medical product development such as drugs to be applied in wound healing as well as skin care use, due to antioxidant peptides action (Antonio Moretta et al. 2021; Golonka et al. 2021). AMPs can also act as inhibitors of angiotensin-converting enzyme 1 (ACE1) and pancreatic lipase, being useful to control metabolic syndrome. The discovery and characterization of AMPs have significantly broadened our understanding of the intricate interplay between host defense mechanisms and microbial pathogenesis.

The unique physicochemical properties of AMPs, including their broad-spectrum activity, rapid bactericidal action, and relatively low propensity for inducing resistance, have propelled them into the spotlight as a promising avenue for combating the growing menace of antimicrobial resistance. Some of the popular AMPs which were FDA approved for clinical use are listed in the Table 1.2 below. This chapter aims to provide a profound understanding of the challenges and future opportunities in the dynamic field of antimicrobial peptide research.

Table 1.2 List of few FDA-approved AMPs used for commercial applications: Clinically validated AMPs with their origin, properties, mode of action, and usage

AMPs/ peptide name	Properties of the peptide	Origin	Antimicrobial activity	Treatment indication	Route of administration	Mechanism of action	References
Daptomycin (Cubicin)	Cyclic lipopeptide	Fungi (*Penicillium chrysogenum*)	Antibacterial (gram-positive bacteria, including methicillin-resistant *Staphylococcus aureus* [MRSA] and vancomycin-resistant *Enterococci* [VRE])	Complicated skin infections	Intravenous injection	Membrane lysis/ depolarization	Matthew Shoemaker et al. (2006)
Gramicidin	Linear pentadecapeptides (15 hydrophobic amino acids)	Bacteria (*Brevibacillus brevis*)	Antibacterial (gram-positive pathogens)	Skin lesions, surface wounds, and eye infections	Topical application only	Channel-forming ionophore on the bacterial membrane	Kessler et al. (2004), David and Rajasekaran (2015)
Nisin	Polycyclic peptide (34 amino acids)	Bacteria (*Lactococcus lactis*)	Antibacterial (gram-positive organisms and spore producers, including lactic acid bacteria, *Listeria monocytogenes*, *S. aureus*, *Bacillus cereus*, etc.)	Bacterial infections		Membrane depolarization	Shin et al. (2016), Field et al. (2023)
Melittin	A small linear basic peptide with no disulfide bridge (26 amino acids)	Honey bee (*Apis mellifera*) venom	Anti-inflammatory applications, anticancer properties	Relieving pain and swelling associated with rheumatoid arthritis and multiple sclerosis	Topical	Membrane disruption	Gajski and Garaj-Vrhovac (2013)

Histatin	Small cationic peptides (rich in histidine amino acid)	Human saliva	Antibacterial and antifungal (mostly drug-resistant opportunistic pathogen *Candida albicans*)	Treatment of oral candidiasis (gingivitis and periodontal diseases)	Topical gels or denture acrylic	Binds to the receptor on fungal cell membrane, enters the cytoplasm, and targets mitochondria by non-lytic loss of ATP from actively respiring cells	Kavanagh and Dowd (2004), Khurshid et al. (2017)
Lactoferricin	Amphipathic, cationic peptide (25 residues)	Mammals (milk protein)	Antimicrobial and antitumor properties and immunomodulatory roles	Oral administration	Oral care and food sector (infant formula and supplements)	Membrane depolarization	Gifford et al. (2005)
Polymyxin E (colistin)	Cyclic lipopeptide	*Bacillus polymyxa* var. *colistinus*	Antimicrobial properties against like *A. baumannii*, *P. aeruginosa*, *K. pneumonia* and caused by *E.coli* and *Salmonella* spp.	Last resort antibiotic option for tackling MDR gram-negative pathogens, cystic fibrosis, and gastrointestinal tract infections	Intravenous and intramuscular administration	Membrane disruption/ immunomodulation	Antimicrobial peptides basics for clinical application (2019), Jiang et al. (2021)
Polymyxin B	Cyclic polypeptide	*Paenibacillus polymyxa*	Antibacterial activity against gram-negative bacteria	Meningitis, periodontitis, wound infection, pneumonia, sepsis, and urinary tract infections	Intravenous and inhale	Bacterial membrane disruption	Arnold et al. (2007), Biswaro et al. (2018)

2 Challenges in the Development of AMPs

While AMPs hold great therapeutic potential in the post-antibiotic era to fight against AMR infections, their development and clinical application are also accompanied by various complexities that have impeded their swift transition from laboratory to clinical use. It was noticed that during clinical trials, certain AMPs were terminated or discontinued citing unfavorable pharmacokinetic profiles, lack of clear advantage, or better efficacy over traditional treatments. A lot of these challenges stemming from both scientific and practical aspects have been addressed earlier in many scientific reports, but in most of the cases, issues related to peptide stability, toxicity, production costs, and regulatory hurdles are the key factors. In this section, a few of such challenges related to AMP development will be discussed in detail.

2.1 Peptide Design and Synthesis

Generally, the AMPs from natural sources are purified from their sources, like insulin from animal pancreatic tissue. The purification of natural AMPs hence poses challenges in terms of quality, quantity, safety, and so on. Maintaining a balance between selectivity and potency of AMPs is a critical step for the clinical validation of AMPs. Currently, peptide synthesis is performed either by chemical methods (solid phase peptide synthesis [SPPS] and liquid phase peptide synthesis [LPPS]) or biological methods (production of recombinant protein expression and fermentation) (Brandenburg et al. 2012). Due to technological advancements, the pharmaceutical-grade peptide synthesis process is much better now; still, the chemical synthesis of AMPs always remains a challenge based on the peptide sequence, structure, length, and composition of the amino acid residues. In addition, the downstream processing and large-scale production of peptide drugs is considered to be very critical, which is generally the most expensive, technically challenged, and laborious part of the production. The manufacturing cost of AMPs is estimated to be around $50–400 per gram of amino acid produced by SPSS technology (Chen and Lu 2020; Yuhui Du et al. 2022). Therefore, establishing scalable and cost-effective peptide production methods (such as fermentation or biotechnological engineering) would make AMP production economically feasible and widely accepted in the future. A sustainable AMP production approach will also contribute to minimizing harmful effects on our environment that are often a byproduct of traditional antibiotic manufacturing.

2.2 Stability and Bioavailability

Many AMPs are susceptible to degradation by endogenous proteases in the human body and enzymes in biological fluids, limiting their therapeutic efficacy and bioavailability in vivo. Formulating AMPS to escape this enzymatic degradation while retaining their antimicrobial activity presents a significant hurdle in peptide-based drug development (Giuliani et al. 2007). Different strategies to enhance the stability of AMPs in physiological conditions, such as peptide modifications, incorporation of nonnatural amino acids, and formulation approaches, are being actively explored to overcome the shortcomings of pharmacokinetics and bioavailability. In general, optimization is performed by reducing peptide length or altering peptide surface properties by substituting amino acid residues. In this regard, computer-assisted AMP design and cutting-edge machine learning tools have been very useful for the accurate prediction of the desired biological activity from the primary peptide structure. Another issue with the bioavailability of AMPs lies with their reduction in antimicrobial activity under physiological salt conditions and high ionic strength owing to the loss of electrostatic interactions between AMPs and cell membranes (Koehbach and Craik 2019). This is evident in the failure of reproducibility of several potential in vitro data, which ultimately resulted in unexpectedly low performance in vivo. It is very pertinent to mention here that peptides have generally lower oral bioavailability. This is due to the instability of AMPs under harsh pH conditions in the gastrointestinal tract, which is further aggravated by their poor penetration ability across the gut epithelium (Pushpanathan et al. 2013). Even though the oral route of administration is widely adapted for the delivery of AMPs, due to its noninvasive and painless administration to patients, another route of administration like intramuscular or intravenous injections may be necessary to maintain the stability of AMP physicochemical and biological characteristics (Conlon et al. 2014).

2.3 Resistance Development

Although AMPs possess a lower propensity for developing microbial resistance compared to conventional antibiotics, it is important to consider that the resistance can be provoked by some pathogens. Considering the mandatory interactions between the AMPs and the microbial membrane for the efficacy of the drug, it is critical that the pathogen does not remodel or reorganize its membrane composition to escape action. Literature on AMP resistance has emphasized some known mechanisms, including the production of degrading enzymes, membrane surface modifications, biofilm formation, capsules, and efflux pumps. For instance, daptomycin was FDA-approved in 2003 for its application against gram-positive pathogens, but soon after its commercialization, resistance to the drug was noticed in many gram-positive bacteria like *Bacillus subtilis* and *S. aureus* that modified their phosphatidylglycerol content of the membrane to restrict daptomycin antimicrobial activity. Similarly, biofilm formation greatly enhances the ability of the pathogen to develop resistance even though it is reported that AMPs do restrict biofilm formation, and this is well evident in the AMP database, where nearly 2.16% of the peptides are designated to be anti-biofilm peptides (Luo et al. 2023). Research data showed promising potential for synergistic combinations of AMPs and various antibiotics tested against each of the *Enterococcus faecium*, *Staphylococcus aureus*, *Klebsiella pneumoniae*, *Acinetobacter baumannii*, *Pseudomonas aeruginosa*, and *Enterobacter* species (ESKAPE) pathogens in animal model, toward reducing AMP-mediated resistance. In addition, synergistic activity of daptomycin with β-lactams antibiotics in combination demonstrated greater efficacy of 90% as compared to 57% efficacy in monotherapy in treating patients with staphylococcal bacteremia infections (Magana et al. 2020). Besides combinatorial therapies with antibiotics, exploring synergistic effects of two, three, or more AMPs in combination would minimize the chance of AMR development in the future and significantly improve the therapeutic value as compared to their efficacy. Hence, understanding the mechanisms behind such AMP resistance and devising strategies to overcome resistance is the bigger challenge in long-term clinical applications of AMPs.

2.4 Specificity and Selectivity

Another drawback of the clinical implementation of AMPs is its broad spectrum of antimicrobial activity, which works like a two-edged sword. While AMPs can target a wide range of pathogens, including bacteria, viruses, fungi, and parasites, this broad specificity could also lead to unintended harm to beneficial microbes in the human microbiome. AMPs usually target the microbial membranes, which are not always highly selective and may be toxic to host cells. Several reports on the hemolytic and cytotoxic effects of AMPs at antimicrobial concentrations question their clinical use (Koehbach and Craik 2019). For example, polymyxins, which were developed for treating MDR gram-negative infections and are currently being used in clinics, were also reported to cause neurotoxicity and nephrotoxicity at the same antimicrobial dosage. Designing AMPs to retain potent antimicrobial activity while reducing side effects such as cytotoxicity to host cells remains an important aspect of its development (Zasloff 2002).

Also, optimizing the formulation, delivery, and dosage of AMPs for various applications can be a challenging task and needs careful monitoring. In the context of the drug delivery platform of AMPs, novel modifications such as topical, systemic, or targeted delivery would reduce adverse side effects and simultaneously enable a controlled release of AMPs. As elucidated in literature, the AMP dosage forms in ongoing clinical trials constitute hydrogel, topical gel or cream, polyvinyl alcohol-based solution for better performance and bioavailability in the wound, or, at the surgical site, oral solutions or mouth rinse (Luo et al. 2023; Zasloff 2002).

Table 1.3 List of some of the AMPs being investigated for clinical use under the clinical trials

Peptide name	Structural features	Route of administration	Treatment or use	Stage of clinical trials	Clinical trial ID	Mechanism of action	References
Aviptadil	A 28-amino-acid vasoactive intestinal peptide (VIP) hormone	Intravenous	Acute respiratory distress syndrome and sepsis	I	NCT04311697	Inhibit cytokine synthesis and block replication of SARS-CoV2 in pulmonary cells	Youssef et al. (2022)
Lytixar (LTX-109)	Chemically synthesized, peptide mimetic drug	Topical hydrogel	Treatment of nasal carriers like MRSA, impetigo, and other complicated gram-positive infections	I/IIa	NCT01158235, NCT01803035	Antibacterial activity by membrane disruption and cell lysis	Nilsson et al. (2015)
PAC113	12-amino-acid histidine–rich peptide derived from histatin 5 found in human saliva	Mouth rinse	Treatment of oral candidiasis, specifically in HIV immunocompromised patients	IIb	NCT00659971	Antifungal activity by membrane disruption	Cheng et al. (2020)
CZEN-002	Synthetic octapeptide from α-melanocyte-stimulating hormone (α-MSH)	Intravaginal gel	Vaginal candidiasis	II[b]		Antifungal activity not by membrane damage, rather by inhibiting pro-inflammatory cytokine production	
Sifuvirtide (SFT)	Synthetic peptide based on the 3D structure of the HIV-1 gp41 fusogenic core conformation	Universal gel for topical use	HIV fusion inhibitor as a treatment option for AIDS	II		Impair the gp41-mediated cell entry or viral fusion	Li et al. (2012)
Omiganan (MBI-226)	Synthetic 12-mer cationic peptide derived from indolicidin	Topical gel	Treatment of atopic dermatitis (AD), genital warts, catheter-related infections, rosacea	II/III	NCT00231153, NCT01784133	Broad microbicidal activity by depolarizing cytoplasmic membrane	Sader et al. (2004)

2.5 Regulatory Hurdles

During the initial period of AMP discovery, most of the research was focused on the identification, characterization, and validation of antimicrobial properties in vitro. There were very little to no specific or robust regulatory guidelines for their approval as new drugs in clinical practice. Also, researchers on AMPs promised their potential in vitro without providing a complete picture of their actual biological activity in a clinically relevant physiological condition (Brandenburg et al. 2012; Nitin et al. 2023). In fact, approvals for any new drugs or antibiotics, including AMPs, can be a lengthy, iterative, and complicated process for both big pharma companies as well as scientists. For any AMP to be clinically approved and be successful pharmaceutically, it has to pass through some regulatory guidelines primarily focusing on validating in vivo activity as reported in vitro, pathological condition, route of administration, and targeted population for each AMPs. Specifically, agencies like the European Medicines Agency (EMA) or U.S. Food and Drug Administration (FDA) approve clinical trials being conducted for safety and immunogenicity profiles to approve for human application (Machado et al. 2023). As observed in Table 1.3, there are numerous examples of AMPs currently being investigated in different developmental phases of clinical trials, with some showing positive efficacy while others being discontinued. It is critical to note here that the success of AMPs in clinical trials and posterior application is variable and most of the time failure is unexpected and unknown. Therefore, navigating through these hurdles of clinical approval, alongside demonstrating better safety and efficacy, poses a significant challenge in AMP development. The main reason behind this is the lack of standardized guidelines for evaluation and comparison among various AMP subclasses and groups (Zhang et al. 2021; Wang et al. 2022). In addition, intellectual property management and product marketing simultaneously during the preclinical phase is a very crucial regulatory factor because this has a huge impact on public perception and acceptance of the new drug. During preclinical evaluation of AMPs, it is highly important to raise awareness and address any issues with their safety, effectiveness, and bioactivity.

3 Future Prospects for AMP Development

As researchers continue to address the shortcomings associated with AMP development, the future is filled with several promising opportunities for application across different fields, from medicine to agriculture, the food sector, water resource management, and beyond. In this section, some future opportunities for further development and utilization of AMPs will be discussed in detail.

3.1 Personalized Medicine

The ability to custom design AMPs for certain patient populations or individualized disease profiles based on their genetic makeup and microbiome holds great promise in the realm of personalized medicine. AMPs could be smartly modified to increase their selectivity while minimizing side effects like cell cytotoxicity, thereby optimizing therapeutic outcomes. Such tailored peptides present a new generation of designer drugs to simultaneously enhance antimicrobial activity and overcome AMP resistance (Patel and Akhtar 2017). For instance, to reduce the hemolytic and cytotoxic effects of a natural broad-spectrum AMP bacaucin, researchers removed its lipid portion and the ring opening of the parent peptide, named bacaucin-1, a derivative to the parent compound, which decreased its overall hydrophobicity. Bacaucin-1 could selectively target antibiotic-resistant *Staphylococcus aureus* infections by a membrane disruption, without lipid sequence (de la Fuente-Nunez et al. 2017). In particular, there is an ease and feasibility associated with AMP development in personalized therapy by using peptide engineering tools. Again, the immunomodulatory properties and antitumor properties of AMPs make them an ideal candidate for their application in personalized immunotherapy for treating cancer patients. In general, AMPs are utilized to enhance the immune response against tumors and infectious agents, paving the way for innovative personalized therapy in oncology and immunology (Santos et al. 2022). AMPs are also known to play a crucial role in improving gut health and homeostasis by balancing the human microbiome. This aspect of the AMP role can be exploited to address gut-related disorders and support gastrointestinal health in an individualized manner.

Similarly, exploring the synergistic effects of AMPs with another bioactive peptide molecule or other antimicrobial agent can enhance treatment efficacy and mitigate the development of resistance. The combinatorial therapeutic approach could offer novel strategies for managing complex and multidrug-resistant infections and, at the same time, will reduce the burden and cost needed to develop another new class of antimicrobial agents for tackling MDR pathogens (Viñas 2021).

3.2 One Health Approach

The rationale behind the "One Health" approach values the interrelatedness of human health to the health of animals and shared environmental health. This includes several emerging and reemerging zoonotic diseases, vector-borne diseases, antibiotic resistance, food safety, food security, environmental contamination, and climate change, to list a few. In this context, AMPs have wide applications in human healthcare settings, veterinary medicine, agriculture, and environmental preservation, which means AMPs can provide a holistic approach to disease prevention and treatment (Wang et al. 2022). Hence, applying AMPs into a "One Health" framework, integrating human, animal, and environmental health, presents an opportunity to combat the spread of antimicrobial resistance across different ecosystems. AMPs can be explored as alternatives in veterinary medicine and agricultural practices. Specifically, plant antimicrobial peptides (PAMPs) were greatly used as plant growth promoters, insecticides, and improved transgenic products that are resistant to pathogens (Bakare et al. 2022). AMPs can be successfully used in agriculture and food industry to reduce the need for chemical pesticides and reliance on traditional antibiotics in animal husbandry and crop protection. Additionally, they can serve as natural preservatives in the food industry, contributing to food safety (Zasloff 2002). AMPs derived from sustainable sources, such as plants and algae, offer eco-friendly alternatives to synthetic peptides and traditional antibiotics. This aligns with the increasing emphasis on sustainability in healthcare and agriculture. Exploring the potential of AMPs in environmental applications, such as water treatment and biofilm control, can contribute to addressing public health concerns related to waterborne pathogens and antimicrobial-resistant microbes in various environmental settings (Koehbach and Craik 2019). Overall, AMPs can offer a very natural and sustainable approach to controlling pathogens in livestock, aquaculture, and plant diseases, thus contributing greatly to food safety and security.

3.3 Technological Advancements

The advancements in nano-biotechnology have facilitated AMP engineering to minimize the undesirable characteristics of both natural and synthetic peptides, improve drug delivery systems, and thereby increase the stability and targeted antimicrobial action. Nanotechnology primarily encapsulates AMPs in two ways: either a passive nano-delivery system that does not possess surface modification of their structure to guide the nano-carrier rather is driven by carrier's size and shape or a directed delivery system that involves surface modifications of the nano-carrier with ligands to allow better interaction at the site of action following targeted delivery and increased bioavailability. In this way, nanoparticles and liposome technologies could facilitate the translation of AMP studies in vitro toward clinical applications (Biswaro et al. 2018).

Leveraging bioinformatics and computational tools for the design and optimization of AMPs can expedite the discovery of novel peptides with enhanced antimicrobial properties and minimal side effects. Computational approaches coupled with synthetic biology enable the prediction of peptide structures, activity, and interactions, facilitating the rational design of engineered AMPs for specific applications (Brandenburg et al. 2012; Nie 2022).

3.4 *Public Health Education and Awareness*

AMPs may play a crucial role in responding to emerging infectious diseases and bioterrorism threats. Developing stockpiles of effective AMP-based therapies is an important preparedness measure. Educating the public and healthcare professionals about the health benefits and multifaceted applications of AMPs can promote awareness and acceptance of these novel antimicrobial agents (Carratala et al. 2020). Building public trust and understanding of the role of AMPs in combating infectious diseases is crucial for their successful integration into mainstream healthcare practices. The increasing awareness of AMPs and their potential benefits is expected to drive growth in the AMP market. Pharmaceutical and biotechnology companies are investing in research and development, creating economic opportunities.

4 Global AMP Market Scenario

As per various reports on the emerging business of antimicrobial peptide production, the market was valued at 4.37 billion USD in 2021 and was expected to rise up to 7.85 billion USD by 2029, with a 4.6% compound annual growth rate (CAGR) from 2022 to 2030 (https://www.marketresearch.com/LP-Information-Inc-v4134/Global-Antimicrobial-Peptides-Growth-32980753/, https://www.researchreportsworld.com/enquiry/request-sample/21492999). North America (United States, Canada, and Mexico) is dominating the peptide therapeutics development market followed by Europe on a global scale with the highest share of sales around 35%. However, it is noteworthy that Asia Pacific holds great promise for AMP market development owing to their increased infectious disease burden and growing health awareness among patients and healthcare workers. Some of the key industrial contributors to this growing AMP market are GenScript Biotech Corporation, Novo Nordisk A/S, InvivoGen, Sanofi, AstraZeneca PLC, GlaxoSmithKline PLC, Merck & Co., Pfizer, Inc., and Takeda Pharmaceutical Company (https://www.grandviewresearch.com/industry-analysis/peptide-antibiotics-market-report). Such market reports provide an integrated as well as holistic view of the current landscape of the AMP market in different geographical regions based on their type (natural or synthetic), route of administration (oral, injection, topical, or others), different types of applications (agricultural, pharmaceutical, cosmetics, food industry, etc.), and evolving trends in industry policies and dynamics.

One of the most significant factors responsible for the growth of the AMP market is rising antibiotic resistance in different kinds of deadly infectious diseases, such as bacterial skin infections, ophthalmic infections, bloodstream infections, and hospital-acquired infections. Undoubtedly, the alarming frequency of AMR cases worldwide and the rising need for an effective and safe alternative to antibiotics are fueling the rapid expansion of the antimicrobial peptide market globally. For instance, according to the 2019 reports on antimicrobial resistance in the USA published by the Centers for Disease Control and Prevention (CDC), more than 2.8 million antimicrobial resistance infections occur every year with more than 35,000 deaths. Moreover, the CDC's Antimicrobial Resistance (AR) Solutions Initiative is focused on developing infrastructure to detect, respond, contain, and prevent such infections across different communities and environments. During 2016–2023, the CDC invested more than $430 million in around 700 innovative antimicrobial resistance projects in more than 60 countries, collaborating with more than 200 public and private institutions to tackle AMR globally.

In 2023, BioCote Limited and Eco Finish formed a strategic merger/partnership to introduce antimicrobial additives targeting different microbes for the better safety of swimmers in residential and commercial pool environments. Similarly, Boehringer Ingelheim, Evotec SE, and bioMérieux in July 2022 formed a collaborative venture to develop the next generation of antimicrobials to combat emerging drug-resistant infections. Such company mergers and collaborations would boost the peptide therapeutics development market. As per the data repository of AMPs (DRAMP) database, currently, there are approximately 96 AMPs in different stages of development as drug candidates. In the category of clinical AMPs, 33 are in the preclinical stage, 8 are in phase III, 6 have completed phase II, 7 have failed in phase III, and one DP178 (T20, enfuvirtide (Fuzeon)), which is an anti-HIV therapeutics developed by Trimeris company, has been approved by the FDA (http://dramp.cpu-bioinfor.org/) (Shi et al. 2022).

5 Conclusion

Undoubtedly, antimicrobial peptides present significant potential as new-generation antimicrobial agents that can mitigate our current globally prevalent AMR health issue. As we march toward the post-antibiotic area, these AMPs hold the greatest potential to be alternatives to conventional antibiotics. The vast diversity of natural and synthetic AMPs provides an ocean of opportunities for researchers, pharma companies, and regulatory bodies to develop these peptides strategically as a smarter antimicrobial therapy for humans and the environment. The lessons learned from new antibiotics development and the co-emergence of microbial resistance should always be considered while structuring these AMPs into preclinical or clinical trials and subsequent applications. While further research is needed to address the knowledge gaps in this growing field of AMPs, the healthcare providers, and regulatory bodies have to come together for a better and more rewarding application of AMPs in substituting traditional therapies.

References

Almaaytah A (2022) Antimicrobial peptides as potential therapeutics: advantages, challenges and recent advances. Farmacia 70(6):991–1003

Antimicrobial peptides basics for clinical application (2019) Springer

Antonio Moretta CS, Petrone AM, Rosanna Salvia AF, Manniello MD, Lucchetti D, Vassallo A, Vogel ASAPFH (2021) Antimicrobial peptides: a new hope in biomedical and pharmaceutical fields. Front Cell Infect Microbiol 11:668632

Apel HBAK (1991) Thionins. Annu Rev Plant Physiol Plant Mol Bioi 42:227–240

Arnold TM, Forrest GN, Messmer KJ (2007) Polymyxin antibiotics for gram-negative infections. Am J Health Syst Pharm 64(8):819–826

Aslam B et al (2018) Antibiotic resistance: a rundown of a global crisis. Infect Drug Resist 11:1645–1658

Bakare OO et al (2022) Plant antimicrobial peptides (PAMPs): features, applications, production, expression, and challenges. Molecules 27(12):3703

Baltzer SA, Brown MH (2011) Antimicrobial peptides: promising alternatives to conventional antibiotics. J Mol Microbiol Biotechnol 20(4):228–235

Binder U et al (2010) The antifungal activity of the Penicillium chrysogenum protein PAF disrupts calcium homeostasis in Neurospora crassa. Eukaryot Cell 9(9):1374–1382

Biswaro LS et al (2018) Antimicrobial peptides and nanotechnology, recent advances and challenges. Front Microbiol 9:855

Brandenburg L-O et al (2012) Antimicrobial peptides: multifunctional drugs for different applications. Polymers 4(1):539–560

Carratala JV et al (2020) Nanostructured antimicrobial peptides: the last push towards clinics. Biotechnol Adv 44:107603

Chaturvedi P, Bhat RAH, Pande A (2018) Antimicrobial peptides of fish: innocuous alternatives to antibiotics. Rev Aquac 12(1):85–106

Chen CH, Lu TK (2020) Development and challenges of antimicrobial peptides for therapeutic applications. Antibiotics (Basel) 9(1):24

Cheng KT et al (2020) The interactions between the antimicrobial peptide P-113 and living Candida albicans cells shed light on mechanisms of antifungal activity and resistance. Int J Mol Sci 21(7):2654

Conlon JM et al (2014) Potential therapeutic applications of multifunctional host-defense peptides from frog skin as anti-cancer, anti-viral, immunomodulatory, and anti-diabetic agents. Peptides 57:67–77

da Cunha NB et al (2017) The next generation of antimicrobial peptides (AMPs) as molecular therapeutic tools for the treatment of diseases with social and economic impacts. Drug Discov Today 22(2):234–248

David JM, Rajasekaran AK (2015) Gramicidin a: a new mission for an old antibiotic. J Kidney Cancer VHL 2(1):15–24

de la Fuente-Nunez C et al (2017) Next-generation precision antimicrobials: towards personalized treatment of infectious diseases. Curr Opin Microbiol 37:95–102

Edqvist J et al (2018) Plant lipid transfer proteins: are we finally closing in on the roles of these enigmatic proteins? J Lipid Res 59(8):1374–1382

Field D et al (2023) After a century of nisin research - where are we now? FEMS Microbiol Rev 47(3):fuad023

Gajski G, Garaj-Vrhovac V (2013) Melittin: a lytic peptide with anticancer properties. Environ Toxicol Pharmacol 36(2):697–705

Ganz T (2003) Defensins: antimicrobial peptides of innate immunity. Nat Rev Immunol 3(9):710–720

Gifford JL, Hunter HN, Vogel HJ (2005) Lactoferricin: a lactoferrin-derived peptide with antimicrobial, antiviral, antitumor and immunological properties. Cell Mol Life Sci 62(22):2588–2598

Giuliani A, Pirri G, Nicoletto S (2007) Antimicrobial peptides: an overview of a promising class of therapeutics. Open Life Sci 2(1):1–33

Global Antimicrobial Resistance and Use Surveillance System (GLASS) report (2021) World Health Organization

Golonka I et al (2021) Antimicrobial and antioxidative activity of newly synthesized peptides absorbed into bacterial cellulose carrier against Acne vulgaris. Int J Mol Sci 22(14):7466

Hegedus N, Marx F (2013) Antifungal proteins: more than antimicrobials? Fungal Biol Rev 26(4):132–145

Holly MK, Diaz K, Smith JG (2017) Defensins in viral infection and pathogenesis. Ann Rev Virol 4(1):369–391

Hou X et al (2022) Peptaibols: diversity, bioactivity, and biosynthesis. Eng Microbiol 2(3):100026

Jiang Y et al (2021) Recent advances in design of antimicrobial peptides and polypeptides toward clinical translation. Adv Drug Deliv Rev 170:261–280

Kavanagh K, Dowd S (2004) Histatins: antimicrobial peptides with therapeutic potential. J Pharm Pharmacol 56(3):285–289

Kessler N et al (2004) The linear pentadecapeptide gramicidin is assembled by four multimodular nonribosomal peptide synthetases that comprise 16 modules with 56 catalytic domains. J Biol Chem 279(9):7413–7419

Khurshid Z et al (2017) Histatin peptides: pharmacological functions and their applications in dentistry. Saudi Pharm J 25(1):25–31

Koehbach J, Craik DJ (2019) The vast structural diversity of antimicrobial peptides. Trends Pharmacol Sci 40(7):517–528

Kosciuczuk EM et al (2012) Cathelicidins: family of antimicrobial peptides. A review. Mol Biol Rep 39(12):10957–10970

Li L et al (2012) Efficacy, stability, and biosafety of sifuvirtide gel as a microbicide candidate against HIV-1. PLoS One 7(5):e37381

Luo X et al (2023) Advancements, challenges and future perspectives on peptide-based drugs: focus on antimicrobial peptides. Eur J Pharm Sci 181:106363

Machado M, Silva S, Costa EM (2023) Are antimicrobial peptides a 21st-century solution for atopic Dermatitis? Int J Mol Sci 24(17):13460

Magana M et al (2020) The value of antimicrobial peptides in the age of resistance. Lancet Infect Dis 20(9):e216–e230

Mahlapuu M, Bjorn C, Ekblom J (2020) Antimicrobial peptides as therapeutic agents: opportunities and challenges. Crit Rev Biotechnol 40(7):978–992

Manniello MD et al (2021) Insect antimicrobial peptides: potential weapons to counteract the antibiotic resistance. Cell Mol Life Sci 78(9):4259–4282

Marx F et al (2008) The Penicillium chrysogenum antifungal protein PAF, a promising tool for the development of new antifungal therapies and fungal cell biology studies. Cell Mol Life Sci 65(3):445–454

Masso-Silva JA, Diamond G (2014) Antimicrobial peptides from fish. Pharmaceuticals (Basel) 7(3):265–310

Matthew Shoemaker D, J.S, Roland WE (2006) A review of daptomycin for injection (Cubicin) in the treatment of complicated skin and skin structure infections. Ther Clin Risk Manag 2(2):169–174

Meyer V, Jung S (2018) Antifungal peptides of the AFP family revisited: are these cannibal toxins? Microorganisms 6(2):50

Mishra B et al (2017) Host defense antimicrobial peptides as antibiotics: design and application strategies. Curr Opin Chem Biol 38:87–96

Nie P (2022) Understanding antimicrobial peptides and combatting antimicrobial resistance problems. Rev Aquac 14:523–524

Nilsson AC et al (2015) LTX-109 is a novel agent for nasal decolonization of methicillin-resistant and -sensitive Staphylococcus aureus. Antimicrob Agents Chemother 59(1):145–151

Nitin B, Nage VNW, Baig MMV (2023) Antimicrobial peptides: current perspectives and future promises. Microbes Infect Dis 5:185
Orivel J et al (2001) Ponericins, new antibacterial and insecticidal peptides from the venom of the ant Pachycondyla goeldii. J Biol Chem 276(21):17823–17829
Patel S, Akhtar N (2017) Antimicrobial peptides (AMPs): the quintessential 'offense and defense' molecules are more than antimicrobials. Biomed Pharmacother 95:1276–1283
Pushpanathan M, Gunasekaran P, Rajendhran J (2013) Antimicrobial peptides: versatile biological properties. Int J Pept 2013:675391
Sader HS et al (2004) Omiganan pentahydrochloride (MBI 226), a topical 12-amino-acid cationic peptide: spectrum of antimicrobial activity and measurements of bactericidal activity. Antimicrob Agents Chemother 48(8):3112–3118
Santos CRGR, Lima LF, dos Reis MCG, D.S.C.a.F.O.L, Cunha NB (2022) Advances and perspectives for antimicrobial peptide and combinatory therapies. Front Bioen Biotechnol 10:1051456
Sher Khan R et al (2019) Plant defensins: types, mechanism of action and prospects of genetic engineering for enhanced disease resistance in plants. 3 Biotech 9(5):192
Shi G et al (2022) DRAMP 3.0: an enhanced comprehensive data repository of antimicrobial peptides. Nucleic Acids Res 50(D1):D488–D496
Shin JM et al (2016) Biomedical applications of nisin. J Appl Microbiol 120(6):1449–1465
Stec B (2006) Plant thionins--the structural perspective. Cell Mol Life Sci 63(12):1370–1385
Viñas JMSM (2021) Future prospects for antimicrobial peptide development: peptidomimetics and antimicrobial combinations. Expert Opin Drug Discov 16(6):601–604
Wang L et al (2022) Therapeutic peptides: current applications and future directions. Signal Transduct Target Ther 7(1):48
Wong LH, Gatta AT, Levine TP (2019) Lipid transfer proteins: the lipid commute via shuttles, bridges and tubes. Nat Rev Mol Cell Biol 20(2):85–101
Wu Q, Patocka J, Kuca K (2018) Insect antimicrobial peptides, a mini review. Toxins (Basel) 10(11):461
Youssef JG et al (2022) The use of IV vasoactive intestinal peptide (Aviptadil) in patients with critical COVID-19 respiratory failure: results of a 60-day randomized controlled trial. Crit Care Med 50(11):1545–1554
Yuhui Du LL, Zheng Y, Liu J, Gong J, Qiu Z, Li Y, Qiao J, Huo Y-X (2022) Incorporation of non-canonical amino acids into antimicrobial peptides: advances, challenges, and perspectives. Appl Environ Microbiol 88(23):e0161722
Yulema Valero EC-P, Meseguer J, Esteba MA, Cuesta A (2013) Biological role of fish antimicrobial peptides. In: Antimicrobial peptides. Nova Science
Zasloff M (2002) Antimicrobial peptides of multicellular organisms. Nature 415:389
Zhang QY et al (2021) Antimicrobial peptides: mechanism of action, activity and clinical potential. Mil Med Res 8(1):48

Chapter 2
Bacterial Antimicrobial Peptides: A Potential Alternative to Antibiotics

Sumeeta Kumari and Alka Kumari

1 Introduction

There is a wide variety of antimicrobial agents currently accessible in the market and widely employed to address various human infections. Particularly, antibiotics are effective drugs used to treat the infections caused by the pathogenic bacteria (Lei et al. 2019). Despite their widespread and sustained use in human and veterinary treatment, as well as in agriculture, particularly in developing nations, they have aided in the emergence and dissemination of drug-resistant microorganisms (Huan et al. 2020). The global increase in resistance to traditional antibiotics has been widely reported by the World Health Organisation (WHO), and it poses a potential and significant risk to public health (Moretta et al. 2021). The WHO has named multiple drug resistance (MDR) as a standout severe concern to development, food security, and health worldwide. According to estimates, roughly 700,000 people globally per year die from illnesses brought on by multidrug-resistant (MDR) bacteria. If nothing is done, this figure might increase to ten million deaths by 2050, surpassing the present monthly number of cancer-related deaths (Alaoui Mdarhri et al. 2022; O'Neill 2016; Lin et al. 2022). Furthermore, multiple studies found that the widespread use of antibiotics for hospitalized COVID-19 patients with no confirmed secondary illness enhanced antimicrobial resistance by pushing the selection of MDR. The Centers for Disease Control and Prevention (CDC) also concluded that antibiotic-resistant diseases are not only still a problem but have also gotten worse in their special report "COVID-19: U.S. Impact on Antimicrobial Resistance" (Alaoui Mdarhri et al. 2022). Therefore, finding new medicines with innovative

S. Kumari (✉)
Microbial Type Culture Collection, CSIR-Institute of Microbial Technology, Chandigarh, India

A. Kumari (✉)
Department of Chemical Engineering, Indian Institute of Technology Delhi, New Delhi, India

P. Baindara, S. M. Mandal (eds.), *Evolution of Antimicrobial Peptides*,
https://doi.org/10.1007/978-3-031-67515-7_2

mechanisms of action and approaches that are less likely to cause bacterial resistance is essential considering the problem of antibiotic resistance (Moretta et al. 2021). In the search for novel antibiotics, antimicrobial peptides (AMPs), alternatively referred to as host defense peptides, have stirred up a lot of interest (Moretta et al. 2021). AMPs have been defined as revolutionarily native antimicrobial weapons (Zasloff 2002). AMPs, which are produced by all organisms, ranging from prokaryotes to humans, play a vital role in innate immunity. AMPs, one of the important components of the innate immune system, provide immediate, nonspecific responses against pathogens (Wang et al. 2016a). AMPs (size <10 kDa) exhibit a wide range of antimicrobial activity against a wide spectrum of pathogens, encompassing bacteria, fungi, protozoa, and even some viruses (Lai and Gallo 2009). Unlike traditional antibiotics, which typically operate through a specified high-affinity antimicrobial target and can direct toward the advancement of resistance in microorganisms, AMPs employ a range of antimicrobial actions, which could be used as a technique to hamper bacteria from progressing resistance (Peschel and Sahl 2006). In addition to actively combating microorganisms, AMPs can also give protection through several pathways, such as the preservation of healthy gut homeostasis and the control of host inflammatory responses (Wang et al. 2016a).

Bacteriocins are a class of antimicrobials that have been generated by a variety of bacteria and recently garnered significance for their potential to fight antibiotic-resistant bacteria (Hassan et al. 2012). Bacteriocins are tiny, heat-stable peptides produced by bacterial sources that are active against other bacteria and countered by a specific immune mechanism possessed by the producer (Hirsch et al. 1951). Bacteriocin-mediated suppression was initially described 80 years ago when Cotter et al. discovered the antagonism between *Escherichia coli* strains (Cotter et al. 2005). AMPs encoded by genes and produced by bacteria are now commonly known as "bacteriocins," which is a change from their former name, "colicins," which represented the original producer organism (Hirsch et al. 1951). Dubos discovered the first AMP gramicidin in 1939 from *Bacillus brevis*, and its antimicrobial activity against infections was tested in mice. Bacillomycin, a peptide family discovered from *Bacillus subtilis* a few years later in 1948, has low antibacterial activity but significant antifungal activity (De Cesare et al. 2020). Several research have shown that bacteriocins have considerable possibility in the pharmaceutical sector as bacteriocinogenic probiotics and as therapeutic agents in the clinic (Hassan et al. 2012). They frequently have very high potency at very low concentrations and particularly target the species related to the producer as compared to eukaryotic AMPs, and that is why bacteriocins are thought to be potential antibacterial agents with applications ranging from food preservation to disease treatment (Nissen-Meyer and Nes 1997; Nes et al. 2007; Guaní-Guerra et al. 2010). Table 2.1 summarizes some of the AMPs approved by the Food and Drug Administration (FDA) for various infections. Bacteriocins kill their targets by a variety of different mechanisms, the most frequent of which are the destruction of target cells through the development of pores and/or the suppression of cell wall synthesis (Hassan et al. 2012). Based on the biochemical and genetic characteristics of bacteriocins produced by gram-positive bacteria, three groups have been identified. These groups include the

Table 2.1 List of few AMPs approved by the Food and Drug Administration (FDA) for various infections (Divyashree et al. 2019)

Generic name	Brand name	Source	FDA approval	Medical use
Vancomycin	Antimicrobial	*Amycolatopsis orientalis*	1954	Gram-positive bacterial infections that are fatal and resistant to other medicines
Bacitracin	Antimicrobial	*Bacillus subtilis*	1984	Infants with pneumonia and emphysema (Staphylococci)
Teicoplanin	Antimicrobial	Semisynthetic glycopeptide	1990	Treatment for diarrhea due to *Clostridium difficile* and pseudomembranous colitis
Daptomycin	Antimicrobial	*Streptomyces roseosporus*	2003	Bacterial infections of the skin and subcutaneous tissues
Enfuvirtide	Antiviral	Synthetic peptide	2003	Antiretroviral medications used to prevent HIV-1 fusion
Telavancin	Antimicrobial	Semisynthetic derivative of vancomycin	2009	Skin and skin structure infections that are complex (cSSSI)
Boceprevir	Antiviral	Synthetic peptide	2011	Genotype 1 of the hepatitis C virus
Telaprevir	Antiviral	Semisynthetic peptide	2011	NS3/4A serine protease from the genotype 1 of the hepatitis C virus
Dalbavancin	Antimicrobial	Semisynthetic lipoglycopeptide	2014	Bacterial infections of the skin and its structure
Oritavancin	Antimicrobial	Semisynthetic glycopeptide	2014	Acute bacterial infections of the skin and skin structure (ABSSSI)

presence or absence of disulfide bonds and monosulfide bonds, molecular weight, thermal stability, proteolytic enzyme stability, amino acid posttranslational modification, and antimicrobial activity of bacteriocins. Bacteriocins of class I, or lantibiotics, form lanthionine and methyllanthionine, typical amino acids that are the end result of posttranslational modification. These amino acids are small (11–19 kDa), heat-stable, and heat-resistant. Bacteriocins in class II (non-lantibiotics) have an amphiphilic helical structure, are small (<10 kDa), and contain peptides with no modified amino acid residues. There are two classes of class III bacteriocins, each comprising a bacteriolytic (IIIa) and a nonlytic mechanism of action (IIIb). Their molecular weight exceeds 30 kDa, and they are thermolabile and unmodified peptides (Hernández-González et al. 2021).

It is likely that the widespread presence of the genetic determinants responsible for bacteriocin synthesis has aided in the rapid spread of this feature among bacteria. Mobile genetic elements, like conjugative plasmids and transposons, contain genetic determinants for AMP production (Jack et al. 1995). Bacteriocins are thought to have an ecological function in nature where they give an advantage to the producer in competition among other bacterial species for shared resources, actively determining the microbial composition in particular habitats (Cotter et al. 2005). This brief chapter will concentrate on the potential of bacterial antimicrobial peptides (bacteriocins), antibiotic resistance, possible alternatives to antibiotics, and AMP applications in different fields.

2 Resistance to Antibiotics and Current Issues

The antibiotic era was inaugurated with the wonder medication penicillin in 1928, and modern medicine has advanced significantly since then (Mulani et al. 2019). Antibiotics underwent a golden era between 1940 and 1965 when many antimicrobials were discovered and introduced into modern medicine (Jadimurthy et al. 2022). Unfortunately, frequent antibiotic usage, self-medication, and hospital-acquired infections have provoked the emergence of antibimicrobial resistance (AMR) (Mulani et al. 2019). Due to the emergence of AMR, it is challenging to achieve universal health care, as it also hampers the attainment of several sustainable development goals, such as improved nutrition, food security, zero hunger, and sustainable agriculture (Denissen et al. 2022). Growing populations result in an increase in the consumption of antibiotics in the food-animal sector, as well as good health, hygiene, and sanitation. The World Health Organization (WHO) developed the global priority pathogen list in 2017 in order to encourage pharmaceutical research and development of new antibiotics. The WHO has latterly enumerated the pathogens in three categories: medium, high, and critical priority status, which urgently necessitate new antibiotics (Denissen et al. 2022). On the WHO list of priorities for pathogens, the multidrug-resistant (bacteria resistant to three or more classes of antibiotics) gram-negative ESKAPE (*Enterococcus faecium*, *Staphylococcus aureus*, *Klebsiella pneumoniae*, *Acinetobacter baumannii*, *Pseudomonas aeruginosa*, and *Enterobacter species*) pathogens are included in the "Priority 1: Critical" pathogen group. The gram-positive ESKAPE pathogens *Enterococcus faecium* (vancomycin-intermediate, vancomycin-resistant, and methicillin-resistant) and *Staphylococcus aureus* (vancomycin-resistant) are included in the "Priority 2: High" pathogen group. "Priority 3: Medium" pathogen group includes *Streptococcus pneumoniae*, *Haemophilus influenzae*, and *Shigella* spp (Jadimurthy et al. 2022). ESKAPE pathogens acquired resistance from genetic mutations and attainment of mobile genetic elements, target altercation, reduction in drug intake, overexpression of efflux pump, development of biofilm formation, production of exopolysaccharide and degrading enzyme against lipopeptides, β-lactams, tetracyclines, and antibiotics inclusive of glycopeptides and carbapenems (Hassan et al. 2012). AMR classification has largely been confined to clinical/hospital environments, but from extra-hospital reservoirs, ESKAPE strains have also been reported to be isolated. Indeed, common ESKAPE pathogen reservoirs include food (fruit, vegetables, and animal products), soil, diverse water sources (surface runoff, streams, drinking water, rivers, and dams, among others), plants, and municipal or sewage waste (Denissen et al. 2022). In recent decades, the occurrence of community-acquired and nosocomial fungal infections has elevated significantly. The risk of fungal infections has escalated in diseases associated with immunosuppression or immunodeficiency, and restricted therapeutic options accessible are emphasizing the search for new alternatives. Unfortunately, only three classes of potent antifungal agents like echinocandins (e.g., caspofungin), polyenes (e.g., Amphotericin), and triazoles (e.g., fluconazole) are available. However, their limited spectrum of activity and widespread use have contributed to the emergence of drug resistance strains of *Aspergillus*, *Candida*, *Fusarium*, and *Cryptococcus* (De Cesare et al. 2020). Nevertheless, several pathogenic fungi, such as *Candida auris*, *Mucorales*, and many molds, are primarily resistant to these drugs and difficult to cure now, highlighting the need for alternative antifungal methods (De Cesare et al. 2020).

Understanding the genesis of diseases linked to AMR, an ecological problem involving intricate interactivity between a variety of microbial populations that affect the health of animals, people, and the environment, requires a multidisciplinary approach. For instance, harmful bacteria may receive ABR genes from environmental reservoirs, changing how these organisms interact with their environment in the future. Consequently, a particular approach is required to assess the establishment of disease (such as vector-borne, zoonotic, and chronic), conduct risk analysis, and implement control measures for these pathogens in the environment (Denissen et al. 2022). Given that humans, food, and the environment contribute to the emergence of AMR, a One Health approach (people, animals, and ecosystems can all be balanced and optimized through a holistic approach) is one of the most important strategies to overcome the threat (Interagency Coordination Group on Antimicrobial Resistance 2019).

AMPs are promising antimicrobial agents with good activity against a diverse range of bacteria, viruses, and fungi. Given the limited alternatives available for treating such infections, their capacity to kill ESKAPE pathogens, particularly those exhibiting pan-drug resistance, presents a clinical advantage over antibiotics (Rima et al. 2021).

3 AMP Synthesis and Structural Properties

Insect, plant, and human innate immune systems utilize AMPs as one of the oldest and most prevalent antimicrobial strategies. However, AMP synthesis is not limited to multicellular organisms. Bacteriocins are ribosomally generated AMPs that are produced by a single bacterium and have antibacterial activity against other bacteria, either from the same species (limited spectrum) or from different genera (broad spectrum) (Bowdish et al. 2005). It has been proposed that a range of 30–99% of archaea and bacteria produce a minimum of one bacteriocin although genome analysis should eventually provide a more precise figure about number (Cotter et al. 2005). Antimicrobial peptides (AMPs) are produced by certain genes and are either constitutive or induced by external stimuli (Lei et al. 2019). They are, in fact, mostly small molecules (<100 amino acids), with a positive net charge (normally ranging from +2 to +11) and a significant section of hydrophobic residues (usually 50%) (Haney et al. 2017). These cationic AMPs are generally short, helical polypeptides with a positive charge, characterized by amino acid sequences of less than 100 residues, and they prominently feature the positively charged amino acids arginine and lysine (Rathinakumar and Wimley 2010). AMPs exhibit an amphipathic structure that renders them soluble in aqueous environments due to hydrophobic and hydrophilic regions. Anionic AMPs, with a negative net charge range from −1 to −7, are a less prevalent class of AMPs (Harris et al. 2009). Most of the AMPs' structural flexibility is due to their amphiphilic nature. Based on their secondary structure, AMPs are usually categorized into four types: helical peptides, sheet peptides, extended peptides, and loop peptides (Moretta et al. 2021; Zasloff 2002). The α-helical AMPs, which include LL-37, magainins, and cecropins, are distinguished by their structural formation, which includes an extraordinarily high proclivity for the formation of α-helix. Human α- and β-defensins, plectasin, and protegrins are examples of β-sheet AMPs that are stabilized by disulfide bridges (2–4) and create relatively stiff structures. Extended AMPs are rich in proline, tryptophan, arginine, and histidine and do not adopt a conventional secondary structure (e.g., indolicidin). The disulfide bridge helps the loop AMPs, like bactenecin, generate loop structures. Structure-based knowledge of AMPs will greatly assist in understanding their mechanisms of action and facilitate the development of more pharmaceutically useful AMPs (Wang et al. 2016a). Several different models attempt to describe the activity of these peptides. Some examples of these theories are the barrel-stave model, toroidal pore model, aggregate model, and carpet model, which come under non-membrane action (Lei et al. 2019; Zhang et al. 2021). In non-membrane action, the AMPs penetrate inside the cell and interact with intracellular substances, which leads to the inhibition of DNA and RNA, protein synthesis, protein folding, cell wall synthesis, enzyme activity, and so on (Zhang et al. 2021).

In the beginning, bacteriocins were divided up into four distinct classes (O'Shea et al. 2013). The fourth class of bacteriocins, which consists of large complexes with lipid or carbohydrate moieties, has been dropped and given the new name "bacteriolysins" (Bariş et al. 2013). Bacteriocins are thus primarily divided into three classes (Liu et al. 2014). Class I bacteriocins are made up of 19–50 amino acids and are significantly posttranslationally changed, leading to the inclusion of nonstandard amino acids like lanthionine, dehydrobutyrine, α-methyllanthionine, labyrinthine, and dehydroalanine. In class I, there exists subclass Ia for lantibiotics, subclass Ib for labyrinthopeptins, and subclass Ic for other classes (sanctibiotics) (Parada et al. 2007; Cuozzo et al. 2001). Class II is comprised of nonmodified, short peptides that are heat-stable, and it can be further split into class IIa (pediocin-like bacteriocins), class IIb (bacteriocins with two unmodified peptides), class IIc (circular bacteriocins), and class IId (noncircular bacteriocins) (linear, unmodified, non-pediocin-like bacteriocins) (Drider et al. 2006; Belguesmia et al. 2011). Class III bacteriocins are the big and heat-labile bacteriocins for which there is only limited data (Kumariya et al. 2019).

4 Use of AMPs as Alternatives to Antibiotics

Antibimicrobial-resistant bacteria (AMR) and the sharp rise in infectious diseases caused by pathogens have drawn attention to a situation that is reminiscent of the time before the development of antibiotics. Remarkably, antimicrobial peptides (AMPs) show promise as effective therapeutic options for infections that are resistant to common antibiotics (Elibe Mba and Innocent Nweze 2022). They are a good contender for antibiotic alternatives because of their capacity to enter bacteria, propensity to change, ability to form secondary structures like beta sheets and helices, and solubility in water, aqueous buffers, and organic solvents (Mazumdar and Adam 2021). They are often referred to as "natural antibiotics." Some AMPs can induce rapid mortality in gram-negative, gram-positive, parasites, fungi, encapsulated viruses, or cancerous cells, typically occurring within a few minutes. AMPs have a low chance of developing resistance, and they can even prevent the growth of bacteria that are resistant to antibiotics (Erdem Büyükkiraz and Kesmen 2022). Since AMPs are often hydrophobic and can electrostatically bind to the surface of bacterial membranes, they can readily pass through and damage the membrane structure, leading to cell death. AMPs can kill pathogens at numerous sites, unlike conventional antibiotics, which have only one target, significantly lowering the formation of drug-resistant microorganisms (Xuan et al. 2023). They possess broad-spectrum antibacterial activities and are now used to treat MDR infections, viral infections, fungal infections, wound healing, treatment for biofilms, and *Mycobacterium tuberculosis* infections (Fig. 2.1). Table 2.2 has highlighted a selection of a few antimicrobial compounds utilized across diverse fields. A comprehensive discussion of the various applications of antimicrobial peptides will be provided below. In general, they have the potential to become one of the most effective replacements for antibiotics in the years to come (Xuan et al. 2023).

4.1 Activity Against Multidrug-Resistant Bacteria and ESKAPE

MDR microorganisms have developed because of the improper usage of antibiotics. The multidrug-resistant strains of MRSA, vancomycin-resistant *Enterococcus* (VRE), multidrug-resistant *P. aeruginosa* (MDR-PA), and multidrug-resistant *A. baumannii* (MDR-AB) are the ones that are most frequently seen in clinics. AMPs have received significant attention recently as a potential antibiotic alternative due to their distinctive antibacterial mechanism and complicated antibacterial effect (Xuan et al. 2023). Biopeptides can affect bacterial membranes or other targets, unlike antibiotics. The methods of AMPs for treating drug-resistant bacterial infections include biofilm permeation, resensitization, intracellular bacteriostatic action, immunological activity control, and biofilm inhibition (Xuan et al. 2023). Natural salivary peptide histatin 5 has a strong anti-biofilm and bactericidal activity

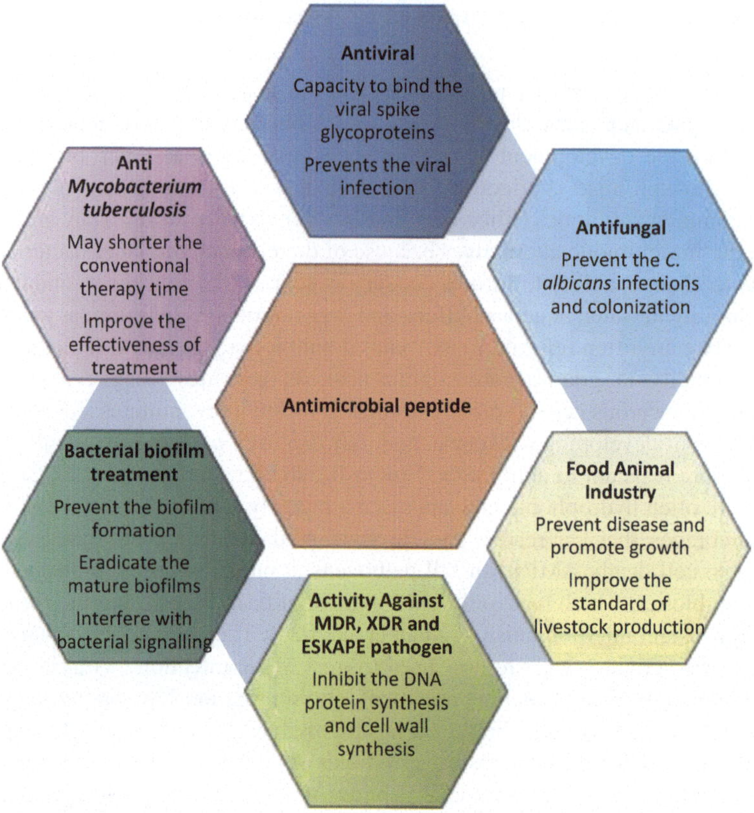

Fig. 2.1 Application of antimicrobial compounds in various fields

toward human ESKAPE (70%) in vitro (Juárez-Rodríguez et al. 2021). Gaglione et al. (2017) evaluated recombinant peptides derived from human ApoB, namely, rpApoBL and rpApoBS, which exhibited wound healing, antibacterial, anti-inflammatory, and antibiofilm properties against MDR strains of *P. aeruginosa* and *S. aureus* in 2017 (Mulani et al. 2019). In contrast, penisin, a lantibiotic, was effective in killing methicillin-resistant *Staphylococcus aureus* (MRSA) and did not cause hemolysis. Contrary to other lantibiotics, it was effective in inhibiting gram-negative bacteria's growth (Baindara et al. 2016a). Bierbaum et al. reported that a lantibiotic mersacidin is a strong inducer of the cell wall stress response of *Staphylococcus aureus* (Sass et al. 2008), and mersacidin effectively inhibits the growth of gram-positive bacteria, including methicillin-resistant *Staphylococcus aureus* strains (MRSA) in vitro and in vivo (Sass et al. 2008).

Staphylococcus aureus's toxicity is thought to be due in part to its ability to create biofilm. Biofilm formation leads to the spread of new bacteria, causing chronic infections that are challenging to treat (Archer et al. 2011). Cationic peptides and negatively charged elements on the exterior of the bacterial envelope are attracted to

Table 2.2 List of a few AMPs used in different fields, their amino acid sequence, and target

AMPs	Amino acid	Target group	References
Cathelicidin-BF	KFFRKLKKSVKKRAKEFFKKPRVIGVSIPF	Gram-negative, gram-positive	Rima et al. (2021)
Defensin	GFGCPLDQMQCHRHCQTITGRSGGYCSGPLKLTCTCYR	Gram-negative, gram-positive	Rima et al. (2021)
Gramicidin S	VXLfPVXLfP	Gram-negative, gram-positive, parasites, fungus, cancer cells, mammalian cells, SARS-CoV-2	Rima et al. (2021)
Polymyxin Colistin	XTXXXILXXT	Gram-negative	Rima et al. (2021)
Citropin-1.1	GLFDVIKKVASVIGGL	Gram-negative, gram-positive, fungus, mammalian cells, mollicute	Rima et al. (2021)
Caerin-1.10	GLLSVLGSVAKHVLPHVVPVIAEKL	Gram-negative, gram-positive, viruses, mammalian cells	Rima et al. (2021)
Murepavadin	LSYXXXXWXXASPP	Gram-negative, gram-positive, cancer cells, mammalian cells	Rima et al. (2021)
Protegrin-1	RGGRLCYCRRRFCVCVGR	*A. baumannii* (XDR and MDR strains from surgical wounds)	de Pontes et al. (2022)
EcDBS1R6	PMKKLFKLLARIAVKIPVW	*E. coli*, *P. aeruginosa*, *K. pneumoniae*, *A. baumannii*	de Pontes et al. (2022)
Iztli peptide 1 (IP-1)	KFLNRFWHWLQLKPGQPMY	*M. tuberculosis*	de Pontes et al. (2022)
B1CTcu5	LIAGLAANFLPQILCKIARKC	*M. tuberculosis*	de Pontes et al. (2022)
ARP788.14	KRWIILGLNKIVRMYSPTSI	*Candida utilis* and in vitro antifungal activity against yeast	de Pontes et al. (2022)
ARP788.13	PPIPVGEIYKRWIILGLNK	In vitro antifungal activity against yeast	de Pontes et al. (2022)

(continued)

Table 2.2 (continued)

AMPs	Amino acid	Target group	References
Ctn[15–34] (the C-terminal fragment of Crotalicidin peptide)	KKRLKKIFKKPMVIGVTIPF	*Candida albicans* biofilms	de Pontes et al. (2022)
pBD-2 (Porcine _-Defensin-2)	DHYICAKKGGTCNFSPCPLFNRIEGTCYSGKAKCCIR	PRV (pseudorabies virus), PRRSV (porcine reproductive and respiratory syndrome virus)	de Pontes et al. (2022)
D-HB43	FAKLLAKLAKKLL	Methicillin-resistant *S. aureus* strains	de Pontes et al. (2022)
Lactoferrin	–	Weanling pigs. Reduced total viable counts of *E. coli* and *Salmonella* in the small intestine	Wang et al. (2016a)
Antimicrobial peptide-A3	FLFSLIRKAIGGLISAFK	Broilers reduced *coliforms* and *Clostridium* spp. counts in feces	Wang et al. (2016a)
Sublancin	MKKFGILKKGCLSLLVLLFLITLIVIKAKTEATKIAAKL KASGELITIADDIAKKKLNKILKAGYGGVGLVTDN FSAVITKQAKQGLSASKKAQNVKDKLLIYVSAG YGPVNNILTVYNNQQPGQLILKIDAGNYKGITVTEA	Broilers reduced *Clostridium perfringens* in the cecum	Wang et al. (2016a)
GF-17	GFKRIVQRIKDFLRNLV	Zika virus (ZIKV)	Yang et al. (2023)

one another by electrostatic bonding. The peptide eventually reaches the cell membrane's surface and achieves membrane permeability via a variety of models (Yasir et al. 2018a). According to the most recent studies, AMPs can also break down cell membranes through unidentified methods. The disruption of quorum sensing, loss of biological membrane potential, alarm system inhibition, degradation of the polysaccharides and biofilm matrix, and inhibition of transporter expression are the five primary suggested mechanisms (Yasir et al. 2018a). In the case of resensitization, researchers suggested that sustained osmosis at subinhibitory concentrations of peptides may improve bacterial membrane permeability and help antibiotics reach targets more often (Marks et al. 2012). Many AMPs eliminate cellular endotoxin, boosting innate immunity for antibacterial actions (Bommineni et al. 2010). Currently, it is understood that several AMPs can prevent free cells from adhering to the biofilm by altering their shape. Another established way for how AMPs undermine quorum sensing is by lowering the expression of genes associated with it, which directly impacts the development of biofilms (Luo and Song 2021). Numerous studies demonstrate that AMPs impact various physiological processes in cells, including DNA and protein synthesis, protein folding, enzyme function, and cell wall formation (Xuan et al. 2023). This characteristic demonstrates the potential for AMPs to take the place of conventional antibiotics to exhibit antibacterial action and prevent drug resistance (Xuan et al. 2023).

4.2 Immunomodulatory Activity

Increasing data suggests that AMPs protect hosts from pathogens via mechanisms other than their direct antibacterial action. AMPs play an important role in innate and adaptive immunity. they have chemotactic activity, modulate anti- and proinflammatory responses, and have direct impacts on immune response (Wang et al. 2016a). AMPs regulate the immune microenvironment's dynamic equilibrium by controlling the section of cytokines such as interleukins, IFNs, tumor necrosis factors (TNFs), chemokines, and immune cell activities such as dendritic cells (DCs), macrophages, monocytes, granulocytes, mast cells, and lymphocytes. Peptides regulate cell surface receptors such as chemokine, cytokine, and G-protein-coupled receptors (GPCRs), Toll-like receptors (TLRs), and intracellular signal pathways like NF-κB, ERK1/2, p38, JNK, MAPK, and phosphoinositide 3-kinase (Zhang et al. 2021). LL-37 is expressed by mast cells, NK cells, neutrophils, and epithelial cells. It attracts immune cells to microbial invasion sites by binding to FPR1 (newly named FPR2) (Zhang et al. 2021). LL-37 suppresses neutrophil-programmed cell death and promotes keratinocyte migration. LL-37 binds to the bone marrow cell's integrin receptor to activate macrophages against gram-positive and gram-negative bacteria (Datta et al. 2023a). As a result, AMPs have an immunomodulatory effect, preventing infection-related tissue damage and maintaining microenvironment homeostasis (Zhang et al. 2021).

4.3 Antiviral Peptides

Enfuvirtide (Enf), a peptide medicine of 36-amino-acid residues, was the first peptide medication against the human immunodeficiency virus (HIV), to be sanctioned for clinical use. Enf prevented the binding of the heptad repeat 1 (HR1) domain with the HR2 domain, which prevented infection. In addition, peptide-based treatments for various viruses have been formulated, including those that cause influenza, hepatitis, herpes, and so on. In 2011, the FDA approved the synthetic peptides telaprevir and boceprevir to treat hepatitis C virus (HCV).

The current coronavirus outbreak, known as severe acute respiratory syndrome coronavirus (SARS-CoV-2) or COVID-19, is the third recorded case of animal-to-human transmission and has been declared a pandemic by the WHO. In recent decades, COVID-19 has been considered one of the most hazardous viral illnesses to evolve into a global pandemic, posing a significant threat to humanity. As a result, discovering an antiviral medicine or a vaccination is the only way for humans to survive COVID-19 infection. Apart from a few nonenveloped viruses, the AMPs' antiviral action was demonstrated against enveloped DNA and RNA viruses. The antiviral efficacy of AMPs is dependent on whether it acts directly on the viral envelope or via adsorption on the viral surface and entrance process (Elnagdy and Alkhazindar 2020).

Electrostatic interactions between AMPs and heparan sulfate and other negatively charged cell surface molecules are facilitated by the AMPs' positively charged residues. Glycosaminoglycan molecules found in heparan sulfate have a close connection to viral attachment. It has been demonstrated that viral infection can be decreased by AMPs that block heparan sulfate. Additionally, the antiviral impact of AMPs is linked to their capacity to prevent virus transmission from one cell to another via tight junctions (cell-to-cell spread) or to prevent the development of large cells (syncytium) (Elnagdy and Alkhazindar 2020).

Gramicidin S, which has a similar action to melittin, has recently been shown to be able to reduce the viral load and enhance viral clearance in Vero cells that are infected with SARS-CoV-2. Gramicidin S can bind with SARS-CoV-2 spike glycoprotein, according to an in silico study, which can prevent viral infection (Yang et al. 2023).

4.4 Food Animal Industry

In conventional livestock and poultry production, antibiotics are frequently used to prevent disease and promote growth (Wang et al. 2016a). Kerr et al. (2001) discovered that expressing a bioactive form of lysostaphin in mouse mammary glands inhibited *S. aureus*, the most common infectious mastitis pathogen. Donovan et al. (2005) were successful in generating transgenic dairy cows that expressed lysostaphin in their mammary epithelium as a step toward mastitis prevention and treatment (Mabrouk 2022). Transgenic AMP expression is a viable solution to some significant issues in animal husbandry, such as mammary gland infection, which has a direct impact on both animal health and milk production. According to studies, adding nisin to broiler chickens' diets modulated their gut microbial ecology and markedly reduced the number of *Bacteroides* and *Enterobacteriaceae* found in the ileum digesta. AMPs have the potential to get around problems with traditional antibiotics and improve the standard of livestock production (Erdem Büyükkiraz and Kesmen 2022).

4.5 Activity Against Mtb

According to the WHO, 10.6 million people were infected with TB in 2021, an increase of 4.5% from 10.1 million the previous year (Robert and Brown 2004). There was also an increase in the number of drug-resistant tuberculosis (Dr-Tb) cases in 2021, with 450,000 new cases of rifampicin-resistant tuberculosis (rr-Tb). Following COVID-19, *Mtb* ranked second as the leading infectious killer. As a result of vaccination and pandemic management, it is forecast that tuberculosis will return to be the top cause of death soon. Extreme drug resistance (XDR) and multi-drug resistance (MDR) *Mtb* strains employ distinct strategies to prevent antibiotics: i. Under hypoxic conditions, cell division increases (Robert and Brown 2004); ii. the cell wall becomes impermeable to antibiotics; iii. biofilm formation (Fux et al. 2005); iv. activation of efflux pumps to escape antibiotics (Te Brake et al. 2018; Ramón-García et al. 2009); and v. persistence is one of the major characteristics of Mtb, as some cells can evade antibiotic treatments even at concentrations far above their MIC (minimum inhibitory concentration) (Parbhoo et al. 2022). The treatment of drug-resistant TB needs to be improved, and antimicrobial peptides may reduce the length of conventional therapy and improve its effectiveness.

4.5.1 *Mtb* Resistance: A Novel Treatment

AMPs are antimicrobial compounds that can be used alone or in combination with other drugs (Mehta et al. 2022; Jacobo-Delgado et al. 2023). A lower minimum inhibitory concentration of synthetic analogs has also been demonstrated than that of natural AMPs (Travis et al. 2000). Different modalities are used by AMPs to exert their effects: (i) bacterial membrane disruption by electroporation or the formation of pores, (ii) cell death induction and interaction with bacterial components, (iii) cell wall biosynthesis inhibition, and (iv) antimicrobial action mediated by innate immune system activation (Silva and Gomes 2017; Kumar et al. 2018; Hancock and Sahl 2006).

AMPs have been studied first on *Mtb bacilli* using cytotoxic T lymphocytes. According to pioneer research published in 1999, granulysin kills *Mtb* dose-dependently after 72 hours. In addition, they described that perforins and granulysin are required for access to intracellular mycobacteria in vitro (Stenger et al. 1998). A more direct way of demonstrating AMP's protective effects on *Mtb* was to use human neutrophil peptide 1 (HNP-1) (Sharma et al. 2000; Miyakawa et al. 1996). According to other studies, HNP-1 application inhibited intracellular mycobacterial growth after 3 days of treatment in macrophages infected with mycobacteria (Sharma et al. 2000). *Mtb* colony-forming units (CFUs) were minimized by 99% when porcine leukocyte protegrin was used, even in clinical isolates (Miyakawa et al. 1996). As a result of an in vitro evaluation of human β-defensin 1 (HBD-1) and protegrin 1 (PG-1) activity, not only was HBD-1 or PG-1 effective against tuberculosis infections, but it was also shown that when combined with isoniazid, HBD-1 or PG-1 significantly reduced *Mtb* growth (Fattorini et al. 2004). AMPs such as, lacticin 3147 and nisin were derived through microbial fermentation and both peptides were reported to inhibit the synthesis of peptidoglycan in nonpathogenic *M. bovis* and *M. smegmatis* and also in clinically relevant mycobacteria (Carroll et al. 2010). The peptides lacticin 3147 and nisin bind to the precursor of peptidoglycan, which is located at the inner membrane of the cell. However, they are not intracellular targets (Rivas-Santiago et al. 2013).

Cathelicidins, lactoferrin (LF), and human defensins are major types of AMPs. Human cathelicidin, LL-37, binds to the microbial membrane outer leaflet, causing its disruption (Dürr et al. 2006). Furthermore, LL-37 has microbicidal action against *Mtb* through modulating the innate immune response (Yang et al. 2020). LL-37 was produced during *Mtb* infection after oral treatment of vitamin D3 and phenylbutyrate as well as upregulation of the vitamin D receptor, resulting in a reduction of *Mtb* intracellular load (Liu et al. 2006; Mily et al. 2015). Notably, LL-37 activation of infected macrophages resulted in the generation of anti-inflammatory cytokines, IL-10, and TGF-β implying immunological regulation in the early stages of *Mtb* infection (Dürr et al. 2006). Additionally, cathelicidin LL-37 lytic activity and chemicals E6 and E2 (peptide-derived), electron microscope studies revealed that these peptides E6 and E2 cause cytoplasmic components condensation in the *Mtb*. The following effect could be caused by the peptide's interaction with membrane (Thye et al. 2012). Among AMP analogs, acyldepsipeptides (ADEPs) appear to be a future generation of tuberculosis antibiotics (Cobongela et al. 2022). ADEPs have antibacterial action against *Mtb* strains by activating the caseinolytic protease ClpP1P2. AMPs' interactions with mycobacterial membranes and various mechanisms of action apart from AMP's lytic capacities are summarized in Table 2.3. AMPs appear to accumulate inside permeabilized bacterial cells after permeabilization, as demonstrated in these studies.

4.6 Antifungal Peptides

There are around 70,000 taxonomically unique fungi known, and some of these could pose major health risks to people (Li et al. 2012). One of the main human fungal pathogens, *Candida albicans*, can cause a variety of diseases, from

Table 2.3 Peptides and peptidomimetics with antimycobacterial activities are reported to interact with the cell membrane of infection-causing *Mycobacterium tuberculosis* (H37Rv and/or clinical isolates)

Peptides	AMPs' origin	Mechanism on *Mtb* membrane	Mechanism of action against *Mtb*	References
Callyaerin A	Marine sponge *Callyspongia aerizusa*		Not determined	Ibrahim et al. (2010), Daletos et al. (2015)
Capreomycin	*Streptomyces capreolus*		Synthesis of protein inhibited	Rastogi et al. (1996), Magnus et al. (1976)
Cyclomarin A	*Streptomyces* sp.		Inhibits ClpC1 action	Kiefer et al. (2019), Barbie and Kazmaier (2016), Schmitt et al. (2011), Vasudevan et al. (2013)
Dermcidin	Human sweat glands	Binds to the bacterial envelope and inhibits the synthesis of cell walls	RNA and protein synthesis inhibition	Li et al. (2009), Banerjee and Gohil (2016), Schittek et al. (2001)
E50-52	*E. faecalis*		E50-liposome complex formation without damage to the cell membrane	Sosunov et al. (2007), Teng et al. (2015)
Ecumicin	*Nonomuraea* sp. MIM5123		Inhibits ClpC1 action	Gao et al. (2015)
G-13	Synthetic	The membrane of the microbial cell is disrupted	Induces apoptosis of mammalian cells	Toro et al. (2006)
GranF2	Synthetic	The membrane of the microbial cell is disrupted	Induces apoptosis of mammalian cell	Toro et al. (2006), Andreu et al. (1999), Linde et al. (2005)
HBD-2	Human epithelial cells and leukocytes		By inhibiting vitamin D pathways, it inhibits bacterial growth	Nickel et al. (2012), Pazgier et al. (2006), Corrales-Garcia et al. (2013), Kisich et al. (2001)
HCL2	Associated with human cytochrome C oxidase subunit-III		Protein from culture filtrate interacts with it	Samuchiwal et al. (2014)

(continued)

Table 2.3 (continued)

Peptides	AMPs' origin	Mechanism on *Mtb* membrane	Mechanism of action against *Mtb*	References
Hepcidin-25	Human liver hepatocytes and macrophages		By preventing recycled iron from being released, this product inhibits bacterial growth	Sow et al. (2007), Sow et al. (2011)
HNP-1	Human polymorphonuclear neutrophil azurophilic granules	Biosynthesis of cell walls is inhibited and microbial cell membranes are disrupted	Inhibits synthesis of DNA	Banerjee and Gohil (2016), Sharma and Khuller (2001), Kalita et al. (2004)
IP-1	Synthetic	Not determined	Secretion of TNF-o and activation of autophagy	Coyotl et al. (2020)
Lassomycin	L. *kentuckyensis*		ATPase is uncoupled from proteolytic activity when it binds to ClpC1P1P2	Gavrish et al. (2014)
Laterosporulin-10	*Brevibacillus* sp.	Disruption of bacterial cell membrane	Not studied	Baindara et al. (2016b)
LL-37	Human neutrophils and macrophages	Pore formation	Immunomodulatory activity	Rivas-Santiago et al. (2008), Kapoor et al. (2011)
LLKKK18	An analog of LL-37, synthetic	Pore formation	Immunomodulatory activity	Silva et al. (2016a, b)
PR-39	Porcine (*Sus scrofa*) leucocytes		Inhibition of protein and DNA synthesis	Linde et al. (2001)
Pro region of hepcidin	Human liver hepatocytes and macrophages		By preventing recycled iron from being released, this product inhibits bacterial growth	Rivas-Santiago et al. (2008)

RN7, RN6, RN3 (1-45)	Produced by human eosinophil secondary granules	Disruption of bacterial cell wall	Intracellular macrophage killing, cell agglutination	Pulido et al. (2013)
llamycin A/rufomycin I	*Streptomyces* sp. (MIM3502) and *S. atratus* (NRRL B-16927)		Inhibits ClpC1 action	Sweatt et al. (2016), Khalil et al. (2014)
SAMPs-Dma	Synthetic		Binding of DNA	Sweatt et al. (2016)
Trichoderin A	*Trichoderma* species derived from marine sponges		An inhibitor of ATP synthase	Pruksakorn et al. (2011)
Viomycin	*Streptomyces puniceus*		Synthesis of protein inhibited	Akbergenov et al. (2011), Stanley et al. (2010)

superficial mucosal infections to hematogenous prominent candidiasis. AMPs, cytokine immunotherapy, monoclonal antibodies, vaccines, and other novel biopharmaceuticals have been developed to treat fungus infections (Nicola et al. 2019). Peptides are gaining attention as promising novel antibiotic agents. Antimicrobial peptides contain pleiotropic properties that allow them to exert potent antifungal activity in addition to broad-spectrum antibacterial activity, which may be effective in combating fungal infections. Fungal cell interference and lysis with the synthesis of the fungal cell wall were originally speculated as the AMP mechanisms of action against fungi (Wang et al. 2016b). *C. albicans* and *Cryptococcus neoformans* cells were efficiently eradicated by the synthetic cathelicidin peptides SMAP-29, indolicidin, BMAP-28, BMAP-27, and protegrin-1 through membrane permeabilization and damage (Benincasa et al. 2006). However, AMPs appear to exert their effects via additional mechanisms, according to a growing corpus of research. Human defensins 2 and 3 could eradicate *C. albicans* in a salt-sensitive and energy-dependent way without generating severe membrane disruption or lysis, according to Vylkova et al. (Vylkova et al. 2007). However, further research is needed to understand the precise mechanisms of destruction. Abiotic surfaces, murine urinary bladders, and oral epidermis can be inhibited by LL-37 by interacting with yeast protein cell wall components and carbohydrates. This is crucial for preventing *C. albicans* infection and colonization by antimicrobial peptides (Tsai et al. 2011).

4.7 Bacterial Biofilm Treatment

The biofilm is an extracellular polymeric matrix embedded in a sessile community of bacteria. It is notoriously hard to eradicate bacteria embedded in biofilms, compared to those in a planktonic state, since they are intrinsically resistant to antimicrobial agents. In recent decades, there has been a dearth of novel antibiotics, compounded by the emergence of resistance among pathogens with a tendency to form biofilms. Therefore, solutions are urgently needed to address these crises. Antimicrobial peptides seem to be strong antibiofilm agents, as they tend to prevent the formation of biofilm and also eradicate matured biofilms formed by multidrug-resistant bacteria. AMPs possess numerous anti-biofilm defenses, including (i) membrane potential degradation, (ii) interference with bacterial cell signaling systems, (iii) exopolysaccharide layer degradation, (iv) alarmone system degradation, and (v) biofilm-forming genes, that are downregulated (Datta et al. 2023b; Yasir et al. 2018b). Table 2.4 summarizes AMPs against the formation of biofilm and its antibacterial properties.

Table 2.4 Highlights of promising AMPs against biofilm formation and potential antibacterial properties

Peptides	Peptides antimicrobial activity	Sequence and properties	Highlights	References
Pexiganan	*Staphylococcus* sp., *Streptococcus* spp., *E. faecium*, *Corynebacterium* sp., *Pseudomonas* sp., *Acinetobacter* sp., *Stenotrophomonas* sp.	GIGKFLKKAKKFGKAFVKILKK-NH2	Anti-biofilm agent with weak activity against CL structures	Maciejewska et al. (2016)
Citropin 1.1	*S. aureus*, *P. aeruginosa*, *S. pyogenes*, *E. coli*, *S. epidermidis*	GLFDVIKKVASVIGGL-NH2	Efficacious antibiofilm agent against strains of *S. aureus*	Maciejewska et al. (2016)
Palm-KK-NH2	*S. aureus*, *P. aeruginosa*, *S. pyogenes*, *E. coli*, *S. epidermidis*	Palm-KK-NH2 (palm–hexadecanoic acid residue)	It works as an anti-biofilm agent against most strains. Combination with standard antibiotics potentiates activity	Maciejewska et al. (2016)
D-Ranalexin	MRSA strains	FLGGLIKIVPAMICAVTKKC	Biofilms are effectively killed in a dose-dependent manner, but it has a high cytotoxic and hemolytic effect	Zapotoczna et al. (2017)
AMP2	*E. coli*, *E. faecalis*, *S. aureus*, *S. epidermidis*, *P. aeruginosa*	KRRWRIWLV	Biofilm area has been reduced by 76%	Bormann et al. (2017)
GH12	*S. mutans*	GLLWHLLHHLLH-NH2	In vitro antimicrobial activity against cariogenic bacteria and their biofilms	Wang et al. (2017)
LyeTxI	*P. gingivalis*, *F. nucleatum*, *A. actinomycetemcomitans*	IWLTALKFLGKNLGKHLALKQQLAKL	It inhibits the growth of periodontopathic bacteria. It has a rapid bactericidal effect and prevents the development of biofilms. Dental applications are possible	Cruz Olivo et al. (2017)

(continued)

Table 2.4 (continued)

Peptides	Peptides antimicrobial activity	Sequence and properties	Highlights	References
L12	*P. aeruginosa*, *S. aureus*, *K. pneumoniae*, *E. coli*	LKKLLKKLLKKL-NH2	The mechanism of action involves pore formation, inhibiting biofilm formation, causing rapid permeabilization of bacterial membranes, suppressing LPS-induced pro-inflammatory mediators, and disrupting drug-resistant biofilms, even at low concentrations of L12.	Khara et al. (2017)
SAAP-148	*E. faecium*, *S. aureus*, *P. aeruginosa*, *A. baumannii*, and *Enterobacter* spp., *K. pneumoniae*	LKRVWKRVFKLLKRYWRQLKKPVR	Because of its broad antibacterial activity against multidrug-resistant strains, biofilms forming strains, and persistent bacteria, this promising peptide fights difficult-to-treat infections	De Breij et al. (2018)
WAM-1	*A. baumannii*	KRGFGKKLRKRLKKFRNSIKKRLKNFNVVIPIPLPG	As a result of its origin in LL37 AMPs, this peptide has a greater effect on preventing biofilm dispersion than LL-37, its parent peptide	Spencer et al. (2018)

5 AMPs: Face of the 2019 SARS-CoV-2 Pandemic

The Coronavirus Disease Pandemic in 2019 has emerged due to SARS-CoV-2, a highly pathogenic virus. Approximately 700 million cases of COVID-19 have been reported, and six million deaths have been reported around the world as of October 2023, according to the WHO report (https://covid19.who.int/). Antimicrobial peptides (AMPs) are commonly recognized as a promising approach for combating pathogenic microorganisms (Wang et al. 2016c). AMPs have been shown to act on the membrane of bacteria, fungi, and viruses as well as modulate higher organisms' innate immune responses. Lactoferrin (LF), an AMP, is an iron-binding glycoprotein found in a variety of mucosal secretions. LF has antiviral effects against a wide variety of animal and human viruses (RNA and DNA) due to its capacity to prevent viral replication. Likewise, it has been demonstrated that LF increases host immunity to viral infection rather than combating the virus once it has infected the host. This is accomplished by inhibiting virus entry into the host cell either directly by binding to the virus particles or by blocking cellular receptors. Several studies have demonstrated that LF has a vital role in the fight against viruses like HSV, HCV, polio, HIV, and rotavirus. The LF antiviral mechanisms, however, differ among viruses and can attach either directly to the host cell receptor or to the virus particle (Superti et al. 1997; Ikeda et al. 2000). In a study, the peptide analog mucroporin-M1 (LFRLIKSLIKRLVSAFK), which differs from the parent peptide mucroporin in four amino acid positions (P6K, G3R, G11R, and G10K), exhibits therapeutic effects against SARS-CoV-2 (Mahendran et al. 2020). Similarly, by blocking protein synthesis without causing any adverse effects, minor modifications to lactococcin G and glycocin F improved their effectiveness against SARS-CoV-2 (Balmeh et al. 2021). In the development of anti-SARS-CoV-2 drugs against SARS-CoV-2 and its variants, lipopeptides like surfactants could also be promising drug agents. The surfactin produced by *Bacillus clausii* TS, a probiotic bacterial strain with a molecular weight similar to surfactin C, exhibits antiviral activity against a variety of enveloped viruses as shown by competitive ELISA assay, molecular docking, dynamics, and simulation experiments, and the producer probiotic strain TS has the potential to act as a dual-active drug entity against SARS-CoV-2 (Baindara et al. 2023). Two other AMPs, caerin 1.6 and caerin 1.10, show a limited affinity for the ACE2 receptor but a very high potential to interact with the SARS-CoV-2 spike surface protein. These peptides are interesting candidates for SARS-CoV-2 therapy due to their selectivity toward viral proteins (Liscano et al. 2020).

6 AMPs as Conjugated or Subunit Vaccine

Despite the development of numerous vaccines against infectious diseases, the swift emergence of new pathogens underscores the pressing requirement for innovative strategies to address infectious diseases (Dinata and Baindara 2023). The successful advancement of novel vaccines relies significantly on adjuvants or subunit vaccines and formulations capable of not only eliciting robust humoral responses against a specific pathogen but also activating effector and memory $CD4^+$ and $CD8^+$ T cells. Nevertheless, most vaccines approved for human use or presently undergoing clinical examination struggle to induce effective cellular responses. Addressing these limitations, recent research has concentrated on formulating combinations of adjuvants not only to enhance antigen-specific immune responses but also to facilitate their modulation (Ebensen et al. 2019). Beyond the direct applications, AMPs can be explored for development as subunit vaccines or utilized as highly immunogenic carrier proteins for antigens that are highly antigenic but non-immunogenic. Currently, immunoinformatic tools enable the study of the physiochemical and immunogenic properties of peptides through molecular dynamic simulations and in silico immune simulations. However, these findings should be substantiated through in vitro and in vivo analyses. Few antimicrobial peptides (AMPs) such as laterosporulin 25, identified as a probiotic peptide through immunoinformatic tools, are proposed to possess defensin-like bacteriocin characteristics and exhibit immunogenic properties. These attributes could potentially facilitate their utilization as carrier proteins or in the development of subunit vaccines (Dinata and Baindara 2023).

7 Limitation of AMPs

While broad-spectrum antimicrobial peptides (AMPs) offer numerous advantages, there are still practical application disadvantages associated with their use (Table 2.5). Nevertheless, there exist constraints in both the production and application of antimicrobial peptides across diverse fields and commercial uses. However, advanced technologies, including refined genetic engineering and purification techniques, need further enhancement to boost the production yield of antimicrobial peptides (Lyu et al. 2023).

Table 2.5 Advantages and disadvantages of AMPs (Rima et al. 2021; Garvey 2023)

Disadvantages	Advantages
Cytotoxicity	Some AMPs synergize with standard antibiotics
Bacterial resistance to specific AMPs may develop	Broad-spectrum antibacterial action against bacteria, viruses, yeast, and fungus
Limited stability	Shorter sequences of amino acids are simpler to synthesize
Brief half-lives	Rapid onset of action
Instability toward protease	Potent
High expense of extraction	Possibility of using adjuvants for vaccines
Toxicity to microbial expression systems during production	Some AMPs are stable and active in a wide pH range
Absence of specificity	Broad-spectrum activity

8 Conclusions

Since AMPs can be used as alternative or complementary approaches for treating microbial infections, they have garnered significant interest in recent years due to their promising potential. Compared with conventional antibiotics, these peptides are a good alternative because of their potency, wide-spectrum activity, variety of natural sources, low accumulation in tissues, and rapid killing ability. The research community is also pursuing AMPs as potential commercially available drugs because they also possess anti-inflammatory, immunomodulating, and antioxidant properties. AMPs were discussed in this chapter, along with their mechanisms of action and sources, along with their antimicrobial activity against several human pathogens. Several potential applications for AMPs in medicine can be explained by their ability to regulate pro-inflammatory reactions, regulate wound healing, and stimulate cell proliferation, via modulating cell migration, chemotaxis, angiogenesis, and cytokine release. To enhance the delivery and effectiveness of AMPs, pharmaceutical companies are investing considerable efforts in developing them as therapeutic agents, and this involves improving the metabolic and chemical stability of AMPs and devising innovative formulation strategies.

References

Akbergenov R, Shcherbakov D, Matt T, Duscha S, Meyer M, Wilson DN, Böttger EC (2011) Molecular basis for the selectivity of antituberculosis compounds capreomycin and viomycin. Antimicrob Agents Chemother 55:4712–4717. https://doi.org/10.1128/AAC.00628-11

Alaoui Mdarhri H, Benmessaoud R, Yacoubi H, Seffar L, Guennouni Assimi H, Hamam M, Boussettine R, Filali-Ansari N, Lahlou FA, Diawara I et al (2022) Alternatives therapeutic approaches to conventional antibiotics: advantages, limitations and potential application in medicine. Preprint at MDPI. https://doi.org/10.3390/antibiotics11121826

Andreu D, Carreño C, Linde C, Boman HG, Andersson M (1999) Identification of an anti-mycobacterial domain in NK-lysin and granulysin. Biochem J 344:845–849. https://doi.org/10.1042/0264-6021:3440845

Archer NK, Mazaitis MJ, William Costerton J, Leid JG, Powers ME, Shirtliff ME (2011) Staphylococcus aureus biofilms: properties, regulation and roles in human disease. Preprint at Taylor and Francis Inc. https://doi.org/10.4161/viru.2.5.17724

Baindara P, Chaudhry V, Mittal G, Liao LM, Matos CO, Khatri N, Franco OL, Patil PB, Korpole S (2016a) Characterization of the antimicrobial peptide penisin, a class Ia novel lantibiotic from Paenibacillus sp. strain A3. Antimicrob Agents Chemother 60:580–591. https://doi.org/10.1128/AAC.01813-15

Baindara P, Singh N, Ranjan M, Nallabelli N, Chaudhry V, Pathania GL, Sharma N, Kumar A, Patil PB, Korpole S (2016b) Laterosporulin10: a novel defensin like class iid bacteriocin from brevibacillus sp. strain SKDU10 with inhibitory activity against microbial pathogens. Microbiology (United Kingdom) 162:1286–1299. https://doi.org/10.1099/mic.0.000316

Baindara P, Chowdhury T, Roy D, Mandal M, Mandal SM (2023) Surfactin-like lipopeptides from Bacillus clausii efficiently bind to spike glycoprotein of SARS-CoV-2. J Biomol Struct Dyn. https://doi.org/10.1080/07391102.2023.2196694

Balmeh N, Mahmoudi S, Fard NA (2021) Manipulated bio antimicrobial peptides from probiotic bacteria as proposed drugs for COVID-19 disease. Inform Med Unlock 23:100515. https://doi.org/10.1016/j.imu.2021.100515

Banerjee DI, Gohil TP (2016) Interaction of antimicrobial peptide with mycolyl transferase in Mycobacterium tuberculosis. Int J Mycobacteriol 5:83–88. https://doi.org/10.1016/j.ijmyco.2015.07.002

Barbie P, Kazmaier U (2016) Total synthesis of Cyclomarin A, a marine Cycloheptapeptide with anti-tuberculosis and anti-malaria activity. Org Lett 18:204–207. https://doi.org/10.1021/acs.orglett.5b03292

Bariş Ö, Güllüce M, Karadayı M, Barış Ö (2013) Bacteriocins: promising natural antimicrobials. https://doi.org/10.13140/2.1.5014.5606

Belguesmia Y, Naghmouchi K, Chihib N-E, Drider D (2011) Class IIa Bacteriocins: current knowledge and perspectives. In: Prokaryotic antimicrobial peptides. Springer, New York, pp 171–195. https://doi.org/10.1007/978-1-4419-7692-5_10

Benincasa M, Scocchi M, Pacor S, Tossi A, Nobili D, Basaglia G, Busetti M, Gennaro R (2006) Fungicidal activity of five cathelicidin peptides against clinically isolated yeasts. J Antimicrob Chemother 58:950–959. https://doi.org/10.1093/jac/dkl382

Bommineni YR, Achanta M, Alexander J, Sunkara LT, Ritchey JW, Zhang G (2010) A fowlicidin-1 analog protects mice from lethal infections induced by methicillin-resistant Staphylococcus aureus. Peptides (N.Y.) 31:1225–1230. https://doi.org/10.1016/j.peptides.2010.03.037

Bormann N, Koliszak A, Kasper S, Schoen L, Hilpert K, Volkmer R, Kikhney J, Wildemann B (2017) A short artificial antimicrobial peptide shows potential to prevent or treat bone infections. Sci Rep 7:1–14. https://doi.org/10.1038/s41598-017-01698-0

Bowdish DME, Davidson DJ, Hancock REW (2005) A re-evaluation of the role of host defence peptides in mammalian immunity. Curr Protein Pept Sci 6(1):35

Carroll J, Draper LA, O'Connor PM, Coffey A, Hill C, Ross RP, Cotter PD, O'Mahony J (2010) Comparison of the activities of the lantibiotics nisin and lacticin 3147 against clinically significant mycobacteria. Int J Antimicrob Agents 36:132–136. https://doi.org/10.1016/j.ijantimicag.2010.03.029

Cobongela SZZ, Makatini MM, Mdluli PS, Sibuyi NRS (2022) Acyldepsipeptide analogues: a future generation antibiotics for tuberculosis treatment. Pharmaceutics 14. https://doi.org/10.3390/pharmaceutics14091956

Corrales-Garcia L, Ortiz E, Castañeda-Delgado J, Rivas-Santiago B, Corzo G (2013) Bacterial expression and antibiotic activities of recombinant variants of human β-defensins on pathogenic bacteria and M. tuberculosis. Protein Expr Purif 89:33–43. https://doi.org/10.1016/j.pep.2013.02.007

Cotter D, Hills C, Ross RP (2005) Bacteriocins: developing innate immunity for food. Nat Rev Microbiol 3:777–788. https://doi.org/10.1038/nrmicro1240

Coyotl EAP, Palacios JB, Muciño G, Moreno-Blas D, Costas M, Montes TM, Diener C, Uribe-Carvajal S, Massieu L, Castro-Obregón S et al (2020) Antimicrobial peptide against mycobacterium tuberculosis that activates autophagy is an effective treatment for tuberculosis. Pharmaceutics 12:1–24. https://doi.org/10.3390/pharmaceutics12111071

Cruz Olivo EA, Santos D, de Lima ME, dos Santos VL, Sinisterra RD, Cortés ME (2017) Antibacterial effect of synthetic peptide LyeTxI and LyeTxI/β-Cyclodextrin association compound against planktonic and multispecies biofilms of periodontal pathogens. J Periodontol 88:e88–e96. https://doi.org/10.1902/jop.2016.160438

Cuozzo SA, Sesma FJM, Pesce AA, Holgado R, Raya RR (2001) Methods for the detection and concentration of Bacteriocins produced by lactic acid bacteria. In: Food microbiology protocols. Methods in biotechnology. Humana Press

Daletos G, Kalscheuer R, Koliwer-Brandl H, Hartmann R, De Voogd NJ, Wray V, Lin W, Proksch P (2015) Callyaerins from the marine sponge Callyspongia aerizusa: cyclic peptides with antitubercular activity. J Nat Prod 78:1910–1925. https://doi.org/10.1021/acs.jnatprod.5b00266

Datta M, Rajeev A, Chattopadhyay I (2023a) Application of antimicrobial peptides as next-generation therapeutics in the biomedical world. Preprint at Taylor and Francis Ltd. https://doi.org/10.1080/02648725.2023.2199572

Datta M, Rajeev A, Chattopadhyay I (2023b) Application of antimicrobial peptides as next-generation therapeutics in the biomedical world. Biotechnol Genet Eng Rev 00:1–39. https://doi.org/10.1080/02648725.2023.2199572

De Breij A, Riool M, Cordfunke RA, Malanovic N, De Boer L, Koning RI, Ravensbergen E, Franken M, Van Der Heijde T, Boekema BK et al (2018) The antimicrobial peptide SAAP-148 combats drug-resistant bacteria and biofilms. Sci Transl Med 10. https://doi.org/10.1126/scitranslmed.aan4044

De Cesare GB, Cristy SA, Garsin DA, Lorenz MC (2020) Antimicrobial peptides: a new frontier in antifungal therapy. MBio 11:1–21. https://doi.org/10.1128/mBio.02123-20

de Pontes JTC, Borges ABT, Roque-Borda CA, Pavan FR (2022) Antimicrobial peptides as an alternative for the eradication of bacterial biofilms of multi-drug resistant bacteria. Preprint at MDPI. https://doi.org/10.3390/pharmaceutics14030642

Denissen J, Reyneke B, Waso-Reyneke M, Havenga B, Barnard T, Khan S, Khan W (2022) Prevalence of ESKAPE pathogens in the environment: antibiotic resistance status, community-acquired infection and risk to human health. Preprint at Elsevier GmbH. https://doi.org/10.1016/j.ijheh.2022.114006

Dinata R, Baindara P (2023) Laterosporulin25: a probiotically produced, novel defensin-like bacteriocin and its immunogenic properties. Int Immunopharmacol 121. https://doi.org/10.1016/j.intimp.2023.110500

Divyashree M, Mani MK, Reddy D, Kumavath R, Ghosh P, Azevedo V, Barh D (2019) Clinical applications of antimicrobial peptides (AMPs): where do we stand now? Protein Pept Lett 27:120–134. https://doi.org/10.2174/0929866526666190925152957

Donovan DM, Kerr DE, Wall RJ (2005) Engineering disease resistant cattle. Preprint. https://doi.org/10.1007/s11248-005-0670-8

Drider D, Fimland G, Héchard Y, McMullen LM, Prévost H (2006) The continuing story of class IIa Bacteriocins. Microbiol Mol Biol Rev 70:564–582. https://doi.org/10.1128/mmbr.00016-05

Dürr UHN, Sudheendra US, Ramamoorthy A (2006) LL-37, the only human member of the cathelicidin family of antimicrobial peptides. Biochim Biophys Acta Biomembr 1758:1408–1425. https://doi.org/10.1016/j.bbamem.2006.03.030

Ebensen T, Delandre S, Prochnow B, Guzmán CA, Schulze K (2019) The combination vaccine adjuvant system Alum/c-di-AMP results in quantitative and qualitative enhanced immune responses post immunization. Front Cell Infect Microbiol 9. https://doi.org/10.3389/fcimb.2019.00031

Elibe Mba I, Innocent Nweze E (2022) Antimicrobial peptides therapy: an emerging alternative for treating drug-resistant bacteria. Yale J Biol Med 95(4):445

Elnagdy S, Alkhazindar M (2020) The potential of antimicrobial peptides as an antiviral therapy against COVID-19. Preprint at American Chemical Society. https://doi.org/10.1021/acsptsci.0c00059

Erdem Büyükkiraz M, Kesmen Z (2022) Antimicrobial peptides (AMPs): a promising class of antimicrobial compounds. Preprint at John Wiley and Sons Inc. https://doi.org/10.1111/jam.15314

Fattorini L, Gennaro R, Zanetti M, Tan D, Brunori L, Giannoni F, Pardini M, Orefici G (2004) In vitro activity of protegrin-1 and beta-defensin-1, alone and in combination with isoniazid, against Mycobacterium tuberculosis. Peptides (NY) 25:1075–1077. https://doi.org/10.1016/j.peptides.2004.04.003

Fux CA, Costerton JW, Stewart PS, Stoodley P (2005) Survival strategies of infectious biofilms. Trends Microbiol 13:34–40. https://doi.org/10.1016/j.tim.2004.11.010

Gaglione R, Dell'Olmo E, Bosso A, Chino M, Pane K, Ascione F, Itri F, Caserta S, Amoresano A, Lombardi A et al (2017) Novel human bioactive peptides identified in Apolipoprotein B: evaluation of their therapeutic potential. Biochem Pharmacol 130:34–50. https://doi.org/10.1016/j.bcp.2017.01.009

Gao W, Kim JY, Anderson JR, Akopian T, Hong S, Jin YY, Kandror O, Kim JW, Lee IA, Lee SY et al (2015) The cyclic peptide ecumicin targeting CLpC1 is active against Mycobacterium tuberculosis in vivo. Antimicrob Agents Chemother 59:880–889. https://doi.org/10.1128/AAC.04054-14

Garvey M (2023) Antimicrobial peptides demonstrate activity against resistant bacterial pathogens. Preprint at Multidisciplinary Digital Publishing Institute (MDPI). https://doi.org/10.3390/idr15040046

Gavrish E, Sit CS, Cao S, Kandror O, Spoering A, Peoples A, Ling L, Fetterman A, Hughes D, Bissell A et al (2014) Lassomycin, a ribosomally synthesized cyclic peptide, kills mycobacterium tuberculosis by targeting the ATP-dependent protease ClpC1P1P2. Chem Biol 21:509–518. https://doi.org/10.1016/j.chembiol.2014.01.014

Guaní-Guerra E, Santos-Mendoza T, Lugo-Reyes SO, Terán LM (2010) Antimicrobial peptides: general overview and clinical implications in human health and disease. Preprint. https://doi.org/10.1016/j.clim.2009.12.004

Hancock REW, Sahl HG (2006) Antimicrobial and host-defense peptides as new anti-infective therapeutic strategies. Nat Biotechnol 24:1551–1557. https://doi.org/10.1038/nbt1267

Haney EF, Mansour SC, Hancock REW (2017) Antimicrobial peptides: An introduction. In: Methods in molecular biology. Humana Press Inc., pp 3–22. https://doi.org/10.1007/978-1-4939-6737-7_1

Harris F, Dennison SR, Phoenix DA (2009) Anionic antimicrobial peptides from eukaryotic organisms. Curr Protein Pept Sci 10(6):585

Hassan M, Kjos M, Nes IF, Diep DB, Lotfipour F (2012) Natural antimicrobial peptides from bacteria: characteristics and potential applications to fight against antibiotic resistance. Preprint. https://doi.org/10.1111/j.1365-2672.2012.05338.x

Hernández-González JC, Martínez-Tapia A, Lazcano-Hernández G, García-Pérez BE, Castrejón-Jiménez NS (2021) Bacteriocins from lactic acid bacteria. A powerful alternative as antimicrobials, probiotics, and immunomodulators in veterinary medicine. Preprint at MDPI AG. https://doi.org/10.3390/ani11040979

Hirsch A, Grinsted E, Chapman HR, Mattick ATR (1951) 446. A note on the inhibition of an anaerobic sporeformer in Swiss-type cheese by a Nisin-producing Streptococcus. Cambridge University Press

Huan Y, Kong Q, Mou H, Yi H (2020) Antimicrobial peptides: classification, design, application and research progress in multiple fields. Preprint at Frontiers Media SA. https://doi.org/10.3389/fmicb.2020.582779

Ibrahim SRM, Min CC, Teuscher F, Ebel R, Kakoschke C, Lin W, Wray V, Edrada-Ebel R, Proksch P (2010) Callyaerins A-F and H, new cytotoxic cyclic peptides from the Indonesian marine sponge Callyspongia aerizusa. Bioorg Med Chem 18:4947–4956. https://doi.org/10.1016/j.bmc.2010.06.012

Ikeda M, Nozaki A, Sugiyama K, Tanaka T, Naganuma A, Tanaka K, Sekihara H, Shimotohno K, Saito M, Kato N (2000) Characterization of antiviral activity of lactoferrin against hepatitis C virus infection in human cultured cells. Virus Res 66:51–63. https://doi.org/10.1016/S0168-1702(99)00121-5

Interagency Coordination Group on Antimicrobial Resistance (2019)

Jack RW, Tagg JR, Ray B (1995) Bacteriocins of gram-positive bacteria. Microbiol Rev 59(2):171

Jacobo-Delgado YM, Rodríguez-Carlos A, Serrano CJ, Rivas-Santiago B (2023) Mycobacterium tuberculosis cell-wall and antimicrobial peptides: a mission impossible? Front Immunol 14:1–18. https://doi.org/10.3389/fimmu.2023.1194923

Jadimurthy R, Mayegowda SB, Nayak SC, Mohan CD, Rangappa KS (2022) Escaping mechanisms of ESKAPE pathogens from antibiotics and their targeting by natural compounds. Biotechnol Rep 34. https://doi.org/10.1016/j.btre.2022.e00728

Juárez-Rodríguez MM, Cortes-López H, García-Contreras R, González-Pedrajo B, Díaz-Guerrero M, Martínez-Vázquez M, Rivera-Chávez JA, Soto-Hernández RM, Castillo-Juárez I (2021) Tetradecanoic acids with anti-virulence properties increase the pathogenicity of Pseudomonas aeruginosa in a Murine Cutaneous Infection Model. Front Cell Infect Microbiol 10. https://doi.org/10.3389/fcimb.2020.597517

Kalita A, Verma I, Khuller GK (2004) Role of human neutrophil peptide-1 as a possible adjunct to antituberculosis chemotherapy. J Infect Dis 190:1476–1480. https://doi.org/10.1086/424463

Kapoor R, Eimerman PR, Hardy JW, Cirillo JD, Contag CH, Barron AE (2011) Efficacy of antimicrobial peptoids against Mycobacterium tuberculosis. Antimicrob Agents Chemother 55:3058–3062. https://doi.org/10.1128/AAC.01667-10

Kerr DE, Plaut K, John BA, Christine MWJ, Lax A, Moore KD, Wells K, Wall RJ (2001) Lysostaphin expression in mammary glands confers protection against Staphylococcal infection in transgenic mice. Nat Biotechnol 19:66–70

Khalil ZG, Salim AA, Lacey E, Blumenthal A, Capon RJ (2014) Wollamides: Antimycobacterial cyclic hexapeptides from an australian soil Streptomyces. Org Lett 16:5120–5123. https://doi.org/10.1021/ol502472c

Khara JS, Obuobi S, Wang Y, Hamilton MS, Robertson BD, Newton SM, Yang YY, Langford PR, Ee PLR (2017) Disruption of drug-resistant biofilms using de novo designed short α-helical antimicrobial peptides with idealized facial amphiphilicity. Acta Biomater 57:103–114. https://doi.org/10.1016/j.actbio.2017.04.032

Kiefer A, Bader CD, Held J, Esser A, Rybniker J, Empting M, Müller R, Kazmaier U (2019) Synthesis of new Cyclomarin derivatives and their biological evaluation towards mycobacterium tuberculosis and plasmodium falciparum. Chem Eur J 25:8894–8902. https://doi.org/10.1002/chem.201901640

Kisich KO, Heifets L, Higgins M, Diamond G (2001) Antimycobacterial agent based on mRNA encoding human β-defensin 2 enables primary macrophages to restrict growth of Mycobacterium tuberculosis. Infect Immun 69:2692–2699. https://doi.org/10.1128/IAI.69.4.2692-2699.2001

Kumar P, Kizhakkedathu JN, Straus SK (2018) Antimicrobial peptides: diversity, mechanism of action and strategies to improve the activity and biocompatibility in vivo. Biomol Ther 8. https://doi.org/10.3390/biom8010004

Kumariya R, Garsa AK, Rajput YS, Sood SK, Akhtar N, Patel S (2019) Bacteriocins: classification, synthesis, mechanism of action and resistance development in food spoilage causing bacteria. Preprint at Academic Press. https://doi.org/10.1016/j.micpath.2019.01.002

Lai Y, Gallo RL (2009) AMPed up immunity: how antimicrobial peptides have multiple roles in immune defense. Preprint. https://doi.org/10.1016/j.it.2008.12.003

Lei J, Sun L, Huang S, Zhu C, Li P, He J, Mackey V, Coy DH, He Q (2019) The antimicrobial peptides and their potential clinical applications. Am J Transl Res 11(7):3919

Li M, Rigby K, Lai Y, Nair V, Peschel A, Schittek B, Otto M (2009) Staphylococcus aureus mutant screen reveals interaction of the human antimicrobial peptide dermcidin with membrane phospholipids. Antimicrob Agents Chemother 53:4200–4210. https://doi.org/10.1128/AAC.00428-09

Li Y, Xiang Q, Zhang Q, Huang Y, Su Z (2012) Overview on the recent study of antimicrobial peptides: origins, functions, relative mechanisms and application. Peptides (N.Y.) 37:207–215. https://doi.org/10.1016/j.peptides.2012.07.001

Lin J, Du F, Long M, Li P (2022) Limitations of phage therapy and corresponding optimization strategies: a review. Preprint at MDPI. https://doi.org/10.3390/molecules27061857

Linde CMA, Hoffner SE, Refai E, Andersson M (2001) In vitro activity of PR-39, a proline-arginine-rich peptide, against susceptible and multi-drug-resistant Mycobacterium tuberculosis. J Antimicrob Chemother 47:575–580. https://doi.org/10.1093/jac/47.5.575

Linde CMA, Grundström S, Nordling E, Refai E, Brennan PJ, Andersson M (2005) Conserved structure and function in the granulysin and NK-lysin peptide family. Infect Immun 73:6332–6339. https://doi.org/10.1128/IAI.73.10.6332-6339.2005

Liscano Y, Oñate-Garzón J, Ocampo-Ibáñez ID (2020) In silico discovery of antimicrobial peptides as an alternative to control sars-cov-2. Molecules 25:11–14. https://doi.org/10.3390/molecules25235535

Liu PT, Stenger S, Li H, Wenzel L, Tan BH, Krutzik SR, Ochoa MT, Schauber J, WU K, Meinken C et al (2006) Toll-like receptor triggering of a vitamin D–mediated human. Science (1979) 311:1770

Liu W, Pang H, Zhang H, Cai Y (2014) Biodiversity of lactic acid bacteria. In: Lactic acid bacteria: fundamentals and practice. Springer Netherlands, pp 103–203. https://doi.org/10.1007/978-94-017-8841-0_2

Luo Y, Song Y (2021) Mechanism of antimicrobial peptides: antimicrobial, anti-inflammatory and antibiofilm activities. Preprint at MDPI. https://doi.org/10.3390/ijms222111401

Lyu Z, Yang P, Lei J, Zhao J (2023) Biological function of antimicrobial peptides on suppressing pathogens and improving host immunity. Preprint at MDPI. https://doi.org/10.3390/antibiotics12061037

Mabrouk DM (2022) Antimicrobial peptides: features, applications and the potential use against covid-19. Preprint at Springer Science and Business Media B.V. https://doi.org/10.1007/s11033-022-07572-1

Maciejewska M, Bauer M, Neubauer D, Kamysz W, Dawgul M (2016) Influence of amphibian antimicrobial peptides and short lipopeptides on bacterial biofilms formed on contact lenses. Materials 9. https://doi.org/10.3390/ma9110873

Magnus PD, Santiago FD, Squires RG, Chemical T, Shiba T (1976) Haynes, g. leclerc 427:1109–1111

Mahendran ASK, Lim YS, Fang CM, Loh HS, Le CF (2020) The potential of antiviral peptides as COVID-19 therapeutics. Front Pharmacol 11:1–5. https://doi.org/10.3389/fphar.2020.575444

Marks LR, Clementi EA, Hakansson AP (2012) The human milk protein-lipid complex HAMLET sensitizes bacterial pathogens to traditional antimicrobial agents. PLoS One 7. https://doi.org/10.1371/journal.pone.0043514

Mazumdar A, Adam V (2021) Antimicrobial peptides-an alternative candidates to antibiotics against Staphylococcus aureus and its antibiotic-resistant strains. J Mol Clin Med 4:1. https://doi.org/10.31083/j.jmcm.2021.01.208

Mehta K, Sharma P, Mujawar S, Vyas A (2022) Role of antimicrobial peptides in treatment and prevention of mycobacterium tuberculosis: a review. Int J Pept Res Ther 28:1–13. https://doi.org/10.1007/s10989-022-10435-9

Mily A, Rekha RS, Kamal SMM, Arifuzzaman ASM, Rahim Z, Khan L, Haq MA, Zaman K, Bergman P, Brighenti S et al (2015) Significant effects of oral phenylbutyrate and Vitamin D3 adjunctive therapy in pulmonary tuberculosis: a randomized controlled trial. PLoS One 10:1–25. https://doi.org/10.1371/journal.pone.0138340

Miyakawa Y, Ratnakar P, Rao AG, Costello ML, Mathieu-Costello O, Lehrer RI, Catanzaro A (1996) In vitro activity of the antimicrobial peptides human and rabbit defensins and porcine leukocyte protegrin against Mycobacterium tuberculosis. Infect Immun 64:926–932. https://doi.org/10.1128/iai.64.3.926-932.1996

Moretta A, Scieuzo C, Petrone AM, Salvia R, Manniello MD, Franco A, Lucchetti D, Vassallo A, Vogel H, Sgambato A et al (2021) Antimicrobial peptides: a new hope in biomedical and pharmaceutical fields. Preprint at Frontiers Media S.A. https://doi.org/10.3389/fcimb.2021.668632

Mulani MS, Kamble EE, Kumkar SN, Tawre MS, Pardesi KR (2019) Emerging strategies to combat ESKAPE pathogens in the era of antimicrobial resistance: a review. Front Microbiol 10. https://doi.org/10.3389/fmicb.2019.00539

Nes IF, Diep DB, Holo H (2007) Bacteriocin diversity in Streptococcus and Enterococcus. Preprint. https://doi.org/10.1128/JB.01254-06

Nickel D, Busch M, Mayer D, Hagemann B, Knoll V, Stenger S (2012) Hypoxia triggers the expression of human β Defensin 2 and antimicrobial activity against mycobacterium tuberculosis in human macrophages. J Immunol 188:4001–4007. https://doi.org/10.4049/jimmunol.1100976

Nicola AM, Albuquerque P, Paes HC, Fernandes L, Costa FF, Kioshima ES, Abadio AKR, Bocca AL, Felipe MS (2019) Antifungal drugs: new insights in research & development. Pharmacol Ther 195:21–38. https://doi.org/10.1016/j.pharmthera.2018.10.008

Nissen-Meyer J, Nes IF (1997) Ribosomally synthsized antimicrobial peptides: their function, structure, biogenesis, and mechanism of action. Arch Microbiol 167:67–77

O'Neill J (2016) Tackling drug-resistant infections globally: final report and recommendations the review on antimicrobial resistance chaired by Jim O'Neill. HM Government and Wellcome Trust, London

O'Shea EF, O'Connor PM, O'Sullivan O, Cotter PD, Ross RP, Hill C (2013) Bactofencin A, a new type of cationic bacteriocin with unusual immunity. MBio 4. https://doi.org/10.1128/mBio.00498-13

Parada JL, Caron CR, Bianchi A, Medeiros P, Soccol CR (2007) Brazilian archives of biology and technology. Bacteriocins from lactic acid bacteria: purification, properties and use as biopreservatives. Braz Arch Biol Technol 50:521–542

Parbhoo T, Schurz H, Mouton JM, Sampson SL (2022) Persistence of Mycobacterium tuberculosis in response to infection burden and host-induced stressors. Front Cell Infect Microbiol 12:1–18. https://doi.org/10.3389/fcimb.2022.981827

Pazgier M, Hoover DM, Yang D, Lu W, Lubkowski J (2006) Human β-defensins. Cell Mol Life Sci 63:1294–1313. https://doi.org/10.1007/s00018-005-5540-2

Peschel A, Sahl HG (2006) The co-evolution of host cationic antimicrobial peptides and microbial resistance. Preprint. https://doi.org/10.1038/nrmicro1441

Pruksakorn P, Arai M, Liu L, Moodley P, Jacobs WR, Kobayashi M (2011) Action-mechanism of trichoderin A, an anti-dormant mycobacterial aminolipopeptide from marine sponge-derived Trichoderma sp. Biol Pharm Bull 34:1287–1290. https://doi.org/10.1248/bpb.34.1287

Pulido D, Torrent M, Andreu D, Nogues MV, Boix E (2013) Two human host defense ribonucleases against mycobacteria, the eosinophil cationic protein (RNase 3) and RNase 7. Antimicrob Agents Chemother 57:3797–3805. https://doi.org/10.1128/AAC.00428-13

Ramón-García S, Martín C, Thompson CJ, Aínsa JA (2009) Role of the Mycobacterium tuberculosis P55 efflux pump in intrinsic drug resistance, oxidative stress responses, and growth. Antimicrob Agents Chemother 53:3675–3682. https://doi.org/10.1128/AAC.00550-09

Rastogi N, Labrousse V, Goh KS (1996) In vitro activities of fourteen antimicrobial agents against drug susceptible and resistant clinical isolates of Mycobacterium tuberculosis and comparative intracellular activities against the virulent H37Rv strain in human macrophages. Curr Microbiol 33:167–175. https://doi.org/10.1007/s002849900095

Rathinakumar R, Wimley WC (2010) High-throughput discovery of broad-spectrum peptide antibiotics. FASEB J 24:3232–3238. https://doi.org/10.1096/fj.10-157040

Rima M, Rima M, Fajloun Z, Sabatier JM, Bechinger B, Naas T (2021) Antimicrobial peptides: a potent alternative to antibiotics. Antibiotics 10. https://doi.org/10.3390/antibiotics10091095

Rivas-Santiago B, Hernandez-Pando R, Carranza C, Juarez E, Contreras JL, Aguilar-Leon D, Torres M, Sada E (2008) Expression of cathelicidin LL-37 during Mycobacterium tuberculosis infection in human alveolar macrophages, monocytes, neutrophils, and epithelial cells. Infect Immun 76:935–941. https://doi.org/10.1128/IAI.01218-07

Rivas-Santiago B, Rivas Santiago CE, Castañeda-Delgado JE, León-Contreras JC, Hancock REW, Hernandez-Pando R (2013) Activity of LL-37, CRAMP and antimicrobial peptide-derived compounds E2, E6 and CP26 against Mycobacterium tuberculosis. Int J Antimicrob Agents 41:143–148. https://doi.org/10.1016/j.ijantimicag.2012.09.015

Robert B, Brown EB (2004) Global tuberculosis report (WHO)

Samuchiwal SK, Tousif S, Singh DK, Kumar A, Ghosh A, Bhalla K, Prakash P, Kumar S, Bhattacharyya M, Moodley P et al (2014) A peptide fragment from the human COX3 protein disrupts association of Mycobacterium tuberculosis virulence proteins ESAT-6 and CFP10, inhibits mycobacterial growth and mounts protective immune response. BMC Infect Dis 14:1–11. https://doi.org/10.1186/1471-2334-14-355

Sass P, Jansen A, Szekat C, Sass V, Sahl HG, Bierbaum G (2008) The lantibiotic mersacidin is a strong inducer of the cell wall stress response of Staphylococcus aureus. BMC Microbiol 8. https://doi.org/10.1186/1471-2180-8-186

Schittek B, Hipfel R, Sauer B, Bauer J, Kalbacher H, Stevanovic S, Schirle M, Schroeder K, Blin N, Meier F et al (2001) Dermcidin: a novel human antibiotic peptide secreted by sweat glands. Nat Immunol 2:1133–1137. https://doi.org/10.1038/ni732
Schmitt EK, Riwanto M, Sambandamurthy V, Roggo S, Miault C, Zwingelstein C, Krastel P, Noble C, Beer D, Rao SPS et al (2011) The natural product Cyclomarin kills mycobacterium tuberculosis by targeting the ClpC1 subunit of the Caseinolytic protease. Angew Chem 123:6011–6013. https://doi.org/10.1002/ange.201101740
Sharma S, Khuller G (2001) DNA as the intracellular secondary target for antibacterial action of human neutrophil peptide-I against Mycobacterium tuberculosis H37Ra. Curr Microbiol 43:74–76. https://doi.org/10.1007/s002840010263
Sharma S, Verma I, Khuller GK (2000) Antibacterial activity of human neutrophil peptide-1 against Mycobacterium tuberculosis H37Rv: in vitro and ex vivo study. Eur Respir J 16:112–117. https://doi.org/10.1034/j.1399-3003.2000.16a20.x
Silva T, Gomes MS (2017) Immuno-stimulatory peptides as a potential adjunct therapy against intra-macrophagic pathogens. Molecules 22:1–12. https://doi.org/10.3390/molecules22081297
Silva JP, Gonçalves C, Costa C, Sousa J, Silva-Gomes R, Castro AG, Pedrosa J, Appelberg R, Gama FM (2016a) Delivery of LLKKK18 loaded into self-assembling hyaluronic acid nanogel for tuberculosis treatment. J Control Release 235:112–124. https://doi.org/10.1016/j.jconrel.2016.05.064
Silva JP, Appelberg R, Gama FM (2016b) Antimicrobial peptides as novel anti-tuberculosis therapeutics. Biotechnol Adv 34:924–940. https://doi.org/10.1016/j.biotechadv.2016.05.007
Sosunov V, Mischenko V, Eruslanov B, Svetoch E, Shakina Y, Stern N, Majorov K, Sorokoumova G, Selishcheva A, Apt A (2007) Antimycobacterial activity of bacteriocins and their complexes with liposomes. J Antimicrob Chemother 59:919–925. https://doi.org/10.1093/jac/dkm053
Sow FB, Florence WC, Satoskar AR, Schlesinger LS, Zwilling BS, Lafuse WP (2007) Expression and localization of hepcidin in macrophages: a role in host defense against tuberculosis. J Leukoc Biol 82:934–945. https://doi.org/10.1189/jlb.0407216
Sow FB, Nandakumar S, Velu V, Kellar KL, Schlesinger LS, Amara RR, Lafuse WP, Shinnick TM, Sable SB (2011) Mycobacterium tuberculosis components stimulate production of the antimicrobial peptide hepcidin. Tuberculosis 91:314–321. https://doi.org/10.1016/j.tube.2011.03.003
Spencer JJ, Pitts RE, Pearson RA, King LB (2018) The effects of antimicrobial peptides WAM-1 and LL-37 on multidrug resistant Acinetobacter baumannii. Pathogen Dis. https://doi.org/10.1093/femspd/fty007/4822137
Stanley RE, Blaha G, Grodzicki RL, Strickler MD, Steitz TA (2010) The structures of the anti-tuberculosis antibiotics viomycin and capreomycin bound to the 70S ribosome. Nat Struct Mol Biol 17:289–293. https://doi.org/10.1038/nsmb.1755
Stenger S, Hanson DA, Teitelbaum R, Dewan P, Niazi KR, Froelich CJ, Ganz T, Thoma-Uszynski S, Melián A, Bogdan C et al (1998) An antimicrobial activity of cytolytic T cells mediated by granulysin. Science 1979(282):121–125. https://doi.org/10.1126/science.282.5386.121
Superti F, Ammendolia MG, Valenti P, Seganti L (1997) Antirotaviral activity of milk proteins: Lactoferrin prevents rotavirus infection in the enterocyte like cell line HT-29. Med Microbiol Immunol 186:83–91. https://doi.org/10.1007/s004300050049
Sweatt SK, Gower BA, Chieh AY, Liu Y, Li L (2016) High-resolution structure of ClpC1-Rufomycin and ligand binding studies provide a framework to design and optimize anti-tuberculosis leads. Physiol Behav 176:139–148. https://doi.org/10.1021/acsinfecdis.8b00276.Additional
Te Brake LHM, De Knegt GJ, De Steenwinkel JE, Van Dam TJP, Burger DM, Russel FGM, Van Crevel R, Koenderink JB, Aarnoutse RE (2018) The role of efflux pumps in tuberculosis treatment and their promise as a target in drug development: unraveling the Black Box. Annu Rev Pharmacol Toxicol 58:271–291. https://doi.org/10.1146/annurev-pharmtox-010617-052438
Teng T, Liu J, Wei H (2015) Anti-mycobacterial peptides: from human to phage. Cell Physiol Biochem 35:452–466. https://doi.org/10.1159/000369711

Thye T, Owusu-Dabo E, Vannberg FO, Van Crevel R, Curtis J, Sahiratmadja E, Balabanova Y, Ehmen C, Muntau B, Ruge G et al (2012) Common variants at 11p13 are associated with susceptibility to tuberculosis. Nat Genet 44:257–259. https://doi.org/10.1038/ng.1080

Toro JC, Hoffner S, Linde C, Andersson M, Andersson J, Grundström S (2006) Enhanced susceptibility of multidrug resistant strains of Mycobacterium tuberculosis to granulysin peptides correlates with a reduced fitness phenotype. Microbes Infect 8:1985–1993. https://doi.org/10.1016/j.micinf.2006.02.030

Travis SM, Anderson NN, Forsyth WR, Espiritu C, Conway BD, Greenberg EP, McCray PB, Lehrer RI, Welsh MJ, Tack BF (2000) Bactericidal activity of mammalian cathelicidin-derived peptides. Infect Immun 68:2748–2755. https://doi.org/10.1128/IAI.68.5.2748-2755.2000

Tsai PW, Yang CY, Chang HT, Lan CY (2011) Human antimicrobial peptide LL-37 inhibits adhesion of Candida albicans by interacting with yeast cell-wall carbohydrates. PLoS One 6. https://doi.org/10.1371/journal.pone.0017755

Vasudevan D, Rao SPS, Noble CG (2013) Structural basis of mycobacterial inhibition by Cyclomarin A. J Biol Chem 288:30883–30891. https://doi.org/10.1074/jbc.M113.493767

Vylkova S, Nayyar N, Li W, Edgerton M (2007) Human β-defensins kill Candida albicans in an energy-dependent and salt-sensitive manner without causing membrane disruption. Antimicrob Agents Chemother 51:154–161. https://doi.org/10.1128/AAC.00478-06

Wang S, Zeng X, Yang Q, Qiao S (2016a) Antimicrobial peptides as potential alternatives to antibiotics in food animal industry. Preprint at MDPI AG. https://doi.org/10.3390/ijms17050603

Wang S, Zeng X, Yang Q, Qiao S (2016b) Antimicrobial peptides as potential alternatives to antibiotics in food animal industry. Int J Mol Sci 17. https://doi.org/10.3390/ijms17050603

Wang G, Li X, Wang Z (2016c) APD3: the antimicrobial peptide database as a tool for research and education. Nucleic Acids Res 44:D1087–D1093. https://doi.org/10.1093/nar/gkv1278

Wang Y, Fan Y, Zhou Z, Tu H, Ren Q, Wang X, Ding L, Zhou X, Zhang L (2017) De novo synthetic short antimicrobial peptides against cariogenic bacteria. Arch Oral Biol 80:41–50. https://doi.org/10.1016/j.archoralbio.2017.03.017

Xuan J, Feng W, Wang J, Wang R, Zhang B, Bo L, Chen ZS, Yang H, Sun L (2023) Antimicrobial peptides for combating drug-resistant bacterial infections. Preprint at Churchill Livingstone. https://doi.org/10.1016/j.drup.2023.100954

Yang B, Good D, Mosaiab T, Liu W, Ni G, Kaur J, Liu X, Jessop C, Yang L, Fadhil R et al (2020) Significance of LL-37 on immunomodulation and disease outcome. Biomed Res Int 2020. https://doi.org/10.1155/2020/8349712

Yang M, Liu S, Zhang C (2023) Antimicrobial peptides with antiviral and anticancer properties and their modification and nanodelivery systems. Preprint at Elsevier BV. https://doi.org/10.1016/j.crbiot.2023.100121

Yasir M, Willcox MDP, Dutta D (2018a) Action of antimicrobial peptides against bacterial biofilms. Preprint at MDPI AG. https://doi.org/10.3390/ma11122468

Yasir M, Willcox MDP, Dutta D (2018b) Action of antimicrobial peptides against bacterial biofilms. Materials 11. https://doi.org/10.3390/ma11122468

Zapotoczna M, Forde É, Hogan S, Humphreys H, O'gara JP, Fitzgerald-Hughes D, Devocelle M, O'Neill E (2017) Eradication of staphylococcus aureus biofilm infections using synthetic antimicrobial peptides. J Infect Dis 215:975–983. https://doi.org/10.1093/infdis/jix062

Zasloff, M. (2002) Antimicrobial peptides of multicellular organisms

Zhang QY, Yan ZB, Meng YM, Hong XY, Shao G, Ma JJ, Cheng XR, Liu J, Kang J, Fu CY (2021) Antimicrobial peptides: mechanism of action, activity and clinical potential. Preprint at BioMed Central Ltd. https://doi.org/10.1186/s40779-021-00343-2

Chapter 3
Antimicrobial Peptide in Oral Health: A Therapeutic Approach

Alka Kumari and Sumeeta Kumari

1 Exploring the Oral Microbiome and Its Role in Oral Infections

An oral cavity is an ecosystem that fosters the interaction and proliferation of a diverse variety of microorganisms, commensals, and pathogens. It is one of the few ecosystems within the human body that is simultaneously dynamic, open, and closed (Hans and Madaan Hans 2014). The oral cavity functions as the gateway to the digestive tract, housing 700 distinct types of microbiota throughout its various regions, including the mouth (Bik et al. 2010; Ding and Schloss 2014; Deo and Deshmukh 2019; Burcham et al. 2020; Griffith et al. 2022; Bowen et al. 2018; Marsh and Devine 2011; Cugini et al. 2019; Vu et al. 2009; Hajishengallis 2015; Vieira Colombo et al. 2015). The oral microbiome of an individual is established as bacteria inoculate the mouth cavity, beginning with their first eating and progressing through certain events, like tooth eruption, that allow specific microorganisms to colonize specific tissue niches (Griffith et al. 2022). Teeth, among the various tissue types within the oral cavity, uniquely serve as opened, non-shedding covering in the body that readily harbors microorganisms, enabling the formation of protective biofilms that support the thriving and proliferation of bacteria (Bowen et al. 2018; Marsh and Devine 2011). Biofilms are composed of bacterial cells entangled within an extracellular matrix that includes extracellular bacterial DNA, proteins, and extracellular polysaccharides (EPS) (Cugini et al. 2019; Vu et al. 2009). The biofilms permit the mouth to serve as a reservoir for potentially harmful bacteria, which

A. Kumari (✉)
Department of Chemical Engineering, Indian Institute of Technology Delhi, New Delhi, India

S. Kumari (✉)
Microbial Type Culture Collection, CSIR-Institute of Microbial Technology, Chandigarh, India

P. Baindara, S. M. Mandal (eds.), *Evolution of Antimicrobial Peptides*,
https://doi.org/10.1007/978-3-031-67515-7_3

can migrate to other regions of the body to cause secondary site infections when shed from their biofilm colony, making it vital to treat the infection in the oral cavity (Hajishengallis 2015; Vieira Colombo et al. 2015). As oral bacteria multiply on hard and soft tissue, they form plaque, a multispecies biofilm (Griffith et al. 2022). When bacteria found in plaque produce acid that harms the tooth, dental caries begin to form. Dental cavities must be avoided by maintaining tooth minerals and managing oral bacteria biofilms.

Streptococcus mutans, *Actinomyces* species, *S. sobrinus*, and *Lactobacilli* are recognized as the key cariogenic bacteria, while *Actinobacillus actinomycetemcomitans* is associated with periodontal disease, particularly among young individuals. However, the primary instigator of cavity formation remains *S. mutans*, a facultative anaerobic gram-positive coccus. In substantial numbers, this bacterium metabolizes sucrose to produce significant amounts of extracellular glucan, a glucose polymer, facilitating its colonization of tooth surfaces. In combination with *S. mutans*, *Actinomyces viscosus*, and *Lactobacillus acidophilus*, a diverse group of bacteria colonizes and collaboratively forms pathogenic plaque biofilms, which significantly contribute to the development of oral infectious diseases. Oral pathogens can be categorized into different groups, including facultative anaerobic bacteria such as *Lactobacillus* species, *Actinomyces* species, *Streptococcus* species, *Veillonella* species, and *Porphyromonas* species.

Additionally, there are obligate anaerobic bacteria like *Fusobacterium* species and *Tannerella* species, along with common aerobic bacteria such as *Pseudomonas* species and *Staphylococcus* species. Consortia of two anaerobes (*Fusobacterium nucleatum* and *Porphyromonas gingivalis*), obligate anaerobe (*Prevotella intermedia*), and facultative anaerobe (*Aggregatibacter actinomycetemcomitans*) are the reason behind periodontal infections. In addition to bacterial infections, another concern is candidiasis or oral thrush, where *Candida albicans*, a typical yeast, gathers on the mouth lining, resulting in creamy white lesions (Luong et al. 2022). Inadequate denture hygiene practices often lead to a common oral cavity issue, creating a favorable environment for *Candida* growth in the gap between the oral mucosa and dentures (Ramage et al. 2012). Moreover, individuals with a collapsed bite due to tooth loss are susceptible to *Candida* growth in the moist areas around the corners of their mouth. Given that *Streptococcus mutans* and *Candida albicans* are both acid-tolerant microorganisms (De Bernardis et al. 1998), a recent study has investigated possible interactions between these genetically distinct organisms in the context of severe caries ecology (Bachtiar and Bachtiar 2018; Kim et al. 2017). Meanwhile, chronic periodontitis is attributed to an overabundance of anaerobic bacteria, including *Treponema denticola*, *Tannerella forsythia*, and *P. gingivalis* (Luong et al. 2022).

During ancient times in China, for alleviating toothache, arsenic trioxide was employed, and this same inorganic compound is now utilized in the treatment of acute promyelocytic leukemia (Siderov and Duggan 2010). Antonie van Leeuwenhoek, a pioneering microscopist and microbiologist, used a microscope to observe the presence of bacteria in his dental plaque (Cotter et al. 2013). Miller proposed that the metabolism of sugars into acids is responsible for the demineralization of teeth (Luong et al. 2022). The bacterium responsible for dental caries was identified as *S. mutans*, which was isolated by Clarke (Bowen et al. 2018). By using a hamster model, Keyes validated the bacterial presence in dental caries, emphasizing the significance of elapsed time as a crucial factor in caries etiology (Keyes 1969). Following these occurrences, antibiotics and different antimicrobial agents have been prescribed to treat and prevent oral diseases (Luong et al. 2022; Simón-Soro and Mira 2015). Clinicians employ antibiotics to address bacterial infections within the oral cavity, including tooth pain caused by sinusitis, dental abscesses, pericoronitis, and acute gingivitis (Griffith et al. 2022; Lockhart et al. 2019).

Apart from these infections, oral cancer, which is the most prevalent form of cancer in the oral cavity, is oral tongue squamous cell carcinoma (OTSCC), which has a high prevalence of invasiveness and metastasis and aggressive clinical behavior (Vierthaler et al. 2020). In the world, oral cancer is the sixth most prevalent cancer. It is prevalent in Europe, Melanesia, and South Central Asia. In Taiwan, oral cancer ranked sixth among all cancers and fourth among males although the 5-year survival rate for patients with this disease was only 60% in the United States in 2011. Oral behaviors that have been linked to an increased risk of oral cancer include chewing betel quid, drinking alcohol, and smoking cigarettes (Lee et al. 2017). Patients with locally advanced disease still have an unsatisfactory prognosis, and patients without disease still have a high relapse rate despite recent improvements in multidisciplinary treatment for oral cancer. Most oral cancers are OSCC, and some researchers have also discovered the microbes associated with oral cancer, which can be explained by DNA damage brought on by inflammation in epithelial cells because of endotoxins generated by microorganisms. Oral cancer and epithelial precursor lesions have been associated with several specific species, including *F. nucleatum*, *Porphyromonas gingivalis*, and *Prevotella intermedia*; *Veillonella*, *Prevotella*, *Fusobacterium*, *Porphyromonas*, *Clostridium*, *Actinomyces*, *Haemophilus*, *Enterobacteriaceae*, and *Streptococcus* spp. (Lee et al. 2017).

In routine clinical practice, clinicians usually do not pinpoint the particular bacterium or bacteria responsible for oral infections through the culturing of pus or exudate. Consequently, dentists commonly prescribe broad-spectrum antibiotics since they do not specifically target the bacteria responsible for the infection (Poveda Roda et al. 2007). With the rising prevalence of antibiotic resistance, the efficacy of current antibiotics diminishes (Meinen et al. 2021). Antibiotics should not be used to treat dental caries since overuse could lead to bacterial resistance and change the flora of the mouth and intestines. Antibiotics also lack remineralizing characteristics to aid in the control of dental caries and are frequently only momentarily beneficial due to changes in the oral environment (Niu et al. 2021). In the field of dentistry, there is a pressing need for more judicious antibiotic prescriptions, as this practice has a profound impact on both systemic health and the growing issue of antibiotic resistance (Koyuncuoglu et al. 2017). Therefore, to manage caries, researchers are looking for alternate antimicrobial drugs to control biofilm. Antimicrobial peptides can target pathogens and protect the mouth's beneficial microflora, according to a groundbreaking study (Niu et al. 2021).

Beyond their antimicrobial function, AMPs also serve as highly effective biological molecules in wound healing, immune activation, and inflammation. They are currently under extensive research for potential clinical applications (Khurshid et al. 2016). These peptides' protective role in preventing microorganisms from entering the mouth cavity leads to an efficient defense against infections. This chapter explores the attributes of the oral cavity, discusses diseases influenced by plaque formation, delves into both natural and synthetic antimicrobial peptides and their role in maintaining oral ecology, and suggests potential applications of these AMPs in preventing and treating conditions related to dental plaque.

2 Oral AMPs

AMPs in the oral cavity contribute to maintaining the equilibrium of oral ecology. Oral disorders such as dental caries, oral mucosal diseases, periodontal diseases, and oral cancer may be treated with antimicrobial peptides (Fig. 3.1). The salivary glands, neutrophils, and epithelial cells in the oral cavity all release these peptides, which have antimicrobial effects on oral pathogenic bacteria and their biofilms.

Antimicrobial resistance is on the rise among oral pathogens, leading to a global search for new antimicrobial treatments. As host defense peptides, AMPs represent a highly promising alternative in development for treating oral pathogens, particularly due to their efficacy against MDR pathogens, highlighting a new way to combat disease-causing bacteria. Oral microbiome AMPs are capable of attacking biofilm-forming bacteria, which may prove useful in treating chronic periodontitis. With their immunomodulatory properties, native oral AMPs can be used to control dysbiosis and inflammation associated with pathogenic bacteria in the oral

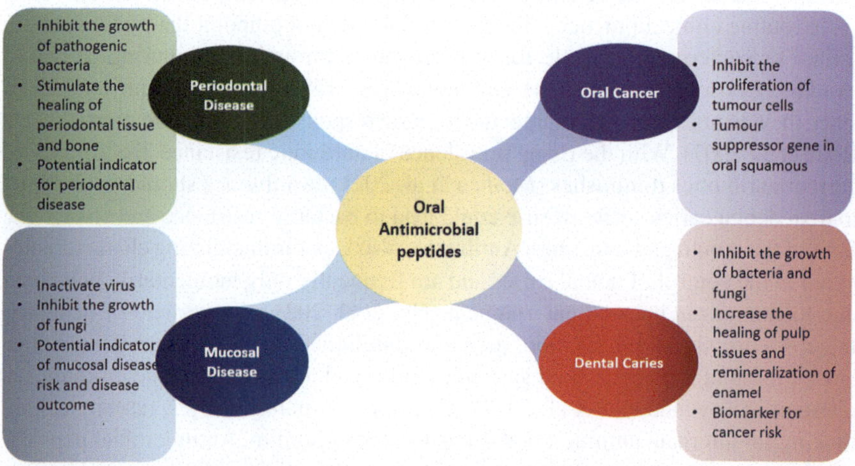

Fig. 3.1 Role of AMPs in oral disease

microbiome, offering potential benefits for dentistry and beyond, an overview has been shown in Fig. 3.2. In the human oral cavity, three classes of AMPs are present: defensins, cathelicidins, and histatins. Each of these distinct pathways overlaps with the others and contributes to sustaining oral health (Izadi et al. 2020).

2.1 *The Defensins Family*

Defensins are characterized as compact, cationic peptides having a low molecular weight (approximately 4–5 kDa) and typically contain approximatelysix to eight cysteine residues, facilitating the formation of three to four intramolecular disulfide bonds (White and Wimley 1995). Defensins have garnered significant research attention because of their widespread expression in the human body and their ability to combat a broad range of microorganisms, including gram-negative and gram-positive bacteria, fungi, and even viruses like herpes simplex (Khurshid et al. 2016; Wang 2014). Human defensins are categorized as alpha (α), beta (β), and theta (θ) defensins, distinguishing them based on factors such as peptide chain length, cellular location, cysteine positioning, and folding patterns (Khurshid et al. 2016; Abiko et al. 2007).

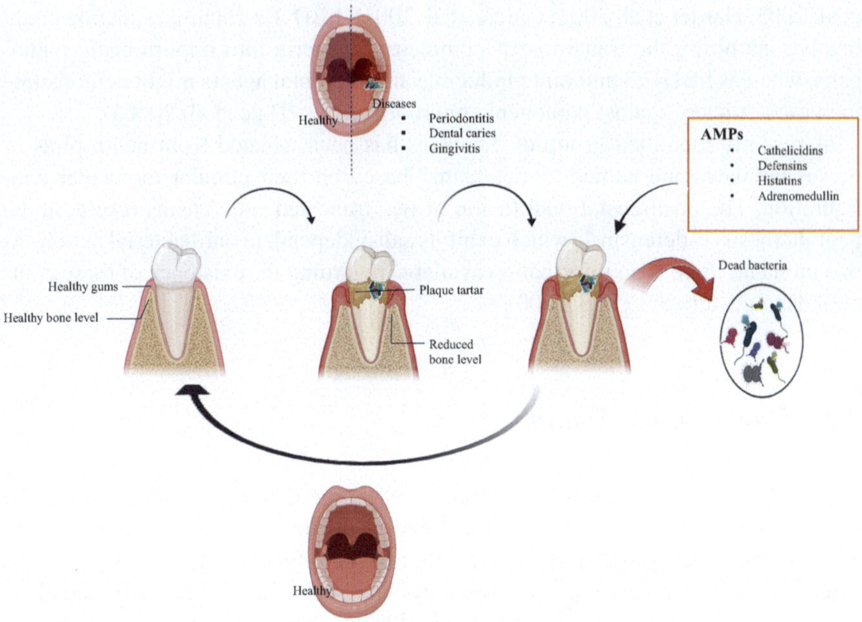

Fig. 3.2 Overview of oro-pathogenic bacteria infecting the tooth and surrounding tissue, showing both the healthy and diseased states and antimicrobial peptides act on the diseased states

The mature α-defensin family has four hNPs, which were isolated from the human neutrophils (Selsted et al. 1985). The hNP-1 exhibits higher activity than hNP-3 and can kill *P. aeruginosa*, *S. aureus*, and *E. coli* (Ganz et al. 1985). Griffith et al. researched and by using the chromatographic method discovered the hNP-4 (33 amino acids), which is active against *Streptococcus faecalis*, *E. coli*, and *Candida albicans* (Wilde et al. 1989). The remaining two members within this family, namely, hNP-5 and hNP-6, are absent in the oral cavity but are found in the enteric system. Among healthy humans, hNP-1 to 3 constitutes the majority in saliva, accounting for approximately 99%, while the presence of hNP-4 is notably lower, approximately 100-fold less. (Gabay et al. 1989). Dale et al. have reported decreased salivary concentrations of α-defensins (specifically hNP-1, hNP-2, and hNP-3) in individuals with dental caries, proposing that these biomarkers could serve as valuable indicators for assessing caries risk within the broader population (Dale et al. 2006a).

The oral cavity expresses three AMPs from the human β-defensin (hBD) family, namely hBD-1, hBD-2, and hBD-3, all of which exhibit a wide-ranging antimicrobial effect. The antibacterial activity of hBD-3 is the highest, whereas hBD-1 has the lowest antimicrobial activity. Healthy gingival epithelial cells express hBD-1 on a constitutive basis, whereas proinflammatory mediators like TNF-, IL-1, and IFN stimulate the production of hBD-2 and 3 (Izadi et al. 2020). Harmful oral bacteria like *P. gingivalis*, *A. actinomycetemcomitans*, *T. forsythensis*, and *T. denticola* and commensal oral bacteria like *F. nucleatum* and its bacterial products both triggered the upregulation of hBD-2 and hBD-3 through NF-B and MAPK pathways (Feucht et al. 2003; Harder et al. 2001; García et al. 2001). hBD-1's potential function could involve inhibiting the transition of commensal bacteria into opportunistic pathogens, whereas hBD-2,3 and other inducible antimicrobial agents might demonstrate increased efficacy against pathogenic microorganisms (Dale et al. 2006a).

A recently identified group of defensins has been isolated from neutrophils of rhesus monkeys and named "θ-defensins" based on their circular molecular configuration. The posttranslational fusion of two truncated α-defensins results in the peptide rhesus θ-defensin 1, which exhibits salt-independent antibacterial action. At this moment, there is no information available regarding the existence of these compounds in the airways (Bals 2000).

2.2 *Human Cathelicidin*

A peptide class consisting of a conserved N-terminal and a variable C-terminal region, cathelicidins were initially identified in bovine neutrophil lysates (Storici et al. 1996). Their primary storage location is the granulocyte and they play an essential role in regulating innate immunity. The N-terminal sequence of cathelicidins is highly homologous (approximately 70%) with the sequence of cathelin, an inhibitor of cathepsin L. Human CAP-18, a precursor produced by cathelicidin-encoding genes (Margherita 2005). By cleaving this propeptide, cathelin and cathelicidin peptides are produced (Murakami et al. 2017). In cathelicidins, the lack of homology between C-terminus residues may account for their diversity in structures, which may include linear, helical, or folded arrangements depending on the arrangement and composition of residues. As of now, the only cathelicidin isolated from humans is LL-37 with 37 amino acids, among 30 identified members. With 37 AAs and 2 leucine residues at its beginning, it was cloned from human cDNA bone marrow (Gudmundsson et al. 1996). Two leucines at the N-terminus and over 50% of AAs in LL-37 are hydrophilic. Many amino acids of LL-37 form intramolecular hydrogen bonds with the α-helical structure, and α-helicity is related to its antibacterial activity (Nizet and Gallo 2003).

In addition to being a cationic antimicrobial peptide, it has a size of 18 kDa, thus being also known as hCAP18 (Nilsson et al. 1999). Chromosome 3 at position 3p21.3 contains the gene coding for LL-37. The gene is known as cathelicidin antimicrobial peptide (CAMP). There are three introns and four exons in the LL-37 gene. A peptide is encoded by the first three exons, which encode the signal sequence and cathelin region, and a mature peptide by the fourth exon (Yang et al. 2004). As LL-37 contains acute phase response factor binding sites, LL-37 is upregulated in inflammation based on its intron and promoter regions (Zanetti et al. 2000).

CAMP, FALL-39, and hCAP-18 (MW = 18 kDa) are other names for this peptide, which is produced by the proteolysis of kallikreins or proteinases in the cytoplasmic granules and lamellar bodies. In the oral cavity, it is mainly derived from granules of neutrophils and, to a lesser extent, from epithelial cells lining the respiratory, gastrointestinal, and urogenital tracts (Dale et al. 2001). The oral cavity expresses LL-37 in buccal mucosa, gingival tissues, and tongue epithelium in response to inflammation (Nilsson et al. 1999). Saliva and GCFs have also been reported to contain this protein (Murakami et al. 2002; Türkoğlu et al. 2009).

With a concentration of 0.14–3 μg/mL in saliva, LL-37 is effective against gram-positive and gram-negative bacteria, which includes *A. actinomycetemcomitans* and *Porphyromonas gingivalis*, two well-established periopathogens (Zigmond et al. 2015). Even in the presence of gingipains, LL-37 can maintain its antimicrobial activity with some salivary components protecting it (Gutner et al. 2009). In gingival fibroblasts exposed to *P. gingivalis* and its products, LL-37 inhibits the inflammatory response (Inomata et al. 2010). LL-37 interacts directly with lipopolysaccharides in bacteria (Larrick et al. 1995). In terms of its mechanism, it works similarly to detergent in that it forms micelles (Dürr et al. 2006) and disrupts anaerobic biofilms. In addition to treating dental caries, LL-37 treats chronic periodontal diseases as well (Neshani et al. 2019; Jain 2017). It is moderately active against *Enterococcus genera*, *Staphylococcus*, *Pseudomonas*, and *Escherichia coli* (Neshani et al. 2019). A higher level of LL-37 promotes wound healing and prevents infection of lesions. The level of LL-37 in saliva is correlated with mucosal lesions in patients with chronic oral inflammation (oral lichen planus) (Greer et al. 2013). LL-37 interacts with peptidoglycans and inner cytoplasmic membranes after having passed through the outer membrane of gram-negative bacteria (Kościuczuk et al. 2012). LL-37 is virtually absent in Kostmann syndrome, a form of periodontal disease exhibited by genetic factors (Pütsep et al. 2002). As a result, LL-37 appears to have significant antimicrobial activity in the oral cavity.

Crystallography and molecular dynamics studies suggest that LL-37 forms a tetramer consisting of antiparallel dimers when exposed to bacterial membranes and detergents (Sancho-Vaello et al. 2020) to form channels lined with peptides stabilized through hydrogen bond interactions between the PE/PG structures and the peptide-lining interfaces. The LL-37 inhibits clinically relevant bacteria at low concentrations, but no biofilm formation steps are disrupted by LL-37 up to 25 μM (Mishra et al. 2016).

2.3 *Histatins*

Several types of histatins are produced in the oral cavity, with the three most prominent being histatin-1, histatin-3, and histatin-5 (Blotnick et al. 2017; HelmerhorsHMt et al. 2001). Due to their low toxicity, histatins play an important role in the maintenance of oral homeostasis, exhibiting antifungal activity against some microbes, including *C. albicans* (the organism that causes oral thrush) (Khurshid et al. 2017). Besides inhibiting fluconazole-resistant *C. albicans* biofilms (Curvelo et al. 2019), histatin-5 also has apoptotic properties against *S. mutans* (Fernández-Presas et al. 2018). Even though histatins are proteolyzed by oral fluid, even after their early degradation, these proteins retain their antimicrobial capabilities, making them an attractive therapeutic target for treating oral disease (Helmerhorst et al. 2006). An antibacterial effect has been shown against oral streptococci using P-113, a fragment of hisatin-5 (residues 4–16) (Huo et al. 2011). In addition to modulating oral inflammation, histatin-5 binds to the hemagglutinin B (HagB) of *P. gingivalis* to incapacitate the chemokine response against the bacteria, which can contribute to the prevention of diseases such as periodontitis (Borgwardt et al. 2014). Additionally, histatin-1 has been shown to facilitate the reattachment of directly attached tooth cells to teeth following soft tissue damage, thereby limiting microbial access to between-tooth junctions (Van Dijk et al. 2018). Considering the prevalence of AMR, especially with fungal infections, histatins can help combat resistance as well as provide a treatment alternative for oral conditions such as candidiasis. We summarized the AMPs' used for the treatment of oral diseases in Table 3.1.

2.4 *Epithelial Antimicrobial*

Epithelial cells naturally produce AMPs and proteins in response to bacterial products and bacteria adhered to the exterior of the oral mucosa and tooth. Antimicrobial peptides (AMPs) play a role in preserving the microbial balance in a healthy mucosal environment. The β-defensin family, the calprotectin (also known as

Table 3.1 An overview of antimicrobial peptides that could be used in treating oral diseases

Antimicrobial peptides	Origin	Sequences	Role	Oral microbes: major targets	References
Cathelicidin (LL37)	Neutrophils, salivary glands and ducts, gingival sulcus, inflamed epithelia	LLGDFFRKSKEKIGKEFKRIVQRIKDFLRNLVPRTES	Primarily antibacterial, Kostmann syndrome (Periodontal disease associated with congenital neutropenia)	*F. nucleatum, S. mutans, Capnocytophaga sputigena*	Ouhara et al. (2005), Dale et al. (2006b)
α-defensins HNP 1–4	Neutrophils, salivary duct cells, sites of inflammation, gingival sulcus	Subtypes differ in sequence, but cysteine residues and cationic nature are conserved.	Antibacterial, antiviral, antifungal. Present in GCF, Kostmann syndrome	*S. mutans* (poor), *C. albicans*, HIV	Raj et al. (2000)
β-defensins hBD1-3	Salivary ducts, epithelia	Subtypes differ in sequence, but cysteine residues and cationic nature are conserved.	Antibacterial, antiviral, antifungal	hBD1: Poor antibacterial *hBD2,* and *hBD3: F. nucleatum, S. mutans, P. gingivalis, S. sanguis, C. albicans,* HIV	Maisetta et al. (2003), Joly et al. (2004)
Histatins	Salivary glands/ducts	Different subtypes have different sequences, but the cationic nature is conserved.	Antifungal, regulate oral hemostasis, andbond metal ions in saliva	*C. albicans*	Oppenheim et al. (1988)

calgranulin) protein, and adrenomedullin are all AMPs expressed in the oral epithelium (Izadi et al. 2020).

2.4.1 Calprotectin

Calprotectin is a multifunctional AMP existing as a heterodimer (S100A8/A9, produced in the cytoplasm of mucosal epithelial cells and neutrophils that is) (Gonzalez et al. 2022). These S100A8 and S100A9 genes are basically from a closely located cluster of 13 genes at location 1q21 on chromosome 1 (Gregory et al. 2006). These genes are organized in an evolutionary conserved manner. Two introns separate the three exons of both peptides. The first exon is untranslated, whereas the second and third exons encode EF-hand motifs at the N- and C-termini, respectively (Kerkhoff et al. 1998). This protein is expressed in monocytes, neutrophils, mucosal keratinocytes, and macrophages and is associated with arachidonic acid metabolism and leukocyte trafficking (Nacken et al. 2003). During infectious and inflammatory diseases, its level increases in saliva, plasma, and synovial fluid (Cuida et al. 1997; Sander et al. 1984). During wound healing and upon stress inducement, the expression of calprotectin is upregulated within epithelial cells (Marionnet et al. 2003). Additionally, its level is elevated through exposure to complement factor C5a and proinflammatory cytokines such as IL-1β, TNF-α, and IL-6 (Benet Bosco Dhas et al. 2012). In gingival crevicular fluid (GCF), calprotectin levels are positively correlated with periodontitis severity (Kido et al. 1999). Calprotectin is derived from both sulcular epithelium and immune cells in GCF. Increased pocket depth and aspartate aminotransferase levels, prostaglandin E2, IL-1β, and collagenase in GCF are associated with increased calprotectin expression (Nakamura et al. 2000). Periodontitis patients show elevated levels of this protein in their pocket epithelium as assessed by immunohistochemistry (Gómez et al. 1995). The gingival crevicular fluid of patients with periodontal disease is elevated with this molecule during inflammatory conditions (Kido et al. 1999). There is a constitutive expression of this gene in stratified oral epithelia and cultured gingival epithelia (Ross and Herzberg 2001); it is also highly sensitive to stress in the epidermis (Marionnet et al. 2003). In response to microbial insults, epithelial cells release calprotectin into the interstitial milieu. Calprotectin is a zinc- and calcium-binding protein, and it functions as an antimicrobial agent by competing with microorganisms for zinc, an essential growth factor. It is expressed constitutively in oral stratified epithelial cells and is increased in periodontal disease (Izadi et al. 2020). The expression of calprotectin provides a shield against bacterial attachment and invasion, potentially enhancing the resistance of gingival cells to being invaded by *P. gingivalis*, a gram-negative pathogen associated with periodontal disease, and *Candida* (Dale et al. 2006a).

2.4.2 Adrenomedullin

With one intramolecular disulfide link, adrenomedullin is a cationic amphipathic peptide with 185 amino acids, which undergoes proteolytic processing and C-terminal amidation to form the mature 52 amino acid adrenomedullin (Khurshid et al. 2016; Dale et al. 2006a). This precursor gene is located on chromosome 11 and encoded by the ADM gene. Adrenomedullin's gene structure and function differ from that of the β-defensins though there are some functional similarities between them. Kitamura et al. (Kitamura et al. 1993) found that adrenomedullin has similar functions to calcitonin gene-related peptide and binds to its receptor-like calcitonin receptor (McLatchie et al. 1998), which is believed to control circulation in a hormone-like manner. Cardiovascular abnormalities result in the death of mice lacking adrenomedullin at mid-gestation (Caron 2001).

Initially known for its vasodilatory properties, adrenomedullin is a multifunctional peptide that has since been acknowledged for its antibacterial activity against both gram-negative and gram-positive bacteria found in the skin, oral cavity, gut, and respiratory tract. The compound has no antifungal properties. Its production and secretion are both constitutively present in oral epithelial cells, and it is induced by TNF, IL1, and active oral microorganisms (Dale et al. 2006a). Adreomedullin levels are approximately twice as high in periodontally unhealthy areas as in healthy areas (Khurshid et al. 2016). Adrenomedullin's antibacterial mechanism is similar to that of other cationic antimicrobial peptides. Although the precise process is unknown, it is thought to increase intramembranous pore development in bacterial cell membranes. When the cell membrane of bacteria is disrupted, essential intracellular activities are halted, leading to cell death (Hans and Madaan Hans 2014).

2.5 *Probiotics and EAPs*

Humans' oral cavity and gastrointestinal tract contain millions of microbes. When the immunity of the host is impaired or the epithelial lining's sensing and defense mechanisms are lost, these microbes can harm the host. Immunomodulatory properties have been demonstrated for these commensal bacteria (Hooper et al. 2001). In addition, they prevent pathogenic microorganisms from colonizing the host (Mazmanian et al. 2005). As a result of finding these endogenous bacteria and understanding their potential benefits, probiotics have been developed. Bacteria administered as probiotics are viable and compete with host bacteria to colonize host sites and promote well-being. Many infectious and inflammatory diseases are treated and prevented using this modality (Reid et al. 2003). The interaction between epithelial tissues and probiotics and/or commensal bacteria and how they altered pathogenic bacteria from nonpathogenic bacteria has always been a source of speculation, including *Lactobacillus* spp., *Streptococcus salivarius*, and *Bifidobacterium* spp. Several studies have examined the interaction between probiotic organisms and epithelium. Both probiotic and pathogenic bacteria are known to initiate different signaling pathways (Chung and Dale 2004; Hasegawa et al. 2007). Three ways by which probiotic bacteria modulate immune function are through altered inhibition of the NF-2 pathway, toll-like receptor signaling, and release of IL-10 cytokine (Grangette et al. 2005; Tien et al. 2006). These mechanisms prevent probiotics from accessing proinflammatory pathways and producing EAPs, which protects them from the host's defense mechanism. Hence, probiotics ensure not only a healthy oral epithelium but also promote epithelial integrity and homeostasis (Cosseau et al. 2008).

2.6 Synthetic Antimicrobial Peptide

Scientists turned to designing synthetic antimicrobial peptides (AMPs) due to several drawbacks associated with natural AMPs. These included the insufficient concentration of natural AMPs in saliva and gingival crevicular fluid (GCF) to reach the minimum inhibitory concentration (MIC) required for inhibiting the growth of various microorganisms. Additionally, using natural AMPs at higher doses led to an escalation in toxicity toward mammalian cells (Aoki et al. 2012). Another challenge arises when utilizing natural AMPs in the context of periodontal diseases, as the relevant pathogens exhibit substantial proteolytic activity. Furthermore, the considerable size of antimicrobial proteins found in saliva renders their purification a complex and costly endeavor (Fjell et al. 2012). Based on both empirical evidence and theoretical discussions, there are doubts surrounding the broad applications of natural human AMPs. The incorporation of these compounds into the human body could elicit the emergence of resistance to our innate defense proteins, potentially compromising our inherent immune response to infections. (Bell and Gouyon 2003). These constraints have driven research efforts toward the development of synthetic antimicrobial peptides. The primary objectives are to enhance their resistance to degradation by the proteolytic activity of bacteria associated with periodontal diseases, increase their tolerance to the salt concentration in saliva, minimize their cytotoxic effects, and enhance their pharmaceutical and pharmacokinetic characteristics (Izadi et al. 2020).

According to the process of synthesis, synthetic peptides are split into three groups: (i) mimesis, (ii) novel design, and (iii) fusion of antibacterial and functional sequences with peptide. During the synthesis of AMP mimesis, the natural antimicrobial peptide human amelogenin served as the model for QP5, which also has a hydrophobic tail with seven residues. Incorporating a hydrophilic tail renders the peptide soluble in water and enables it to bind with calcium (Lv et al. 2015). The function-based design uses two synthesis techniques: new designs and the fusion of peptides with antimicrobial and functional sequences. GH12 synthetic peptide was developed by Tu et al. utilizing α-helical wheel projection based on the α-helical folding principles. The hydrophobic residue leucine and cationic histidine was used to create an amphipathic α-helix. To assist the peptides, adhere to the surface of the lipid bilayer, they also added tryptophan and a glycine to the first position (Tu et al.

2016). In the process of synthesizing peptides, it is common to truncate or modify a sequence by combining antibacterial and functional segments. C16G2 (Eckert et al. 2006), IMB-2 (Mai et al. 2011), and C11H (Huo et al. 2018) are precise target sequences designed to target *S. mutans*, and they are derived from shortened fragments of the competence-stimulating peptide found in *S. mutans* (Niu et al. 2021).

2.7 Specifically Targeted Antimicrobial Peptides (STAMPs)

As previously stated, plaque is a balanced microbial population, and disrupting this balance allows pathogenic species to establish themselves (Eckert et al. 2006). Antimicrobial agents cannot distinguish between resident and pathogenic organisms, causing them to have a lethal impact on both. This leads to a disruption in the ecological balance of the microbial community (plaque) within the oral cavity (Eckert et al. 2006; Qiu et al. 2003). Hence, scientists developed STAMPs with the aim of selectively eliminating the pathogenic species within plaque, rather than eradicating the entire plaque structure (Kaplan et al. 2011). STAMPs are composed of three distinct regions: the killing region, which essentially serves as the antimicrobial peptide, a flexible tri-glycine linker, and a species-specific binding region. The killing region is responsible for its antimicrobial properties, while the binding region targets a particular species and facilitates the delivery of the antimicrobial peptide to a specific pathogen, minimizing its impact on the other residents within the plaque community (Izadi et al. 2020; Qiu et al. 2003).

3 Antibiofilm Peptides for Oral Biofilm Prevention

Oral bacteria can only grow in situ if they adhere to a surface since the continuous flow of host secretions prevents them from growing before, they are swallowed. As a result, oral bacteria have evolved to form biofilms on soft epithelial tissue and hard tooth surfaces, which often contain multiple bacteria strains and are responsible for most dental caries, endodontic infections, and periodontal diseases. There are two types of oral biofilm associated with teeth: supragingival biofilms (on enamel surfaces exposed to the air) and subgingival biofilms (under the gumline and within the sulcus or periodontal pocket) (Kolenbrander et al. 2010). Antimicrobial agents are highly resistant to oral biofilms in humans. Antibiofilm peptides serve as host-defense molecules that inhibit biofilm invasions and infections. Because of their specific targeting of bacterial biofilms, these peptides have garnered significant attention as potential disinfecting agents in recent decades. Thus, gram-negative and gram-positive bacterial pathogens and fungi are prevented from forming biofilms or destroying preexisting ones. In 1928, nisin was identified as a cationic peptide from *Lactobacillus lactis* and was found to have high antimicrobial activity at a pH range of 2–6 and room temperature (Pepperney and Chikindas 2011). Commercial polylysine is produced for food applications derived from *Streptomyces albulus* 346 (Yoshida and Nagasawa 2003). In dental research, nisin and polylysine inhibit *Streptococcus mutans* in vitro in a synergistic manner (Badaoui Najjar et al. 2009). Synthetic peptides with helical structures have a great deal of potential since they successfully inhibit biofilm formation on dental caries-related strains. KSL (KKVVFKVKFK) exhibits broad antibacterial activity due to its helical structure. In a dose-dependent manner, KSL targets bacterial membranes through electrostatic interactions and causes destabilization (Leung et al. 2009). A few oral bacteria strains have been tested for antibiofilm activity using KSL, including *S. sobrinus*, *S. mutans*, and *Streptococcus sanguinis* (Concannon et al. 2003) with MICs ranging from 3 to 100 μg/mL in vitro. As demonstrated by Tu et al. (2016) in vitro, the modified peptide GH12 exhibited strong and rapid antimicrobial activity against oral *Streptococci*, especially *S. sobrinus*, *S. mutans*, and *S. salivarius* with MICs ranging from 6.7 to 32.0 μg/mL. It has been studied that lacticin 3147, L-K6, and chrysophsin-1 combat *S. mutans* biofilms (Datta et al. 2023). Human saliva contains BPIFA2, which inhibits the growth of *S. gordonii* biofilms. Oral biofilms containing multispecies are also effectively treated with AMPs. Multispecies oral biofilms can be inhibited by peptide 1018 (Wang et al. 2015a). Biofilm is formed on the teeth surface by multispecies bacteria that cause periodontal disease (Frédéric et al. 2018). A study by Wang et al. (Wang et al. 2015b) found that antibiofilm peptides such as Nal-P-113 were useful in treating this disease. The LL-37 inhibitory activity was demonstrated against *P. gingivalis*, *Fusobacterium nucleatum*, *Treponema denticola*, and *S. mutans* (Rosen et al. 2012). Combining nisin with sodium fluoride hampers the development of biofilms by *S. mutans* as nisin prompts fluoride ions to enter the bacterial cytoplasm and exert an antibacterial effect by forming transmembrane pores. The actions of human β-defensin 1, 3, and 5 inhibited the growth of *Eikenella corrodens*, *F. nucleatum*, and *Tannerella forsythia* (Paulo et al. 2018).

4 Alternative to Human AMP from Plants and Bacteria for the Treatment of Oral Disease

Plant-derived peptides demonstrating antimicrobial activity can be employed as an alternative to human AMPs for preventing or treating infectious diseases. The suggested potential application of hydroxychavicol, an extract from piper betel plant leaves, as an agent for oral health care, is attributed to its inhibitory activity against oral microorganisms. Synthetically generated AMPs may also have an excellent potential for clinical application in addition to natural peptides (Stotz et al. 2009). Liu et al. presented an intriguing, cost-effective method employing plant-produced antimicrobial peptides (PMAMPs), such as retrocyclin or protegrin, for topical applications in biofilm control. After a sole topical application on a surface mimicking a tooth, PMAMPs rapidly eradicated the *Streptococcus mutans* pathogen and hindered biofilm formation (Liu et al. 2016).

Initially, the probiotics were isolated from dairy and the gut of animals and humans for improved gastrointestinal and immune system function. Many classical probiotics like *Lactobacilli* and bifidobacterium were acidogenic, leading to the initiation of dental caries and progression and increasing the acidogenicity of oral biofilms. For these reasons, healthy oral bacteria were suggested to be used for the treatment of oral diseases. Many healthy oral streptococcal isolated are *Streptococcus dentisani*, *S. salivarius*, *S. oligofermentans*, and *Streptococcus* A12, and they inhibit the growth of pathogenic bacteria by producing the bacteriocin, protease, and hydrogen peroxide (Ferrer et al. 2020).

S. dentisani was isolated from the dental plaque of a caries-free individual in 2011. Its metagenomic analysis revealed the overexpression of the gene, which codes for 11 AMP, and they already been reported before for the inhibition of *S. mutans* and *Fusobacterium nucleatum* (Ferrer et al. 2020). It is possible to use this strain as a potential probiotic against tooth decay and its antimicrobial peptide against oral pathogens (López-López et al. 2017). There are very few studies regarding their use for oral health, consequences, and in vivo efficacies for human health, and clinical studies are aiming toward probiotics efficacies for human oral health (Ferrer et al. 2020).

5 AMPs Currently in Clinical Trials for Oral Disease Treatment

Presently, five antimicrobial peptides, including PAC-113 (histatin 5 based), histatin (natural AMP), dusquetide (SGX942) (IDR-1 analog), C16G2 (synthetic AMP), and iseganan (IB-367) (protegrin I analog), are undergoing clinical trials for the treatment of oral diseases (Table 3.1) (Izadi et al. 2020). GERM CLEAN, developed by Scisyn in Chengdu, China, is a new oral spray featuring a newly synthesized polypeptide as its active ingredient (Wang et al. 2023). As per the manufacturer's instruction, this spray inhibits the growth of *S. mutans*, a dual biofilm formed by *S. mutans* and *C. albicans* (Wang et al. 2023; Xiong et al. 2020). As described in the literature, in ongoing clinical trials, various formulations are being investigated, including topical gel and hydrogel, topical cream, polyvinyl alcohol-based solution for wound bed administration, hyaluronic acid–based hydrogel for surgical site administration, oral solutions, and mouth rinse (Moretta et al. 2021). In the previously mentioned oral AMPs undergoing clinical trials, there is a possibility of using AMPs like C16G2, iseganan, histatin, and dusquetide for mouthwash and oral rinse solutions for the treatment of oral diseases, especially in HIV patients or patients undergoing chemotherapy (Izadi et al. 2020) (Table 3.2).

Table 3.2 AMPs under clinical trials for oral disease treatment

Name	Source	DRAMp ID	Developmental stage	Treatment	References
Dusquetide (SGX942)	IDR-1 analog	20773	Phase III	Oral mucositis in head-neck chemotherapy patients	Kudrimoti et al. (2016), North et al. (2016)
C16G2	Synthetic AMP	20760	Phase II	Tooth decay pathogen *S. mutans* (oral rinse, and gel)	Guo et al. (2015)
Iseganan (IB-367)	Protegrin I analog	18059	Phase III (failure)	Polymicrobial infection linked to mouth rinse-induced oral mucositis in chemotherapy patients	Trotti et al. (2004)
PAC-113	histatin 5 based	18081	Phase IIb	Oral candidiasis in HIV patients, gingivitis (mouth rinse)	Mickels et al. (2001), Paquette et al. (2002)
Histatin	Natural AMP	18061	Phase II-III	Oral candidiasis (mouthwash)	Kavanagh and Dowd (2010)

6 Limitations of AMPs in Oral Health

The utilization of AMPs in oral health comes with specific limitations in addition to their application in treating oral diseases. AMPs offer distinct advantages for oral health, including broad-spectrum activity (antibacterial, anti-inflammatory, antifungal, and antiviral) and the rapid elimination of pathogens, potentially minimizing the development of induced resistance. Drawbacks associated with AMPs in oral health include susceptibility to proteolysis and the manifestation of local and systemic toxicity. Their efficacy diminishes due to sensitivity to salt, serum, and pH variations (Gupta et al. 2018). The oral application is constrained by the enzymatic degradation of peptides in the pre-systemic phase and their limited penetration into the intestinal mucosa. The synthesis, purification, delivery, and in vivo effectiveness of antimicrobial peptides present substantial challenges, as indicated by the scarcity of antimicrobial peptides undergoing clinical trials (Gorr 2009). An encouraging substitute for natural or synthetic peptides involves utilizing peptide mimetics, which emulate the biological activity of antimicrobial peptides but in a chemical form that is easier to produce and deliver (Gorr 2009).

7 Conclusion

Commensal and pathogenic bacterial species are both present in the oral microbiota, but pH, immune system, and nutritional factors influence their balance. Dental caries and periodontitis can develop with inadequate oral care, contributing to dysbiosis and its onset. Conventional antibiotics can be used in conjunction with these treatments, although resistant strains of bacteria have been found to form polymicrobial biofilms commonly associated with these diseases. Antibiotics are widely misused and mismanaged in human health and livestock, which exacerbates the problem. Thus, in both oral health and beyond, alternative antibiotic treatments are being sought to address the increasing AMR threats. Because pathogens and host AMPs coevolve, AMPs have the potential to be used as therapeutic regimens due to their potential efficacy. Oral cavity components often contain native AMPs that are antibacterial. There are several appealing attributes for further research of AMPs targeting oropathogenic bacteria, including their native origin (as a result, they are less invasive to the host) and their chemical specificity (allowing them to target pathogenic bacteria). AMPs such as C16G2 demonstrate substantial antibacterial activity in clinical trials, which indicates that they are a promising alternative or adjunct to other treatments, even if they do not selectively target pathogenic bacteria. Therapeutic benefits can be achieved through oral EAPs in a variety of ways. A variety of products can be produced using EAPs such as gels, mouthwashes, and gum paints that can be applied to periodontal tissues to prevent and reverse periodontitis. The use of EAPs as local drug delivery agents can reduce the effects of periopathogenic bacteria. This chapter aims to summarize a series of AMPs that have antimicrobial activity against bacteria implicated in oral infectious diseases, such as defensins, LL-37, and histatins. Although more research is needed to optimize their therapeutic applications, AMPs are likely to achieve a better understanding of their role in dentistry with subsequent trials. To prevent the emergence and advancement of oral infections and combat dysbiosis, AMPs can be helpful alongside mechanical removal, traditional antibiotics, and other interventions.

References

Abiko Y, Saitoh M, Nishimura M, Yamazaki M, Sawamura D, Kaku T (2007) Role of β-defensins in oral epithelial health and disease. Preprint. https://doi.org/10.1007/s00795-007-0381-8

Aoki W, Kuroda K, Ueda M (2012) Next generation of antimicrobial peptides as molecular targeted medicines. Preprint. https://doi.org/10.1016/j.jbiosc.2012.05.001

Bachtiar EW, Bachtiar BM (2018) Relationship between candida albicans and streptococcus mutans in early childhood caries, evaluated by quantitative PCR [version 1; peer review: 2 approved]. F1000Res 7. https://doi.org/10.12688/F1000RESEARCH.16275.1

Badaoui Najjar M, Kashtanov D, Chikindas ML (2009) Natural antimicrobials ε-poly-l-lysine and Nisin A for control of Oral microflora. Probiot Antimicrob Prot 1:143–147. https://doi.org/10.1007/s12602-009-9020-0

Bals R (2000) Epithelial antimicrobial peptides in host defense against infection. Respir Res 1:141–150. https://doi.org/10.1186/rr25

Bell G, Gouyon PH (2003) Arming the enemy: the evolution of resistance to self-proteins. Preprint at Society for General Microbiology. https://doi.org/10.1099/mic.0.26265-0

Benet Bosco Dhas D, Vishnu Bhat B, Bahubali Gane D (2012) Role of calprotectin in infection and inflammation. Curr Pediatr Res 16:83–94

Bik EM, Long CD, Armitage GC, Loomer P, Emerson J, Mongodin EF, Nelson KE, Gill SR, Fraser-Liggett CM, Relman DA (2010) Bacterial diversity in the oral cavity of 10 healthy individuals. ISME J 4:962–974. https://doi.org/10.1038/ismej.2010.30

Blotnick E, Sol A, Bachrach G, Muhlrad A (2017) Interactions of histatin-3 and histatin-5 with actin. BMC Biochem 18:1–13. https://doi.org/10.1186/s12858-017-0078-0

Borgwardt DS, Martin AD, Van Hemert JR, Yang J, Fischer CL, Recker EN, Nair PR, Vidva R, Chandrashekaraiah S, Progulske-Fox A et al (2014) Histatin 5 binds to Porphyromonas gingivalis hemagglutinin B (HagB) and alters HagB-induced chemokine responses. Sci Rep 4. https://doi.org/10.1038/srep03904

Bowen WH, Burne RA, Wu H, Koo H (2018) Oral biofilms: pathogens, matrix, and polymicrobial interactions in microenvironments. Preprint at Elsevier Ltd. https://doi.org/10.1016/j.tim.2017.09.008

Burcham ZM, Garneau NL, Comstock SS, Tucker RM, Knight R, Metcalf JL, Miranda A, Reinhart B, Meyers D, Woltkamp D et al (2020) Patterns of Oral microbiota diversity in adults and children: A crowdsourced population study. Sci Rep 10. https://doi.org/10.1038/s41598-020-59016-0

Caron KM (2001) Extreme hydrops fetalis and cardiovascular abnormalities in mice lacking a functional Adrenomedullin gene. Proc Natl Acad Sci 98:615–619. https://doi.org/10.1073/pnas.021548898

Chung WO, Dale BA (2004) Innate immune response of oral and foreskin Keratinocytes: utilization of different signaling pathways by various bacterial species. Infect Immun 72:352–358. https://doi.org/10.1128/IAI.72.1.352-358.2004

Concannon SP, Crowe TD, Abercrombie JJ, Molina CM, Hou P, Sukumaran DK, Raj PA, Leung KP (2003) Susceptibility of oral bacteria to an antimicrobial decapeptide. J Med Microbiol 52:1083–1093. https://doi.org/10.1099/jmm.0.05286-0

Cosseau C, Devine DA, Dullaghan E, Gardy JL, Chikatamarla A, Gellatly S, Yu LL, Pistolic J, Falsafi R, Tagg J et al (2008) The commensal Streptococcus salivarius K12 downregulates the innate immune responses of human epithelial cells and promotes host-microbe homeostasis. Infect Immun 76:4163–4175. https://doi.org/10.1128/IAI.00188-08

Cotter PD, Ross RP, Hill C (2013) Bacteriocins-a viable alternative to antibiotics? Preprint. https://doi.org/10.1038/nrmicro2937

Cugini C, Shanmugam M, Landge N, Ramasubbu N (2019) The role of exopolysaccharides in oral biofilms. Preprint at SAGE Publications Inc. https://doi.org/10.1177/0022034519845001

Cuida M, Halse AK, Johannessen AC, Tynning T, Jonsson R (1997) Indicators of salivary gland inflammation in primary Sjögren's syndrome. Eur J Oral Sci 105:228–233. https://doi.org/10.1111/j.1600-0722.1997.tb00205.x

Curvelo JAR, De Moraes DC, Dos Anjos CA, Portela MB, Soares RMA (2019) Histatin 5 and human lactoferrin inhibit biofilm formation of a fluconazole resistant candida albicans clinical isolate. An Acad Bras Cienc 91:1–7. https://doi.org/10.1590/0001-3765201920180045

Dale BA, Kimball JR, Krisanaprakornkit S, Roberts F, Robinovitch M, O'Neal R, Valore EV, Ganz T, Anderson GM, Weinberg A (2001) Localized antimicrobial peptide expression in human gingiva. J Periodontal Res 36:285–294. https://doi.org/10.1034/j.1600-0765.2001.360503.x

Dale BA, Tao R, Kimball JR, Jurevic RJ (2006a) Oral antimicrobial peptides and biological control of caries. BMC Oral Health. https://doi.org/10.1186/1472-6831-6-S1-S13

Dale BA, Tao R, Kimball JR, Jurevic RJ (2006b) Oral antimicrobial peptides and biological control of caries. BMC Oral Health 6:1–7. https://doi.org/10.1186/1472-6831-6-S1-S13

Datta M, Rajeev A, Chattopadhyay I (2023) Application of antimicrobial peptides as next-generation therapeutics in the biomedical world. Biotechnol Genet Eng Rev 00:1–39. https://doi.org/10.1080/02648725.2023.2199572

De Bernardis F, Muhlschlegel FA, Muhlschlegel M, Cassone A, Fonzi WA (1998) The pH of the Host Niche controls gene expression in and virulence of Candida albicans. Infect Immun 66(7):3317

Deo PN, Deshmukh R (2019) Oral microbiome: unveiling the fundamentals. Preprint at Wolters Kluwer Medknow Publications. https://doi.org/10.4103/jomfp.JOMFP_304_18

Ding T, Schloss PD (2014) Dynamics and associations of microbial community types across the human body. Nature 509:357–360. https://doi.org/10.1038/nature13178

Dürr UHN, Sudheendra US, Ramamoorthy A (2006) LL-37, the only human member of the cathelicidin family of antimicrobial peptides. Biochim Biophys Acta Biomembr 1758:1408–1425. https://doi.org/10.1016/j.bbamem.2006.03.030

Dyke TV, Paquette D, Grossi S, Braman V, Massaro J, D'Agostino R, Dibart S, Friden P (2002) Clinical and microbial evaluation of a histatin-containing mouthrinse in humans with experimental gingivitis: a phase-2 multi trial study. J Clin Periodontol 28:404–410. https://doi.org/10.1034/j.1600-051x.2002.290212.x

Eckert R, He J, Yarbrough DK, Qi F, Anderson MH, Shi W (2006) Targeted killing of Streptococcus mutans by a pheromone-guided "smart" antimicrobial peptide. Antimicrob Agents Chemother 50:3651–3657. https://doi.org/10.1128/AAC.00622-06

Fernández-Presas AM, Márquez Torres Y, García González R, Reyes Torres A, Becker Fauser I, Rodríguez Barrera H, Ruíz García B, Toloza Medina R, Delgado Domínguez J, Molinarí Soriano JL (2018) Ultrastructural damage in Streptococcus mutans incubated with saliva and histatin 5. Arch Oral Biol 87:226–234. https://doi.org/10.1016/j.archoralbio.2018.01.004

Ferrer MD, López-López A, Nicolescu T, Perez-Vilaplana S, Boix-Amorós A, Dzidic M, Garcia S, Artacho A, Llena C, Mira A (2020) Topic application of the probiotic streptococcus dentisani improves clinical and microbiological parameters associated with oral health. Front Cell Infect Microbiol 10. https://doi.org/10.3389/fcimb.2020.00465

Feucht EC, DeSanti CL, Weinberg A (2003) Selective induction of human beta-defensin mRNAs by Actinobacillus actinomycetemcomitans in primary and immortalized oral epithelial cells. Oral Microbiol Immunol 18(6):359. https://doi.org/10.1046/j.0902-0055.2002.00097.x

Fjell CD, Hiss JA, Hancock REW, Schneider G (2012) Designing antimicrobial peptides: form follows function. Preprint at Nature Publishing Group. https://doi.org/10.1038/nrd3591

Frédéric LJ, Michel B, Selena T (2018) Oral microbes, biofilms and their role in periodontal and peri-implant diseases. Materials 11:1–17. https://doi.org/10.3390/ma11101802

Gabay JE, Scottt RW, Campanelli D, Griffitht J, Wildet C, Marrat MN, Seeger M, Nathan CF (1989) Antibiotic proteins of human polymorphonuclear leukocytes (bacteria/lysosomes/neutrophils). Proc Natl Acad Sci U S A 86(14):5610

Ganz T, Selsted ME, Szklarek D, Harwig SSL, Daher K, Bainton DF, Lehrer RI (1985) Defensins natural peptide antibiotics of human neutrophils. J Clin Invest 76(4):1427

García JRC, Jaumann F, Schulz S, Krause A, Rodríguez-Jiménez J, Forssmann U, Adermann K, Klüver E, Vogelmeier C, Becker D et al (2001) Identification of a novel, multifunctional β-defensin (human β-defensin 3) with specific antimicrobial activity: its interaction with plasma membranes of Xenopus oocytes and the induction of macrophage chemoattraction. Cell Tissue Res 306:257–264. https://doi.org/10.1007/s004410100433

Gómez RS, Langer P, Pelka M, von den Driesch P, Johannessen AC, Simon M (1995) Variational expression of functionally different macrophage markers (27E10, 25F9, RM3/1) in normal gingiva and inflammatory periodontal disease. J Clin Periodontol 22:341–346. https://doi.org/10.1111/j.1600-051X.1995.tb00159.x

Gorr SU (2009) Antimicrobial peptides of the oral cavity. Periodontol 51:152–180. https://doi.org/10.1111/j.1600-0757.2009.00310.x

Grangette C, Nutten S, Palumbo E, Morath S, Hermann C, Dewulf J, Pot B, Hartung T, Hols P, Mercenier A (2005) Enhanced antiinflammatory capacity of a Lactobacillus plantarum mutant synthesizing modified teichoic acids. Proc Natl Acad Sci USA 102:10321–10326. https://doi.org/10.1073/pnas.0504084102

Greer A, Zenobia C, Darveau RP (2013) Defensins and LL-37: A review of function in the Gingival epithelium. Periodontol 2000(63):67–79. https://doi.org/10.1111/prd.12028

Gregory SG, Barlow KF, McLay KE, Kaul R, Swarbreck D, Dunham A, Scott CE, Howe KL, Woodfine K, Spencer CCA et al (2006) The DNA sequence and biological annotation of human chromosome 1. Nature 441:315–321. https://doi.org/10.1038/nature04727

Griffith A, Mateen A, Markowitz K, Singer SR, Cugini C, Shimizu E, Wiedman GR, Kumar V (2022) Alternative antibiotics in dentistry: antimicrobial peptides. Preprint at MDPI. https://doi.org/10.3390/pharmaceutics14081679

Gudmundsson GH, Agerberth B, Odeberg J, Bergman T, Olsson B, Salcedo R (1996) The human gene FALL39 and processing of the cathelin precursor to the antibacterial peptide LL-37 in granulocytes. Eur J Biochem 238:325–332. https://doi.org/10.1111/j.1432-1033.1996.0325z.x

Guo L, McLean JS, Yang Y, Eckert R, Kaplan CW, Kyme P, Sheikh O, Varnum B, Lux R, Shi W et al (2015) Precision-guided antimicrobial peptide as a targeted modulator of human microbial ecology. Proc Natl Acad Sci USA 112:7569–7574. https://doi.org/10.1073/pnas.1506207112

Gupta S, Bhatia G, Sharma A, Saxena S (2018) Host defense peptides: an insight into the antimicrobial world. Preprint at Wolters Kluwer Medknow Publications. https://doi.org/10.4103/jomfp.JOMFP_113_16

Gutner M, Chaushu S, Balter D, Bachrach G (2009) Saliva enables the antimicrobial activity of LL-37 in the presence of proteases of Porphyromonas gingivalis. Infect Immun 77:5558–5563. https://doi.org/10.1128/IAI.00648-09

Hajishengallis G (2015) Periodontitis: from microbial immune subversion to systemic inflammation. Preprint at Nature Publishing Group. https://doi.org/10.1038/nri3785

Hans M, Madaan Hans V (2014) Epithelial antimicrobial peptides: Guardian of the oral cavity. Preprint at Hindawi Publishing Corporation. https://doi.org/10.1155/2014/370297

Harder J, Bartels J, Christophers E, Schröder JM (2001) Isolation and characterization of human β-Defensin-3, a novel human inducible peptide antibiotic. J Biol Chem 276:5707–5713. https://doi.org/10.1074/jbc.M008557200

Hasegawa Y, Mans JJ, Mao S, Lopez MC, Baker HV, Handfield M, Lamont RJ (2007) Gingival epithelial cell transcriptional responses to commensal and opportunistic oral microbial species. Infect Immun 75:2540–2547. https://doi.org/10.1128/IAI.01957-06

HelmerhorsHMt EJ, Van'T Hof W, Breeuwer P, Veerman ECI, Abee T, Troxler RF, Amerongen AVN, Oppenheim FG (2001) Characterization of Histatin 5 with respect to Amphipathicity, hydrophobicity, and effects on cell and mitochondrial membrane integrity excludes a Candidacidal mechanism of pore formation. J Biol Chem 276:5643–5649. https://doi.org/10.1074/jbc.M008229200

Helmerhorst EJ, Alagl AS, Siqueira WL, Oppenheim FG (2006) Oral fluid proteolytic effects on histatin 5 structure and function. Arch Oral Biol 51:1061–1070. https://doi.org/10.1016/j.archoralbio.2006.06.005

Hooper LV, Wong MH, Thelin A, Hansson L, Falk PG, Gordon JI (2001) Molecular analysis of commensal host-microbial relationships in the intestine. Science 1979(291):881–884. https://doi.org/10.1126/science.291.5505.881

Huo L, Zhang K, Ling J, Peng Z, Huang X, Liu H, Gu L (2011) Antimicrobial and DNA-binding activities of the peptide fragments of human lactoferrin and histatin 5 against Streptococcus mutans. Arch Oral Biol 56:869–876. https://doi.org/10.1016/j.archoralbio.2011.02.004

Huo L, Huang X, Ling J, Liu H, Liu J (2018) Selective activities of STAMPs against streptococcus mutans. Exp Ther Med 15:1886–1893. https://doi.org/10.3892/etm.2017.5631

Inomata M, Into T, Murakami Y (2010) Suppressive effect of the antimicrobial peptide LL-37 on expression of IL-6, IL-8 and CXCL10 induced by Porphyromonas gingivalis cells and extracts in human gingival fibroblasts. Eur J Oral Sci 118:574–581. https://doi.org/10.1111/j.1600-0722.2010.00775.x

Izadi N, Keikha M, Ghazvini K, Karbalaei M (2020) Oral antimicrobial peptides and new therapeutic strategies for plaque-mediated diseases. Preprint at Elsevier Inc. https://doi.org/10.1016/j.genrep.2020.100811

Jain H (2017) Cationic antimicrobial peptide: LL-37 and its role in periodontitis. Front Biol (Beijing) 12:116–123. https://doi.org/10.1007/s11515-017-1432-8

Johnstone KF, Herzberg MC, (2022) Antimicrobial peptides: defending the mucosal epithelial barrier. Front Oral Health 3:01–19. https://doi.org/10.3389/froh.2022.958480

Joly S, Maze C, McCray PB, Guthmiller JM (2004) Human β-Defensins 2 and 3 demonstrate strain-selective activity against oral microorganisms. J Clin Microbiol 42:1024–1029. https://doi.org/10.1128/JCM.42.3.1024-1029.2004

Kaplan CW, Sim JH, Shah KR, Kolesnikova-Kaplan A, Shi W, Eckert R (2011) Selective membrane disruption: mode of action of C16G2, a specifically targeted antimicrobial peptide. Antimicrob Agents Chemother 55:3446–3452. https://doi.org/10.1128/AAC.00342-11

Kavanagh K, Dowd S (2010) Histatins: antimicrobial peptides with therapeutic potential. J Pharm Pharmacol 56:285–289. https://doi.org/10.1211/0022357022971

Kerkhoff C, Klempt M, Sorg C (1998) Novel insights into structure and function of MRP8 (S100A8) and MRP14 (S100A9). Biochim Biophys Acta, Mol Cell Res 1448:200–211. https://doi.org/10.1016/S0167-4889(98)00144-X

Keyes PH (1969) Present and future measures for dental caries control. J Am Dent Assoc 79:1395–1404. https://doi.org/10.14219/jada.archive.1969.0037

Khurshid Z, Naseem M, Sheikh Z, Najeeb S, Shahab S, Zafar MS (2016) Oral antimicrobial peptides: types and role in the oral cavity. Preprint at Elsevier BV. https://doi.org/10.1016/j.jsps.2015.02.015

Khurshid Z, Najeeb S, Mali M, Moin SF, Raza SQ, Zohaib S, Sefat F, Zafar MS (2017) Histatin peptides: pharmacological functions and their applications in dentistry. Saudi Pharm J 25:25–31. https://doi.org/10.1016/j.jsps.2016.04.027

Kido JI, Nakamura T, Kido R, Ohishi K, Yamauchi N, Kataoka M, Nagata T (1999) Calprotectin in gingival crevicular fluid correlates with clinical and biochemical markers of periodontal disease. J Clin Periodontol 26:653–657. https://doi.org/10.1034/j.1600-051X.1999.261004.x

Kim D, Sengupta A, Niepa THR, Lee BH, Weljie A, Freitas-Blanco VS, Murata RM, Stebe KJ, Lee D, Koo H (2017) Candida albicans stimulates Streptococcus mutans microcolony development via cross-kingdom biofilm-derived metabolites. Sci Rep 7. https://doi.org/10.1038/srep41332

Kitamura K, Sakata J, Kangawa K, Kojima M, Matsuo H, Eto T (1993) Cloning and characterization of cDNA encoding a precursor for human adrenomedullin. Preprint. https://doi.org/10.1006/bbrc.1993.1881

Kolenbrander PE, Palmer RJ, Periasamy S, Jakubovics NS (2010) Oral multispecies biofilm development and the key role of cell-cell distance. Nat Rev Microbiol 8:471–480. https://doi.org/10.1038/nrmicro2381

Kościuczuk EM, Lisowski P, Jarczak J, Strzałkowska N, Jóźwik A, Horbańczuk J, Krzyżewski J, Zwierzchowski L, Bagnicka E (2012) Cathelicidins: family of antimicrobial peptides. A review. Mol Biol Rep 39:10957–10970. https://doi.org/10.1007/s11033-012-1997-x

Koyuncuoglu CZ, Aydin M, Kirmizi NI, Aydin V, Aksoy M, Isli F, Akici A (2017) Rational use of medicine in dentistry: do dentists prescribe antibiotics in appropriate indications? Eur J Clin Pharmacol 73:1027–1032. https://doi.org/10.1007/s00228-017-2258-7

Kudrimoti M, Curtis A, Azawi S, Worden F, Katz S, Adkins D, Bonomi M, Elder J, Sonis ST, Straube R et al (2016) Dusquetide: A novel innate defense regulator demonstrating a significant and consistent reduction in the duration of oral mucositis in preclinical data and a randomized, placebo-controlled phase 2a clinical study. J Biotechnol 239:115–125. https://doi.org/10.1016/j.jbiotec.2016.10.010

Larrick JW, Hirata M, Balint RF, Lee J, Zhong J, Wright SC (1995) Human CAP18: A novel antimicrobial lipopolysaccharide-binding protein. Infect Immun 63:1291–1297. https://doi.org/10.1128/iai.63.4.1291-1297.1995

Lee WH, Chen HM, Yang SF, Liang C, Peng CY, Lin FM, Tsai LL, Wu BC, Hsin CH, Chuang CY et al (2017) Bacterial alterations in salivary microbiota and their association in oral cancer. Sci Rep 7. https://doi.org/10.1038/s41598-017-16418-x

Leung KP, Abercrombie JJ, Campbell TM, Gilmore KD, Bell CA, Faraj JA, DeLuca PP (2009) Antimicrobial peptides for plaque control. Adv Dent Res 21:57–62. https://doi.org/10.1177/0895937409335627

Liu Y, Kamesh AC, Xiao Y, Sun V, Hayes M, Daniell H, Koo H (2016) Topical delivery of low-cost protein drug candidates made in chloroplasts for biofilm disruption and uptake by oral epithelial cells. Biomaterials 105:156–166. https://doi.org/10.1016/j.biomaterials.2016.07.042

Lockhart PB, Tampi MP, Abt E, Aminoshariae A, Durkin MJ, Fouad AF, Gopal P, Hatten BW, Kennedy E, Lang MS et al (2019) Evidence-based clinical practice guideline on antibiotic use for the urgent management of pulpal- and periapical-related dental pain and intraoral swelling: A report from the American Dental Association. J Am Dent Assoc 150:906–921.e12. https://doi.org/10.1016/j.adaj.2019.08.020

López-López A, Camelo-Castillo A, Ferrer MD, Simon-Soro Á, Mira A (2017) Health-associated niche inhabitants as oral probiotics: the case of Streptococcus dentisani. Front Microbiol 8. https://doi.org/10.3389/fmicb.2017.00379

Luong AD, Buzid A, Luong JHT (2022) Important roles and potential uses of natural and synthetic antimicrobial peptides (AMPs) in oral diseases: cavity, periodontal disease, and thrush. Preprint at MDPI. https://doi.org/10.3390/jfb13040175

Lv X, Yang Y, Han S, Li D, Tu H, Li W, Zhou X, Zhang L (2015) Potential of an amelogenin based peptide in promoting reminerlization of initial enamel caries. Arch Oral Biol 60:1482–1487. https://doi.org/10.1016/j.archoralbio.2015.07.010

Mai J, Tian XL, Gallant JW, Merkley N, Biswas Z, Syvitski R, Douglas SE, Ling J, Li YH (2011) A novel target-specific, salt-resistant antimicrobial peptide against the cariogenic pathogen Streptococcus mutans. Antimicrob Agents Chemother 55:5205–5213. https://doi.org/10.1128/AAC.05175-11

Maisetta G, Batoni G, Esin S, Luperini F, Pardini M, Bottai D, Florio W, Giuca MR, Gabriele M, Campa M (2003) Activity of human β-defensin 3 alone or combined with other antimicrobial agents against oral bacteria. Antimicrob Agents Chemother 47:3349–3351. https://doi.org/10.1128/AAC.47.10.3349-3351.2003

Margherita Z (2005) The role of Cathelicidins in the innate immune defense of mammals. structure. Curr Issues Mol Biol 7:179–196.

Marionnet C, Bernerd F, Dumas A, Verrecchia F, Mollier K, Compan D, Bernard B, Lahfa M, Leclaire J, Medaisko C et al (2003) Modulation of gene expression induced in human epidermis by environmental stress in vivo. J Invest Dermatol 121:1447–1458. https://doi.org/10.1111/j.1523-1747.2003.12629.x

Marsh PD, Devine DA (2011) How is the development of dental biofilms influenced by the host? In. J Clin Periodontol:28–35. https://doi.org/10.1111/j.1600-051X.2010.01673.x

Mazmanian SK, Cui HL, Tzianabos AO, Kasper DL (2005) An immunomodulatory molecule of symbiotic bacteria directs maturation of the host immune system. Cell 122:107–118. https://doi.org/10.1016/j.cell.2005.05.007

McLatchie LM, Fraser NJ, Main MJ, Wise A, Brown J, Thompson N, Solari R, Lee MG, Foord SM (1998) RAMPS regulate the transport and ligand specificity of the calcitonin- receptor-like receptor. Nature 393:333–339. https://doi.org/10.1038/30666

Meinen A, Reuss A, Willrich N, Feig M, Noll I, Eckmanns T, Al-Nawas B, Markwart R (2021) Antimicrobial resistance and the spectrum of pathogens in dental and oral-maxillofacial infections in hospitals and dental practices in Germany. Front Microbiol 12. https://doi.org/10.3389/fmicb.2021.676108

Mishra B, Golla RM, Lau K, Lushnikova T, Wang G (2016) Anti-staphylococcal biofilm effects of human cathelicidin peptides. ACS Med Chem Lett 7:117–121. https://doi.org/10.1021/acsmedchemlett.5b00433

Moretta A, Scieuzo C, Petrone AM, Salvia R, Manniello MD, Franco A, Lucchetti D, Vassallo A, Vogel H, Sgambato A et al (2021) Antimicrobial peptides: A new hope in biomedical and pharmaceutical fields. Preprint at Frontiers Media SA. https://doi.org/10.3389/fcimb.2021.668632

Murakami M, Ohtake T, Dorschner RA, Gallo RL (2002) Cathelicidin antimicrobial peptides are expressed in salivary glands and saliva. J Dent Res 81:845–850. https://doi.org/10.1177/154405910208101210

Murakami M, Kameda K, Tsumoto H, Tsuda T, Masuda K, Utsunomiya R, Mori H, Miura Y, Sayama K (2017) TLN-58, an additional hCAP18 processing form, found in the lesion vesicle of Palmoplantar Pustulosis in the skin. J Invest Dermatol 137:322–331. https://doi.org/10.1016/j.jid.2016.07.044

Nacken W, Roth J, Sorg C, Kerkhoff C (2003) S100A9/S100A8: myeloid representatives of the S100 protein family as prominent players in innate immunity. Microsc Res Tech 60:569–580. https://doi.org/10.1002/jemt.10299

Nakamura T, Kido J, Kido R, Ohishi K, Yamauchi N, Kataoka M, Nagata T (2000) The association of calprotectin level in Gingival Crevicular fluid with Gingival index and the activities of collagenase and aspartate aminotransferase in adult periodontitis patients. J Periodontol 71:361–367. https://doi.org/10.1902/jop.2000.71.3.361

Neshani A, Zare H, Akbari Eidgahi MR, Kamali Kakhki R, Safdari H, Khaledi A, Ghazvini K (2019) LL-37: review of antimicrobial profile against sensitive and antibiotic-resistant human bacterial pathogens. Gene Rep 17:100519. https://doi.org/10.1016/j.genrep.2019.100519

Nilsson MF, Sandstedt B, Sørensen O, Weber G, Borregaard N, Ståhle-Bäckdahl M (1999) The human cationic antimicrobial protein (hCAP18), a peptide antibiotic, is widely expressed in human squamous epithelia and colocalizes with interleukin-6. Infect Immun 67:2561–2566. https://doi.org/10.1128/iai.67.5.2561-2566.1999

Niu JY, Yin IX, Wu WKK, Li QL, Mei ML, Chu CH (2021) Antimicrobial peptides for the prevention and treatment of dental caries: a concise review. Preprint at Elsevier Ltd. https://doi.org/10.1016/j.archoralbio.2020.105022

Nizet V, Gallo RL (2003) Cathelicidins and innate defense against invasive bacterial infection. Scand J Infect Dis 35:670–676. https://doi.org/10.1080/00365540310015629

North JR, Takenaka S, Rozek A, Kielczewska A, Opal S, Morici LA, Finlay BB, Schaber CJ, Straube R, Donini O (2016) A novel approach for emerging and antibiotic resistant infections: innate defense regulators as an agnostic therapy. J Biotechnol 226:24–34. https://doi.org/10.1016/j.jbiotec.2016.03.032

Oppenheim FG, Xu T, McMillian FM, Levitz SM, Diamond RD, Offner GD, Troxler RF (1988) Histatins, a novel family of histidine-rich proteins in human parotid secretion. Isolation, characterization, primary structure, and fungistatic effects on Candida albicans. J Biol Chem 263:7472–7477. https://doi.org/10.1016/s0021-9258(18)68522-9

Ouhara K, Komatsuzawa H, Yamada S, Shiba H, Fujiwara T, Ohara M, Sayama K, Hashimoto K, Kurihara H, Sugai M (2005) Susceptibilities of periodontopathogenic and cariogenic bacteria to antibacterial peptides, β-defensins and LL37, produced by human epithelial cells. J Antimicrob Chemother 55:888–896. https://doi.org/10.1093/jac/dki103

Paquette DW, Simpson DM, Friden P, Braman V, Williams RC (2002) Safety and clinical effects of topical histatin gels in humans. J. Clin Periodontol 2002 Dec;29(12):1051-8. https://doi.org/10.1034/j.1600-051x.2002.291201.x

Paulo B, Fabio T, da Costa Joao, Ferreira Rita VR (2018) Human antimicrobial peptides in bodily fluids: current knowledge and therapeutic perspectives in the postantibiotic era. Med Res Rev 38:101–146. https://doi.org/10.1002/med

Pepperney A, Chikindas ML (2011) Antibacterial peptides: opportunities for the prevention and treatment of dental caries. Probiot Antimicrob Prot 3:68–96. https://doi.org/10.1007/s12602-011-9076-5

Poveda Roda R, Vicente Bagán J, Bielsa MS, Carbonell Pastor E (2007) E186 Antibiotic use in dental practice. A review. Med Oral Patol Oral Cir Bucal 12:186–192

Pütsep K, Carlsson G, Boman HG, Andersson M (2002) Deficiency of antibacterial peptides in patients with morbus Kostmann: an observation study. Lancet 360:1144–1149. https://doi.org/10.1016/S0140-6736(02)11201-3

Qiu XQ, Wang H, Lu XF, Zhang J, Li SF, Cheng G, Wan L, Yang L, Zuo JY, Zhou YQ et al (2003) An engineered multidomain bactericidal peptide as a model for targeted antibiotics against specific bacteria. Nat Biotechnol 21:1480–1485. https://doi.org/10.1038/nbt913

Raj PA, Antonyraj KJ, Karunakaran T (2000) Large-scale synthesis and functional elements for the antimicrobial activity of defensins. Biochem J 347:633–641. https://doi.org/10.1042/0264-6021:3470633

Ramage G, Coco B, Sherry L, Bagg J, Lappin DF (2012) In vitro Candida albicans biofilm induced proteinase activity and SAP8 expression correlates with in vivo denture stomatitis severity. Mycopathologia 174:11–19. https://doi.org/10.1007/s11046-012-9522-2

Reid G, Jass J, Sebulsky MT, McCormick JK (2003) Potential uses of probiotics in clinical practice. Clin Microbiol Rev 16:658–672. https://doi.org/10.1128/CMR.16.4.658-672.2003

Rosen G, Sela MN, Bachrach G (2012) The antibacterial activity of LL-37 against Treponema denticola is dentilisin protease independent and facilitated by the major outer sheath protein virulence factor. Infect Immun 80:1107–1114. https://doi.org/10.1128/IAI.05903-11

Ross KF, Herzberg MC (2001) Calprotectin expression by gingival epithelial cells. Infect Immun 69:3248–3254. https://doi.org/10.1128/IAI.69.5.3248-3254.2001

Sancho-Vaello E, Gil-Carton D, François P, Bonetti EJ, Kreir M, Pothula KR, Kleinekathöfer U, Zeth K (2020) The structure of the antimicrobial human cathelicidin LL-37 shows oligomerization and channel formation in the presence of membrane mimics. Sci Rep 10:1–16. https://doi.org/10.1038/s41598-020-74401-5

Sander J, Fagerhol MK, Bakken JS, Dale I (1984) Plasma levels of the leucocyte l1 protein in febrile conditions: relation to aetiology, number of leucocytes in blood, blood sedimentation reaction and c-reactive protein. Scand J Clin Lab Invest 44:357–362. https://doi.org/10.3109/00365518409083820

Selsted ME, Harwig SSL, Ganz T, Schilling JW, Lehrer R (1985) Primary structures of three human neutrophil defensins. J Clin Invest 76(4):1436

Siderov J, Duggan J (2010) Arsenic trioxide associated toothache. J Oncol Pharm Pract 16:127–128. https://doi.org/10.1177/1078155209338727

Simón-Soro A, Mira A (2015) Solving the etiology of dental caries. Preprint at Elsevier Ltd. https://doi.org/10.1016/j.tim.2014.10.010

Storici P, Tossi A, Lenarčič B, Romeo D (1996) Purification and structural characterization of bovine cathelicidins, precursors of antimicrobial peptides. Eur J Biochem 238:769–776. https://doi.org/10.1111/j.1432-1033.1996.0769w.x

Stotz HU, Thomson JG, Wang Y (2009) Plant defensins: defense, development and application. Plant Signal Behav 4(11):1010

Tien M-T, Girardin SE, Regnault BA, Lionel Le Bourhis M-A, Dillies J-YC, Bourdet-Sicard R, Sansonetti PJ, Pe T, Dron (2006) Anti-inflammatory effect of Lactobacillus casei on Shigella-infected human intestinal epithelial cells. J Immunol 176(2):1228. https://doi.org/10.4049/jimmunol.176.2.1228

Trotti A, Garden A, Warde P, Symonds P, Langer C, Redman R, Pajak TF, Fleming TR, Henke M, Bourhis J et al (2004) A multinational, randomized phase III trial of Iseganan HCL oral solution for reducing the severity of oral mucositis in patients receiving radiotherapy for head-and-neck malignancy and have equity ownership in the company. Int J Radiat Oncol Biol Phys 58:674–681. https://doi.org/10.1016/S0630-3016(04)01627-4

Tu H, Fan Y, Lv X, Han S, Zhou X, Zhang L (2016) Activity of synthetic antimicrobial peptide GH12 against oral streptococci. Caries Res 50:48–61. https://doi.org/10.1159/000442898

Türkoğlu O, Emingil G, Kütükçüler N, Atilla G (2009) Gingival Crevicular fluid levels of cathelicidin LL-37 and Interleukin-18 in patients with chronic periodontitis. J Periodontol 80:969–976. https://doi.org/10.1902/jop.2009.080532

Van Dijk IA, Veerman ECI, Reits EAJ, Bolscher JGM, Stap J (2018) Salivary peptide histatin 1 mediated cell adhesion: A possible role in mesenchymal-epithelial transition and in pathologies. Biol Chem 399:1409–1419. https://doi.org/10.1515/hsz-2018-0246

Vieira Colombo AP, Magalhães CB, Hartenbach FARR, Martins do Souto R, Maciel da Silva-Boghossian C (2015) Periodontal-disease-associated biofilm: A reservoir for pathogens of medical importance. Microb Pathog 94:27–34. https://doi.org/10.1016/j.micpath.2015.09.009

Vierthaler M, Rodrigues PC, Sundquist E, Siponen M, Salo T, Risteli M (2020) Fluctuating role of antimicrobial peptide hCAP18/LL-37 in oral tongue dysplasia and carcinoma. Oncol Rep 44:325–338. https://doi.org/10.3892/or.2020.7609

Vu B, Chen M, Crawford RJ, Ivanova EP (2009) Bacterial extracellular polysaccharides involved in biofilm formation. Preprint. https://doi.org/10.3390/molecules14072535

Wang G (2014) Human antimicrobial peptides and proteins. Preprint at MDPI AG. https://doi.org/10.3390/ph7050545

Wang Z, De La Fuente-Núñez C, Shen Y, Haapasalo M, Hancock REW (2015a) Treatment of oral multispecies biofilms by an anti-biofilm peptide. PLoS One 10:1–16. https://doi.org/10.1371/journal.pone.0132512

Wang HY, Cheng JW, Yu HY, Lin L, Chih YH, Pan YP (2015b) Efficacy of a novel antimicrobial peptide against periodontal pathogens in both planktonic and polymicrobial biofilm states. Acta Biomater 25:150–161. https://doi.org/10.1016/j.actbio.2015.07.031

Wang Y, Xiong K, Chen X, Chi Y, Han Q, Zou L (2023) The remineralization effect of GERM CLEAN on early human enamel caries lesions in vitro. Sci Rep 13. https://doi.org/10.1038/s41598-023-31405-1

White SH, Wimley WC (1995) Structure, function, and membrane integration. Curr Opin Struct Biol 5(4):521

Wilde CG, Griffith JE, Marra MN, Snable JL, Scott RW (1989) Purification and characterization of human neutrophil peptide 4, a novel member of the defensin family. J Biol Chem 264(19):11200

Xiong K, Chen X, Hu H, Hou H, Gao P, Zou L (2020) Antimicrobial effect of a peptide containing novel oral spray on Streptococcus mutans. Biomed Res Int 2020. https://doi.org/10.1155/2020/6853652

Yang D, Biragyn A, Hoover DM, Lubkowski J, Oppenheim JJ (2004) Multiple roles of antimicrobial defensins, cathelicidins, and eosinophil-derived neurotoxin in host defense. Annu Rev Immunol 22:181–215. https://doi.org/10.1146/annurev.immunol.22.012703.104603

Yoshida T, Nagasawa T (2003) ε-poly-L-lysine: microbial production, biodegradation and application potential. Appl Microbiol Biotechnol 62:21–26. https://doi.org/10.1007/s00253-003-1312-9

Zanetti M, Gennaro R, Scocchi M, Skerlavaj B (2000) Structure and biology of Cathelicidins. In: The biology and pathology of innate immunity mechanisms, p 203–218. https://doi.org/10.1016/0165-6147(93)90081-T

Zigmond M, Yefenof E, Stabholz A, Shapira J, Merrick J, Chaushu S (2015) Salivary LL-37 secretion in individuals with down syndrome. J Dent Res 85:933–936. https://doi.org/10.1177/154405910608501012

Chapter 4
Guardians of the Epidermis: Interplay Between Skin Microbiota and Antimicrobial Peptides

Sk Aftabul Alam and Santi M. Mandal

1 Introduction

The human skin contains a diverse collection of living microorganisms known as the skin microbiota. Growing evidence indicates that this microbiota, characterized by its wide array of microbial species, is essential for maintaining the balance and stability of the skin (Byrd et al. 2018). The human epidermis establishes a robust protective barrier featuring a dense mechanical structure comprising the cornified envelope, keratin, and lipid layers, effectively preventing microbial invasion (Évora et al. 2021). Beyond its mechanical role, neighboring cells within the epidermis actively contribute to immune defense by secreting molecules to combat exogenous pathogens. The initial defense response is termed the "innate immune" response of the skin. Innate immunity refers to the immediate, nonspecific defense mechanisms that an organism possesses against pathogens. It is the first line of defense and acts rapidly upon exposure to a pathogen. The innate immune system comprises physical barriers, chemical defenses, and cellular elements working together to prevent and eradicate potential threats. Innate immunity acts immediately upon encountering a pathogen, providing a quick response before the adaptive immune system can fully engage. Innate immune mechanisms are not pathogen-specific and can target a wide range of microbes, contributing to the overall defense against various threats (Marshall et al. 2018).

Various forms of barriers are integral to innate immunity, encompassing physical, chemical, and cellular defenses. Cells such as neutrophils and macrophages are adept at engulfing and digesting pathogens. They are attracted to the site of infection

S. A. Alam
Department of Botany, Netaji Mahavidyalaya, Arambagh, Hooghly, India

S. M. Mandal (✉)
Department of Bioscience and Biotechnology, Indian Institute of Technology Kharagpur, Kharagpur, WB, India

P. Baindara, S. M. Mandal (eds.), *Evolution of Antimicrobial Peptides*,
https://doi.org/10.1007/978-3-031-67515-7_4

by signals released by damaged cells or microbial products. Natural killer cells can recognize and destroy infected or abnormal cells, providing a defense against viruses and cancer (Mandal and Viswanathan 2015). In addition to cellular components, mucous membranes in the respiratory and gastrointestinal tracts provide a protective barrier, which traps pathogens, and cilia move them out of the body. The skin is a physical barrier that serves as the body's first line of defense. It consists of multiple layers of cells and is coated with oils and other substances that create an inhospitable environment for many pathogens (Pessemier et al. 2021).

AMPs, also referred to as host defense peptides (HDPs), are generally compact (10–50 amino acids). These molecules are positively charged and amphiphilic, with a significant portion, usually around 50%, comprising hydrophobic residues. These characteristics are vital for their essential roles in innate immunity (Herman and Herman 2019; Yamasaki and Gallo 2008; Mahlapuu et al. 2020). The human body has developed defense mechanisms to regulate the microbiome associated with humans, and this involves generating AMPs. These peptides form a group of small, internally produced compounds released through sweat and sebaceous secretions (Gläser et al. 2005).

Although there is diversity in their sequences, numerous AMPs demonstrate wide-ranging antimicrobial efficacy against bacteria, fungi, and viruses (Rademacher et al. 2021). These peptides are characterized by their cationic nature, enabling them to bind to negatively charged bacterial membrane molecules such as lipopolysaccharide and lipoteichoic acid (Kristian et al. 2005). These are postulated to initially adhere to biological membranes, subsequently employing their amphipathic conformation to integrate into the lipid bilayer, disrupting lipid chain interactions in the process (Henriques et al. 2006).

The antimicrobial capabilities of AMPs suggest a probable direct interaction with the microbiota, AMPs have developed as the potential regulators and modulators of the microbiota (Rademacher et al. 2021). In opposition to the gastrointestinal tract, the influence of AMPs on the skin microbiota is still largely uncharted. However, a growing body of research is reinforcing the hypothesis that AMPs originating from the skin may play a pivotal role in shaping the cutaneous microbiota.

2 Skin Microbiota: Composition and Diversity

The skin hosts a varied assembly of bacteria, fungi, and viruses, collectively referred to as the skin microbiota, which, much like their counterparts in the gut, play crucial roles in safeguarding the skin against potential pathogenic invaders. Being the largest organ in the body, the skin accommodates beneficial microorganisms and serves as a robust physical barrier, actively thwarting the entry of pathogens. When this barrier is compromised or the equilibrium between commensal and pathogenic microorganisms is disrupted, it can lead to skin diseases or even systemic health issues. The human skin microbiome encompasses microorganisms, their genetic elements, and interactions with the environment. The major bacterial phyla include

Proteobacteria (11–16%), Firmicutes (24–34%), Actinobacteria (36–51%), and Bacteroidetes (6–9%) (Carmona-Cruz et al. 2022).

Regions with sebaceous glands were primarily inhabited by lipophilic species of *Propionibacterium*, whereas bacteria adapted to humid conditions, such as *Staphylococcus* (Firmicutes) and *Corynebacterium* (Actinobacteria) species, were notably abundant in moist areas like elbow creases and feet. In regions marked by oiliness, the skin accommodates a less diverse community, with *Cutibacterium* (Actinobacteria) species prevailing as the primary isolates. In the central body and arm regions, the prevailing fungal species belonged to the genus *Malassezia*. Conversely, foot sites displayed a more varied assortment of fungal species, encompassing *Aspergillus* spp., *Malassezia* spp., *Rhodotorula* spp., *Cryptococcus* spp., and *Epicoccum* spp., constituting approximately 80% of the overall fungal flora (Findley et al. 2013; Carmona-Cruz et al. 2022).

Viruses within the skin microbiota have not been extensively studied, with cutaneous β and γ human papillomaviruses commonly observed on the skin's surface, but they are less abundant compared to the bacteriophages associated with the bacterial flora residing on the skin (Carmona-Cruz et al. 2022). A thorough examination demonstrated that bacterial colonization depends on the physiological attributes of the skin site, with particular bacteria associated with dry, moist, and sebaceous microenvironments. Staphylococci predominantly inhabit the aerobic habitat on the skin and likely utilize urea derived from sweat as a nitrogen source. The role of corynebacteria, known for their meticulous and slow growth in culture, was underestimated as skin microorganisms until recently. The breakdown of apocrine sweat by corynebacteria, staphylococci, and other microorganisms in the axillary vault contributes to the unique malodor linked with human sweat (Decréau et al. 2003; Emter and Natsch 2008). The skin microbiome performs essential functions in sustaining skin homeostasis, protecting against invading pathogens, and regulating the immune system. Dysbiosis refers to an imbalance among microbial communities in specific body areas, potentially playing a role in the initiation or advancement of diseases. Disruptions in the balance of microorganisms, known as dysbiosis, may be implicated in skin conditions like dermatitis, alopecia areata, psoriasis, and acne (Sanford and Gallo 2013).

The skin's main function is to act as a physical barrier, shielding the body from potentially dangerous foreign objects or poisonous substances. Analyzing the microbiome through metagenomics sequencing across diverse areas of the human body sites provides important insights into how a variety of factors, such as ecology, environment, genetic predisposition, host immunity, and host lifestyle, influence the microbial diversity of the skin (Cogen et al. 2008). Recent advancements in our comprehension of skin function and its microbiome reveal a robust symbiotic interdependence between the microbiota present on the skin and the immune system of the host. Dysbiosis or disruptions in the balance of the microbiome, along with additional factors affecting surface microbiota, can influence the regulation and balance of keratinocytes, potentially compromising the function of the skin barrier.

Microorganisms within the skin are distributed from the epidermis to the deeper dermis layers, with approximately 25% thriving in the deeper layers of the dermis and sebaceous glands. Each person carries an exclusive and distinctive characteristic of the skin microbiota, formed during childhood and maintained throughout adulthood. *Staphylococcus epidermidis* stands out as the most frequently isolated microorganism from human epithelia and plays a crucial role as an integral member of the resident microflora on the skin (Otto 2009). This microorganism displays a flexible association with its host, showcasing the capability to create robust biofilms that present challenges for removal. The complexity arises from navigating the extracellular matrix and is exacerbated by the emergence of antibiotic resistance and immune resistance (Mah and O'Toole 2001). The extracellular matrix functions as a physical barrier, limiting the diffusion of numerous antibiotics and chemicals while also serving as a mechanical barrier that hinders the passage of immune cells.

Growing awareness underscores the significance of fungi and their interplay with the immune system, shaping immune balance and having an impact on the onset of diseases. Disturbances in the chemical composition of the host epidermis, such as alterations in pH and pathological sweat secretion, heighten the pathogenicity of *Malassezia* spp. Consequently, the release of enzymes such as lipases, phospholipases, and bioactive indoles occurs, ultimately modifying the purpose of the epithelial barrier and contributing to immune deregulation and the initiation of diseases (Xu et al. 2007).

Exploring the intricate relationship between the microbiome and its host prompts an examination of how the microbiome influences both the cutaneous and systemic functions of the host, particularly those regulated by the immune system. Skin cells contribute to the generation and metabolism of various substances, including steroid hormones and neurotransmitters, some of which are released through sweat and sebum. When exposed to cutaneous microbes, these substances can impact the growth, virulence, and adhesion of the microbes. Studies conducted in experimental settings have suggested a connection between elevations in local substance P triggered by stress (associated with conditions like eczema, acne, and barrier dysfunction) and alterations in the skin microbiota (N'Diaye et al. 2016). Emphasizing the interaction between the microbiome and the host immune system as a potential foundation, we could consider utilizing this understanding to regulate claims made by cosmetic products, thereby establishing guidelines that ensure a safe and transparent delineation of their topical effects on the body.

The composition and diversity of skin microbiota can vary among individuals and across different body regions. The epidermis of the skin is home to a diverse array of microorganisms, including bacteria, fungi, viruses, and mites. Different regions of the body such as the hands, face, feet, torso, and sebaceous (oil-producing) areas have different microbial compositions due to variations in environmental conditions. Each person has a unique skin microbiota profile influenced by genetic and environmental factors and sometimes also influenced by hygiene practices, the use of skin care products, and exposure to antibiotics. The skin microbiota is thought to contribute to skin health by helping maintain a balanced microbial community, pH, and homeostasis.

3 AMPs in Skin

Innate AMPs are molecules that have remained unchanged across evolutionary processes, displaying a combination of structural and functional variations. These diverse molecules are present in almost all organisms. AMPs are acknowledged for their protective role in skin health, some studies have postulated that AMPs might also play an active role in the development of various skin diseases. AMPs can serve dual functions, functioning not just as agents that combat microbes but also as molecules with the potential to either promote or inhibit inflammation. This characteristic is frequently compared to a double-edged sword. Harder et al. (1997) documented the successful extraction of AMP from human skin for the first time by isolating it from scaly areas in individuals with psoriasis. Recognized for its structural similarity to cationic β-defensins found in bovine epithelia, with a molecular weight ranging from 3 to 5 kDa, this newly identified peptide was named human β-defensin 2 (HBD-2) (Harder et al. 1997). Defensins, a group of small cationic peptides, are classified into three subfamilies: α-defensins, β-defensins, and circular θ-defensins. Typically, they display β-sheet structures and incorporate six cysteine residues, forming three intramolecular disulfide bridges. The distinction between α- and β-defensins is determined by the configuration of these disulfide bridges (Lehrer 2004). Both immune cells and epithelial cells in tissues generate defensins that interface with the external environment. Notably, HBD-2 has demonstrated significant effectiveness against gram-negative bacteria like *Pseudomonas aeruginosa* and *Escherichia coli* while also exhibiting activity against *Candida albicans* (Harder et al. 1997).

Discovered in psoriatic skin in 2001, HBD-3 (5 kDa), another β-defensin, is an AMP characterized by a wide range of activity. It showcases potency at low concentrations and retains its effectiveness even in environments with elevated salt concentrations (Harder et al. 2001). The *CAMP* gene encodes an antimicrobial peptide that can be induced. This gene generates an 18 kDa precursor, which is subjected to proteolytic processing to produce a 37-amino acid peptide referred to as LL-37. This cathelicidin is distinctive to humans and is specifically produced in the skin (Lehrer and Ganz 2002). Cathelicidin comprises two segments: a cathelin domain located at the N-terminal and an antimicrobial domain located at the C-terminal. The *CAMP* gene exhibits its main activity in myeloid bone marrow cells, and the proprotein hCAP18 is stored within granules unique to neutrophils. Individuals, including both mice and humans, who lack the *CAMP* gene, are prone to bacterial infections in diverse tissues (Campbell et al. 2012).

In opposition to LL-37, HBD-2, and HBD-3, which exhibit limited expression in normal epidermis, Gläser et al. recognized the S100 protein psoriasin (or S100 A7) as a plentiful AMP in undamaged skin. This AMP displays antimicrobial activity, particularly against *E. coli* and various fungi (Hein et al. 2015).

Harder and Schröder reported the discovery of a new AMP, named ribonuclease 7 (RNase 7), obtained from the outermost layer of healthy skin known as the stratum corneum. This AMP exhibited potent antimicrobial efficacy against a broad spectrum of pathogenic microorganisms. Structurally similar to the ribonuclease A superfamily, it was designated as ribonuclease 7 (RNase 7). This lysine-rich protein, weighing 14.5 kDa, exhibits a wide range of antimicrobial activities and possesses potent ribonuclease capabilities. Its presence can also be additionally triggered by pro-inflammatory cytokines and bacterial presence (Harder and Schröder 2002).

All the aforementioned AMPs are cationic, possessing a positive charge that facilitates affinity to the bacterial surface carrying a negative charge. Nevertheless, dermcidin (DCD), an AMP solely secreted by sweat glands, diverges from this pattern by carrying a negative charge. This 47-amino-acid peptide also demonstrates antimicrobial activity through membrane interface with numerous microorganisms (Schittek et al. 2001).

Other peptides and proteins with antimicrobial properties have been identified in the epidermis layer. Illustrations encompass protease inhibitors such as secretory leukocyte proteinase inhibitor (SLPI) and elafin, both of which exhibit defensin-like antimicrobial activity (Sallenave 2002). Peptides derived from hornerin, known as cationic intrinsically disordered antimicrobial peptides (CIDAMPs), display microbicidal activity against a range of pathogens, including *Escherichia coli*, *Pseudomonas aeruginosa*, *Staphylococcus aureus*, and *Candida albicans* (Gerstel et al. 2018), calprotectin complex with MRP-8 and MRP-14 having antimicrobial activity against *Candida albicans* (Sohnle et al. 2000).

AMPs possess the ability to fix and neutralize endotoxins of bacteria, induce angiogenesis, and facilitate improved wound curing (Mahlapuu et al. 2020). Such AMPs exhibit inherent activity related to chemotaxis, improve the directional movement of immune cells in response to chemical signals, and activate the differentiation and development of dendritic cells. Recently, AMPs have attracted attention as potential candidates for drugs aimed at treating infectious diseases and as innovative immunomodulatory therapies. Additionally, several AMPs demonstrate immunomodulatory activity, indirectly aiding in the clearance of pathogens within the host (Yeung et al. 2011).

4 Factors Influencing the Expression of AMPs

Immune and epithelial cells, consistently encountering environmental microbes, generate and release AMPs. Crucial for fortifying barriers, deficiencies in AMPs elevate vulnerability to infections. Apart from compromising the structural integrity of microbial membranes, AMPs also interact with lipopolysaccharides (LPSs), serve as a chemoattractant to attract immune cells, and engage with cellular receptors. As a result, they impact the expression of chemokines and cytokines. A review of the literature reveals that HBD-1 is consistently explicated in different tissues, whereas inflammatory stimuli are required to induce the production of HBD-2, HBD-3, and HBD-4. Bacteria, IL-1β, LPS, TNF-α, or phorbol-myristate-acetate (PMA) that cause infections are some examples of these stimuli (Chen et al. 2006).

Two forms of vitamin D are available for human consumption: vitamin D_2, also referred to as ergocalciferol, sourced from fungi, and vitamin D_3, known as cholecalciferol, derived from animal origins or produced in the skin. The activation of the human *CAMP* gene is triggered by $1,25(OH)_2D_3$ and equivalent counterparts found in different cell lines and primary cells (Gombart et al. 2007). Schauber et al. (2007) also reported that explicit cathelicidin is similarly elevated in keratinocytes through a comparable pathway (Toll-like receptor activation) following skin injury. As per Wang et al. (2010), it has been demonstrated that $1,25(OH)_2D_3$ significantly enhances the expression of *NOD2/CARD15/IBD1* in epithelial cells and primary human monocytic. Treating normal keratinocytes of humans with $1,25(OH)_2D_3$ led to a dose-dependent elevation in HBD-3 mRNA levels (Dai et al. 2010) (Table 4.2).

Butyrate is a short fatty acid generated in the colon through the fermentation of dietary fiber by anaerobic bacteria. It is present in several foods and acts as a histone deacetylase inhibitor (HDACi). In adult patients, *Shigella* infections have been noted to reduce LL-37 and HBD-1 levels in the colon. Furthermore, during *Shigella* infection, there is a decrease in rabbit CAP18 levels in the colon surface epithelium; nevertheless, the administration of sodium butyrate orally has been shown to restore CAP18 levels (Islam et al. 2001; Raqib et al. 2006). Due to its disagreeable scent, butyrate is rarely employed in clinical trials. Instead, 4-phenylbutyrate (PBA), a derivative of butyrate that is odorless and palatable, is frequently employed in clinical settings.

Enhancing the endogenous expression of cathelicidin and defensins could be particularly beneficial in infection treatment. Clinical studies have revealed diminished hCAP18 levels in asthma patients. In hemodialysis patients, an increase in circulating hCAP18 levels at the commencement of treatment was associated with a noteworthy reduction in 1-year mortality. Although there was a slight connection with $1,25(OH)_2D_3$ levels, there was no correlation with 25(OH)D levels (Gombart et al. 2009).

The expression of AMPs in the body is regulated by various factors, and their induction is a crucial component of the innate immune response. Inflammatory signals like cytokines, interleukins (IL-1 and IL-6), and tumor necrosis factor alpha (TNF-α) can stimulate AMP expression (Zhang et al. 2023). Moreover, their expression can be induced in response to tissue damage. The need for heightened antimicrobial defense in damaged areas contributes to AMP upregulation. The major bioactive molecules like glucocorticoid hormone can modulate AMP expression. Cortisol, a stress hormone, also influences AMP production (Wang and Jin 2021). Genetic factors, mostly transcription factors, are the main regulator of AMP expression (Mandal et al. 2018). Pattern recognition receptors (PRRs), Toll-like receptors (TLRs), and NOD-like receptors (NLRs) recognize specific microbial patterns and initiate signaling cascades and are also regulators of AMP production. Especially, micronutrients like vitamin D and zinc are important for the proper functioning of the immune system, including AMP expression (Benson et al. 2023). Understanding the complex regulatory network controlling AMP expression is crucial for developing therapeutic strategies to enhance innate immune responses and combat infections. The interplay of these factors highlights the dynamic nature of the host–microbe interactions that shape the immune defense mechanisms in the skin. Microbial exposure is the first line of regulator of AMP expression. The presence of pathogens or microbes, including bacteria, viruses, and fungi, can trigger the expression of AMPs. The innate immune system recognizes conserved microbial structures, such as lipopolysaccharides (LPSs) and peptidoglycans, leading to the activation of AMP production. The microbes-mediated infection or some akin inflammatory diseases are discussed below as their role in AMP production for disease management.

5 Action as Safeguard

Nevertheless, the mechanisms through which certain AMPs can effectively permeate bacterial cells are still subjects of the ongoing investigation. A comprehensive grasp of the AMP mode of action is essential for the formulation of effective drug candidates. The majority of AMPs exert a straight microbicidal impact by compromising the fidelity of the target organism's membrane. Additionally, they possess the ability to traverse the microbial membrane to reach intracellular targets (Hancock and Sahl 2006).

Few peptides, such as proline-rich AMPs, may bind to the bacterial surface, undergo translocation into the cell by forming transient pores, and ultimately exert their effects on intracellular targets (Otvos 2002). Notably, the swift action of AMPs in exerting antibacterial efficacy is a common characteristic. Instances comprise the Os(3–12)NH2 derived from tick defensin, HDP chicken cathelicidin-2 (CATH-2), and the peptide designed through computational methods, PaDBS1R1, all showcasing microbicidal activity within an exceptionally brief time frame of 5–10 minutes upon exposure (Schneider et al. 2016; Ismail et al. 2019; Irazazabal et al. 2018). Both gram-positive and gram-negative microorganisms possess cytoplasmic membranes containing phospholipid molecules. The interaction between AMPs and the bacterial surface is intensified by the existence of teichoic acids in the cell wall of gram-positive microorganisms and the LPS layer in the outer membrane of gram-negative microorganisms, introducing extra electronegative charge (Ebenhan et al. 2014; Malmsten 2016).

AMPs can induce destabilization of the bacterial cell membrane by altering its width, either decreasing or increasing it (Grage et al. 2016). Alternatively, they can bring about the clustering of the head groups of phospholipids within the plasma membrane, following the charged lipid clustering model (Epand 2019). AMPs can initiate molecular electroporation by generating a transmembrane potential, resulting in the formation of pores and the movement of molecules across the membrane (Lai and Kaznessis 2018). Alternatively, such peptides can enhance membrane permeability by forming complexes with tiny organic anions, facilitating their transport through the membrane (Rokitskaya et al. 2011). Additionally, AMPs might be able to enhance the production of non-bilayer intermediates, which would allow the peptide to be internalized into the membrane. Subsequently, the interruption of the intermediate occurs, followed by the release of the peptide (Haney et al. 2010).

Furthermore, studies have suggested that AMPs can translocate across membranes without the formation of pores, potentially involving receptor-mediated processes. Once these molecules traverse bacterial membranes, they can target intracellular macromolecules and bioprocesses, such as inhibiting DNA replication and transcription (Friedrich et al. 2001). Peptides induce positive curvature, promoting the creation of toroidal pores and micelles. In contrast, negative curvature stress might lead to the generation of non-bilayer intermediates, and the induction of cubic lipid phases by AMP could lead to the formation of porous structures in the membrane (Haney et al. 2010; Schmidt and Wong 2013). Moreover, it has been demonstrated that such peptides may deactivate bacterial chaperones responsible for protein folding, resulting in bactericidal effects through the inhibition of protein synthesis (Graf and Wilson 2019).

Numerous instances of AMPs with intracellular targets exist: indolicidin fixes to DNA, thereby preventing DNA replication (Cardoso et al. 2019); teixobactin hinders the synthesis of the cell wall by attaching to the peptidoglycan precursor known as lipid II (Chiorean et al. 2020); the Lser-PRP2 peptide from larvae is suggested to impede protein folding by interacting with DnaK, one of the bacterial chaperone (Cytryńska et al. 2020); and the proline-rich peptide Bac5 is reported to block translation by binding to ribosomes (Mardirossian et al. 2018) (Figs. 4.1 and 4.2).

Thus, AMPs serve as a crucial component of the body's innate immune system, acting as natural safeguards against microbial threats. Antimicrobial peptides are versatile and dynamic elements of the innate immune system, providing a frontline defense for humans. Their multifaceted roles contribute to the overall protection and maintenance of health in various tissues and organs within the body.

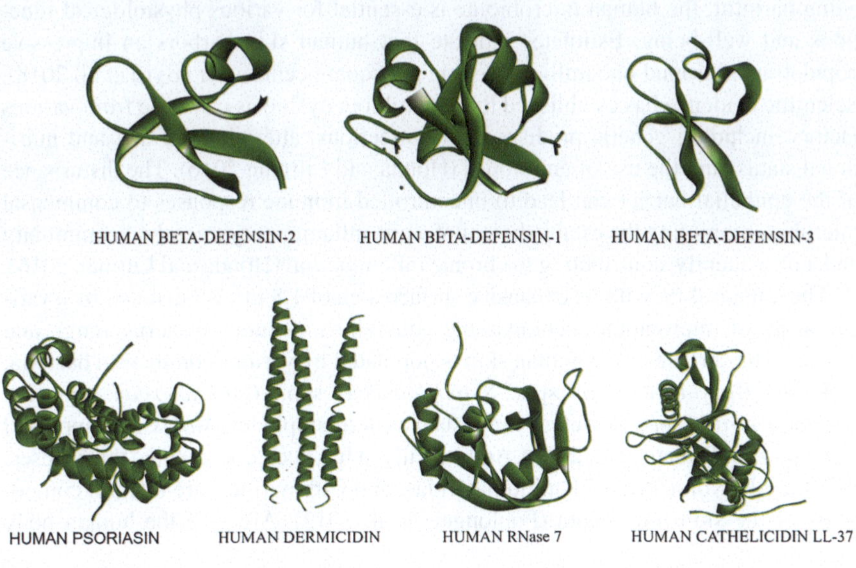

Fig. 4.1 PDB structure of AMPs

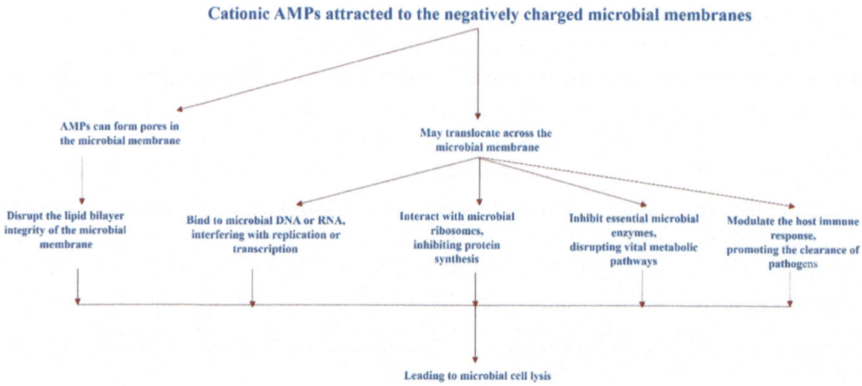

Fig. 4.2 The schematic illustration of most of the skin peptides and their possible action

6 AMP Expression and Their Contribution to Skin Diseases

While AMPs play a vital role in safeguarding the skin, numerous studies have proposed that these peptides might actively participate in the development of various skin diseases. A growing body of evidence substantiates the pathophysiological role of AMPs in conditions such as psoriasis vulgaris, acne vulgaris, atopic dermatitis, hidradenitis suppurativa, rosacea, and wound healing. Recent research on the human microbiome indicates that these microbes are predominantly beneficial, playing a crucial role in maintaining overall health. They demonstrate sensitivity to both genetic factors and environmental influences. The findings suggest that contrary to being harmful, the human microbiome is essential for various physiological functions and well-being. Estimates indicate that human skin harbors an impressive population of around one million bacteria per square centimeter (Byrd et al. 2018). Scientific evidence has established that microbiota dysbiosis can arise from various factors, including genetic predisposition, infections, alterations in diet and nutritional status, and the use of antibiotics (Honda and Littman 2016). The disturbance of the epithelial barrier can lead to uncontrolled immune responses to commensal microbes, promoting the establishment of a pro-inflammatory microbial community and, consequently, contributing to chronic inflammation (Honda and Littman 2016).

The human skin, with its expansive surface area of 1.8 m^2, is inhabited by a various array of microbiota, encompassing generally innocuous bacteria, fungi, and viruses. The surface of the human skin is populated by various commensal bacteria, including *Corynebacterium* spp., *Staphylococcus* spp., *Cutibacterium* spp., and *Actinobacterium* spp., as well as fungi such as dermatophytes, *Malassezia* spp., and *Candida* spp. (Balato et al. 2018). Additionally, it has been documented that viruses, including Polyomaviridae, Papillomaviridae, and Circoviridae, are integral components of the skin microbiota (Foulongne et al. 2012). Areas of the human body

Table 4.1 AMPs produced by skin microbiota and their effects on microorganisms

List of AMPs	Structure	Source	Killing effects on Gram +ve bacteria	Killing effects on Gram -ve bacteria	Killing effects on Virus	Killing effects on Fungi
Beta defensin 1		Skin, sweat glands, urinary tract, respiratory tract, gastrointestinal tract	No	Yes	No	Yes
Beta defensin 2		Skin, gastrointestinal tract, respiratory tract, urinary tract	No	Yes	Yes	Yes
Beta defensin 3		Skin, respiratory tract, gastrointestinal tract	Yes	Yes	Yes	Yes
Cathelicidin		Skin, gastrointestinal tract, respiratory tract	Yes	Yes	Yes	Yes
Dermcidin		Sweat glands	Yes	Yes	Not reported	Not reported
Psoriasin 1		Skin, urinary tract, respiratory tract	Yes	Yes	Not reported	Yes
RNase A7		Skin, urinary tract, respiratory tract	Yes	Yes	Not reported	Yes

characterized by elevated humidity, such as sweat glands or the armpit, are frequently colonized by staphylococci or corynebacteria; in contrast, drier or more lipid-rich regions, such as hair follicles or sebaceous glands, are predominantly inhabited by *Cutibacterium acnes* (Kong et al. 2012) (Table 4.1).

6.1 Atopic Dermatitis

Atopic dermatitis (AD) is a persistent inflammatory skin disorder that impacts around 10–20% of the overall population. It is an intricate, multifactorial condition influenced by a combination of genetic predisposition and environmental triggers (Sandilands et al. 2007; Balato et al. 2018). This condition entails a decrease in gram-negative bacteria and irregular colonization of the skin by *Staphylococcus aureus*. Notably, research has revealed a noteworthy rise in the prevalence of

Table 4.2 Domain and sequence analysis of the skin-abundant peptides

Accession number; sequence character	Sequence	Domain architecture
AAC51728.1; beta defensin 1 (*Homo sapiens*)	MRTSYLLLFTLCLLLSEMASGGNFLTGLGH RSDHYNCVSSGGQCLYSACPIFTKIQGTCYRGKAKCK	
CAB65126.1; beta defensin 2 (*Homo sapiens*)	MRVLYLLFSFLFIFLMPLPGV FGGIGDPVTCLKSGAICHPVFCPRRYKQIGTCGLPGTKCCKKP	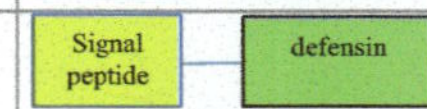
CAC03097.1; beta defensin 3 (*Homo sapiens*)	MRIHYLLFALLFLFLVPVPGHGGIINTLQKYYCRV RGGRCAVLSCLPKEEQIGKCSTRGRKCCRRKK	
NP_004336.4; cathelicidin (*Homo sapiens*)	MKTQRDGHSLGRWSLVLLLLGLVMPLAIIAQV LSYKEAVLRAIDGINQRSSDANLYRLLDLDPRP TMDGDPDTPKPVSFTVKETVCPRTTQQSPEDCD FKKDGLVKRCMGTVTLNQARGSFDISCDKDNK RFALLGDFFRKSKEKIGKEFKRIVQRIKDFLRNLVPRTES	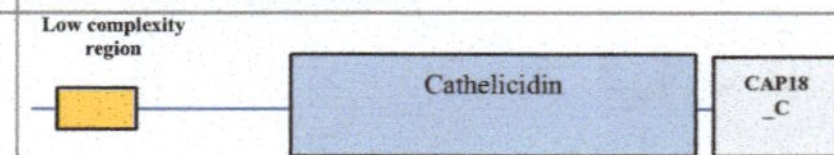
KAI2566128.1; dermcidin (*Homo sapiens*)	MRFMTLLFLTALAGALVCAYDPEAASAPGSGN PCHEASAAQKENAGEDPGLARQAPKPRKQRSSLLEK GLDGAKKAVGGLGKLGKDAVEDLESVGKGAVHDVKDVLDSVL	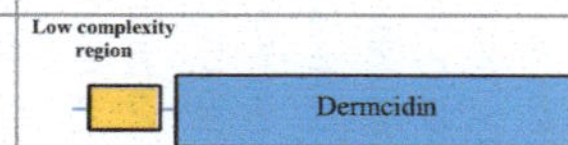
EAW53327.1; psoriasin 1 (*Homo sapiens*)	MSNTQAERSIIGMIDMFHKYTRRDDKIDKPSLLTM MKENFPNFLSACDKKGTNYLADVFEKKDKNEDKKIDFSEF LSLLGDIATDYHKQSHGAAPCSGGSQ	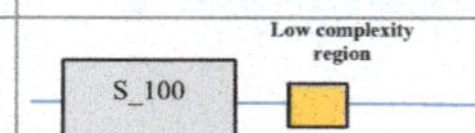
AAH74960.1; ribonuclease (RNase A7) (*Homo sapiens*)	MAPARAGFCPLLLLLLLGLWVAEIPVSAKPKG MTSSQWFKIQHMQPSPQACNSAMKNINKHTKRCK DLNTFLHEPFSSVAATCQTPKIACKNGDKNCH QSHGPVSLTMCKLTSGKYPNCRYKEKRQNKSYVV ACKPPQKKDSQQFHLVPVHLDRVL	

Staphylococcus aureus, particularly in untreated individuals, in inflamed regions, and during flare-ups compared to xerotic (dry) areas (Kong et al. 2012). During flare-ups in affected atopic dermatitis (AD) skin, there is a noticeable decrease in the abundance of other species, including *Propionibacterium*, *Corynebacterium*, and *Streptococcus* (Kong et al. 2012). Significant distinctions have been observed between young children and adolescents concerning atopic dermatitis (AD). In non-lesional AD skin, the diversity of the microbiome was noted to be considerably higher in young children compared to adolescents and adults (Balato et al. 2018). Studies have shown that unlike in healthy skin, various AMPs were activated in affected skin. However, when compared to skin affected by psoriasis, their increase was relatively less pronounced (Harder et al. 2013).

6.2 *Psoriasis Vulgaris*

Psoriasis (Pso) is a persistent inflammatory condition that affects approximately 2–4% of the worldwide population. Chang et al. reported a close association between the proportion of *Staphylococcus aureus* and psoriatic lesions, with a lower ratio of *Staphylococcus epidermidis* and *Cutibacterium acnes* compared to healthy skin (Chang et al. 2018). *Prevotella* and *Staphylococcus* showed a notable association with lesional skin, whereas *Anaerococcus* and *Cutibacterium* were linked to non-lesional skin in individuals with psoriasis. There were no notable differences in the types of aerobic bacteria identified through skin cultures between individuals with psoriasis and those without any skin conditions among the healthy controls. Various AMPs, including S100-proteins, LL-37, HBDs, and RNase 7, are activated in both lesional and non-lesional skin in comparison to a healthy control group (Niyonsaba et al. 2017).

6.3 *Acne Vulgaris*

Acne is a prevalent inflammatory epidermis disorder that impacts 70–80% of adolescents, leading to significant psychological and social consequences. The pathogenesis of acne involves inflammation, abnormal keratinization, increased sebum production, and colonization with *Cutibacterium acnes*. In areas of the affected skin, there has been reported augmented expression of S100 A7, HBD-2, human neutrophil peptides (HNPs) 1–3, and granulysin (Harder et al. 2013). Conversely, it has been demonstrated that the expression of dermcidin (DCD) is diminished in the sweat of individuals with acne (Nakano et al. 2015).

6.4 Hidradenitis Suppurativa

Hidradenitis suppurativa (or HS) is a persistent inflammatory skin condition marked by abscesses, recurrent nodules, fistulae, and scarring, primarily affecting the intertriginous regions (Balato et al. 2018). The manifestation of HS is influenced by a combination of factors, including follicular occlusion, follicular rupture, an immune response resembling a foreign body reaction, a distinctive genetic profile, and environmental elements such as microbial colonization. AMPs were found to exhibit heightened expression in HS lesions when compared to unaffected, non-lesional skin (Schlapbach et al. 2009). Emelianov et al. (2012) documented the elevated appearance of psoriasin, LL-37, and HBD-3 protein in an immunohistological investigation of individuals with HS (Table 4.3).

Table 4.3 List of AMPs associated with different skin diseases

Skin disease	AMP	Changes in skin microbiome	References
Atopic dermatitis	↑S100 A7, ↑RNase 7, ↑HBD-2-3, ↓LL-37	↑*Staphylococcus epidermidis* ↑*Staphylococcus aureus* ↑ *Staphylococcus haemolyticus* ↓*Corynebacterium* ↓*Propionibacterium*	Oh et al. (2013), Kong et al. (2012), Gonzalez et al. (2016)
Psoriasis vulgaris	↑S100 A7, ↑HBD-2-3, ↑RNase 7, ↑LL-37	↑*Propionibacterium* ↑*Corynebacterium* ↑ *Streptococcus* ↑*Staphylococcus aureus* ↓Firmicutes	Alekseyenko et al. (2013), Gao et al. (2008)
Hidradenitis suppurativa	↑LL-37, ↑psoriasin, ↑cathelicidin, ↑HBD-3	↓*S. aureus* ↑*S. aureus* ↑Coagulasi-negative staphylococci ↑*Porphyromonas* and *Peptoniphilus* ↑Corynebacterium	Ring and Emtestam (2016), Sartorius et al. (2012), Ring et al. (2017)
Acne vulgaris	↑S100 A7, ↑HBD-2, ↑HNP 1-3, ↑granulysin, ↓DCD	↑*Staphylococcus epidermidis* ↑*Propionibacterium Acnes* ↓Actinobacteria ↑Proteobacteria and Firmicutes	Fitz-Gibbon et al. (2013), Dreno et al. (2017), Christensen et al. (2016)

7 Common Usage in Pharmaceuticals

The extensive and rapid bactericidal effectiveness of AMPs, coupled with a perceived low likelihood of resistance development, positions them as promising contenders for antimicrobial medications (Mahlapuu et al. 2016). Currently, only a small subset of AMPs has received approval for medical applications as anti-infectives, and cyclic peptides, particularly gramicidins and polymyxins, have been extensively researched since the 1950s (Zavascki et al. 2007). Gramicidin is predominantly applied topically to address infections in superficial injuries, as well as infections in the ocular, nasal, and throat regions. Daptomycin exemplifies a cyclic AMP that has been more recently incorporated into clinical practice for treating complex skin and skin structure infections instigated by gram-positive bacteria, particularly *Staphylococcus aureus*.

Pexiganan, a synthetic peptide composed of 22 amino acids and derived from the *Xenopus peptide* magainin, exhibits effectiveness against a broad spectrum of microorganisms, encompassing gram-positive and gram-negative bacteria, along with fungi. This includes activity against various strains that are resistant to multiple drugs. The topical application of pexiganan has undergone evaluation in two phase III medical hearings for addressing minor infections in diabetic foot ulcers (Sierra et al. 2017).

DPK-060, derived in structure from the human protein kininogen, shows a broad range of antimicrobial capabilities. The effectiveness of DPK-060 gel has been assessed in clinical phase II trials for managing infections in acute external otitis (NCT01447017) and atopic dermatitis (NCT01522391) (Håkansson et al. 2019). LL-37 is a peptide consisting of 37 amino acids, formed through the proteolytic cleavage of the 18-kDa hCAP18, and was initially found to be abundant at the wound margin in acute conditions but absent in chronic wounds. This observation provides early evidence indicating that this peptide may contribute to the usual curing of wounds (Heilborn et al. 2003).

Omiganan, also known as CLS001 or MBI-226, is a synthetic peptide consisting of 12 amino acids. It is an analog of indolicidin, initially derived from neutrophils found in cattle. Omiganan exhibits effectiveness against a broad spectrum of microorganisms, encompassing both gram-positive and gram-negative bacteria as well as fungi. Clinical trials have explored omiganan as a localized intervention for catheter infections, atopic dermatitis, acne vulgaris, rosacea, and genital warts.

The peptide, PXL01, derived from human lactoferrin, a glycoprotein that binds to iron and is found in both milk and mucosal secretions, demonstrated effectiveness during a phase II clinical trial for reducing postoperative adhesions in individuals undergoing surgical intervention for the repair of hand flexor tendons (Wiig et al. 2014). In addition to the straight administration of AMPs, as outlined previously, various efforts are currently underway to augment the intrinsic synthesis of

AMPs. For instance, vitamin D3, also known as 1,25-dihydroxyvitamin D3, has been concerned with directly enhancing the expression of numerous AMPs, including defensin b2, hCAP18, and neutrophil gelatinase-associated lipocalin. The hCAP18 gene stands out as the most extensively studied target for vitamin D3, as demonstrated by its significant induction by 1,25-dihydroxyvitamin D3 [1,25$(OH)_2D_3$] in various human cell types in vitro studies (Jeffery et al. 2018). Furthermore, administering oral vitamin D supplements has been exposed to increased levels of hCAP18/LL-37 in both lesional and non-lesional skin biopsies collected from individuals with atopic dermatitis (Hata et al. 2008). Liu et al. (2006) documented that microorganisms such as *Mycobacterium tuberculosis* induce the expression of VDR (vitamin D receptor) and vitamin D-1-hydroxylase genes, leading to enhanced production of 1,25$(OH)_2D_3$. Subsequently, this upregulates the expression of hCAP18.

8 Conclusion

AMPs present compelling alternatives to conventional therapies as they act as anti-infectives and immunomodulatory agents. Their mechanisms of action are less prone to inducing resistance compared to traditional antibiotics. A survey of literature indicates that manipulating diet or incorporating nutritional supplements has the potential to enhance and regulate immune function. Increasing indications propose that nutrients and the metabolic by-products generated from the microbial breakdown of dietary components could be pivotal in influencing the innate immune response by controlling the activation of AMP genes. As numerous AMPs are presently advancing through late-stage clinical development across various therapeutic domains, the upcoming years hold the potential to substantiate the therapeutic advantages of these innovative candidates and result in the market authorization of several new drugs based on AMPs. Current rigorous research, both in nonclinical and clinical settings, is underway to assess AMP as potential antibiotics for the next generation.

References

Alekseyenko AV, Perez-Perez GI, De Souza A, Strober B, Gao Z, Bihan M, Li K, Methé BA, Blaser MJ (2013) Community differentiation of the cutaneous microbiota in psoriasis. Microbiome 1(1):1–17

Balato A, Cacciapuoti S, Di Caprio R, Marasca C, Masarà A, Raimondo A, Fabbrocini G (2018) Human microbiome: composition and role in inflammatory skin diseases. Arch Immunol Ther Exp 67:1–18

Benson R, Unnikrishnan MK, Kurian SJ, Velladath SU, Rodrigues GS, Chandrashekar Hariharapura R, Muraleedharan A, Bangalore Venkateshiah D, Banerjee B, Mukhopadhyay C, Johnson AS (2023) Vitamin D attenuates biofilm-associated infections via immunomodulation and cathelicidin expression: a narrative review. Expert Rev Anti-Infect Ther 21(1):15–27

Byrd AL, Belkaid Y, Segre JA (2018) The human skin microbiome. Nat Rev Microbiol 16(3):143–155

Campbell Y, Fantacone ML, Gombart AF (2012) Regulation of antimicrobial peptide gene expression by nutrients and by-products of microbial metabolism. Eur J Nutr 51:899–907

Cardoso MH, Meneguetti BT, Costa BO, Buccini DF, Oshiro KG, Preza SL, Carvalho CM, Migliolo L, Franco OL (2019) Non-lytic antibacterial peptides that translocate through bacterial membranes to act on intracellular targets. Int J Mol Sci 20(19):4877

Carmona-Cruz S, Orozco-Covarrubias L, Sáez-de-Ocariz M (2022) The human skin microbiome in selected cutaneous diseases. Front Cell Infect Microbiol 12:834135

Chang HW, Yan D, Singh R, Liu J, Lu X, Ucmak D, Lee K, Afifi L, Fadrosh D, Leech J, Vasquez KS (2018) Alteration of the cutaneous microbiome in psoriasis and potential role in Th17 polarization. Microbiome 6(1):1–27

Chen H, Xu Z, Peng L, Fang X, Yin X, Xu N, Cen P (2006) Recent advances in the research and development of human defensins. Peptides 27(4):931–940

Chiorean S, Antwi I, Carney DW, Kotsogianni I, Giltrap AM, Alexander FM, Cochrane SA, Payne RJ, Martin NI, Henninot A, Vederas JC (2020) Dissecting the binding interactions of teixobactin with the bacterial cell-wall precursor lipid II. Chembiochem 21(6):789–792

Christensen GJ, Scholz CF, Enghild J, Rohde H, Kilian M, Thürmer A, Brzuszkiewicz E, Lomholt HB, Brüggemann H (2016) Antagonism between *Staphylococcus epidermidis* and *Propionibacterium acnes* and its genomic basis. BMC Genomics 17(1):1–14

Cogen AL, Nizet V, Gallo RL (2008) Skin microbiota: a source of disease or defence? Br J Dermatol 158(3):442–455

Cytryńska M, Rahnamaeian M, Zdybicka-Barabas A, Dobslaff K, Zuechner T, Sacheau G, Innis CA, Vilcinskas A (2020) Proline-rich antimicrobial peptides in medicinal maggots of Lucilia sericata interact with bacterial DnaK but do not inhibit protein synthesis. Front Pharmacol 11:532

Dai X, Sayama K, Tohyama M, Shirakata Y, Hanakawa Y, Tokumaru S, Yang L, Hirakawa S, Hashimoto K (2010) PPARγ mediates innate immunity by regulating the 1α, 25-dihydroxyvitamin D3 induced hBD-3 and cathelicidin in human keratinocytes. J Dermatol Sci 60(3):179–186

De Pessemier B, Grine L, Debaere M, Maes A, Paetzold B, Callewaert C (2021) Gut–skin axis: current knowledge of the interrelationship between microbial dysbiosis and skin conditions. Microorganisms 9(2):353

Decréau RA, Marson CM, Smith KE, Behan JM (2003) Production of malodorous steroids from androsta-5, 16-dienes and androsta-4, 16-dienes by Corynebacteria and other human axillary bacteria. J Steroid Biochem Mol Biol 87(4–5):327–336

Dreno B, Martin R, Moyal D, Henley JB, Khammari A, Seité S (2017) Skin microbiome and acne vulgaris: Staphylococcus, a new actor in acne. Exp Dermatol 26(9):798–803

Ebenhan T, Gheysens O, Kruger HG, Zeevaart JR, Sathekge MM (2014) Antimicrobial peptides: their role as infection-selective tracers for molecular imaging. BioMed Res Int 2014:867381

Emelianov VU, Bechara FG, Gläser R, Langan EA, Taungjaruwinai WM, Schröder JM, Meyer KC, Paus R (2012) Immunohistological pointers to a possible role for excessive cathelicidin (LL-37) expression by apocrine sweat glands in the pathogenesis of hidradenitis suppurativa/acne inversa. Br J Dermatol 166(5):1023–1034

Emter R, Natsch A (2008) The sequential action of a dipeptidase and a β-lyase is required for the release of the human body odorant 3-methyl-3-sulfanylhexan-1-ol from a secreted Cys-Gly-(S) conjugate by Corynebacteria. J Biol Chem 283(30):20645–20652

Epand RM (2019) Anionic lipid clustering model. In: Antimicrobial peptides: basics for clinical application, pp 65–71

Évora AS, Adams MJ, Johnson SA, Zhang Z (2021) Corneocytes: relationship between structural and biomechanical properties. Skin Pharmacol Physiol 34(3):146–161

Findley K, Oh J, Yang J, Conlan S, Deming C, Meyer JA, Schoenfeld D, Nomicos E, Park M, Kong HH (2013) Topographic diversity of fungal and bacterial communities in human skin. Nature 498(7454):367–370

Fitz-Gibbon S, Tomida S, Chiu BH, Nguyen L, Du C, Liu M, Elashoff D, Erfe MC, Loncaric A, Kim J, Modlin RL (2013) *Propionibacterium acnes* strain populations in the human skin microbiome associated with acne. J Invest Dermatol 133(9):2152–2160

Friedrich CL, Rozek A, Patrzykat A, Hancock RE (2001) Structure and mechanism of action of an indolicidin peptide derivative with improved activity against gram-positive bacteria. J Biol Chem 276(26):24015–24022

Gao Z, Tseng CH, Strober BE, Pei Z, Blaser MJ (2008) Substantial alterations of the cutaneous bacterial biota in psoriatic lesions. PLoS One 3(7):e2719

Gerstel U, Latendorf T, Bartels J, Becker A, Tholey A, Schröder JM (2018) Hornerin contains a linked series of ribosome-targeting peptide antibiotics. Sci Rep 8(1):16158

Gläser R, Harder J, Lange H, Bartels J, Christophers E, Schröder JM (2005) Antimicrobial psoriasin (S100A7) protects human skin from Escherichia coli infection. Nat Immunol 6(1):57–64

Gombart AF, O'Kelly J, Saito T, Koeffler HP (2007) Regulation of the CAMP gene by 1, 25 (OH) 2D3 in various tissues. J Steroid Biochem Mol Biol 103(3–5):552–557

Gombart AF, Bhan I, Borregaard N, Tamez H, Camargo CA Jr, Koeffler HP, Thadhani R (2009) Low plasma level of cathelicidin antimicrobial peptide (hCAP18) predicts increased infectious disease mortality in patients undergoing hemodialysis. Clin Infect Dis 48(4):418–424

Gonzalez ME, Schaffer JV, Orlow SJ, Gao Z, Li H, Alekseyenko AV, Blaser MJ (2016) Cutaneous microbiome effects of fluticasone propionate cream and adjunctive bleach baths in childhood atopic dermatitis. J Am Acad Dermatol 75(3):481–493

Graf M, Wilson DN (2019) Intracellular antimicrobial peptides targeting the protein synthesis machinery. In: Antimicrobial peptides: basics for clinical application, pp 73–89

Grage SL, Afonin S, Kara S, Buth G, Ulrich AS (2016) Membrane thinning and thickening induced by membrane-active amphipathic peptides. Front Cell Dev Biol 4:65

Håkansson J, Ringstad L, Umerska A, Johansson J, Andersson T, Boge L, Rozenbaum RT, Sharma PK, Tollbäck P, Björn C, Saulnier P (2019) Characterization of the in vitro, ex vivo, and in vivo efficacy of the antimicrobial peptide DPK-060 used for topical treatment. Front Cell Infect Microbiol 9:174

Hancock RE, Sahl HG (2006) Antimicrobial and host-defense peptides as new anti-infective therapeutic strategies. Nat Biotechnol 24(12):1551–1557

Haney EF, Nathoo S, Vogel HJ, Prenner EJ (2010) Induction of non-lamellar lipid phases by antimicrobial peptides: a potential link to mode of action. Chem Phys Lipids 163(1):82–93

Harder J, Schröder JM (2002) RNase 7, a novel innate immune defense antimicrobial protein of healthy human skin. J Biol Chem 277(48):46779–46784

Harder J, Bartels J, Christophers E, Schröder JM (1997) A peptide antibiotic from human skin. Nature 387(6636):861–861

Harder J, Bartels J, Christophers E, Schröder JM (2001) Isolation and characterization of human μ-Defensin-3, a novel human inducible peptide antibiotic. J Biol Chem 276(8):5707–5713

Harder J, Tsuruta D, Murakami M, Kurokawa I (2013) What is the role of antimicrobial peptides (AMP) in acne vulgaris? Exp Dermatol 22(6):386–391

Hata TR, Kotol P, Jackson M, Nguyen M, Paik A, Udall D, Kanada K, Yamasaki K, Alexandrescu D, Gallo RL (2008) Administration of oral vitamin D induces cathelicidin production in atopic individuals. J Allergy Clin Immunol 122(4):829–831

Heilborn JD, Nilsson MF, Sørensen O, Ståhle-Bäckdahl M, Kratz G, Weber G, Borregaard N (2003) The cathelicidin anti-microbial peptide LL-37 is involved in re-epithelialization of human skin wounds and is lacking in chronic ulcer epithelium. J Invest Dermatol 120(3):379–389

Hein KZ, Takahashi H, Tsumori T, Yasui Y, Nanjoh Y, Toga T, Wu Z, Grötzinger J, Jung S, Wehkamp J, Schroeder BO (2015) Disulphide-reduced psoriasin is a human apoptosis-inducing broad-spectrum fungicide. Proc Natl Acad Sci 112(42):13039–13044

Henriques ST, Melo MN, Castanho MA (2006) Cell-penetrating peptides and antimicrobial peptides: how different are they? Biochem J 399(1):1–7

Herman A, Herman AP (2019) Antimicrobial peptides activity in the skin. Skin Res Technol 25(2):111–117

Honda K, Littman DR (2016) The microbiota in adaptive immune homeostasis and disease. Nature 535(7610):75–84

Irazazabal LN, Porto WF, Fensterseifer IC, Alves ES, Matos CO, Menezes AC, Felício MR, Gonçalves S, Santos NC, Ribeiro SM, Humblot V (2018) Fast and potent bactericidal membrane lytic activity of PaDBS1R1, a novel cationic antimicrobial peptide. Biochim Biophys Acta 1861(1):178–190

Islam D, Bandholtz L, Nilsson J, Wigzell H, Christensson B, Agerberth B, Gudmundsson GH (2001) Downregulation of bactericidal peptides in enteric infections: a novel immune escape mechanism with bacterial DNA as a potential regulator. Nat Med 7(2):180–185

Ismail NO, Odendaal C, Serem JC, Strömstedt AA, Bester MJ, Sayed Y, Neitz AW, Gaspar AR (2019) Antimicrobial function of short amidated peptide fragments from the tick-derived OsDef2 defensin. J Pept Sci 25(12):e3223

Jeffery, L. E., Henley, P., Marium, N., Filer, A., Sansom, D. M., Hewison, M., & Raza, K. (2018). Decreased sensitivity to 1,25-dihydroxyvitamin D3 in T cells from the rheumatoid joint. Journal of autoimmunity, 88: 50–60

Kong HH, Oh J, Deming C, Conlan S, Grice EA, Beatson MA, Nomicos E, Polley EC, Komarow HD, Murray PR, Turner ML (2012) Temporal shifts in the skin microbiome associated with disease flares and treatment in children with atopic dermatitis. Genome Res 22(5):850–859

Kristian SA, Datta V, Weidenmaier C, Kansal R, Fedtke I, Peschel A, Gallo RL, Nizet V (2005) D-alanylation of teichoic acids promotes group a streptococcus antimicrobial peptide resistance, neutrophil survival, and epithelial cell invasion. J Bacteriol 187(19):6719–6725

Lai PK, Kaznessis YN (2018) Insights into membrane translocation of protegrin antimicrobial peptides by multistep molecular dynamics simulations. ACS omega 3(6):6056–6065

Lehrer RI (2004) Primate defensins. Nat Rev Microbiol 2(9):727–738

Lehrer RI, Ganz T (2002) Cathelicidins: a family of endogenous antimicrobial peptides. Curr Opin Hematol 9(1):18–22

Liu PT, Stenger S, Li H, Wenzel L, Tan BH, Krutzik SR, Ochoa MT, Schauber J, Wu K, Meinken C, Kamen DL (2006) Toll-like receptor triggering of a vitamin D-mediated human antimicrobial response. Science 311(5768):1770–1773

Mah TFC, O'Toole GA (2001) Mechanisms of biofilm resistance to antimicrobial agents. Trends Microbiol 9(1):34–39

Mahlapuu M, Håkansson J, Ringstad L, Björn C (2016) Antimicrobial peptides: an emerging category of therapeutic agents. Front Cell Infect Microbiol 6:194

Mahlapuu M, Björn C, Ekblom J (2020) Antimicrobial peptides as therapeutic agents: opportunities and challenges. Crit Rev Biotechnol 40(7):978–992

Malmsten M (2016) Interactions of antimicrobial peptides with bacterial membranes and membrane components. Curr Top Med Chem 16(1):16–24

Mandal A, Viswanathan C (2015) Natural killer cells: in health and disease. Hematol Oncol Stem Cell Ther 8(2):47–55

Mandal SM, Manna S, Mondal S, Ghosh AK, Chakraborty R (2018) Transcriptional regulation of human defense peptides: a new direction in infection control. Biol Chem 399:1277–1284

Mardirossian M, Barrière Q, Timchenko T, Müller C, Pacor S, Mergaert P, Scocchi M, Wilson DN (2018) Fragments of the nonlytic proline-rich antimicrobial peptide Bac5 kill *Escherichia coli* cells by inhibiting protein synthesis. Antimicrob Agents Chemother 62(8):10–1128

Marshall JS, Warrington R, Watson W, Kim HL (2018) An introduction to immunology and immunopathology. Allergy, Asthma Clin Immunol 14(2):1–10

Nakano T, Yoshino T, Fujimura T, Arai S, Mukuno A, Sato N, Katsuoka K (2015) Reduced expression of dermcidin, a peptide active against *Propionibacterium acnes*, in sweat of patients with acne vulgaris. Acta Derm Venereol 95(7):783–786

N'Diaye A, Mijouin L, Hillion M, Diaz S, Konto-Ghiorghi Y, Percoco G, Chevalier S, Lefeuvre L, Harmer NJ, Lesouhaitier O, Feuilloley MG (2016) Effect of substance P in *Staphylococcus aureus* and *Staphylococcus epidermidis* virulence: implication for skin homeostasis. Front Microbiol 7:506

Niyonsaba F, Kiatsurayanon C, Chieosilapatham P, Ogawa H (2017) Friends or foes? Host defense (antimicrobial) peptides and proteins in human skin diseases. Exp Dermatol 26(11):989–998

Oh J, Freeman AF, Park M, Sokolic R, Candotti F, Holland SM, Segre JA, Kong HH, NISC Comparative Sequencing Program (2013) The altered landscape of the human skin microbiome in patients with primary immunodeficiencies. Genome Res 23(12):2103–2114

Otto M (2009) *Staphylococcus epidermidis*—the 'accidental' pathogen. Nat Rev Microbiol 7(8):555–567

Otvos L Jr (2002) The short proline-rich antibacterial peptide family. Cellular and Molecular Life Sciences CMLS, 59, pp.1138-1150. Proc Natl Acad Sci 104(47):18730–18735

Rademacher F, Gläser R, Harder J (2021) Antimicrobial peptides and proteins: interaction with the skin microbiota. Exp Dermatol 30(10):1496–1508

Raqib R, Sarker P, Bergman P, Ara G, Lindh M, Sack DA, Nasirul Islam KM, Gudmundsson GH, Andersson J, Agerberth B (2006) Improved outcome in shigellosis associated with butyrate induction of an endogenous peptide antibiotic. Proc Natl Acad Sci 103(24):9178–9183

Ring HC, Emtestam L (2016) The microbiology of hidradenitis suppurativa. Dermatol Clin 34(1):29–35

Ring HC, Thorsen J, Saunte DM, Lilje B, Bay L, Riis PT, Larsen N, Andersen LOB, Nielsen HV, Miller IM, Bjarnsholt T (2017) The follicular skin microbiome in patients with hidradenitis suppurativa and healthy controls. JAMA Dermatol 153(9):897–905

Rokitskaya TI, Kolodkin NI, Kotova EA, Antonenko YN (2011) Indolicidin action on membrane permeability: carrier mechanism versus pore formation. Biochim Biophys Acta 1808(1):91–97

Sallenave JM (2002) Antimicrobial activity of antiproteinases. Biochem Soc Trans 30(2):111–115

Sandilands A, Terron-Kwiatkowski A, Hull PR, O'Regan GM, Clayton TH, Watson RM, Carrick T, Evans AT, Liao H, Zhao Y, Campbell LE (2007) Comprehensive analysis of the gene encoding filaggrin uncovers prevalent and rare mutations in ichthyosis vulgaris and atopic eczema. Nat Genet 39(5):650–654

Sanford JA, Gallo RL (2013) Functions of the skin microbiota in health and disease. In: Seminars in immunology, vol 25, No. 5. Academic Press, pp 370–377

Sartorius K, Killasli H, Oprica C, Sullivan A, Lapins J (2012) Bacteriology of hidradenitis suppurativa exacerbations and deep tissue cultures obtained during carbon dioxide laser treatment. Br J Dermatol 166(4):879–883

Schauber J, Dorschner RA, Coda AB, Büchau AS, Liu PT, Kiken D, Helfrich YR, Kang S, Elalieh HZ, Steinmeyer A, Zügel U (2007) Injury enhances TLR2 function and antimicrobial peptide expression through a vitamin D–dependent mechanism. J Clin Invest 117(3):803–811

Schittek B, Hipfel R, Sauer B, Bauer J, Kalbacher H, Stevanovic S, Schirle M, Schroeder K, Blin N, Meier F, Rassner G (2001) Dermcidin: a novel human antibiotic peptide secreted by sweat glands. Nat Immunol 2(12):1133–1137

Schlapbach C, Yawalkar N, Hunger RE (2009) Human β-defensin-2 and psoriasin are overexpressed in lesions of acne inversa. J Am Acad Dermatol 61(1):58–65

Schmidt NW, Wong GC (2013) Antimicrobial peptides and induced membrane curvature: geometry, coordination chemistry, and molecular engineering. Curr Opinion Solid State Mater Sci 17(4):151–163

Schneider VA, Coorens M, Ordonez SR, Tjeerdsma-van Bokhoven JL, Posthuma G, van Dijk A, Haagsman HP, Veldhuizen EJ (2016) Imaging the antimicrobial mechanism (s) of cathelicidin-2. Sci Rep 6(1):32948

Sierra JM, Fusté E, Rabanal F, Vinuesa T, Viñas M (2017) An overview of antimicrobial peptides and the latest advances in their development. Expert Opin Biol Ther 17(6):663–676

Sohnle PG, Hunter MJ, Hahn B, Chazin WJ (2000) Zinc-reversible antimicrobial activity of recombinant calprotectin (migration inhibitory factor—related proteins 8 and 14). J Infect Dis 182(4):1272–1275

Wang H, Jin H (2021) Update on the aetiology and mechanisms of generalized pustular psoriasis. Eur J Dermatol 31:602–608

Wang TT, Dabbas B, Laperriere D, Bitton AJ, Soualhine H, Tavera-Mendoza LE, Dionne S, Servant MJ, Bitton A, Seidman EG, Mader S (2010) Direct and indirect induction by 1, 25-dihydroxyvitamin D3 of the NOD2/CARD15-defensin β2 innate immune pathway defective in Crohn disease. J Biol Chem 285(4):2227–2231

Wiig ME, Dahlin LB, Friden J, Hagberg L, Larsen SE, Wiklund K, Mahlapuu M (2014) PXL01 in sodium hyaluronate for improvement of hand recovery after flexor tendon repair surgery: randomized controlled trial. PLoS One 9(10):e110735

Xu J, Saunders CW, Hu P, Grant RA, Boekhout T, Kuramae EE, Kronstad JW, DeAngelis YM, Reeder NL, Johnstone KR, Leland M (2007) Dandruff-associated *Malassezia* genomes reveal convergent and divergent virulence traits shared with plant and human fungal pathogens. Proc Natl Acad Sci U S A 104(47):1873

Yamasaki K, Gallo RL (2008) Antimicrobial peptides in human skin disease. Eur J Dermatol 18(1):11–21

Yeung AT, Gellatly SL, Hancock RE (2011) Multifunctional cationic host defence peptides and their clinical applications. Cell Mol Life Sci 68:2161–2176

Zavascki AP, Goldani LZ, Li J, Nation RL (2007) Polymyxin B for the treatment of multidrug-resistant pathogens: a critical review. J Antimicrob Chemother 60(6):1206–1215

Zhang Z, Zhou Y, Zhang H, Du X, Cao Z, Wu Y, Liu C, Sun Y (2023) Antibacterial activity and mechanisms of trohepc2-22, a derived peptide of hepcidin2 from golden pompano (*Trachinotus ovatus*). Int J Mol Sci 24:9251

Chapter 5
Insect Antimicrobial Peptides in Host Defense and Their Therapeutic Application

Roy Dinata, Chettri Arati, Bose Manikandan, Giri Abinash, Buragohain Pori, Laskar Saeed-Ahmed, Rema Momin Bidanchi, Vikas Kumar Roy, and G. Gurusubramanian

1 Introduction

The improper and excessive use of antibiotics has led to the emergence of antibiotic resistance, a significant concern for global public health as recognized by the World Health Organisation. The pursuit of discovering alternative medications has proven to be a genuine difficulty (Martens and Demain 2017). This phenomenon, along with the scarcity of novel molecules, poses significant challenges in the field of healthcare (Payne et al. 2007; Tommasi et al. 2015). The ESKAPE pathogens, including *Enterococcus faecium*, *Staphylococcus aureus*, *Klebsiella pneumoniae*, *Acinetobacter baumannii*, *Pseudomonas aeruginosa*, and *Enterobacter species*, are highly resistant to the majority of antibiotics currently in use (Ghosh & Saha 2020). These pathogens have significantly contributed to the rise of hospital-acquired infections. In addition, the World Health Organisation (WHO) has recently revised its priority list, which consists of 12 bacterial diseases that require the development of new medicines. The World Health Organisation (WHO) categorizes its list into three distinct groups based on the level of urgency in requiring new antibiotics: critical, high, and medium priority. The category of utmost importance includes the gram-negative species *P. aeruginosa* and *A. baumannii*, which are bacteria resistant to carbapenem. These bacteria are commonly linked to serious and potentially fatal diseases such as bloodstream infections and pneumonia. The high-priority list includes S. aureus methicillin-resistant (MRSA), as well as vancomycin-intermediate and vancomycin-resistant species (Beceiro et al. 2013; World Health Organization 2017). The escalating rate of antibiotic resistance poses a particularly formidable

B. Manikandan

R. Dinata · C. Arati · G. Abinash · B. Pori · L. Saeed-Ahmed · R. M. Bidanchi · V. K. Roy · G. Gurusubramanian (✉)
Department of Zoology, Mizoram University, Aizawl-796004, Mizoram, India

P. Baindara, S. M. Mandal (eds.), *Evolution of Antimicrobial Peptides*,
https://doi.org/10.1007/978-3-031-67515-7_5

problem in the management of tropical diseases. The clinical progression of ulcers, diabetic foot infections, postsurgical infections, and burn wounds can be further complicated by chronic skin and soft tissue infections. In cases where antibiotic-resistant pathogens are present, the clinical outcome of these complications tends to deteriorate progressively. Similarly, the gram-negative bacterium *A. baumannii* has been identified as the causative agent of various diseases that are resistant to antibiotics, including wounds, skin, urinary tract infections, pneumonia, and bacteremia (Neshani et al. 2020). Infections caused by bacteria in the lower respiratory tract, which are frequently linked to bronchiectasis, are a prevalent and escalating chronic respiratory condition. These infections are connected with lung illness in individuals not only with cystic fibrosis (CF) but also with chronic obstructive pulmonary disease. The clinical evolution of bronchiectasis that is resistant to antibiotics can lead to a deterioration of the health state. This is caused by the development of an uncontrollable cycle of infection and inflammation, which cannot be managed with currently available medications (Varga et al. 2015; De Soyza and Aksamit 2016; McShane et al. 2018). Lung infections linked to bronchiectasis can progress to respiratory failure and mortality. Furthermore, the diminished standard of living and escalated healthcare expenses can exacerbate patient adherence (Gollan et al. 2019). The effective control of bacterial infections relies on the synergistic interaction between the host immune response and the use of antibiotics. Therefore, weaknesses in the host's immune system and/or decreased effectiveness of antibiotics caused by the existence of resistant microorganisms may unfortunately lead to a transition to a persistent infection. Chronic bacterial infections are linked to higher rates of illness and death caused by the infection itself, as well as an increased likelihood of the disease spreading throughout the body, which is a dangerous condition that is challenging to treat, particularly when antibiotic-resistant pathogens are present. Hence, in addition to the reckless utilization of antibiotics, the corresponding problem of resistance to the most frequently employed compounds is a worldwide apprehension (Bush and Page 2017). The pursuit of discovering alternative medications has proven to be a significant problem, as indicated by references (Dian et al. 2020; Haisma et al. 2016).

AMPs are small proteins associated with immunity that possess the ability to combat bacteria, viruses, fungi, or parasites produced by a large number of organisms, including bacteria, plants, insects, and mammals. In insects, antimicrobial peptides are produced by cells and tissues involved in the host's innate immune response, such as hemocytes or the fat body. These peptides have the potential to serve as effective substitutes for traditional antibiotics, particularly in light of the increasing problem of antimicrobial resistance (Bulet et al. 2004; Hancock and Sahl 2006; Vale et al. 2014). Antimicrobial peptides (AMPs) have a role in several defense-related mechanisms, such as pathogen elimination, endotoxin binding and neutralization, and regulation of immunological responses during infection (Brogden 2005; Makarova et al. 2016). In insects, the activation of antimicrobial peptides (AMPs) leads to a temporary immune response in the body fluids, known as the hemolymph. This response is characterized by increased concentrations of AMPs and lasts longer than the initial cellular responses. It is believed to serve as a secondary defense mechanism against persistent infections. AMPs, previously recognized for their defensive role against viruses, have now been found to also play a role in regulating endosymbionts (Masson et al. 2016). AMPs are widely present in eukaryotes although they have been extensively researched in insects (Bulet and Stocklin 2005; Yi et al. 2014). The non-ribosomally synthesized antimicrobial peptides (AMPs) produced by bacteria and fungi are distinct from the ribosomally synthesized AMPs that contribute to the innate immunity of multicellular animals. The term AMP often refers specifically to these ribosomally produced molecules (Wiesner and Vilcinskas 2010). The Antimicrobial Peptide Database (APD) is an excellent resource for a comprehensive compilation of known AMPs. It can be accessed at http://aps.unmc.edu/AP. This chapter explored the wide range of antimicrobial peptides obtained from insects and their potential medical uses, with a particular focus on their activity as antimicrobials.

2 Mechanism of Action of Insect AMPs Against Microbes

Bacteria exhibit a lower propensity to develop resistance to AMPs compared to conventional antibiotics. These peptides can physically rupture microbial cell membranes, resulting in the elimination of a wide range of disease-causing bacteria. Therefore, the microbiological membrane is commonly regarded as the principal target of antimicrobial peptides (AMPs) (Chernysh et al. 2015; Shen et al. 2018). Furthermore, the remarkable ability of these peptides to break cell membranes makes them highly suitable candidates for combination treatments with traditional antibiotics (Hollmann et al. 2018). AMPs can enhance the permeability of antibiotic molecules into the cytoplasm of microorganisms, allowing them to effectively interact with their intended targets. Antimicrobial peptides (AMPs) eliminate bacteria in many ways, such as disrupting the bacterial membrane, interfering with bacterial metabolism, and targeting cytoplasmic components (Ongey et al. 2018; Patocka et al. 2019; Shen et al. 2018; Jozefiak and Engberg 2017; Bechinger and Gorr 2017). The main connection between an antimicrobial peptide (AMP) and the target bacterium is established through either an electrostatic or hydrophobic interaction, which is highly influenced by the lipid composition of the bacterial membrane.

AMPs can interact with the cell membrane surface to modify the membrane's permeability to kill the microbes (Shen et al. 2018). Upon interaction with the cell membrane, AMPs influence the osmotic pressure balance by the establishment of a transmembrane potential (Fig. 5.1) (Patocka et al. 2019). Essentially, the antibacterial activity of the AMPs is directly linked to the way they interact with the membrane (Fig. 5.1). Currently, there are at least four generally employed modes of action to explain the membrane activity of antimicrobial peptides (AMPs): barrel-stave, carpet, toroidal pore, and disordered toroidal pore (Shen et al. 2018; Jozefiak and Engberg 2017). To achieve the antibacterial effect, a minimum concentration threshold is necessary for all antimicrobial effects.

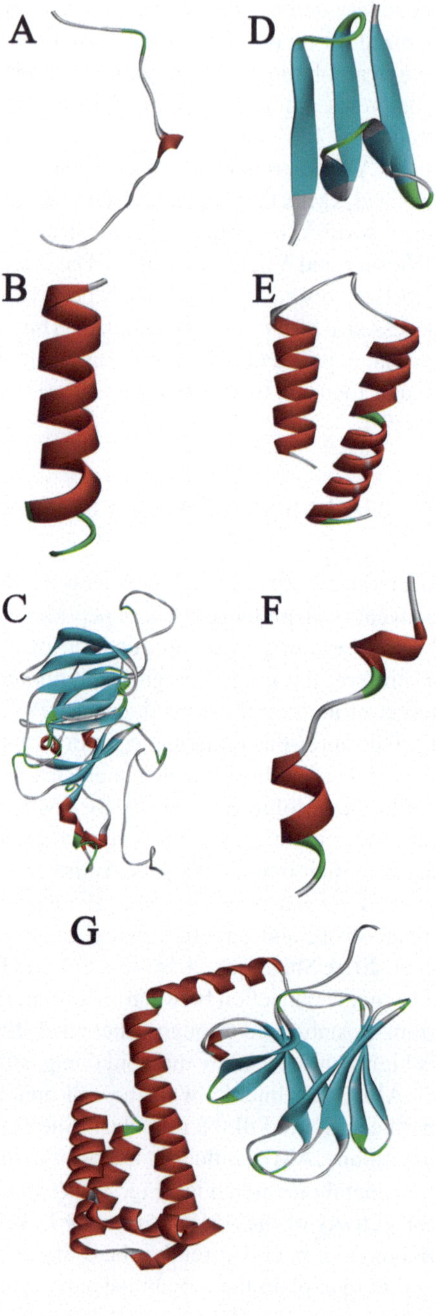

Fig. 5.1 Three-dimensional structure of insect antimicrobial peptides. (**a**) abaecin, (**b**) apamin, (**c**) apisimin, (**d**) hymenoptaecin, (**e**) melittin, (**f**) cecropin, and (**g**) metchnikowin

3 Insect AMPs

AMPs are essential elements of the insect's innate and cellular immune system. They are produced in the fat body or hemocytes in response to immune challenges posed by viruses and parasites. Subsequently, they are released into the hemolymph. AMPs, or antimicrobial peptides, perform a crucial role as peptides that defend the host in insects. This is one of the ancient evolutionary mechanisms in insects that may be rapidly generated and transferred to the site of infection (Dinata et al. 2023). Antimicrobial peptides originating from insects are discussed below (Fig. 5.1 and Table 5.1).

3.1 Cecropins

The cecropins were initially extracted from the hemolymph of the *Hyalophora cecropia*, a kind of enormous silk moth. This is how the term "cecropin" was coined. The structure of these peptides is primarily determined by a diverse collection of antibacterial and poisonous peptides obtained from different species of lepidopteran and dipteran insects. These peptides play a significant role in the innate immunity of insects. Cecropins are compact proteins consisting of approximately 35 amino acid residues. They possess the ability to combat both gram-positive and gram-negative bacteria (Ouyang et al. 2015). The primary insect cecropins (A, B, and D) are composed of 35–37 amino acid residues, excluding cysteine (Hultmark et al. 1982; Van et al. 1985). Cecropins possess the ability to disrupt the integrity of bacterial cell membranes, leading to their lysis. Additionally, they can impede the process of proline uptake and induce membrane permeability. This has been supported by studies (Moore et al. 1996; Bechinger and Lohner 2006). In addition to being called bactericidin, lepidopteran, and sarcotoxin, insect cecropins are known by various additional names (Bechinger and Lohner 2006). Cecropin D is derived from *H. cecropia* and exhibits similarity to cecropin A and cecropin B (Hultmark et al. 1982). Cecropin D is detected in the hemolymph after cecropin A or cecropin B following a bacterial infection (Gudmundsson et al. 1991). The expression of a genetically-modified cecropin D in *Pichia pastoris* was achieved successfully, demonstrating antibacterial properties. Activity exhibited against both gram-positive and gram-negative bacteria (Guo et al. 2012). Phosphorylation of the C-terminal lysine residue of cecropin D enhances its antibacterial action (Park et al. 2013). Cecropin D has been found to effectively prevent the infection and replication of porcine reproductive and respiratory syndrome virus (PRRSV) in laboratory conditions (Liu et al. 2015). Cecropin P1 is an antimicrobial peptide derived from *Ascaris suum*, a parasitic worm found in the intestines of pigs. Cecropin P1 demonstrated potent growth inhibition against enterotoxigenic *Escherichia coli*, with a minimum inhibitory concentration (MIC) of 1 mg/mL (Hou et al. 2011). An analysis of the tertiary structure reveals that cecropin P1 can form α-helical structures with the C-terminal

Table 5.1 Insect AMPs, amino acid sequences, and their therapeutic application

Insect AMPs	Amino acid sequences	Therapeutic application	References
Cecropin A	GGLKKLGKKLEGVGKRVFKASEKALPVAVGIKALG-NH2	Against both gram-positive and gram-negative bacteria and fungi, anti-inflammatory activity, antitumor activity	Brady et al. (2019)
Cecropin B	KWKVFKKIEKMGRNIRNGIVKAGPAIAVLGEAKAL-NH2		
Cecropin B1	KWKVFKKIEKMGRNIRNGIVKAGPKWKVFKKIEK-NH2		
Cecropin B3	AIAVLGEAKALMGRNIRNGIVKAGPAIAVLGEAKAL-NH2		
Cecropin C	GWLKKLGKRIERIGQHTRDATIQGLGIAQQAANVAATAR-NH2		
Cecropin D	WNPFKELEKVGQRVRDAVISAGPAVATVAQATALAK-NH2		
Cecropin P1	SWLSKTAKKLENSAKKRISEGIAIAIQGGPR-NH2		
Attacins A-C	AGALTINSDGTSGAVVKVPI-NH2	Against gram-negative bacteria	Yi et al. (2014)
Attacins E-F	DAHGALTLNSDGTSGAVVKVPFAGNDLNI-NH2		
Lebocin 1	DLRFLYPRGKLPVPTPPPFNPKPIYIDMGNRY-NH2	Against gram-positive bacteria	
Drosocin	GKPRPYSPRPTSHPRPIRV-NH2	Against gram-positive bacteria	Manniello et al. (2021)
Diptericin A	DDMTMKPTPPPQYPLNLQGGGGGQSGDGFGFAVQGHQKVW TSDNGRHEIGLNGGYGQHLGGPYGNSEPSWKVGSTYTYRFP NF-NH2	Active against gram-negative bacteria	Mylonakis et al. (2016)
Metchnikowin	HRHEGPIFNTRPSPFNPNEPRPGPIY-NH2	Against both gram-positive and gram-negative bacteria and fungi	Mylonakis et al. (2016)

Ponericin G1	GWKDWAKKAGGWLKKKGPGMAKAALKAAMQ-NH2	Against both gram-positive and gram-negative bacteria and fungi	Mylonakis et al. (2016)
Ponericin G2	GWKDWLKKGKEWLKAKGPGIVKAALQAATQ-NH2		
Ponericin G3	GWKDWLNKGKEWLKKKGPGIMKAALKAATQ-NH2		
Ponericin G4	DFKDWMKTAGEWLKKKGPGILKAAMAAAT-NH2		
Ponericin G5	GLKDWVKIAGGWLKKGPGILKAAMAAATQ-NH2		
Ponericin G6	GLVDVLGKVGGLIKKLLP-NH2		
Ponericin G7	GLVDVLGKVG GLIKKLLPG-NH2		
Ponericin W1	WLGSALKIGAKLLPSVVGLFKKKKQ-NH2		
Ponericin W2	WLGSALKIGAKLLPSVVGLFQKKKK-NH2		
Ponericin W3	GIWGTLAKIGIKAVPRVISMLKKKKQ-NH2		
Ponericin W4	GIWGTALKWGVKLLPKLVGMAQTKKQ-NH2		
Ponericin W5	FWGALIKGAAKLIPSVVGLFKKKQ-NH2		
Ponericin W6	FIGTALGIASAIPAIVKLFK-NH2		
Ponericin L1	LLKELWTKMKGAGKAVLGKI-NH2		
Ponericin L2	LLKELWTKIKGAGKAVLGKIKGLL-NH2		
Pyrrhocoricin	VDKGSYLPRPTPPRPIYNRN-NH2	Active against gram-negative bacteria	Mylonakis et al. (2016)
Persulcatusin	GFGCPFNQGACHRHCRSIGRRGGYCAGLFKQTCTCYSR-NH2	Against gram-positive bacteria	Mylonakis et al. (2016)
Abaecin	MKVVIFIFALLATICAAFAYVPLPNVPQPGRRPFPTF PGQGPFNPKIKWPQGY	Antibacterial activity	Casteels et al. (1990)
Apidaecin	MKNFALAILVVTFVVAVFGNTNLDPPTRPTRLRREAE PEAEPGNNRPVYIPQPRPPHPRLRREAEPEAEPGNNRPVYIPQP RPPHPRLRREAEPEAEPGNNRPVYIPQPRPPHPRLRREAEPEAE PGNNRPVYIPQPRPPHPRI	Antibacterial activity	Casteels et al. (1989)

(continued)

Table 5.1 (continued)

Insect AMPs	Amino acid sequences	Therapeutic application	References
Defensin	MKIYFIVGLLFMAMVAIMAAPVEDEFEPLEHFENEER ADRHRRVTCDLLSFKGQVNDSACAANCLSLGKAGGHCEKGV CICRKTSFKDLWDKRFG	Antibacterial activity	Casteels-Josson et al. (1994)
Hymenoptaecin	MKFIVLVLFCAVAYVSAQAELEPEDTMDYIPTRFRRQERGSIVI QGTKEGKSRPSLDIDYKQRVYDKNGMTGDAYGGLNIRPGQPS RQHAGFEFGKEYKNGFIKGQSEVQRGPGGRLSPYFGINGGFRF	Antibacterial activity	Casteels et al. (1993)
Jellin	EPFKISIHL	Antimicrobial activity	Fontana et al. (2004)
Apisimin	MSKIVAVVVLAAFCVAMLVSDVSAKTSISVKGESNVD VVSQINSLVSSIVSGANVSAVLLAQTLVNILQILIDANVFA	Antibacterial activity	Bíliková et al. (2002)
Royalisin/ Defensin	MKIYFIVGLLFMAMVAIMAAPVEDEFEPLEHFENEER ADRHRRVTCDLLSFKGQVNDSACAANCLSLGKAGGHCEKGV CICRKTSFKDLWDKRFG	Growth and development and antiaging	Fujiwara et al. (1990), Bíliková et al. (2002)
Tertiapin	ALCNCNRIIIPHMCWKKCGKK	Antibacterial activity	Ko et al. (2020)
Melectin	GFLSILKKVLPKVMAHMK	Antibacterial activity	Cerovský et al. (2008)
Adolapin		Anti-arthritic, analgesic, anti-inflammatory, antipyretic, antinociceptive	Son et al. (2007)
Apamin	MISMLRCIYLFLSVILITSYFVTPVMPCNCKAPETAL CARRCQQHG	Blood-brain barrier drug-delivery shuttles, antifungal, antiviral, anti-inflammatory, analgesic, and neurodegenerative disorders	

Melittin	MKFLVNVALVFMVVYISYIYAAPEPEPAPEPEAEADA EADPEAGIGAVLKVLTTGLPALISWIKRKRQQG	Anticancer, antitumor, antiangiogenesis, antifungal, antiparasitic, antimicrobial, anti-arthritic, anti-inflammatory, antipsychotic, anti-atherosclerosis, antibacterial, antiviral, neuroprotective, and anti-multidrug resistant	Gevod and Birdi (1984), Ma et al. (2017)
Secapin	MKNYSKNATHLITVLLFSFVVILLIIPSKCEAVSNDMQPLEARS ADLVPEPRYIIDVPPRCPPGSKFIKNRCRVIVP	Antibacterial activity	Gauldie et al. (1978)

region (Lys15–Gly29) in the lipopolysaccharide (LPS) of the outer membrane of gram-negative bacteria (Baek et al. 2016). The bactericidal effect of cecropin P1 on *E. coli* is influenced by the degree of initial permeabilization caused by the peptide (Arcidiacono et al. 2009). The antibacterial action of cysteine-terminus-modified cecropin P1 (CP1C) is reduced due to the inclusion of a polyethylene glycol (PEG) linker, which hinders CP1C from interacting with the bilayer (Wu et al. 2018). Cecropin P1 exhibits a notable inhibitory impact on the human fungal infection *Candida albicans* (Zakharchenko et al. 2017). Cecropin P1 hinders the attachment of PRRSV, hence inhibiting its activity (Guo et al. 2014). Cecropin P1 hinders the release of viral particles and weakens the occurrence of virus-induced cell death (Guo et al. 2014). At present, researchers have created biosensors that utilize cecropin P1. By immobilizing the peptide through various termini, multiple functions and activities can be achieved (Wang et al. 2014). Lucilin, a cecropin consisting of 36 amino acids, has been found as a partial genetic sequence in maggots of the species *Lucilia sericata* (Téllez and Castaño-Osorio 2014). The fusion protein, GWLK-Lucilin-CPD-His8, exhibits promising efficacy against *E. coli* bacteria that are resistant to multiple drugs (MDR) (Téllez and Castaño-Osorio 2014). *Musca domestica* cecropin has the potential to kill bacteria, including clinical isolates of *E. coli* (Lu et al. 2012).

3.2 Attacin

Attacins are proteins that are rich in glycine and are classified as part of the AMP group. The presence of attacks was initially observed in *Hyalophora cecropia* (Hultmark et al. 1983; Kockum et al. 1984). They exhibit efficacy against gram-negative bacteria (Carlsson et al. 1991). Attacins A–F are antibacterial proteins that are intimately associated with and derived from the hemolymph of immunized pupae. Attacins function by inhibiting the production of the primary outer membrane proteins in dividing gram-negative bacteria. This disruption compromises the structural integrity of the cell wall, resulting in the bacteria growing in elongated chains (Carlsson et al. 1998). Attacins are a type of inducible immune protein P5 that exhibits antibacterial activity. Attacins possess potent bactericidal activity against *E. coli* and other gram-negative bacteria. Attacins, along with cecropin and lysozyme, constitute the third antibacterial protein found in the humoral immune system of *H. cecropia* (Carlsson et al. 1998). Attacin and attacin-related proteins are obtained from many insect species, including *Bombyx mori*, *Glossina morsitans* (tse-tse fly), *Heliothis virescens*, *Trichoplusia ni*, *Samia cynthia ricini* (wild silkmoth), and *Musca domestica* (housefly) (Dushay et al. 2000; Geng et al. 2004). Attacins A–D are classified as a basic group based on their amino acid composition, while attacins E and F include acidic residues.

3.3 Lebocin

Lebocins are a type of antibacterial peptide composed of 32 amino acids. They were discovered by Hara and Yamakawa (1995a, b) in the hemolymph of the silkworm *Bombyx mori*, which had been immunized with *E. coli* (Hara and Yamakawa 1995a, b). Lebocin is a peptide that is rich in proline and contains O-glycosylation, as indicated by Liu et al. (2000). Overall, 41% of the amino acid sequence of lebocin is the same as abaecin, which is a significant antibacterial peptide consisting of 34 amino acids (YVPLPNVPQPGRRPFPTFPGQGPFNPKIKWPQGY-NH2) found in the honeybee *Apis mellifera* (Casteels et al. 1990). The amino acid sequence of lebocin 1 is DLRFLYPRGKLPVPTPPPFNPKPIYIDMGNRY-NH2. The main structure of lebocins 1 and 2 only varies in their sugar component. Lebocin 3 possesses an identical structure to lebocin 2, with the exception that residue 16 is leucine instead of proline.

3.4 Dorsocin

Drosocin is a peptide synthesized by the fruit fly *Drosophila melanogaster* (Bulet et al. 1993). The peptide consists of 19 amino acids, namely GKPRPYSPRPTSHPRPIRV-NH2. O-glycosylation of the peptide is necessary to achieve optimal biological activity (McManus et al. 1999). Glycosylation is a significant posttranslational alteration for a certain family of proline-rich AMPs (Lele et al. 2015). Drosocin exhibits primary resistance against gram-negative bacteria, in addition to the gram-positive bacterium *M. luteus*. The action of drosocin (Bulet et al. 1996) is abolished when the first five N-terminal residues are deleted. Drosocin that has undergone glycosylation exhibits activity against both *E. coli* and fungus (Imler and Bulet 2005). Apidaecin IB and drosocin have substantial similarities in their sequence and interaction mechanism; however, they do not possess any ability to generate pores (Gobbo et al. 2002). Apidaecins play a crucial role in the honeybee's immune response to protect against microbial invasion (Gobbo et al. 2002). The N-terminal alteration of apidaecins enhances both the binding to unknown intracellular target(s) and the efficacy of cell penetration (Matsumoto et al. 2010). The antibacterial action against *Pseudomonas aeruginosa* is enhanced by the presence of repetitions of N-terminal Ile-Orn- and Trp-Orn-motif structures (Bluhm et al. 2016).

3.5 Diptericins

Diptericins are a group of antibacterial peptides that are rich in glycine and have a molecular weight of approximately 8 kilodaltons. They are derived from hemolymph proteins found in dipteran insects and consist of roughly 82 amino acids (Cudic et al. 1999). The dipteran *Phormia terranovae* larvae that were immunized have yielded the isolation of diptericins A–C (Dimarcq et al. 1988). Diptericin is a substance additionally observed in *Drosophila melanogaster* (Reichhart et al. 1992), *Sarcophaga peregrina* (a type of flesh fly) (Ishikawa et al. 1992), and *Mayetiola destructor* (Hessian fly). The most common peptide in this peptide family is diptericin A (DDMTMKPTPPPQYPLNLQGGGGGQSGDGFGFAVQGHQK VWTSDNGRHEIGLNGGYGQHLGGPYGNSEPSWKVGSTYTYRFPNF-NH2). This peptide exhibits activity exclusively against a restricted range of gram-negative bacteria, namely, *E. coli* K12, *Erwinia hericola* T, and *Erwinia carotovora* 113. Diptericin A primarily targets the cytoplasmic membrane of actively dividing bacteria (Keppi et al. 1989). Prolixicin is a newly discovered antimicrobial peptide belonging to the diptericins family. This peptide has been extracted from the hemipteran bug *Rhodnius prolixus*. Prolixicin is a peptide consisting of 21 amino acids (Ursic-Bedoya et al. 2011). Additionally, it possesses two potential phosphorylation sites although no glycosylation sites have been detected. Sequence research indicates that a specific part of prolixicin is associated with the diptericin/attacin family of antimicrobial peptides (AMPs). The production of this peptide occurs in the midgut tissues following bacterial infection of the hemolymph (Ursic-Bedoya et al. 2011).

3.6 Metchnikowin

In 1995, Russian scientists discovered a unique 26-residue proline-rich immune-inducible linear peptide (HRHEGPIFNTRPSPFNPNEPRPGPIY-NH2) from *D. melanogaster*. This peptide stood out due to its exceptional ability to combat gram-positive bacteria and fungi. Nevertheless, this peptide does not exhibit any efficacy against gram-negative bacteria (Levashina et al. 1995). The peptide was designated as metchnikowin as a tribute to *Élie Metchnikoff*, the pioneer of this research area. The Oregon Drosophila strain contains two isoforms of metchnikowin, which exhibit a single amino acid difference (His instead of Arg) (Levashina et al. 1998). Research has been conducted on the target of metchnikowin. Metchnikowin interacts with the fungal enzyme β(1,3)-glucanosyltransferase Gel1 (FgBGT), a crucial enzyme in the formation of the fungal cell wall (Moghaddam et al. 2017). Metchnikowin specifically interacts with the iron-sulfur subunit (SdhB) of succinate-coenzyme Q reductase (SQR). In Moghaddam et al.'s study (Moghaddam et al. 2017), it was found that metchnikowin reduced the activity of *Fusarium graminearum* mitochondrial SQR by up to 52% by inhibiting its SDH activity.

3.7 *Ponericins*

Ponericins are peptides obtained from the venom of the predatory ant *Pachycondyla goeldii*. Orivel and associates (2001) conducted isolation of members belonging to the subfamily Ponerinae and characterized their amino acid sequences. The major structure similarities of ponericins allow them to be categorized into three families: ponericins G, W, and L. Ponericins have significant sequence homology with various known peptides: ponericin G shares a resemblance with cecropin-like peptides, ponericin W shows similarity with gaegurins and melittin, and ponericin L has similarity with dermaseptins. Ponericins exhibit both hemolytic and insecticidal actions, specifically targeting cricket larvae. Ponericins have α-helical conformations within cellular membranes (Johnson and Copello 2010).

3.8 *Pyrrhocoricin*

Cociancich et al. (1994) discovered pyrrhocoricin, an antimicrobial peptide that is rich in proline, in the hemolymph of the sap-sucking bug *Pyrrhocoris apterus*. The 20 amino acid peptide (VDKGSYLPRPTPPRPIYNRN-NH2) binds to the heat shock protein DnaK, which is associated with antimicrobial action (Kragol et al. 2001). Pyrrhocoricin can attach to and enhance the ATPase activity of the molecular chaperone DnaK (Kragol et al. 2002; Chesnokova et al. 2004). Boxell et al. (2008) showed that pyrrhocoricin can serve as a delivery mechanism for transducing peptides through the cell membrane of the parasite *Cryptosporidium parvum*. Efficient transduction enhances the process of confirming the target's validity. Additionally, it will facilitate the transportation of medications composed of peptides to this significant human pathogen. The process of forming a ring structure in pyrrhocoricin is crucial for its antibacterial properties, as stated in the original study (Rosengren et al. 2004).

3.9 *Persulcatusin*

Persulcatusin has been detected within the midgut of *Ixodes persulcatus*. The amino acid sequence is GFGCPFNQGACHRHCRSIGRRGGYCAGLFKQTCTCYSR-NH2 (Saito et al. 2009). The full molecular architecture of persulcatusin remains undiscovered at present. The similarity between this tick AMP and other known tick AMPs ranges from 71% to 88% (Tsuji et al. 2007; Rudenko et al. 2007). The structural integrity of persulcatusin is upheld by three S–S bonds, which play a crucial role in the stability and development of the α-helix and β-sheet structures (Miyoshi et al. 2016, 2017). This peptide can suppress the development of methicillin-sensitive *S. aureus*

(MSSA) and methicillin-resistant *S. aureus* (MRSA) with the MIC of 0.156–1.25 μg/mL and 0.625–2.5 μg/mL, respectively. Recently, persulcatusin has demonstrated significant antibacterial action against multidrug-resistant (MDR) *S. aureus* strains, including vancomycin-resistant *S. aureus* (VRSA) (Miyoshi et al. 2017). Persulcatusin had a more potent antibacterial effect against MRSA compared to other antimicrobial peptides (AMPs) (Miyoshi et al. 2016).

3.10 Moricin

Moricin, an α-helix antimicrobial peptide (AMP), exhibits cationic and amphipathic properties. It has charged amino acid residues at regular intervals of three to four amino acids, which contribute to its antimicrobial activity against bacteria and certain yeast strains. Moricin is composed of 42 amino acid residues and was initially extracted from the hemolymph of *B. mori*. It shows activity against gram-positive bacteria *S. aureus* (Hara and Yamakawa 1995a, b). Twelve genes that encode moricin have been identified in *B. mori*. These genes are classified into two subfamilies based on their sequence similarity. Among the twelve moricin genes, four are classified under subfamily BmmorA (A1–A4), while the remaining eight are classified under subfamily BmmorB (B1–B8) (Cheng et al. 2006).

3.11 Gloverins

Gloverins are antimicrobial peptides (AMPs) that are rich in glycine and have a molecular weight ranging from 8 to 30 kDa. They were initially discovered in the hemolymph of pupae belonging to the gigantic silk moth species *H. gloveri* (Axen et al. 1997). Gloverins have a pliable random coil configuration when dissolved in water. Gloverins derived from several insects exhibit activity against bacteria, fungi, and viruses but do not show activity against *E. coli* strains that possess smooth lipopolysaccharides (LPSs). According to reports, the activity of LPS requires the binding of gloverins to both the inner portion and the lipid A region. Gloverins undergo a conformational change when they enter the hydrophilic areas of the LPS layer and interact with the negatively charged hydrophobic regions composed of lipid A (Axen et al. 1997; Yi et al. 2013). BmGlvs attaches to the rough lipopolysaccharide (LPS), causing a structural change in the peptide from a disordered coil to an α-helix. This transformation is thought to be the primary cause of pore creation on the bacterial cell membrane (Yi et al. 2013). The binding of gloverin to the surface of microbes is necessary for its conformational change and antibacterial action.

Four genes that encode gloverins (namely, Bmglv1, Bmglv2, Bmglv3, and Bmglv4) have been discovered in the silkworm species *B. mori*. The activation of all four gloverin genes was observed in response to *E. coli*, *B. subtilis*, and *Salmonella abortusequi*. However, the expression of gloverin genes was diminished upon exposure to *S. aureus* (Kaneko et al. 2007). The variations in the structure and content of bacterial cell walls across various bacterial strains could potentially explain the varied expression of gloverins (Kaneko et al. 2007). BmGloverin2 (BmGlv2), along with other antimicrobial peptides (AMPs) found in silkworms, has been found to exhibit a synergistic effect in inhibiting the growth of *B. bassiana* fungi (Lü et al. 2017). Additionally, it has been observed that BmGLv2 suppressed the growth of two gram-negative bacteria (*E. coli* JM109 and *Pseudomonas putida*) by increasing the permeability of their cell membranes (Wang et al. 2018). This caused a disturbance in the ion gradient between the cytoplasm and the external environment, ultimately resulting in the death of the bacterial cells.

3.12 *Apidaecin*

Apidaecin refers to a collection of proline-rich peptides, consisting of 18–20 amino acid residues that are synthesized by honeybees and stingless bees (Casteels et al. 1989). Apidaecin-type peptides generated by insects are often shorter than those produced by mammals. This peptide is a member of the short proline-rich antimicrobial peptides (AMPs) family, and it exhibits both sequence similarity and sequence motif patterns with drosocin, pyrrhocoricin, formaecin, lebocin, and abaecin (Otvos 2002; Li et al. 2006). Four variations of apidaecins, namely, HbIa (GNNRPVYIPQPRPPHPRI), HbIb (GNNRPVYIPQPRPPHPRL), HbII (GNNRPIYIPQPRPPHPRL), and HbIII (GNNRPIYISQPRPPHPRL), were identified in the hemolymph of *A. mellifera* infected with *Escherichia coli* (Li et al. 2006). The peptide consists of both conserved and variable domains, which are responsible for its antimicrobial effectiveness and its ability to target certain microbes, respectively. Apidaecin exhibits defensive activity against the gram-negative bacteria through two distinct mechanisms of action. First, it inhibits the DnaK protein, which is part of the bacterial chaperone network. This inhibition blocks the ATPase activity of DnaK, preventing it from facilitating chaperone-assisted protein folding and ribosomal biogenesis (Li et al. 2006). The second mechanism involves the inhibition of translation by apidaecin through the binding of apidaecin with peptide chain release factors to the A-site of ribosomes, which then inhibits the termination phase of translation (Matsumoto et al. 2017).

3.13 *Hemolin*

Hemolin is a protein with a similar structure to immunoglobulin that is only present in Lepidoptera, which includes moths and butterflies. The initial detection occurred in immunocompromised pupae of *Hyalophora cecropia* (Faye et al. 1975) and *Manduca sexta* (Ladendorff and Kanost 1991). Hemolin possesses a crystal structure that resembles a horseshoe, as seen by its four domains. Hemolin acts as a pattern recognition molecule in the silkworm, *Bombyx mori*, and activates opsonization and melanization in silkworms upon microbial infections. Hemolin also has an affinity toward both bacterial and viral cell components (Hoffmann 1995).

3.14 Hymenoptaecin

The *A. mellifera* hymenoptaecin is a glycine-rich cationic antimicrobial peptide (AMP) that consists of 93 amino acids. It also contains a 2-pyrrolidone-5-carboxylic acid at the N-terminus. This information was reported by Casteels et al. in 1993 and Ratzka et al. in 2012. The bee hymenoptaecins consist of a signal peptide, a prosequence, and a mature peptide. The propeptide of bee hymenoptaecins exhibits similarities with the wasp *Nasonia vitripennis* and encodes several peptide precursors (Casteels-Josson et al. 1994). Thirteen distinct hymenoptaecin peptides were discovered in adult workers of *A. cerana* (Xu et al. 2009). The haemolymph of wild *A. mellifera* treated with mild infection of pathogens produced only apidaecin and hymenoptaecin genes, while abaecin and defensin were not present (Casteels-Josson et al. 1994).

3.15 Defensin

The family of cysteine-rich cationic antimicrobial peptides (AMPs) is composed of molecules containing 36–51 amino acids. These peptides have a specific structure consisting of an α-helix, two antiparallel β-strands, and a loop at the amino end. The stability of this structure is maintained by three disulfide bridges (Bonmatin et al. 1992; Cornet et al. 1995). Defensin serves as the primary defense mechanism in bees and is comprised of two distinct forms, namely, defensin 1 (5.5 KDa) and defensin 2 (4.8 KDa), which are regulated by the Toll and Imd signal pathways. Defensin 1 is generated in the salivary glands and is also present in the royal jelly and honey, which are associated with social immunity. On the other hand, defensin 2 is created in the fat body and hemolymph of bees and is connected to individual immunity. The defensin antimicrobial peptide (AMP) of *A. mellifera* consists of 43 amino acids (VTCDVLSWQSKWLSINHSACAIRCLAQRRKGGSCRNG VCICRK), while other bee species have 51 amino acids due to variations in the amino acid sequence of their defensin peptides. The open reading frame of *A. cerana* defensin possesses an extra glycine amino acid that is amidated at the carboxy terminus, in contrast to *A. mellifera* defensin (Eipper et al. 1983; Casteels-Josson et al. 1994). The defensin structure exhibits a resemblance to that of *Phormia terranovae*, and a degree of similarity between defensins 1 and 2 has been identified as 55.8% (Il'isov et al. 2012). Three variations of defensin 1 were identified in honeybees: one variation was found in the hemolymph, while the other two variations were found in royal jelly. These two variations in royal jelly are known as royalisin (Ro-F) and Ro-K, with molecular weights of 5.525 KDa and 5.515 KDa, respectively. The discovery of these isoforms was reported by Fujiwara et al. in 1990 and Klaudiny et al. in 2005. The hemolymph isoform 1 defensin variation differs from the royalysin defensin variants by two amino acid changes (Klaudiny et al. 2005). The genes defensin 1 and 2 exhibit notable differences in terms of their length (2012 and 1950 base pairs), intron–exon structure (two introns located between positions 773-1345 and 1525-180 and one intron between positions 947 and 1283), pre-propeptide region (defensin 2 is longer than defensin 1), and mature peptide (51 and 43 amino acids, with defensin 1 lacking a C-terminal elongation). In their study, Yoshiyama and Kimura (2010) demonstrated a strong evolutionary correlation (93%) between *Apis cerana* japonica defensin (AcjDef2) and *Apis mellifera* defensin (AmDef). Defensin exhibits specific antimicrobial activity against *Paenibacillus larvae*, *Melissococcus pluton*, *Ascosphaera apis*, *Nosema apis*, *N. ceranae*, *Aspergillus flavus*, *A. niger*, *Candida albicans*, *Aurobasidium pullulans*, *Varroa destructor*, *V. jacabsoni*, *Acarapis woodi*, and *Tropilaelaps clareae*, as reported by Bilikova et al. (2002), Bachanova et al. (2002), Aronstein and Saldivar (2005), Klee et al. (2007), Antunez et al. (2009), Higes et al. (2009), and Aronstein et al. (2010). The antipathogenic activity of honeybee defensins can vary due to changes in the processing of defensin precursors, resulting in the production of different molecular variants. This is supported by the presence of 29 different defensin cDNA genes that code for seven distinct defensin peptides, as reported by Il'iasov et al. (2012) and Xu et al. (2009).

3.16 Apamin

Apamin is an antimicrobial peptide (AMP) present in bee venom at a concentration of 2–3%. Its chemical structure is H-Cys-Asn-Cys-Lys-Ala-Pro-Glu-Thr-Ala-Leu--Cys-Ala-Arg-Arg-Cys-Gln-Gln-His-NH2. The substance is a neurotoxic peptide of 18 amino acid residues, arranged in a tight structure held together by two disulfide links. The polypeptide is rigid because it contains two disulfide bridges connecting Cys1 to Cys11 and Cys3 to Cys15, as well as seven hydrogen bonds (Son et al. 2007; Gu et al. 2020). It has various pharmacological effects such as blocking calcium-activated potassium channels; regulating cell growth through signal transduction pathways; reducing neurodegeneration; treating neurofibromatosis, acute pancreatitis, liver fibrosis, atopic dermatitis, atherosclerosis, rheumatoid arthritis, multiple sclerosis, cognitive deficit, and dementia; and inhibiting Th2-related cytokines, TNF-α, and IFN-γ (Castle et al. 1989; Aufschnaiter et al. 2020).

3.17 Melittin

The primary antimicrobial peptide (AMP) found in bee venom consists of 26 amphipathic amino acids (H-Gly-Ile-Gly-Ala-Val-Leu-Lys-Val-Leu-Thr-Thr-Gly-Leu-Pro-Ala-Leu-Ile-Ser-Trp-Ile-Lys-Arg-Lys-Arg-Gln-Gln-NH2), with a concentration of 40–60%. This AMP does not contain any disulfide bridges and has six positive charges located in the C-terminal region, which contribute to the stability of its α-helix structure. There are no negative charges present in this peptide. The polypeptide is a tiny molecule that loses its structure when exposed to water and regains its structure without any external assistance when placed in nonpolar substances (Abd El-Wahed et al. 2019). The N-terminal segment of melittin is mostly hydrophobic and does not possess lytic activity, whereas the C-terminal section is hydrophilic, highly basic, and responsible for the lytic action (Terwilliger 1982; Raghuraman and Chattopadhyay 2007). Due to its amphipathic character, both the monomeric and tetrameric forms of melittin are capable of dissolving in water. Melittin disrupts both bacterial and eukaryotic cell membranes by forming pores and causing lysis. Cell membrane undergoes transient (Lee et al. 2013) and stable pore development (Pino-Angeles and Lazaridis 2018), allowing for the orientation and permeability of ions and big molecules (such as glucose) accordingly. Melittin exhibits a wide range of biological activities, including the release of pro-inflammatory cytokines, activation of G-protein-coupled receptor-mediated opening of transient receptor potential channels, inhibition of protein kinase C, Ca^{2+}/calmodulin-dependent protein kinase II, myosin light-chain kinase, Na^{+}/K^{+}-ATPase (synaptosomal membrane), and blockade of transport pumps (Na^{+}-K^{+}-ATPase and the H^{+}-K^{+}-ATPase) (Jin & Lu 1998; Abd El-Wehbe et al. 2019). It also possesses antinociceptive and hemolytic properties (Wehbe et al. 2019); activates Phospholipase A2 (PLA2) (Keith et al. 2011); exhibits anticancer effects (Gajski and Garaj-Vrhovac 2013; Zarrinnahad et al. 2018), antitumor activity (Duffy et al. 2020), antidiabetic effects (Hossen et al. 2017), anti-biofilm properties (Picoli et al. 2017), and anti-inflammatory properties (Lee et al. 2016); and displays antibacterial, antifungal, and antiviral activities (Doublet et al. 2017; Memariani et al. 2019, 2020). Additionally, it has been found to possess anti-arthritic effects (Lee and Bae 2016) and plays a role in the immune response of bees to infectious diseases (Chen et al. 2016).

3.18 Melectin

The AMP consists of 18 amino acids and has a net charge of +5. It has an amphipathic and α-helical structure, with a molecular weight of 2038.3 Da. The sequence contains hydrophobic and basic amino acid residues, as well as a proline (H-GFLSILKKVLPKVMAHMK-NH2). Melectin, despite its short amino acid sequence, exhibits significant antimicrobial action against both gram-positive

(*Staphylococcus aureus*) and gram-negative bacteria (*Pseudomonas aeruginosa*, *Salmonella typhimurium*, *Klebsiella pneumoniae*, and *E. coli*). It also demonstrates broad-spectrum potency, cell selectivity, low cytotoxicity in mammalian cells, salt-resistant properties, and rapid bactericidal action compared to melittin (which has 26 amino acids). Furthermore, melectin is effective against drug-resistant bacteria, such as *S. aureus* MRSA 254/366/660, *P. aeruginosa* 1034/3543/5018, and *E. coli* CCARM 1229/1238 (Cerovský et al. 2008; Ko et al. 2020). Melectin exhibits selective targeting of bacterial cells by inducing a distortion in the α-helical structure (Cerovský et al. 2008). It also triggers degranulation of peritoneal mast cells in rats (Niederhafner et al. 2010) while demonstrating low levels of hemolytic activity and antitumor properties (Liang et al. 2021).

3.19 Adolapin

Adolapin is a constituent of honeybee venom, comprising around 0.1–1% of the dry venom (Orsolic 2012; Pucca et al. 2019; Aufschnaiter et al. 2020). The bee venom contains 2–5% of the antimicrobial peptides (AMPs). One specific AMP, which is a basic AMP with a molecular weight of 11.5 kDa and consists of 103 amino acids, exhibits significant pharmacological properties. These properties include anti-inflammatory, antinociceptive, analgesic, and antipyretic effects, as well as the inhibition of COX1, prostaglandin synthase, and lipoxygenase systems in human platelets. Additionally, it increases the c-GMP level in the rat spleen and brain while decreasing the c-AMP level in the rat spleen. These findings have been reported by Shkenderov and Koburova in 1982, Koburova in 1984 and 1985, Son et al. in 2007, Ram et al. in 2014, Nipate et al. in 2015, Jung et al. in 2015, and Cherniack and Govorushko in 2018.

3.20 Secapin

The molecule in concern is a nontoxic polypeptide found in bee venom, comprising approximately 0.5–2% of the venom's dry weight. It consists of 25 amino acids arranged in the sequence YIIDVPPRCPPGSKFIKNRCRVIVP-NH2 and contains a significant amount of proline. Additionally, there is a disulfide link between the Cys9 and Cys20 residues inside the polypeptide (Aufschnaiter et al. 2020). Several isoforms of secapin (including secapin, secapin-1, secapin-2, and secapin-3) have been identified in *A. mellifera* (Matysiak et al. 2011; Meng et al. 2012; Mourelle et al. 2014), as well as AcSecapin-1 in *A. cerana* (Lee et al. 2016). The AMP has three isoforms with the following amino acid sequences: secapin (YIIDVPPRCP PGSKFIKNRCRVIVP-NH2), secapin-1 (YIINVPPRCPPGSKFVKNKCRV IVP-NH2), and secapin-2 (YIIDVPPRCPPGSKFV HKRCRVIVP-NH2). The variations between secapin-1 and secapin-2 isoforms were due to substitutions in amino acids 17–19. Secapin-1 exhibited greater sequence homology with *A. dorsata* secapin, *A. florea* secapin, and *A. mellifera* secapin-3 compared to secapin. Secapin-1 functions as a peptide that resembles a serine protease. It possesses properties that prevent the breakdown of fibrin, the breakdown of elastin, and the growth of microorganisms. Additionally, it demonstrates inhibitory actions against trypsin, chymotrypsin, serine proteases produced by microorganisms, elastases, and plasmin (Lee et al. 2016). Secapin-2 does not cause hemolytic activity, mast cell degranulation, or polymorphonuclear leukocytes (PMNL) chemotaxis. However, it does affect hyperalgesia and edema, which are mediated by leukotrienes (Mourelle et al. 2014).

3.21 Tertiapin

Tertiapin AMP is composed of 21 amino acids, two disulfide bridges, and a carboxy-terminal residue that has been amidated. It constitutes a small portion of bee venom (0.1% of dry venom) and is classified as a neurotoxic peptide, similar to apamin. It can modulate potassium channels, as demonstrated by Drici et al. (2000) and Swiatły-Błaszkiewicz et al. (2020). In humans, it inhibits inward rectifier potassium channels (Cornara et al. 2017) and mitigates atrioventricular conduction abnormalities (Drici et al. 2000).

3.22 Cardiopep

Cardiopep, a polypeptide with a molecular weight of 1940 Da, consists of 0.7% bee venom and is composed of low molecular weight peptides that make up 50% of its content. According to Shipman and Cole (1969), cardiopep is comparatively less harmful than phospholipase A, melittin, and entire bee venom, with an LD50 value of 15 mg/kg in mice. In contrast, phospholipase A has an LD50 value of 1–5 mg/kg, melittin has an LD50 value of 2–5 mg/kg, and whole bee venom has an LD50 value of 3–5 mg/kg. In addition to being less toxic, it possesses beta-adrenergic-like stimulant and anti-arrhythmic characteristics (Vick et al. 1974).

4 Pharmacological Properties and Therapeutic Potential of Insects AMPs

Previous investigations have identified potential effective medical applications for antimicrobial peptides (AMPs) (Baltzer and Brown 2011; Kumar et al. 2018). The mechanisms of action vary and include inhibiting gene expression or protein synthesis (such as targeting ribosomal proteins or RNA polymerase or directly binding to DNA), inhibiting cell wall synthesis (such as targeting the corresponding enzymes or lipid phosphatidylethanolamines), or causing the removal of bacterial cell surface proteins (Lakshmaiah Narayana and Chen 2015).

Honeybees secrete defensin 1 in their hypopharyngeal gland, which is later transferred to honey. Defensin 1 was discovered in Revamil honey and is accountable for its antibacterial effect on gram-positive bacteria (Valanne et al. 2011). Apalbumin 1, obtained from Ziziphus honey, effectively suppressed the release of reactive oxygen species (ROS) in zymosan-activated human neutrophils and murine macrophages (Bíliková et al. 2009). It also inhibited the formation of nitric oxide (NO), phagocytosis in LPS-activated murine macrophages, and the production of TNF-α in human monocytic cells (Kwakman and Zaat 2012).

Bee venom exhibits various beneficial properties, including anti-inflammatory, antibacterial, antiviral, anticancer, antimutagenic, antinociceptive, and radioprotective activities (Table 5.1). It has been utilized for the management of skin disorders, diseases of the blood circulation system, immune-related deficiencies, tumors, back discomfort, rheumatism, and arthritis (Ma et al. 2017; Kim et al. 2020). The components found in bee venom include mast cell-degranulating peptide (MCDP), melittin, adolapin, melectin, tertiapin, secarpin, minimine, procamine, and Apamin (Ullah et al. 2023).

Melittin, an AMP present in bee venom, has demonstrated efficacy in treating prostate, liver, kidney, and cervical cancers (Zhang et al. 2018). Apitoxin and its constituents are mostly known for their anti-inflammatory, anticancer, anti-arthritic, and neuroprotective properties. Melittin possesses anti-inflammatory qualities and has been utilized in the treatment of several disorders, including dermatitis, neuritis, liver inflammation, atherosclerosis, and arthritis (Abdela and Jilo 2016). Additionally, it has demonstrated potent antibacterial effects against methicillin-resistant *S. aureus*, as well as anticancer characteristics and the ability to trigger apoptosis in human ovarian cancer cells (SKOV3 and PA-1). Administering melittin stimulates the body to create cortisol and functions as a powerful anti-inflammatory agent derived from BV29.

Several AMPs have shown immune-stimulating actions by promoting cell migration, cell proliferation, or the secretion of cytokines and chemokines (Kruse and Kristensen 2008). Harmoniasin, extracted from the harlequin ladybird *Harmonia axyridis*, and other AMPs obtained from insects have the potential to serve as promising candidates in the development of drugs to fight cancer (Kim et al. 2013; Jin & Weinberg 2019).There is an increasing amount of evidence demonstrating that insect-derived antimicrobial peptides (AMPs) can suppress infections that affect humans. Some examples of bacteria that are susceptible to certain conditions or treatments include multidrug-resistant *A. baumannii*, *Bacillus coagulans*, *Citrobacter freundii*, *Enterobacter aerogenes*, *E. coli*, *Francisella tularensis*, *K. pneumonia*, *Legionella dumoffei*, *Listeria monocytogenes*, *Proteus vulgaris*, *S. aureus*, and *Streptococcus sanguinis* (Mukherjee et al. 2011; Vonkavaara et al. 2013). Fungi that are susceptible to insect antimicrobial peptides (AMPs) include *Aspergillus fumigatus*, *Alternaria* spp., *Botrytis cinerea*, *C. albicans*, *Cryptococcus neoformans*, *Fusarium* spp., *Neurospora crassa*, *Pichia pastoris*, and *Trichoderma viridae* (Po¨ppel et al. 2014). It is worth mentioning that different insect antimicrobial peptides (AMPs) have demonstrated the ability to hinder viruses. For example, a derivative of mellitin called hecate, found in honeybees (*Apis mellifera*), is effective against herpes simplex virus 1 (HSV-1) (Kruse and Kristensen 2008). Additionally, two alloferons from the blowfly *Calliphora vicina* have shown activity against human influenza viruses A and B (Chernysh et al. 2002). There is ongoing discourse over the capacity of bacteria to develop resistance against antimicrobial peptides (AMPs). Some researchers claim that AMPs, like defensins, have maintained their ability to kill bacteria for millions of years. They have additionally observed no development of resistance to AMPs under laboratory conditions (Wilmes and Sahl 2014).

In addition, exposure to antimicrobial peptides (AMPs) does not induce stress responses or lead to higher mutation rates in the bacteria that are treated (Rodriguez-Rojas et al. 2014). Conversely, various mechanisms have been found that can facilitate bacterial resistance against antimicrobial peptides (AMPs) (Joo et al. 2016). Gramicidins, which are obtained from bacteria, were the first commercially available peptide antibiotics (Prenner et al. 1999). Several studies have indicated that antimicrobial peptides (AMPs) derived from insects are promising prospects for the development of alternative antibiotics. Despite global endeavors, there are currently no insect-derived antimicrobial peptides (AMPs) available for commercial use. Their advancement as antibiotics necessitates the resolution of certain challenges, which will be succinctly discussed. Producing sufficient quantities of insect-derived antimicrobial peptides (AMPs) for assessing their strong action in both laboratory and live animal experiments is economically challenging (Hilpert et al. 2005). Nevertheless, the expenses associated with creating short peptides (up to around 80 residues) have significantly reduced in recent times (Vlieghe et al. 2010). This cost reduction has made it possible to produce insect-derived antimicrobial peptides (AMPs) that are too tiny to cause an immune response or allergic reactions. To produce larger insect antimicrobial peptides (AMPs) that have intricate three-dimensional structures held together by intramolecular disulfide bonds, we require cost-effective methods for expressing these peptides in a different organism. Insect cell lines, in particular, have shown great potential as a viable option for producing functional recombinant peptides derived from insects (Kollewe and Vilcinskas 2013). The creation of cost-effective systems for producing proteins using insect cells has become a significant problem in the field of insect biotechnology (Druzinec et al. 2013). While recent advancements have been groundbreaking, the manufacturing costs of these systems remain above those of traditional pharmaceuticals (Mu¨ller et al. 2015). The stability of medications directly affects their bioavailability. The susceptibility of insect antimicrobial peptides (AMPs) to host proteases varies significantly. In general, linear peptides are more susceptible to degradation by proteases compared to antimicrobial peptides (AMPs) that have an intramolecular structure stabilized by disulfide bonds, like defensins. The antibacterial properties of the latter hinder their creation in bacteria, which cannot often synthesize correctly folded functional peptides. These constraints can be surmounted through the utilization of sophisticated insect cell-based expression systems (Kollewe and

Vilcinskas 2013; Druzinec et al. 2013; Mu¨ller et al. 2015). Developing new antibiotics employing insect-derived antimicrobial peptides (AMPs) as leads has led to the emergence of a technique, including the production of functional analogs or peptidomimetics that are more resistant to hydrolysis by host proteases (Baghian et al. 1997; Berthold et al. 2013). In addition, very small quantities of specific antimicrobial peptides (AMPs) are needed for systemic administration when they are utilized to enhance the vulnerability of bacteria to traditional antibiotics (Berthold et al. 2013). As a result, researchers are exploring the combined use of insect-derived antimicrobial peptides (AMPs) and antibiotics as a potential strategy to incorporate these natural compounds into treatments. The aforementioned challenges in using insect-derived antimicrobial peptides (AMPs) have led to their focus on ectopic applications that require less rigorous preclinical and clinical studies (Steinstraesser et al. 2008). An excellent illustration is the utilization of therapeutic maggots from *L. sericata* to produce AMPs. These AMPs are then included in hydrogels to evaluate the effectiveness of synthetic alternatives in wound dressings. Additionally, they are used as cosmetic ingredients to repel dermatological pathogens. This research is supported by references (Rahnamaeian and Vilcinskas 2015). Additional antimicrobial peptides (AMPs) obtained from insects, such as the insect metalloproteinase inhibitor (IMPI) found in *Galleria mellonella*, are currently being researched for their potential use in treating chronic wounds. Ongoing research is currently investigating the potential use of insect-derived antimicrobial peptides (AMPs) in the treatment of eye, lung, and urogenital infections (Brandt 2014; Lo and Lange 2015). For instance, studies have shown that a defensin derived from *G. mellonella* or antimicrobial peptides (AMPs) found in medicinal maggots can effectively combat the microorganisms responsible for lung infections (Palusin'ska-Szysz et al. 2012). Recombinant versions of antimicrobial peptides (AMPs) generated from insects can be administered to the lungs by attaching them to inhalable microparticles. It is anticipated that using many AMPs with synergistic effects will decrease the necessary dosage for therapeutic purposes.

5 Conclusion

AMPs have gained heightened interest from researchers, health experts, and the pharmaceutical sector due to their pharmacological and therapeutic properties. Insect AMPs are small proteins that exhibit a wide spectrum of antibacterial and immunomodulatory effects against many pathogenic microorganisms. Furthermore, honeybee AMPs also demonstrate enhanced effectiveness, strong selectivity, reduced drug interference, low toxicity, drug transport mechanisms, biological variety, and direct offensive properties. Multidrug resistance is a complex and diverse phenomenon, with multiple facets and dimensions. It is also the second most common cause of mortality globally. AMP research has been motivated by the pressing necessity to get novel antimicrobials. AMPs are regarded as promising antimicrobial agents for the development of a new generation of antimicrobials. As a result, multiple pharmaceutical industries are currently performing suitable clinical trials to create possible medicinal medications derived from AMPs. More than 60 peptide medications have previously been introduced to the market, and several innovative therapeutic peptides are currently undergoing preclinical and clinical development. The utilization of therapeutic AMPs derived from insect sources can also impact the development of innovative strategies for addressing human health concerns. These AMPs could serve as viable alternatives to traditional antibiotics. Nevertheless, the progress of their research must confront the constraints linked to the utilization of peptide-based medications.

References

Abd El-Wahed AA, Khalifa SA, Sheikh BY, Farag MA, Saeed A, Larik FA et al (2019) Bee venom composition: from chemistry to biological activity. Stud Nat Prod Chem 60:459–484

Abdela N, Jilo K (2016) Bee venom and its therapeutic values: a review. Life Sci Technol Sci Commun 44:18–22

Antúnez K, Martín-Hernández R, Prieto L, Meana A, Zunino P, Higes M (2009) Immune suppression in the honey bee (Apis mellifera) following infection by Nosema ceranae (Microsporidia). Environ Microbiol 11(9):2284–2290

Arcidiacono S, Soares JW, Meehan AM, Marek P, Kirby R (2009) Membrane permeability and antimicrobial kinetics of cecropin P1 against Escherichia coli. J Pept Sci 15(6):398–403

Aronstein K, Saldivar E (2005) Characterization of a honey bee Toll related receptor gene Am18w and its potential involvement in antimicrobial immune defense. Apidologie 36(1):3–14

Aronstein KA, Murray KD, Saldivar E (2010) Transcriptional responses in honey bee larvae infected with chalkbrood fungus. BMC Genomics 11:391

Aufschnaiter A, Kohler V, Khalifa S, Abd El-Wahed A, Du M, El-Seedi H, Büttner S (2020) Apitoxin and its components against cancer, neurodegeneration and rheumatoid arthritis: limitations and possibilities. Toxins 12(2):66

Axén A, Carlsson A, Engström A, Bennich H (1997) Gloverin, an antibacterial protein from the immune hemolymph of Hyalophora pupae. Eur J Biochem 247(2):614–619. https://doi.org/10.1111/j.1432-1033.1997.00614.x

Bachanová K, Klaudiny J, Kopernický J, Šimúth J (2002) Identification of honeybee peptide active against Paenibacillus larvae larvae through bacterial growth-inhibition assay on polyacrylamide gel. Apidologie 33(3):259–269

Baek MH, Kamiya M, Kushibiki T, Nakazumi T, Tomisawa S, Abe C et al (2016) Lipopolysaccharide-bound structure of the antimicrobial peptide cecropin P1 determined by nuclear magnetic resonance spectroscopy. J Pept Sci 22(4):214–221

Baghian A, Jaynes J, Enright F, Kousoulas KG (1997) An amphipathic alpha-helical synthetic peptide analogue of melittin inhibits herpes simplex virus-1 (HSV-1)-induced cell fusion and virus spread. Peptides 18(2):177–183

Baltzer SA, Brown MH (2011) Antimicrobial peptides: promising alternatives to conventional antibiotics. J Mol Microbiol Biotechnol 20(4):228–235

Beceiro A, Tomás M, Bou G (2013) Antimicrobial resistance and virulence: a successful or deleterious association in the bacterial world. Clin Microbiol Rev 26(2):185–230

Bechinger B, Gorr SU (2017) Antimicrobial peptides: mechanisms of action and resistance. J Dent Res 96(3):254–260

Bechinger B, Lohner K (2006) Detergent-like actions of linear amphipathic cationic antimicrobial peptides. Biochim Biophys Acta 1758(9):1529–1539

Berthold N, Czihal P, Fritsche S, Sauer U, Schiffer G, Knappe D, Alber G, Hoffmann R (2013) Novel apidaecin 1b analogs with superior serum stabilities for treatment of infections by gram-negative pathogens. Antimicrob Agents Chemother 57(1):402–409
Bílíková K, Hanes J, Nordhoff E, Saenger W, Klaudiny J, Simúth J (2002) Apisimin, a new serine-valine-rich peptide from honeybee (Apis mellifera L.) royal jelly: purification and molecular characterization. FEBS Lett 528(1–3):125–129
Bílíková K, Mirgorodskaya E, Bukovská G, Gobom J, Lehrach H, Simúth J (2009) Towards functional proteomics of minority component of honeybee royal jelly: the effect of post-translational modifications on the antimicrobial activity of apalbumin2. Proteomics 9(8):2131–2138
Bluhm ME, Schneider VA, Schäfer I, Piantavigna S, Goldbach T, Knappe D et al (2016) N-terminal Ile-Orn-and Trp-Orn-motif repeats enhance membrane interaction and increase the antimicrobial activity of apidaecins against Pseudomonas aeruginosa. Front Cell Dev Biol 4:39
Bonmatin JM, Bonnat JL, Gallet X, Vovelle F, Ptak M, Reichhart JM, et al. (1992) Two-dimensional 1H NMR study of recombinant insect defensin A in water: resonance assignments, secondary structure and global folding. J Biomol NMR 2:235–256
Boxell A, Lee SHC, Jefferies R, Watt P, Hopkins R, Reid S et al (2008) Pyrrhocoricin as a potential drug delivery vehicle for Cryptosporidium parvum. Exp Parasitol 119(2):301–303
Brady D, Grapputo A, Romoli O, Sandrelli F (2019) Insect Cecropins, antimicrobial peptides with potential therapeutic applications. Int J Mol Sci 20(23):5862. https://doi.org/10.3390/ijms20235862
Brandt CR (2014) Peptide therapeutics for treating ocular surface infections. J Ocul Pharmacol Ther 30(9):691–699
Brogden KA (2005) Antimicrobial peptides: pore formers or metabolic inhibitors in bacteria. Nat Rev Microbiol 3(3):238–250
Bulet P, Stöcklin R (2005) Insect antimicrobial peptides: structures, properties and gene regulation. Protein Pept Lett 12(1):3–11
Bulet P, Dimarcq JL, Hetru C, Lagueux M, Charlet M, Hegy G, Van Dorsselaer A, Hoffmann JA (1993) A novel inducible antibacterial peptide of Drosophila carries an O-glycosylated substitution. J Biol Chem 268(20):14893–14897
Bulet P, Urge L, Ohresser S, Hetru C, Otvos L Jr (1996) Enlarged scale chemical synthesis and range of activity of drosocin, an O-glycosylated antibacterial peptide of Drosophila. Eur J Biochem 238(1):64–69
Bulet P, Stöcklin R, Menin L (2004) Anti-microbial peptides: from invertebrates to vertebrates. Immunol Rev 198:169–184
Bush K, Page MGP (2017) What we may expect from novel antibacterial agents in the pipeline with respect to resistance and pharmacodynamic principles. J Pharmacokinet Pharmacodyn 44(2):113–132
Carlsson A, Engström P, Palva ET, Bennich H (1991) Attacin, an antibacterial protein from Hyalophora cecropia, inhibits synthesis of outer membrane proteins in Escherichia coli by interfering with omp gene transcription. Infect Immun 59(9):3040–3045
Carlsson A, Nyström T, de Cock H, Bennich H (1998) Attacin-an insect immune protein-binds LPS and triggers the specific inhibition of bacterial outer-membrane protein synthesis. Microbiology 144(8):2179–2188
Casteels P, Ampe C, Jacobs F, Vaeck M, Tempst P (1989) Apidaecins: antibacterial peptides from honeybees. EMBO J 8(8):2387–2391
Casteels P, Ampe C, Riviere L, Van Damme J, Elicone C, Fleming M, Jacobs F, Tempst P (1990) Isolation and characterization of abaecin, a major antibacterial response peptide in the honeybee (Apis mellifera). Eur J Biochem 187(2):381–386
Casteels P, Ampe C, Jacobs F, Tempst P (1993) Functional and chemical characterization of Hymenoptaecin, an antibacterial polypeptide that is infection-inducible in the honeybee (Apis mellifera). J Biol Chem 268(10):7044–7054

Casteels-Josson K, Zhang W, Capaci T, Casteels P, Tempst P (1994) Acute transcriptional response of the honeybee peptide-antibiotics gene repertoire and required post-translational conversion of the precursor structures. J Biol Chem 269(46):28569–28575
Castle NA, Haylett DG, Jenkinson DH (1989) Toxins in the characterization of potassium channels. Trends Neurosci 12(2):59–65
Cerovský V, Hovorka O, Cvacka J, Voburka Z, Bednárová L, Borovicková L, Slaninová J, Fucík V (2008) Melectin: a novel antimicrobial peptide from the venom of the cleptoparasitic bee Melecta albifrons. Chembiochem Eur J Chem Biol 9(17):2815–2821
Chen J, Guan SM, Sun W, Fu H (2016) Melittin, the major pain-producing substance of bee venom. Neurosci Bull 32(3):265–272
Cheng T, Zhao P, Liu C, Xu P, Gao Z, Xia Q, Xiang Z (2006) Structures, regulatory regions, and inductive expression patterns of antimicrobial peptide genes in the silkworm Bombyx mori. Genomics 87(3):356–365. https://doi.org/10.1016/j.ygeno.2005.11.018
Cherniack EP, Govorushko S (2018) To bee or not to bee: the potential efficacy and safety of bee venom acupuncture in humans. Toxicon 154:74–78
Chernysh S, Kim SI, Bekker G, Pleskach VA, Filatova NA, Anikin VB, Platonov VG, Bulet P (2002) Antiviral and antitumor peptides from insects. Proc Natl Acad Sci USA 99(20):12628–12632
Chernysh S, Gordya N, Suborova T (2015) Insect antimicrobial peptide complexes prevent resistance development in bacteria. PLoS One 10(7):e0130788
Chesnokova LS, Slepenkov SV, Witt SN (2004) The insect antimicrobial peptide, L-pyrrhocoricin, binds to and stimulates the ATPase activity of both wild-type and lidless DnaK. FEBS Lett 565(1–3):65–69
Cociancich S, Dupont A, Hegy G, Lanot R, Holder F, Hetru C et al (1994) Novel inducible antibacterial peptides from a hemipteran insect, the sap-sucking bug Pyrrhocoris apterus. Biochem J 300(2):567–575
Cornara L, Biagi M, Xiao J, Burlando B (2017) Therapeutic properties of bioactive compounds from different honeybee products. Front Pharmacol 8:412
Cornet B, Bonmatin JM, Hetru C, Hoffmann JA, Ptak M, Vovelle F (1995) Refined three-dimensional solution structure of insect defensin A. Structure 3(5):435–448
Cudic M, Bulet P, Hoffmann R, Craik DJ, Otvos L Jr (1999) Chemical synthesis, antibacterial activity and conformation of diptericin, an 82-mer peptide originally isolated from insects. Eur J Biochem 266(2):549–558
De Soyza A, Aksamit T (2016) Ciprofloxacin dry powder for inhalation in non-cystic fibrosis bronchiectasis. Expert Opin Orphan Drugs 4(8):875–884
Dian Permana A, Mir M, Utomo E, Donnelly RF (2020) WITHDRAWN: bacterially sensitive nanoparticle-based dissolving microneedles of doxycycline for enhanced treatment of bacterial biofilm skin infection: a proof of concept study. Int J Pharm:119220. Advance online publication
Dinata R, Arati C, Manikandan B, Abinash G, Nisa N, Bhanushree B, et al. (2023) Pharmacological and therapeutic potential of honey bee antimicrobial peptides: bee antimicrobial peptides. Indian J Biochem Biophy 60(5):365–384
Dimarcq JL, Keppi E, Dunbar B, Lambert J, Reichhart JM, Hoffmann D et al (1988) Insect immunity: purification and characterization of a family of novel inducible antibacterial proteins from immunized larvae of the dipteran Phormia terranovae and complete amino-acid sequence of the predominant member, diptericin A. Eur J Biochem 171(1–2):17–22
Doublet V, Poeschl Y, Gogol-Döring A, Alaux C, Annoscia D, Aurori C, Barribeau SM, Bedoya-Reina OC, Brown MJ, Bull JC, Flenniken ML, Galbraith DA, Genersch E, Gisder S, Grosse I, Holt HL, Hultmark D, Lattorff HM, Le Conte Y, Manfredini F et al (2017) Unity in defence: honeybee workers exhibit conserved molecular responses to diverse pathogens. BMC Genomics 18(1):207
Drici MD, Diochot S, Terrenoire C, Romey G, Lazdunski M (2000) The bee venom peptide tertiapin underlines the role of I(KACh) in acetylcholine-induced atrioventricular blocks. Br J Pharmacol 131(3):569–577

Druzinec D, Salzig D, Brix A, Kraume M, Vilcinskas A, Kollewe C, Czermak P (2013) Optimization of insect cell based protein production processes - online monitoring, expression systems, scale up. Adv Biochem Eng Biotechnol 136:65–100

Duffy C, Sorolla A, Wang E, Golden E, Woodward E, Davern K, Ho D, Johnstone E, Pfleger K, Redfern A, Iyer KS, Baer B, Blancafort P (2020) Honeybee venom and melittin suppress growth factor receptor activation in HER2-enriched and triple-negative breast cancer. NPJ Precis Oncol 4:24

Dushay MS, Roethele JB, Chaverri JM, Dulek DE, Syed SK, Kitami T, Eldon ED (2000) Two attacin antibacterial genes of Drosophila melanogaster. Gene 246(1–2):49–57

Eipper BA, Mains RE, Glembotski CC (1983) Identification in pituitary tissue of a peptide alpha-amidation activity that acts on glycine-extended peptides and requires molecular oxygen, copper, and ascorbic acid. Proc Natl Acad Sci USA 80(16):5144–5148

Faye I, Pye A, Rasmuson T, Boman HG, Boman IA (1975) Insect immunity. 11. Simultaneous induction of antibacterial activity and selection synthesis of some hemolymph proteins in diapausing pupae of Hyalophora cecropia and Samia cynthia. Infect Immun 12(6):1426–1438. https://doi.org/10.1128/iai.12.6.1426-1438.1975

Fontana R, Mendes MA, de Souza BM, Konno K, César LM, Malaspina O, Palma MS (2004) Jelleines: a family of antimicrobial peptides from the Royal Jelly of honeybees (Apis mellifera). Peptides 25(6):919–928

Fujiwara S, Imai J, Fujiwara M, Yaeshima T, Kawashima T, Kobayashi K (1990) A potent antibacterial protein in royal jelly. Purification and determination of the primary structure of royalisin. J Biol Chem 265(19):11333–11337

Gajski G, Garaj-Vrhovac V (2013) Melittin: a lytic peptide with anticancer properties. Environ Toxicol Pharmacol 36(2):697–705

Gauldie J, Hanson JM, Shipolini RA, Vernon CA (1978) The structures of some peptides from bee venom. Eur J Biochem 83(2):405–410

Geng H, An CJ, Hao YJ, Li DS, Du RQ (2004) Yi chuan xue bao =. Acta Genet Sin 31(12):1344–1350

Gevod VS, Birdi KS (1984) Melittin and the 8-26 fragment. Differences in ionophoric properties as measured by monolayer method. Biophys J 45(6):1079–1083

Ghosh ANS, Saha S (2020) Survey of drug resistance associated gene mutations in Mycobacterium tuberculosis, ESKAPE and other bacterial species. Sci Rep 10(1):8957

Gobbo M, Biondi L, Filira F, Gennaro R, Benincasa M, Scolaro B, Rocchi R (2002) Antimicrobial peptides: synthesis and antibacterial activity of linear and cyclic drosocin and apidaecin 1b analogues. J Med Chem 45(20):4494–4504

Gollan B, Grabe G, Michaux C, Helaine S (2019) Bacterial Persisters and infection: past, present, and progressing. Ann Rev Microbiol 73:359–385

Gu H, Han SM, Park KK (2020) Therapeutic effects of Apamin as a bee venom component for non-neoplastic disease. Toxins 12(3):195

Gudmundsson GH, Lidholm DA, Asling B, Gan R, Boman HG (1991) The cecropin locus. Cloning and expression of a gene cluster encoding three antibacterial peptides in Hyalophora cecropia. J Biol Chem 266(18):11510–11517

Guo C, Huang Y, Zheng H, Tang L, He J, Xiang L et al (2012) Secretion and activity of antimicrobial peptide cecropin D expressed in Pichia pastoris. Exp Ther Med 4(6):1063–1068

Guo C, Huang Y, Cong P, Liu X, Chen Y, He Z (2014) Cecropin P1 inhibits porcine reproductive and respiratory syndrome virus by blocking attachment. BMC Microbiol 14(1):1–11

Haisma EM, Göblyös A, Ravensbergen B, Adriaans AE, Cordfunke RA, Schrumpf J, Limpens RW, Schimmel KJ, den Hartigh J, Hiemstra PS, Drijfhout JW, El Ghalbzouri A, Nibbering PH (2016) Antimicrobial Peptide P60.4Ac-containing creams and gel for eradication of methicillin-resistant Staphylococcus aureus from cultured skin and airway epithelial surfaces. Antimicrob Agents Chemother 60(7):4063–4072

Hancock RE, Sahl HG (2006) Antimicrobial and host-defense peptides as new anti-infective therapeutic strategies. Nat Biotechnol 24(12):1551–1557

Hara S, Yamakawa M (1995a) A novel antibacterial peptide family isolated from the silkworm, Bombyx mori. Biochem J 310(2):651–656

Hara S, Yamakawa M (1995b) Moricin, a novel type of antibacterial peptide isolated from the silkworm, Bombyx mori. J Biol Chem 270(50):29923–29927. https://doi.org/10.1074/jbc.270.50.29923

Higes M, Martín-Hernández R, Garrido-Bailón E, González-Porto AV, García-Palencia P, Meana A et al (2009) Honeybee colony collapse due to Nosema ceranae in professional apiaries. Environ Microbiol Rep 1(2):110–113

Hilpert K, Volkmer-Engert R, Walter T, Hancock RE (2005) High-throughput generation of small antibacterial peptides with improved activity. Nat Biotechnol 23(8):1008–1012

Hoffmann JA (1995) Innate immunity of insects. Curr Opin Immunol 7(1):4–10. https://doi.org/10.1016/0952-7915(95)80022-0

Hollmann A, Martinez M, Maturana P, Semorile LC, Maffia PC (2018) Antimicrobial peptides: interaction with model and biological membranes and synergism with chemical antibiotics. Front Chem 6:204

Hossen MS, Gan SH, Khalil MI (2017) Melittin, a potential natural toxin of crude bee venom: probable future arsenal in the treatment of diabetes mellitus. J Chem 2017:1

Hou Z, Wang W, Liu Z, Liu G, Souffrant WB, Yin Y (2011) Effect of lactoferrincin B and cecropin P1 against enterotoxigenic Escherichia coli in vitro. J Food Agric Environ 9(2 part 1):271–274

Hultmark D, Engström Å, Bennich H, Kapur R, Boman HG (1982) Insect immunity: isolation and structure of cecropin D and four minor antibacterial components from Cecropia pupae. Eur J Biochem 127(1):207–217

Hultmark D, Engström A, Andersson K, Steiner H, Bennich H, Boman HG (1983) Insect immunity. Attacins, a family of antibacterial proteins from Hyalophora cecropia. EMBO J 2(4):571–576

Il'iasov RA, Gaĭfullina LR, Saltykova ES, Poskriakov AV, Nikolenko AG (2012) Defensins in the honeybee antinfectious protection. Zhurnal Evoliutsionnoi Biokhimii i Fiziologii 48(5):425–432

Imler JL, Bulet P (2005) Antimicrobial peptides in Drosophila: structures, activities and gene regulation. Chem Immunol Allergy 86:1–21

Ishikawa M, Kubo T, Natori S (1992) Purification and characterization of a diptericin homologue from Sarcophaga peregrina (flesh fly). Biochem J 287(2):573–578

Jin W, Lu Z (1998) A novel high-affinity inhibitor for inward-rectifier K+ channels. Biochemistry 37(38):13291–13299

Jin G, Weinberg A (2019, April) Human antimicrobial peptides and cancer. In: Seminars in cell & developmental biology, vol 88. Academic Press, pp 156–162

Johnson SR, Copello JA (2010) A biochemical characterization of the major peptides from the venom of the giant neotropical hunting ant Dinoponera australis. Biophys J 98(3):108a

Joo HS, Fu CI, Otto M (2016) Bacterial strategies of resistance to antimicrobial peptides. Philos Trans R Soc B Biol Sci 371(1695):20150292

Jozefiak A, Engberg RM (2017) Insect proteins as a potential source of antimicrobial peptides in livestock production. A review. J Anim Feed Sci 26(2):87–99

Jung SY, Lee KW, Choi SM, Yang EJ (2015) Bee venom protects against rotenone-induced cell death in NSC34 motor neuron cells. Toxins 7(9):3715–3726

Kaneko Y, Furukawa S, Tanaka H, Yamakawa M (2007) Expression of antimicrobial peptide genes encoding Enbocin and Gloverin isoforms in the silkworm, Bombyx mori. Biosci Biotechnol Biochem 71(9):2233–2241. https://doi.org/10.1271/bbb.70212

Keith DJ, Eshleman AJ, Janowsky A (2011) Melittin stimulates fatty acid release through non-phospholipase-mediated mechanisms and interacts with the dopamine transporter and other membrane-spanning proteins. Eur J Pharmacol 650(2–3):501–510

Keppi E, Pugsley AP, Lambert J, Wicker C, Dimarcq JL, Hoffmann JA, Hoffmann D (1989) Mode of action of diptericin A, a bactericidal peptide induced in the hemolymph of Phormia terranovae larvae. Arch Insect Biochem Physiol 10(3):229–239

Kim IW, Lee JH, Kwon YN, Yun EY, Nam SH, Ahn MY et al (2013) Anticancer activity of a synthetic peptide derived from harmoniasin, an antibacterial peptide from the ladybug Harmonia axyridis. Int J Oncol 43(2):622–628

Kim JY, Lee SJ, Maeng YI, Leem J, Park KK (2020) Protective effects of bee venom against endotoxemia-related acute kidney injury in mice. Biology 9(7):154

Klaudiny J, Albert Š, Bachanová K, Kopernický J, Šimúth J (2005) Two structurally different defensin genes, one of them encoding a novel defensin isoform, are expressed in honeybee Apis mellifera. Insect Biochem Mol Biol 35(1):11–22

Klee J, Besana AM, Genersch E, Gisder S, Nanetti A, Tam DQ et al (2007) Widespread dispersal of the microsporidian Nosema ceranae, an emergent pathogen of the western honey bee, Apis mellifera. J Invertebr Pathol 96(1):1–10

Ko SJ, Park E, Asandei A, Choi JY, Lee SC, Seo CH et al. (2020) Bee venom-derived antimicrobial peptide melectin has broad-spectrum potency, cell selectivity, and salt-resistant properties. Sci rep 10(1):10145

Koburova K (1984) Antipyretic effect of a polypeptide from bee venom-adolapin. Eksp Med Morfol 23:143–148

Kockum K, Faye I, Hofsten PV, Lee JY, Xanthopoulos KG, Boman HG (1984) Insect immunity. Isolation and sequence of two cDNA clones corresponding to acidic and basic attacins from Hyalophora cecropia. EMBO J 3(9):2071–2075

Kollewe C, Vilcinskas A (2013) Production of recombinant proteins in insect cells. Am J Biochem Biotechnol 9:255

Kragol G, Lovas S, Varadi G, Condie BA, Hoffmann R, Otvos L (2001) The antibacterial peptide pyrrhocoricin inhibits the ATPase actions of DnaK and prevents chaperone-assisted protein folding. Biochemistry 40(10):3016–3026

Kragol G, Hoffmann R, Chattergoon MA, Lovas S, Cudic M, Bulet P et al (2002) Identification of crucial residues for the antibacterial activity of the proline-rich peptide, pyrrhocoricin. Eur J Biochem 269(17):4226–4237

Kruse T, Kristensen HH (2008) Using antimicrobial host defense peptides as anti-infective and immunomodulatory agents. Expert Rev Anti-Infect Ther 6(6):887–895

Kumar P, Kizhakkedathu JN, Straus SK (2018) Antimicrobial peptides: diversity, mechanism of action and strategies to improve the activity and biocompatibility in vivo. Biomol Ther 8(1):4

Kwakman PH, Zaat SA (2012) Antibacterial components of honey. IUBMB life 64(1):48–55

Ladendorff NE, Kanost MR (1991) Bacteria-induced protein P4 (hemolin) from Manduca sexta: a member of the immunoglobulin superfamily which can inhibit hemocyte aggregation. Arch Insect Biochem Physiol 18(4):285–300. https://doi.org/10.1002/arch.940180410

Lakshmaiah Narayana J, Chen JY (2015) Antimicrobial peptides: possible anti-infective agents. Peptides 72:88–94

Lee G, Bae H (2016) Anti-inflammatory applications of melittin, a major component of bee venom: detailed mechanism of action and adverse effects. Molecules 21(5):616

Lee KS, Kim BY, Yoon HJ, Choi YS, Jin BR (2016) Secapin, a bee venom peptide, exhibits anti-fibrinolytic, anti-elastolytic, and anti-microbial activities. Dev Comp Immunol 63:27–35

Lee MT, Sun TL, Hung WC, Huang HW (2013) Process of inducing pores in membranes by melittin. Proc Natl Acad Sci 110(35):14243–14248

Lele DS, Talat S, Kumari S, Srivastava N, Kaur KJ (2015) Understanding the importance of glycosylated threonine and stereospecific action of Drosocin, a Proline rich antimicrobial peptide. Eur J Med Chem 92:637–647

Levashina EA, Ohresser S, Bulet P, Reichhart JM, Hetru C, Hoffmann JA (1995) Metchnikowin, a novel immune-inducible proline-rich peptide from Drosophila with antibacterial and antifungal properties. Eur J Biochem 233(2):694–700

Levashina EA, Ohresser S, Lemaitre B, Imler JL (1998) Two distinct pathways can control expression of the gene encoding the Drosophila antimicrobial peptide metchnikowin. J Mol Biol 278(3):515–527

Li WF, Ma GX, Zhou XX (2006) Apidaecin-type peptides: biodiversity, structure–function relationships and mode of action. Peptides 27(9):2350–2359
Liang X, Yan J, Lu Y, Liu S, Chai X (2021) The antimicrobial peptide melectin shows both antimicrobial and antitumor activity via membrane interference and DNA binding. In: Drug design, development and therapy, vol Volume 15, pp 1261–1273
Liu G, Kang D, Steiner H (2000) Trichoplusia ni lebocin, an inducible immune gene with a downstream insertion element. Biochem Biophys Res Commun 269(3):803–807
Liu X, Guo C, Huang Y, Zhang X, Chen Y (2015) Inhibition of porcine reproductive and respiratory syndrome virus by Cecropin D in vitro. Infect Genet Evol 34:7–16
Lo JCY, Lange D (2015) Current and potential applications of host-defense peptides and proteins in urology. BioMed Res Int 2015:189016
Lu X, Shen J, Jin X, Ma Y, Huang Y, Mei H et al (2012) Bactericidal activity of Musca domestica cecropin (Mdc) on multidrug-resistant clinical isolate of Escherichia coli. Appl Microbiol Biotechnol 95:939–945
Lü D, Geng T, Hou C, Qin G, Gao K, Guo X (2017) Expression profiling of Bombyx mori gloverin2 gene and its synergistic antifungal effect with cecropin A against Beauveria bassiana. Gene 600:55–63. https://doi.org/10.1016/j.gene.2016.11.011
Ma R, Mahadevappa R, Kwok HF (2017) Venom-based peptide therapy: insights into anti-cancer mechanism. Oncotarget 8(59):100908
Makarova O, Rodriguez-Rojas A, Eravci M, Weise C, Dobson A, Johnston P, Rolff J (2016) Antimicrobial defence and persistent infection in insects revisited. Philos Trans R Soc B Biol Sci 371(1695):20150296
Manniello MD, Moretta A, Salvia R, Scieuzo C, Lucchetti D, Vogel H, Sgambato A, Falabella P (2021) Insect antimicrobial peptides: potential weapons to counteract the antibiotic resistance. Cell Mol Life Sci 78(9):4259–4282. https://doi.org/10.1007/s00018-021-03784-z
Martens E, Demain AL (2017) The antibiotic resistance crisis, with a focus on the United States. J Antibiot 70(5):520–526
Masson F, Zaidman-Rémy A, Heddi A (2016) Antimicrobial peptides and cell processes tracking endosymbiont dynamics. Philos Trans R Soc B Biol Sci 371(1695):20150298
Matsumoto KI, Orikasa Y, Ichinohe K, Hashimoto S, Ooi T, Taguchi S (2010) Flow cytometric analysis of the contributing factors for antimicrobial activity enhancement of cell-penetrating type peptides: case study on engineered apidaecins. Biochem Biophys Res Commun 395(1):7–10
Matsumoto KI, Yamazaki K, Kawakami S, Miyoshi D, Ooi T, Hashimoto S, Taguchi S (2017) In vivo target exploration of apidaecin based on Acquired Resistance induced by Gene Overexpression (ARGO assay). Sci Rep 7(1):12136
Matysiak J, Schmelzer CE, Neubert RH, Kokot ZJ (2011) Characterization of honeybee venom by MALDI-TOF and nanoESI-QqTOF mass spectrometry. J Pharm Biomed Anal 54(2):273–278
McManus AM, Otvos L, Hoffmann R, Craik DJ (1999) Conformational studies by NMR of the antimicrobial peptide, drosocin, and its non-glycosylated derivative: effects of glycosylation on solution conformation. Biochemistry 38(2):705–714
McShane PJ, Weers JG, Tarara TE, Haynes A, Durbha P, Miller DP, Mundry T, Operschall E, Elborn JS (2018) Ciprofloxacin dry powder for inhalation (ciprofloxacin DPI): technical design and features of an efficient drug-device combination. Pulm Pharmacol Ther 50:72–79
Memariani H, Memariani M, Shahidi-Dadras M, Nasiri S, Akhavan MM, Moravvej H (2019) Melittin: from honeybees to superbugs. Appl Microbiol Biotechnol 103(8):3265–3276
Memariani H, Memariani M, Moravvej H, Shahidi-Dadras M (2020) Melittin: a venom-derived peptide with promising anti-viral properties. Eur J Clin Microbiol Infect Dis 39(1):5–17
Meng Y, Yang XX, Sheng YX, Zhang JL (2012) A novel peptide from Apis mellifera and solid-phase synthesis of its analogue. Chin Chem Lett 23(10):1161–1164
Miyoshi N, Saito T, Ohmura T, Kuroda K, Suita K, Ihara K, Isogai E (2016) Functional structure and antimicrobial activity of persulcatusin, an antimicrobial peptide from the hard tick Ixodes persulcatus. Parasit Vectors 9(1):1–11

Miyoshi N, Isogai E, Hiramatsu K, Sasaki T (2017) Activity of tick antimicrobial peptide from Ixodes persulcatus (persulcatusin) against cell membranes of drug-resistant Staphylococcus aureus. J Antibiot 70(2):142–146

Moghaddam MRB, Gross T, Becker A, Vilcinskas A, Rahnamaeian M (2017) The selective antifungal activity of Drosophila melanogaster metchnikowin reflects the species-dependent inhibition of succinate–coenzyme Q reductase. Sci Rep 7(1):8192

Moore AJ, Beazley WD, Bibby MC, Devine DA (1996) Antimicrobial activity of cecropins. J Antimicrob Chemother 37(6):1077–1089

Mourelle D, Brigatte P, Bringanti LDB, De Souza BM, Arcuri HA, Gomes PC et al (2014) Hyperalgesic and edematogenic effects of Secapin-2, a peptide isolated from Africanized honeybee (Apis mellifera) venom. Peptides 59:42–52

Mukherjee K, Mraheil MA, Silva S, Müller D, Cemic F, Hemberger J et al (2011) Anti-listeria activities of Galleria mellonella hemolymph proteins. Appl Environ Microbiol 77(12):4237–4240

Müller H, Salzig D, Czermak P (2015) Considerations for the process development of insect-derived antimicrobial peptide production. Biotechnol Prog 31(1):1–11

Mylonakis E, Podsiadlowski L, Muhammed M, Vilcinskas A (2016) Diversity, evolution and medical applications of insect antimicrobial peptides. Philos Trans R Soc Lond Ser B Biol Sci 371(1695):20150290. https://doi.org/10.1098/rstb.2015.0290

Neshani A, Sedighian H, Mirhosseini SA, Ghazvini K, Zare H, Jahangiri A (2020) Antimicrobial peptides as a promising treatment option against Acinetobacter baumannii infections. Microb Pathog 146:104238

Niederhafner P, Bednárová L, Buděšínský M, Šafařík M, Ehala S, Ježek J et al (2010) Melectin MAPs: the influence of dendrimerization on antimicrobial and hemolytic activity. Amino Acids 39:1553–1561

Nipate SS, Hurali PB, Ghaisas MM (2015) Evaluation of anti-inflammatory, anti-nociceptive, and anti-arthritic activities of Indian Apis dorsata bee venom in experimental animals: biochemical, histological, and radiological assessment. Immunopharmacol Immunotoxicol 37(2):171–184

Ongey EL, Pflugmacher S, Neubauer P (2018) Bioinspired designs, molecular premise and tools for evaluating the ecological importance of antimicrobial peptides. Pharmaceuticals 11(3):68

Orivel J, Redeker V, Le Caer JP, Krier F, Revol-Junelles AM, Longeon A et al (2001) Ponericins, new antibacterial and insecticidal peptides from the venom of the ant Pachycondyla goeldii. J Biol Chem 276(21):17823–17829

Oršolić N (2012) Bee venom in cancer therapy. Cancer Metastasis Rev 31:173–194

Otvos L Jr (2002) The short proline-rich antibacterial peptide family. Cell Mol Life Sci 59:1138–1150

Ouyang L, Xu X, Freed S, Gao Y, Yu J, Wang S et al (2015) Cecropins from Plutella xylostella and their interaction with Metarhizium anisopliae. PLoS One 10(11):e0142451

Palusińska-Szysz M, Zdybicka-Barabas A, Pawlikowska-Pawlęga B, Mak P, Cytryńska M (2012) Anti-Legionella dumoffii activity of Galleria mellonella defensin and apolipophorin III. Int J Mol Sci 13(12):17048–17064

Park SI, An HS, Chang BS, Yoe SM (2013) Expression, cDNA cloning, and characterization of the antibacterial peptide cecropin D from Agrius convolvuli. Anim Cells Syst 17(1):23–30

Patocka J, Nepovimova E, Klimova B, Wu Q, Kuca K (2019) Antimicrobial peptides: Amphibian host defense peptides. Curr Med Chem 26(32):5924–5946

Payne DJ, Gwynn MN, Holmes DJ, Pompliano DL (2007) Drugs for bad bugs: confronting the challenges of antibacterial discovery. Nat Rev Drug Discov 6(1):29–40

Picoli T, Peter CM, Zani JL, Waller SB, Lopes MG, Boesche KN et al (2017) Melittin and its potential in the destruction and inhibition of the biofilm formation by Staphylococcus aureus, Escherichia coli and Pseudomonas aeruginosa isolated from bovine milk. Microb Pathog 112:57–62

Pino-Angeles A, Lazaridis T (2018) Effects of peptide charge, orientation, and concentration on melittin transmembrane pores. Biophys J 114(12):2865–2874

Pöppel AK, Koch A, Kogel KH, Vogel H, Kollewe C, Wiesner J, Vilcinskas A (2014) Lucimycin, an antifungal peptide from the therapeutic maggot of the common green bottle fly Lucilia sericata. Biol Chem 395(6):649–656

Prenner EJ, Lewis RN, McElhaney RN (1999) The interaction of the antimicrobial peptide gramicidin S with lipid bilayer model and biological membranes. Biochim Biophys Acta 1462(1–2):201–221

Pucca MB, Cerni FA, Oliveira IS, Jenkins TP, Argemí L, Sørensen CV et al (2019) Bee updated: current knowledge on bee venom and bee envenoming therapy. Front Immunol 10:2090

Raghuraman H, Chattopadhyay A (2007) Melittin: a membrane-active peptide with diverse functions. Biosci Rep 27(4–5):189–223

Rahnamaeian M, Vilcinskas A (2015) Short antimicrobial peptides as cosmetic ingredients to deter dermatological pathogens. Appl Microbiol Biotechnol 99:8847–8855

Ram SKM, Jayapal N, Nanaiah P, Aswal GS, Ramnarayan BK, Taher SM (2014) The therapeutic benefits of bee venom. Int J Curr Microbiol App Sci 3:377–381

Ratzka C, Förster F, Liang C, Kupper M, Dandekar T, Feldhaar H, Gross R (2012) Molecular characterization of antimicrobial peptide genes of the carpenter ant Camponotus floridanus, vol 7, p e43036

Reichhart JM, Meister M, Dimarcq JL, Zachary D, Hoffmann D, Ruiz C et al (1992) Insect immunity: developmental and inducible activity of the Drosophila diptericin promoter. EMBO J 11(4):1469–1477

Rodríguez-Rojas A, Makarova O, Rolff J (2014) Antimicrobials, stress and mutagenesis. PLoS Pathog 10(10):e1004445

Rosengren KJ, Göransson U, Otvos L Jr, Craik DJ (2004) Cyclization of pyrrhocoricin retains structural elements crucial for the antimicrobial activity of the native peptide. Peptide Sci Origin Res Biomol 76(5):446–458

Rudenko N, Golovchenko M, Grubhoffer L (2007) Gene organization of a novel defensin of Ixodes ricinus: first annotation of an intron/exon structure in a hard tick defensin gene and first evidence of the occurrence of two isoforms of one member of the arthropod defensin family. Insect Mol Biol 16(4):501–507

Saito Y, Konnai S, Yamada S, Imamura S, Nishikado H, Ito T et al (2009) Identification and characterization of antimicrobial peptide, defensin, in the taiga tick, Ixodes persulcatus. Insect Mol Biol 18(4):531–539

Shen W, He P, Xiao C, Chen X (2018) From Antimicrobial Peptides to Antimicrobial Poly (α-amino acid) s. Adv Healthc Mater 7(20):1800354

Shipman WH, Cole LJ (1969) A surfactant bee venom fraction: separation on a newly devised constant flow rate chromatographic column and detection by changes in effluent drop volume. Anal Biochem 29(3):490–497

Shkenderov S, Koburova K (1982) Adolapin-a newly isolated analgetic and anti-inflammatory polypeptide from bee venom. Toxicon 20(1):317–321

Son DJ, Lee JW, Lee YH, Song HS, Lee CK, Hong JT (2007) Therapeutic application of anti-arthritis, pain-releasing, and anti-cancer effects of bee venom and its constituent compounds. Pharmacol Ther 115(2):246–270

Steinstraesser L, Koehler T, Jacobsen F, Daigeler A, Goertz O, Langer S et al (2008) Host defense peptides in wound healing. Mol Med 14(7):528–537

Świątły-Błaszkiewicz A, Mrówczyńska L, Matuszewska E, Lubawy J, Urbański A, Kokot ZJ et al (2020) The effect of bee venom peptides melittin, tertiapin, and apamin on the human erythrocytes ghosts: a preliminary study. Meta 10(5):191

Téllez GA, Castaño-Osorio JC (2014) Expression and purification of an active cecropin-like recombinant protein against multidrug resistance Escherichia coli. Protein Expr Purif 100:48–53

Terwilliger TC (1982) The structure of melitting. II. Interpretation of the structure. J Biol Chem 257:6016–6022

Tommasi R, Brown DG, Walkup GK, Manchester JI, Miller AA (2015) ESKAPEing the labyrinth of antibacterial discovery. Nat Rev Drug Discov 14(8):529–542

Tsuji N, Battsetseg B, Boldbaatar D, Miyoshi T, Xuan X, Oliver JH Jr, Fujisaki K (2007) Babesial vector tick defensin against Babesia sp. parasites. Infect Immun 75(7):3633–3640

Ullah A, Aldakheel FM, Anjum SI, Raza G, Khan SA, Gajger IT (2023) Pharmacological properties and therapeutic potential of honey bee venom. Saudi Pharm J 31(1):96–109

Ursic-Bedoya R, Buchhop J, Joy JB, Durvasula R, Lowenberger C (2011) Prolixicin: a novel antimicrobial peptide isolated from Rhodnius prolixus with differential activity against bacteria and Trypanosoma cruzi. Insect Mol Biol 20(6):775–786

Valanne S, Wang JH, Rämet M (2011) The Drosophila toll signaling pathway. J Immunol 186(2):649–656

Vale N, Aguiar L, Gomes P (2014) Antimicrobial peptides: a new class of antimalarial drugs? Front Pharmacol 5:275

Van Hofsten P, Faye I, Kockum K, Lee JY, Xanthopoulos KG, Boman IA et al (1985) Molecular cloning, cDNA sequencing, and chemical synthesis of cecropin B from Hyalophora cecropia. Proc Natl Acad Sci 82(8):2240–2243

Varga JJ, Barbier M, Mulet X, Bielecki P, Bartell JA, Owings JP et al (2015) Genotypic and phenotypic analyses of a Pseudomonas aeruginosa chronic bronchiectasis isolate reveal differences from cystic fibrosis and laboratory strains. BMC Genomics 16:1–27

Vick JA, Shipman WH, Brooks R Jr (1974) Beta adrenergic and anti-arrhythmic effects of cardiopep, a newly isolated substance from whole bee venom. Toxicon 12(2):139–142

Vlieghe P, Lisowski V, Martinez J, Khrestchatisky M (2010) Synthetic therapeutic peptides: science and market. Drug Discov Today 15(1–2):40–56

Vonkavaara M, Pavel STI, Hölzl K, Nordfelth R, Sjöstedt A, Stöven S (2013) Francisella is sensitive to insect antimicrobial peptides. J Innate Immun 5(1):50–59

Wang Z, Han X, He N, Chen Z, Brooks CL III (2014) Molecular structures of C-and N-terminus cysteine modified cecropin P1 chemically immobilized onto maleimide-terminated self-assembled monolayers investigated by molecular dynamics simulation. J Phys Chem B 118(21):5670–5680

Wang Q, Guo P, Wang Z, Liu H, Zhang Y, Jiang S, Han W, Xia Q, Zhao P (2018) Antibacterial mechanism of Gloverin2 from silkworm, Bombyx mori. Int J Mol Sci 19(8):2275. https://doi.org/10.3390/ijms19082275

Wehbe R, Frangieh J, Rima M, El Obeid D, Sabatier JM, Fajloun Z (2019) Bee venom: overview of main compounds and bioactivities for therapeutic interests. Molecules 24(16):2997

WHO (2017) Global priority list of antibiotic-resistant bacteria to guide research, discovery, and development of new antibiotics

Wiesner J, Vilcinskas A (2010) Antimicrobial peptides: the ancient arm of the human immune system. Virulence 1(5):440–464

Wilmes M, Sahl HG (2014) Defensin-based anti-infective strategies. Int J Med Microbiol 304(1):93–99

Wu Q, Patočka J, Kuča K (2018) Insect antimicrobial peptides: a mini review. Toxins 10(11):461

Xu P, Shi M, Chen XX (2009) Antimicrobial peptide evolution in the Asiatic honey bee Apis cerana. PLoS One 4(1):e4239

Yi HY, Chowdhury M, Huang YD, Yu XQ (2014) Insect antimicrobial peptides and their applications. Appl Microbiol Biotechnol 98:5807–5822

Yi HY, Deng XJ, Yang WY, Zhou CZ, Cao Y, Yu XQ (2013) Gloverins of the silkworm Bombyx mori: structural and binding properties and activities. Insect Biochem Mol Biol 43(7):612–625

Yoshiyama M, Kimura K (2010) Characterization of antimicrobial peptide genes from Japanese honeybee Apis cerana japonica (Hymenoptera: Apidae). Appl Entomol Zool 45(4):609–614

Zakharchenko NS, Belous AS, Biryukova YK, Medvedeva OA, Belyakova AV, Masgutova GA et al (2017) Immunomodulating and Revascularizing activity of Kalanchoe pinnata synergize with fungicide activity of biogenic peptide cecropin P1. J Immunol Res 2017:1

Zarrinnahad H, Mahmoodzadeh A, Hamidi MP, Mahdavi M, Moradi A, Bagheri KP, Shahbazzadeh D (2018) Apoptotic effect of melittin purified from Iranian honey bee venom on human cervical cancer HeLa cell line. Int J Pept Res Ther 24:563–570

Zhang S, Liu Y, Ye Y, Wang XR, Lin LT, Xiao LY et al (2018) Bee venom therapy: potential mechanisms and therapeutic applications. Toxicon 148:64–73

Chapter 6
Herbal Antimicrobial Peptides: A Medicinal and Nutraceutical Approach

Varisha Anjum, Pritya Jha, Areefa Anjum, Vishal Dixit, Uday Bagale, Irina Potoroko, and Sweta Joshi

1 Introduction

Antibiotic resistance as newer resistance mechanism has emerged and spread across the universe concerning healthcare professionals. Various clinically used antibiotics have been associated with resistant bacterial species (Dijksteel et al. 2021). Drug discovery of new effectual antimicrobial agents has become critical for hostility of antibiotic resistance that is modeled as the greatest threat to global conditions (Afroz et al. 2020). For instance, immediately following the 1940 discovery of penicillin-resistant *Staphylococcus aureus* (Dijksteel et al. 2021), in addition to penicillin, fewer pathogen bacteria also developed resistance to cephalosporins, novel carbapenems, and semisynthetic penicillin (Kumarasamy et al. 2010). Additionally, the issue made worse by regulatory authorities is the deterioration of the approval of new antibiotics. Antimicrobial peptides (AMPs) as a substitute to antibiotics have led to transnational efforts to develop resistance bacteria as they have less risk (Wimley and Hristova 2011). These peptides are a varied set of naturally developing

V. Anjum (✉) · U. Bagale · I. Potoroko
Department of Food and Biotechnology, South Ural State University, Chelyabinsk, Russia

P. Jha
Department of Pharmacy, Banasthali Vidyapeeth, Banasthali, Rajasthan, India

A. Anjum
Department of Ilmul Advia, School of Unani Medical Education and Research, New Delhi, India

V. Dixit
School of Pharmaceutical Sciences, Lingaya's Vidyapeeth, Nachouli, Faridabad, Haryana, India

S. Joshi
Department of Food Technology, School of Interdisciplinary Science and Technology, Jamia Hamdard, New Delhi, India

P. Baindara, S. M. Mandal (eds.), *Evolution of Antimicrobial Peptides*,
https://doi.org/10.1007/978-3-031-67515-7_6

peptides with an intrinsic defense mechanism that protects against a variety of plants, animals, microbes, and pathogens, including bacteria, viruses, yeast, and fungi (Beisswenger and Bals 2005). AMPs exhibit a nonspecific, multi-target, and quick action to deactivate these pathogens, causing the sluggish or restricted advent of opposition (Wimley and Hristova 2011). Moreover, few AMPs interact synergistically with antibiotics naturally, which might prevent the emergence of bacteria resistant to antibiotics (Wu et al. 2020).

Plants produce a large amount of secondary active metabolites, such as tannins, terpenoids, quinines, phenols, and AMPs, through metabolic processes derived from the basic metabolic pathway, which enable the creation of an effective defense mechanism (Isah 2019). The produced secondary metabolites act as a defense against bacteria, insects, and fungi. Secondary metabolites contain UV-absorbing molecules that prevent intense damage to the leaf from sunlight and have a variety of actions (antifungal, antibiotic, and antiviral) against pathogens (Sharma et al. 2021). However, antimicrobial peptides (AMPs) have become more popular, and results show a lesser tendency for microorganisms to develop resistance (Crabbé et al. 2019).

Antimicrobial peptides (AMPs) are cationic peptides that can interact with DNA and RNA; block protein transport, enzymes, or ion channels; and act as regulators of steroid hormones (Kenny and Marmion 2019). The involvement of residues like cysteine, which causes the formation of multiple bridges of disulfide, led to the evolution of herbal AMPs (HAPs) differently as compared to other AMPs. The various characteristics, namely, molecular weight between 2 and 6 kDa with 2 or 6 intramolecular disulfide bonds, amphipathicity, and positive charge, of HAPs compared to other peptides obtained from insects, microbes, and animals have shown a relation to defensive mechanisms (Wang et al. 2020). Peptides rich in cysteine cross-braced using disulfide bridges of cysteine-rich HAPs, which are isolated from leaves, flowers, roots, seeds, and stems of various plant species, are components of barrier defense mechanisms (Magana et al. 2020). The changes in the family were classified according to the presence or absence of cysteine, conserved secondary and tertiary structure or sequence similarity, compact structure, isolation with bioprocess precursors from ribosomes, and N- and C-terminal prodomain (Okafor and Okeke 2017).

AMPs typically have broad-spectrum antimicrobial action and the power to eradicate germs that are resistant to a variety of drugs (Wang et al. 2016). The cellular and molecular mechanisms underpinning the AMP activity entail causing alterations in cytoplasmic targets, like cell molecular synthesis, or on membrane-associated targets (Haney et al. 2019). Additionally, they engage in immunomodulatory functions, including enlisting and activating cells of the innate and adaptive immune systems; other AMPs help in protein biosynthesis and inhibit hydrolase (Finkina et al. 2019). The antiproliferative properties of herbal AMPs are due to the presence of chemical components (Parisi et al. 2019), which are used for the discovery of biological plant protection products and new drugs (De Souza Cândido et al. 2014). With different research, it was found that AMPs present in every plant with a specific taxonomy like genus and family have several diversities of molecular peptides that belong to different families structurally. These can be obtained underground (Shen et al. 2019) or above ground, and they can also be obtained from generative (Astafieva et al. 2015) or vegetative sources. Finkina et al. (2018) showed that a large number and wide variety of AMPs were isolated from seeds. Hence, fruits and seeds as a source have shown greater interest in diversity AMPs.

2 Biological Basis of Herbal Antimicrobial Peptides (HAPs)

During their life cycles, plants in natural ecosystems frequently coexist with a wide range of microbes and pests. Over the course of their evolution, plants have developed sophisticated Defense mechanisms to shield them from the harm caused by diseases and pests. As a result, they are now able to repel dangerous nematodes, bacteria, insects, and fungi (Iqbal et al. 2019). These protective mechanisms against the pathogens include both chemical and physical barriers to prevent the spread and growth of the pathogen, such as an intricate network of phytohormones and sophisticated cell recognition systems, countless transcription pathways, varied proteins, and secondary active metabolites (Campos et al. 2018). AMPs are the most prevalent and well-known chemical defenses that have helped plants evolve and withstand biotic stressors (Campos et al. 2018; Kulaeva et al. 2020).

The physiologically active compounds are generated by a wide range of species as a crucial component of innate immunity. The main function of AMPs is to act as cytotoxic host defense against invasive pathogen bacteria, and in higher species like mammals and plants, they also function as immune modulators (Zanetti 2004). Due to its wide spectrum of activity, lower toxicity, and less development of resistance by the target cells, AMP is regarded as a favorable and prospective medication candidate for the future. The AMPs were found to have a variety of helices, looping strands, extended structures, and additional disulfide bridges in their secondary structures. AMPS must exist in such a wide range of structural configurations for them to have such a broad range (Hancock 2001).

AMPs are tiny-molecule polypeptides that are typically created by ribosomes. To produce mature polypeptides, larger protein precursors must first be broken down, and then additional post-translational modifications must be made. Additionally, non-ribosomal peptide synthetases can be used to create various AMPs (Mukherjee et al. 2011). Numerous AMPs have broad-spectrum antibiotic action against parasites, fungi, enveloped viruses, and pathogenic bacteria of gram-negative and gram-positive origin (Havenga et al. 2019). In humans and other higher creatures, AMPs play a crucial role in innate immunity in response to infections with pathogenic bacteria. They also serve as many species' initial line of defense against infections with several dangerous bacteria (Tang et al. 2018; Zahedifard et al. 2020).

Every aspect of existence has been mined for AMPs. It has been discovered that microbes, plants, insects, amphibians, and mammals all contain more than 3000 AMPs. Plants are a promising source of AMPs, which are peptides with strong antibacterial activity against diseases that affect both humans and plants. Most experts agree that AMPs can be created in the flowers, leaves, stems, roots, and seeds of plants (Tam et al. 2015). Research on HAPs has advanced quickly since the initial publication of the thionin peptide from *Triticum aestivum* (Hughes et al. 2000). HAPs are thought to have a considerable impact on plant growth and development in addition to helping plants defend themselves against pathogens (de Zélicourt et al. 2007).

All classes of life contain natural antimicrobial peptides (NAMPs), which belong to different peptide or protein families and are crucial for both the defense barrier and innate immune responses (Diamond et al. 2009). AMPs have shown potential as sources of novel antibiotics and kill viruses, bacteria, fungi, and malignant cells, serving as broad-spectrum antibiotics (Ageitos et al. 2017). Over the past 20 years, there has been an increase in interest in AMP research due to its potential activity and prospects, and as a result, the number of known NAMPs is rising every day (Mahlapuu et al. 2016). NAMPs have more distinctive characteristics than traditional antibiotics. They offer a promising opportunity for the creation of new antibacterial lead compounds.

NAMPs' extraordinary qualities in both structure and function might enable them to take the place of traditional antibiotics. Additionally, NAMPs have been discovered to be stable under a variety of circumstances and to have little to no toxicity in humans, offering them a promising opportunity for the creation of novel antibacterial leads (Bondaryk et al. 2017). NAMPs mostly originate from natural sources, which include plants, bacteria, fungi, archaea, protists, and mammals. These natural resources may be used in ethnomedical treatments for infections, providing crucial hints for the development of novel NAMPs. According to recent studies, more than 1500 NAMPs have been found in nature, and some of these NAMPs, such as kalata B1 and B2, pexiganan, novexatin, thionins, omiganan, and thionein, etc., are now undergoing clinical or preclinical trials (Gründemann et al. 2019). Additionally, certain HAPs continue to be helpful in the treatment of several deadly human illnesses despite the pathogenic bacteria's rising resistance to standard antibiotics. As a result, the use of AMPs in the treatment of illnesses is garnering more and more attention in the medical community.

3 Herbal Sources of Antimicrobial Peptides

Plants serve as a source of a variety of biologically active chemicals. Some of them have been use in both agriculture and medicine (Dhama et al. 2018). All categories of chemicals with a plant origin are of interest, but antimicrobial peptides (AMPs) are of special focus. The tiny molecules that makeup AMPs range in molecular weight from 2 to 10 kDa, have amphiphilic characteristics, and are typically positively charged at physiological and neutral pH levels (Tang et al. 2018). Meanwhile, the main and secondary architectures of AMPs differ significantly. However, it is crucial to keep in mind that most HAPs have a limited spatial configuration that is made possible by intramolecular disulfide bonds (Campos et al. 2018). This also does so in addition to offering temperature, enzyme, and chemical agent stability (Campos et al. 2018). Since the beginning of time, these plant secondary metabolites have been a valuable source of bioactive ingredients utilized in conventional medications to treat infectious illnesses. Reports state that plants are a major source of different AMPs. Previous research has identified several AMPs from different plant sections, using details about their traditional or ethnomedical uses to treat disease. This was how plants and people interacted when they were used as infection-fighting ethnomedicine. Ethnobotany is essential for the discovery and development of new drugs, as evidenced by the identification of the phytoconstituent artemisinin from the Artemisia species (Kayser (2018). A natural chemical library with a variety of structures can also be found through ethnobotanical study.

According to their charge, HAPs may be divided into anionic and cationic charged peptides. However, according to Hammami et al. (2009), the categorizations of HAPs are frequently based on the similarity of sequence, presence of tertiary chemical structure, and motifs of cysteine. According to Wang et al. (2008), the class of NAMPs known as cathelicidins has been found in a variety of animal taxa, including mammals, birds, fish, reptiles, and amphibians. Numerous cathelicidin-related antimicrobial peptides (CRAMPs) with strong antibacterial activity have been found in various snake venoms. Cathelicidins found in snake venom were the subject of a review by Barros et al. (2019) of both recently and previously published research. The chemistry, characterization, and pharmacological activity—particularly antibacterial and anti-biofilm efficacy—furthermore to the mode of action of cathelicidins were all discussed by the authors. They also emphasized the potential of cathelicidins in the creation of fresh antibiotics to combat germs that are resistant to them. Thionins are a group of amphipathic proteins from plants with small molecular weights that were first identified in the proportion of barley seeds and wheat that could be extracted from organic solvents (Hughes et al. 2000). They have 3.0–4.0 disulfide bond linkages, 6.0–8.0 cysteines, and 45–48 amino acid residues (5 kDa) in them (Stec 2006). Numerous species of plants have shown presence of thionins (Hughes et al. 2000). They may be characterized as cyclic peptides because they have an S–S bond between the C- and N-termini, which causes them to form a ring structural topology (Tam et al. 2015).

4 Structure–Activity Relationship (SAR) of Herbal Antimicrobial Peptides (HAPs)

HAPs are small-molecular polypeptides that are produced by ribosomes from mature protein precursors, further maturing by undergoing additional post-translational modifications (Li and Rebuffat 2020). The structural makeup of HAPs varied between species but shared similar characteristics, including shorter lengths, positive charges, and hydrophilic and hydrophobic regions. HAPs have multiple roles, structures, specific targets, and expression patterns, making them more challenging and complex (Santos-Silva et al. 2020). HAPs have unique SAR and are classified into different families, including thionins, hevein-like peptides, snakins, defensins, lipid transfer proteins, α- hairpinins, knottin-type peptides, and cyclotides (Li et al. 2021). HAPs recognize different targets by changing the non-cysteine residue sequence because of their ability to tolerate hypervariable sequences with a conserved framework. The disulfide bonds present in HAPs increase resistance to thermal, enzymatic, and chemical degradation and shield the quaternary and tertiary chemical structures (Maan et al. 2020). HAPs also withstand abiotic stressors due to the presence of chemical barrier systems in plants (Sharma and Gupta 2020). HAPs have a broad range of antioxidation, antiviral, antibacterial, antifungal, proteinase inhibitory, and chitinase activities to produce a waxy layer of cuticles and trichomes as physical barriers to prevent the penetration and spread of pathogens or chemical barriers to stop the growth of pathogens using cell complexes, transcriptional pathways of phytohormone networks, and secondary metabolites (Aweya et al. 2021). The antimicrobial potential and cell selectivity of HAPs are demonstrated by the presence of net charge, amino acid residue, amphipathicity, hydrophobicity, and structural propensity (Wang et al. 2019).

5 Extraction and Isolation of HAPs

In 1970, Okada and Yoshizumi extracted and isolated the first-ever AMP from a plant, from the endosperm of barley (*Hordeum vulgare*) (Okada and Yoshizumi 1970). Homogenization, extraction, concentration, and purification of the plant extract are the three primary phases of HAP's isolation. The resulting extract is typically fractionated using several liquid chromatographic techniques (Fig. 6.1). The whole plant—tubers, seeds, roots, bulbs, leaves, flowers, and fruits—was used for the extraction of herbal AMPs. During homogenization, some plant parts withstand mechanical strain. The parts to be used for extraction are dried with liquid nitrogen or lyophilized before peeling. The disintegration technique is one of the processes used due to the physical properties of plant samples.

Fruits are powdered in extraction buffer using a blender, plant frozen parts are crumbled with liquid nitrogen, and seeds are ground in a coffee mill. Few plants or plant parts accrue substantial low-molecular-weight and high-molecular-weight metabolite content in the root containing inulin, fatty oil, or stored proteins in seeds and tannins in stems and leaves. Whenever needed, supplementary steps like defatting (Egorov et al. 2005), removing excess protein (Evaristus et al. 2018), or prefractionating (Chan and Ng 2013), followed by heating (dos Santos et al. 2017), may sometimes be used in the extraction process. For extraction buffers, diluted acids, salts, and buffer solutions were used in an aqueous medium. The first herbal peptide of microbial origin was purothionins, which were extracted with a sulfuric acid solution (Okada and Yoshizumi 1970). In previous research, AMP extraction from *Bufo bufo* gargarizans, also known as Asian toad stomachs, reached histidine

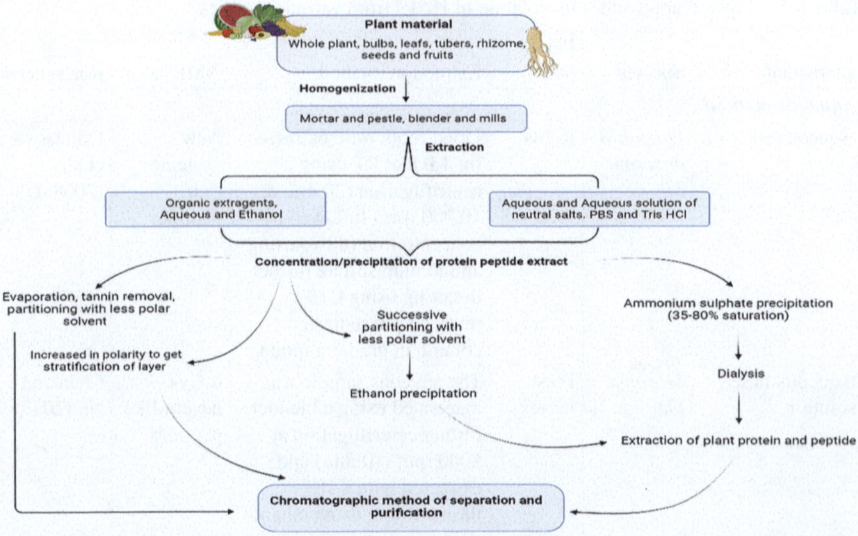

Fig. 6.1 Multiple approaches for extraction and isolation of HAPs

and glycine-rich peptides from plant origin. Presently, there are two ways of extracting AMP: the primary technique for extraction was the use of water and aqueous solutions containing salts, buffers, and acid solutions, whereas another method uses organic-based solutions, including hydroalcoholic solutions. The best-targeted approaches to extracting AMP are shown in Table 6.1.

6 Targeted Approach for Herbal Antimicrobial Isolation

The Research Lab of the Russian Academy of Sciences proposed an optimized pattern as shown in Fig. 6.2. The proposed extractant was acetic acid (10%). The following approaches were used for the extraction of HAPs belonging to various families: thionins (Vasilchenko et al. 2016), defensins (Odintsova et al. 2008b), α-hairpinins (Nolde et al. 2011), and hevein-like peptides (Odintsova et al. 2009). The peptides are partially characterized as not belonging to core families and were extracted and isolated by following steps (Astafieva et al. 2013).

7 Plant Preparation and Extraction

As per the prescribed method, a blender and coffee mill are used for powdering plant material. The acetic acid (10%) was used as an extractant along with the addition of combined proteinase inhibitors. These were added in the ratio of 1/10 w:v

Table 6.1 Targeted approaches in isolation of HAPs from various families

Extractant	Species	Plant part	Extraction method	AMP	References
Aqueous method					
Aqueous solution	*Potentilla anserina*	Roots	Dried roots were extracted for 1.0 h at RT using centrifugation (20 min & 10,000 rpm) followed by concentration (80%) using ammonium sulfate further desalting using C18 reversed-phase flash column in gradient mode.	New cysteine-rich peptides	Odintsova et al. (2008a)
Aqueous-based solution	*Moringa oleifera*	Fresh leaves	The aqueous sample was macerated using a blender further centrifugation at 8000 rpm (10 min) and extracted with C18 flash-column using ethanol as eluent.	8-Cys hevein-like peptides	Chan and Ng (2013)
Acid solution					
Sulfuric acid solution (50 mM)	*Nicotiana alata*	Flowers	Powdered flowers using liquid nitrogen extracted with sulfuric acid for 1 h further filtered solution was centrifuged and pH adjusted to 7.8 followed by concentrating using ammonium sulfate. Dissolved precipitate separated by gel-filtration method using Sephadex G-50.	Defensin	Egorov et al. (2005)
TFA (1% v/v), HCl (1 M) formic acid, and NaCl (5% v/v and 1% w/v, respectively)	*Capsella bursa-pastoris*	Roots	Sample macerated with 1:4 w:v of extraction mixture and centrifuged, separated by means of reversed-phase C18 of cartridge Sep-Pack followed by washing in 80% acetonitrile with 0.1% TFA.	Gly/His-rich peptides	García-Olmedo et al. (1998)

(continued)

Table 6.1 (continued)

Extractant	Species	Plant part	Extraction method	AMP	References
Buffer solution					
Tris–HCl (0.1 M; pH 7.2)	*Phaseolus limensis*	Seeds	Seeds were priorly homogenated by buffer followed by centrifugation (20 min; 12,000 rpm) then further fractionated using two-step ammonium sulfate precipitation. The saturated solution to 20% and final to 85% saturation and dissolving all in 0.01 M Tris–HCl buffer (pH 7.2; 100 mL), dialyzing against the buffer and subjected to separation further.	Peptide defensin	Evaristus et al. (2018)
Na2HPO4 (10 mM), NaH2PO4 (15 mM), KCl (100 mM), EDTA (5%), pH 5.4	*Capsicum anuum*	Seedless fruits	The fruits were extracted for 2 h and saturated with ammonium sulfate, redissolving in distilled water and centrifuged, freeze-dried, and further fractionated by the chromatographic method.	Thionin-like peptides	Kul'Ko et al. (2016)
Organic solutions					
EtOH/MeOH 20%/water (50%)	*Viola odorata*	Aerial parts	The dried part was finely powdered and extracted for 6 h using a mixture of 1:20 w:v ratio.	Cyclotydes	Barber et al. (1988)
Hydroalcoholic solution (50%)	*Viscum album*	White–green parts	The crushed plant sample was extracted with the hydroalcoholic solution and concentrated and partitioned using cyclohexane, dichloromethane, and ethyl acetate. Further aqueous phase was concentrated with ethanol and centrifuged at 2000 g further supernatant was pooled.	Thionins	Limas et al. (1990)

and extracted further by constant rigorous stirring, followed by centrifugation at 4700 rpm (10 min). When the large amounts are taken into concentration, the resulting supernatant is 1.5–2.0 times concentrated using a rotary evaporator.

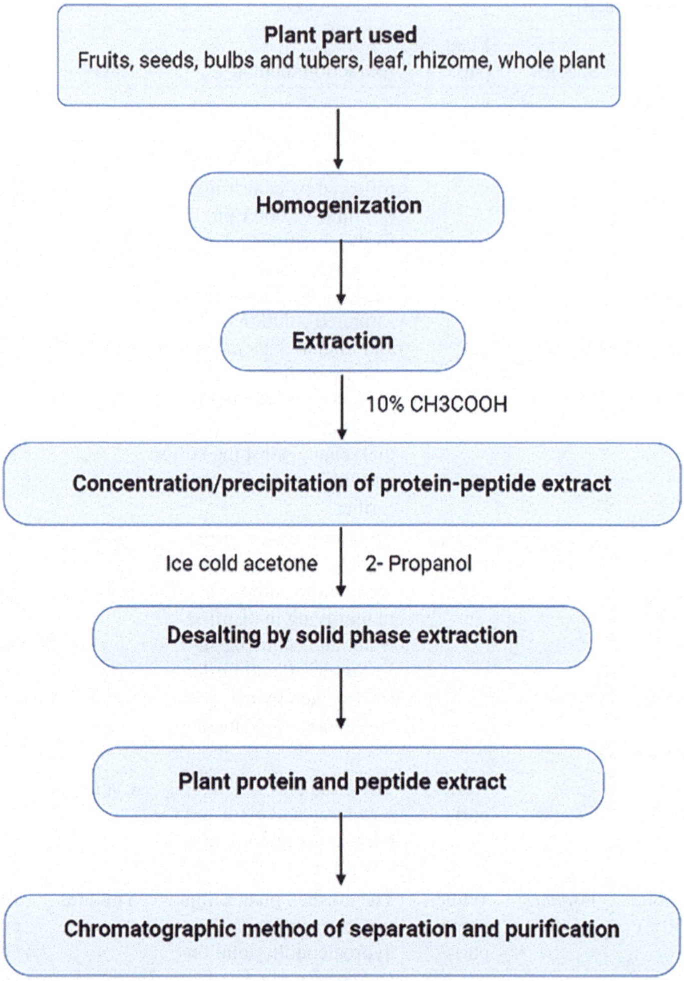

Fig. 6.2 Optimized targeted approach for isolation of HAPs

7.1 Precipitation

The saturation method was used for organic solvent precipitation. The filtrate is mixed gently with a 1:7 ratio of cold acetone (−70 °C) previously stored at +4 °C (6–8 h). The suspended part was centrifuged (10 min, 4700 rpm) for collecting separated precipitation fraction, if necessary, and then further removed. The obtained fraction dried at ±25 °C. The dried extract was resuspended and purified by performing solid-phase extraction in the C18 column using HPLC. A vacuum centrifuge is used to evaporate desalted extract having protein peptide and then lyophilized. The resulting lyophilisate can be further fractionated and employed in antimicrobial activity tests. It is significant to note that the methodology that is being presented allows dialysis to be avoided because it forgoes the use of sulfated ammonia in the salting-out process in favor of precipitation using an organic solvent. Thus, extraction time was significantly reduced by preventing substance loss.

7.2 Fractionation

Factors such as the kind of biomaterial being utilized determine which steps are involved in the fractionation process. The samples are fractionated in the first phase of affinity chromatography using a solid phase of heparin-sepharose sorbent and gradually increasing NaCl concentrations. As a result, the mixture's components can be distinguished by negative and high cation charges. It involves desalting in the solid state. Further size exclusion chromatography is used as the next step; sometimes this step is removed when enough proteins in the reserved form are not present in the plant; else, this step becomes important when plant material contains enough quantity of total protein in seeds. The end step involves reverse-phase chromatography after the fractionation of the mixture.

8 Applications of Herbal Antimicrobial Peptides (HAPs)

8.1 Application of Herbal Antimicrobial Peptides in Medicine

The HAPs molecular mechanisms in many diseases, their significance to the pharmaceutical industry, the difficulties in deploying HAPs as marketable treatments, and HAPs in clinical trials are some of the major obstacles to the validation of HAPs in trials (Divyashree et al. 2020). HAPs can operate as immune modulators in addition to their potential therapeutic uses as antifungal, anticancer, antimitogenic, antiviral, and anti-inflammatory agents (Table 6.2). Without a doubt, HAPs have the potential to displace traditional medications to increase their proportion of the worldwide drug market.

Due to their low toxicity, excellent selectivity, and strong binding to their targets due to their vast chemical space and native amino acid side chain polymorphisms, the existing databases currently contain 7700 antimicrobial peptides and HAPs inclusive (Rempel et al. 2020). The speedy growth of resistant genes in nature is concerning since it results in significant loss of lives worldwide and has unavoidable consequences that could devastate the economies of many nations (Mwangi et al. 2019). Plants are typically attacked by microbes throughout their life cycle because of their sessile properties. Protective agents like HAPs, which successfully manage functional and bacterial infection, earned the tag of future-generation antibiotics in response to this endless barrage of biotic factors (Xia et al. 2018). The study

Table 6.2 Some examples of herbal antimicrobial peptides used in medicine

S. No	HAPs	Source	No. of AA residues (S–S bond)	Mechanism	References
1.	Snakin-1, Snakin-2, Snakin 1/StSN1	*Solanum tuberosum*	88 (6), 104 (6), 88 (6)	Antimicrobial potential in both types of bacteria, i.e., gram −ve and gram +ve rapidly aggregating; Biochemical stability in abiotic and biotic stressors, plant hormone supporting cross-talk focusing on SA and ABA, respectively. Anti-yeast, antibacterial, and antifungal action	Deng et al. (2021), Shwaiki et al. (2020), Almasia et al. (2017)
2.	Trypsin inhibitor-2, Trypsin inhibitor-1	*Momordica cochichinensis*	29 (3), 38 (3)	Enhance biological activity by reducing flexibility, improving stability and vulnerability to proteolytic assault	Le et al. (2019), Swedberg et al. (2016)
3.	1/AMP1_ COCNU (P86705); 2/ AMP 2 and 2/ AMP 3 (P86706)	*Cocos nucifera*	9 (0), 11 (0), 8 (0)	Antibacterial effect	Anaya et al. (2020), Raghubeer et al. (2020)
4.	Anticancer peptide 1/ Cr-ACP1	*Cycas revoluta*	9 (0)	Apoptosis and proliferation of cells by inhibition in cancer cells	Chauhan et al. (2021)
5.	Pharbitis nil/ Pn-AMP1, Pn-AMP2 (P81591)	*Ipomoea nil*	41 (5), 41 (5)	Binds to chitin and shows antifungal properties against chitin-containing fungal infections. prevents the growth of gram-positive bacteria like *B. subtilis*; shows antimicrobial, antibiotic, and fungicide function	Das et al. (2019), Rogozhin et al. (2018)
6.	1/Fa-AMP1; 1/ Fa-AMP2 (P0DKH7)	*Fagopyrum esculentum*	40 (5)	Positive action against gram −ve and gram +ve bacteria and plant-harmful fungus	Majid and Priyadarshini (2020)
7.	1/Ac-AMP1 (Q9S8Z6)	*Amaranthus caudatus*	29 (3)	Binding by Chitin	Phazang et al. (2020)

(continued)

Table 6.2 (continued)

S. No	HAPs	Source	No. of AA residues (S–S bond)	Mechanism	References
8.	Ar-AMP (Q5I2B2)	*Amaranthus retroflexus*	89 (3)	Prevent the development of fungal infections and alter their morphology	Bakare et al. (2022)
9.	1.1a/AMP-1.1a (E1UYT9), 3/ SmAMP3 (C0HJU5)	*Stellaria media*	167 (7), 35 (3)	Antifungal	Caceres-Delpiano et al. (2020)
10.	Cowpea-thionin/ Cp- thionin II	*Vigna unguiculata*	46 (4)	Antibacterial and antifungal action	Schmidt et al. (2019)
11.	Thionin 2.4	*Arabidopsis thaliana*	134 (3)	Antifungal and antibacterial	Hao et al. (2020)
12.	Cycloviolacin O2	*Viola odorata*	30 (3)	Antibacterial and antifungal activities	Parsley et al. (2018)
13.	PAFP-S	*Phytolacca americana*	65 (3)	Antifungal	Das et al. (2019)
14.	1/Mj-AMP1, 2/ Mj-AMP2	*Mirabilis jalapa*	61 (3), 63 (3)	Antifungal	Prasad et al. (2019)
15.	Antifungal peptide 1/ EAFP1 and 1/ EAFP2	*Eucommia ulmoides*	41 (5)	Antifungal	Wong et al. (2019)
16.	10/Ha-AP10	*Helianthus annuus*	116 (4)	Antifungal	Rauf et al. (2020)
17.	Mung bean nsLTP	*Phaseolus mungo*	91 (4)	Antifungal and antibacterial action	Zhou et al. (2022)

conducted by Oliveira and coworkers showed HAPs Mo-CCBP3 first isolated from seeds of *Moringa oleifera* produced 03 chemically synthesized AMPs (Mo-CBP3-PepI, Mo-CBP3-PepII, and Mo-CBP3-PepIII). The pharmacology of synthetic PAMP Mo-CBP3-PepIII in mammalian cells showed no toxic activity or hemolysis, and further ability of permeabilization of the plasma membrane was retained in *Staphylococcus aureas* (Oliveira et al. 2019).

Another study was also conducted and showed a similar pattern where the extracted cysteine-rich AMPs from the Fabaceae family from the species *Bauhinia rufescens* showed antimicrobial activity. AMPs are effective against gram-negative bacteria and distorted cell membranes, as demonstrated by several studies (Nganso et al. 2020). It has been suggested by numerous researchers that HAPs have demonstrated a protective effect against cancer treatment. It has also shown its application in various fields of the medical industry, controlling the growth of pathogenic organisms in animals and humans (Salas et al. 2015). To find potential HAPs and assess therapeutic benefits, fewer studies have been conducted. The HAPs (BRS-P19) isolated from *B. refescens* were investigated, and various researchers have discovered other additional uses for HAPs besides their use against microbial infection and cancer treatment (Afroz et al. 2020). The researcher also investigated the therapeutic inhibitory effect on venom phospholipase A2 obtained from snake venom. Further research needs to be conducted to explore the potential therapeutic benefits of HAPs derived from plants (Table 6.3).

8.2 Application of Herbal Antimicrobial Peptides (HAPs) in Drug Design Using Computational Studies

Through structural interaction and predictions using targeted protein complexes, many techniques were utilized to investigate the PPI of hefty contact surfaces. With CADD (computer-assisted design techniques), in silico modeling was used to produce more potent, economical, and broad-spectrum HAPs to address challenging issues.

To combat different multidrug-resistant diseases, this involves the conversion of the primary sequence to the peptide structure (Santos-Silva et al. 2020). One of the strategies in the development of the bacteriocin peptide having 20 amino acids by Fields and Freed utilized machine learning and a simple biophysical property that can penetrate the pathogen membrane and have hemolytic, cytotoxic, and antibacterial effects. The roles of CADD in identifying prospective therapeutic candidates in pharmacology research, such as HAPs, with their economic cost and limited use of animal models, were explored by Brogi et al. (2020). This work aided pharmacologists and marketers in rationalizing new and secure medications while repositioning commercially available pharmaceuticals. Porto and his teammates (2018) used an algorithm genetically to fortify guava with glycine, producing arginine-rich helical and guavanin peptides (Corrêa et al. 2019). These peptides distorted membranes and caused hyperpolarization to have bactericidal effects at lower concentrations. In addition, Li et al. (2021) focused on employing various HAP types, the variables influencing the antimicrobial function, their mode of action, and possible applications in breeding and the food and fodder industries, as well as in medicinal and agricultural areas.

The advanced research was conducted on a sophisticated CADD approach that can avoid the difficult problems of relative primary sequence to subsequent peptide structures while simultaneously delivering more effective, more potent, economical, and broad-spectrum peptides as potent future-generation antibiotics. The conclusion drawn is that the majority of AMPs have little direct antibiotic activity against microorganisms (Karwal et al. 2020). From a rumen metagenomics dataset, Oyama and his colleagues (2019) discovered two novel (hg2 and hg4) linear HAPs with antibacterial efficacy alongside S. aureus having multidrug resistance. To visualize, analyze, and explain their actions, they used a classified design model and a

Table 6.3 Clinical trials were conducted on some HAPs with their mechanism

Peptide name	MW (g/mol)	Routes	Description	Clinical trial ID	Phase trial	Goal	Mode of action	References
Nisin	3354	Oral	Polycyclic lantibiotic	NCT02928042; NCT02467972	III	Gram-positive bacteria	Depolarization of cell membrane	Prince et al. (2016)
Gramicidin	1882	Topical	Polycyclic peptide	NCT00534391	III	Infected wounds and ulcers	Membrane disruption/ Immune-modulation	David and Rajasekaran (2015)
Polymyxin E (Colistin)	1155	Intravenous	Cyclic polypeptide	NCT01292031; NCT02573064	III	*A. baumannii*/ pneumonia	Membrane disruption/ immunomodulation	Yu et al. (2015)
Daptomycin	1621	Intravenous	Lipopeptide	NCT01922011; NCT02972983	III	Skin infection/ bacteremia	Membrane disruption/ immunomodulation	Taylor and Palmer (2016)
Melittin	2846.5	Intradermal	a-helical peptide	NCT02364349, NCT01526031	I/II	Inflammation	Membrane disruption/ immunomodulation	Lee and Bae (2016)
Murepavadin (POL7080)	1553.8	Intravenous	Analog of Protegrin	EUCTR2017-003933-27-EE	II	*P. aeruginosa*, *K. pneumoniae*	Binding to LptD	Srinivas et al. (2010)
Neuprex R (rBPI21)	~21,000	Intravenous	Derivative of BPI	NCT00462904	III	Pediatric Meningococcaemia	Membrane disruption	Schultz and Weiss (2007)
Surotomycin (CB-315)	1680.8	Oral	Cyclic lipopeptide	NCT01597505	III	*C. difficile*	Membrane disruption	Alam et al. (2015)
Pexiganan (MSI-78)	2477.2	Topical	Analog of Magainin	NCT00563394; NCT00563433	III	Diabetic foot ulcers	Membrane disruption/ immunomodulation	Gottler and Ramamoorthy (2009)

(continued)

Table 6.3 (continued)

Peptide name	MW (g/mol)	Routes	Description	Clinical trial ID	Phase trial	Goal	Mode of action	References
XOMA-629 (XMP-629)	1158.4	Topical	Derivative of BPI		III	Impetigo/acne rosacea	Immunomodulation	Easton et al. (2009)
Omiganan (MBI-226)	1779.2	Topical	Derivative of Indolicidin	NCT00231153; NCT00608959	III	Antisepsis/catheter infection	Membrane disruption/ Immunomodulation	Rubinchik et al. (2009)
NVB-302	1754.0	Oral	Lantibiotic	ISRCTN40071144	I	*C. difficile*	Cell wall synthesis inhibition	Crowther et al. (2013)
OP-145	3093.8	Ear drops	Derivative of LL-37	ISRCTN84220089	I/II	Chronic middle ear infection	Membrane disruption/ immunomodulation	Malanovic et al. (2015)
P113 (PAC-113)	1564.8	Mouth rinse	Fragment of Histatin-5	NCT00659971	II	Oral candidiasis	Membrane disruption/ immunomodulation	Woong et al. (2008)
LTX-109	788.1	Topical	Synthetic tripeptide	NCT01803035; NCT01158235	I/II	MRSA/impetigo	Membrane disruption	Sivertsen et al. (2014)
EA-230	415.5	Intravenous	Oligopeptide	NCT03145220	II	Sepsis	Immunomodulation	Van Groenendael et al. (2018)
SGX942 (Dusquetide)	553.7	Oral rinse	Analog of IDR-1	NCT03237325	III	Oral mucositis	Immunomodulation	Kudrimoti et al. (2016)
hLF1-11	1373.7	Intravenous	Fragment of human lactoferrin	NCT00430469	I/II	Bacterial/fungal Infections	Membrane disruption/ immunomodulation	Welling et al. (2018)
Novexatin (NP213)	1093.3	Topical	Cyclic Cationic Peptide	NCT02933879	II	Fungal nail infection	Membrane disruption	Mercer et al. (2020)

Peptide name	MW (g/mol)	Routes	Description	Clinical trial ID	Phase trial	Goal	Mode of action	References
Ramoplanin (NTI-851)	2568.1	Oral	Glycolipodepsipeptide	–	III	*C. difficile*, VRE	Inhibition of cell wall synthesis	Fulco and Wenzel (2006)
p2TA (AB103)	1037.2	Intravenous	Synthetic peptide	–	III	Necrotic tissue infection	Immunomodulation	Bulger et al. (2014)
D2A21	2775.4	Topical	Synthetic peptide	–	III	Burn wound infections	Membrane disruption	Muchintala et al. (2020)
Melamine	2786.6	Topical	Chimeric peptide	–	II/III	Contact lenses Microbials	Membrane disruption	Yasir et al. (2020)
Mel4	2347.8	Topical	Derivative of Melamine	ACTRN1261500072556	II/II	Contact lenses microbials	Membrane disruption	Yasir et al. (2019)
Delmitide (RDP58)	1228.6	Topical	Derivative of HLA	ISRCTN84220089	II	Inflammatory bowel disease	Immunomodulation	Travis et al. (2005)
DPK-060	2503.2	Ear drops	Derivative of Kininogen	NCT01447017	II	Acute external otitis	Membrane disruption/ immunomodulation	Håkansson et al. (2019)
PXL01	3061.6	Topical	Analog of Lactoferrin	NCT01022242	II	Postsurgical adhesions	Immunomodulation	Edsfeldt et al. (2017)
AP-214	2433.9	Intravenous	Derivative of a-MSH	NCT00903604	II	Postsurgical organ failure	Membrane disruption/ immunomodulation	Doi et al. (2008)
PMX-30063 (Brilacidin)	936.9	Intravenous	Defensin mimetic	NCT01211470; NCT02052388	II	Acute bacterial skin infection	Membrane disruption/ immunomodulation	Mensa et al. (2014)
Wap-8294A2 (Lotilibcin)	1562.8	Topical	Produced by Lysobacter species		I/II	Gram-positive bacteria	Membrane disruption	Itoh et al. (2018)

specific extraction technique employing chemical forms for amino acid identification. In contrast to human cell lipids, the peptide MRSA (methicillin-resistant *S. aureus*) in hg4 suggests that it may provide a better treatment candidate for MDR for bacterial infection. The process of designing drugs in silico is essential, and it has the potential to improve HAP's drug design and optimization procedures. To overcome functional constraints, in silico modeling may estimate the peptide structure without resolving the 3D structure, examine the physical–chemical characteristics, and screen peptide libraries for novel targets. Continual updating of these computational methods is possible based on data from experimental research. An in silico technique called molecular dynamics simulation (MD) examines the physical motion of molecules and atoms that are permitted to interact for a set amount of time. With the help of this method, molecular pathogen interactions, like lipid–peptide interactions and the progression of newer antibiotic molecules, can be revealed (Palmer et al. 2021).

8.3 Application of Herbal Antimicrobial Peptides in Food and Nutraceuticals

HAPs, apart from having antimicrobial actions, are utilized as food additives and have an important role in regulating the growth of plants and their development (Li et al. 2021). Antimicrobial resistance has grown to be a significant global health issue in recent years due to the increased interest in nonantibiotic prevention of recurring microbial illnesses. Various tactics have been created to lessen the worrisome rise in resistant microorganisms (Cai et al. 2016). Nutraceuticals can be considered an alternate therapeutic approach because they are prepared products that have been particularly tailored to meet dietary needs and provide preventative healthcare. These foods, invented in 1989 by Dr. Stephen De Felice, have several health advantages, such as the ability to treat and prevent disease. The study of nutrition has progressed from predicting dietary deficits to concentrating on chronic illnesses. Between food supplements and nutraceuticals, there is no apparent distinction although new work stresses reframing the notion to take efficacy, safety, and toxicity into account.

Food products are nutrient-dense substances that are needed to sustain life, provide energy, and promote growth. Safety and efficacy studies must come first, then epidemiological goals must be set to distinguish between foods, dietary supplements, and nutraceuticals. Food supplements can make up for micro- or macronutrient deficits, but nutraceuticals require substantial scientific backing (Santini and Novellino 2018). Additionally, nutritional supplements provide several benefits over synthetic medications. The high production costs of contemporary medications, for instance, can be decreased by the usage of nutraceuticals. Setting up a system for mass plant production, maintaining the infrastructure needed for the collection of natural goods, and curating raw extracts all require significantly reduced costs. As a result, it is extremely viable to implement such a system in low- to middle-income nations where high infrastructure costs are a barrier to the research and development of innovative treatments. Additionally, by lowering the price of therapies to make them more accessible, such low manufacturing costs might successfully address the issue of medication access disparity, particularly for patients living in rural parts of low-income nations (Paudel et al. 2022).

Nutraceuticals are synthetic compounds or materials (typically concentrated in the form of powder, capsule, or pill) that have the potential to improve human health and wellness, including the prevention and treatment of disease. Additionally, phytochemicals were described by López-Gutiérrez et al. (2015) as non-nutritive substances in plants that can prevent illness. Because they include a wide range of phytochemicals—some of which have not yet been discovered—with positive health effects and disease-prevention abilities, plant-based nutraceuticals have captured the interest of the majority of researchers in recent years. According to Vats et al. (2012), plants have historically been employed as antimicrobials. Therefore, considering that the prevalence of bacteria resistance to antibiotics has been on the rise, plants might be considered alternatives to the treatment of infectious disorders. In the majority of rural parts of sub-Saharan Africa, bambara groundnut (BGN) is primarily grown for its seed (Karunaratne et al. 2010).

Despite its nutritional importance, BGN has reportedly been overlooked by researchers, according to Udeh et al. (2020). This may be due to a lack of understanding of the phytochemical makeup of the plant and its related nutraceutical qualities, such as its antioxidant and antibacterial capabilities. Locally cultivated BGN, especially in rural areas of sub-Saharan Africa, provides a source of protein, carbs, and nutrients (functional food) as well as income for households (Mwale et al. 2007). Many dietary proteins' structures contain short peptide sequences that have functional characteristics (Singh and Kumar 2019). These peptides can be created by hydrolyzing dietary proteins using a variety of proteases, or they can be released during animal gastric digestion, microbial fermentation, industrial food processing, or animal gastric digestion (Sánchez and Vázquez 2017). Chemical processes can also be used to produce them. Bioactive peptides need to pass the gastrointestinal barrier and resist enzyme breakdown to have a favorable impact on health (Toldrá et al. 2018). According to Gurumallesh et al. (Toldrá et al. 2018), proteases are one of the major classes of enzymes that catalyze the breakdown of peptide bonds in proteins and peptides. By hydrolyzing the Actinopyga lecanora protein with bromelain, Ghanbari and Ebrahimpour (Ghanbari and Ebrahimpour 2018) were able to produce antibacterial peptides that were effective against *Pseudomonas* sp., *E. coli*, and *Staphylococcus aureus*. According to Mudgil et al. (2019), who used chymotrypsin, bromelain, and a microbial protease to break down the proteins of quinoa and amaranth, bromelain was found to be more effective than the other two proteases in producing antimicrobial peptides. Similar results were found by Borrajo et al. (2020), who found that the hydrolysis of porcine liver to produce antimicrobial peptides can be effectively achieved with bromelain. In the literature, antimicrobial peptides from milk, marine, and plant sources have been reported to be generated via enzymatic hydrolysis (Bhandari et al. 2020).

9 Challenges of HAPs

There are still a lot of problems with the effective use of HAPs to bind and identify their specific target. One of these was attributed by Tang and his colleague (2018) to the difficulty involved in the sanitization of certain HAPs by a variety of assessment, documentation, and purification procedures. The issue is made worse by the fact that practically every plant organ can make HAPs. In addition, Rogozhin and teammates (Rogozhin et al. 2018) conducted research and found that more HAPs were produced in wild-type grains than cultivated variants because of the fragment sequence variability of high C-terminal and high percent of hydrophobic amino acids, posing a greater challenge to yield more active HAPs in larger amounts. In addition, because artificial farming employs transgene techniques on a commercial scale, it can be challenging to express HAP because of different growth parameters. However, a singular amino acid alteration in AMP modification can lessen the natural breakdown of HAPs using proteases in leaves. However, expected transgenic functions can be achieved by using the proteases found in the internal fluid of the leaf. Another issue with HAPs is their selectivity, which restricts their ability to quickly connect to or kill target cells by interfering with the integrity of the plasma membrane. To address this issue and create further unique PAMPs from the existing ones, further research on the link between peptide structure and activity is necessary (Bakare et al. 2021).

10 Advantages and Disadvantages of HAP

A. HAPs, like any medical strategy, have both advantages and disadvantages. Unlike conventional antibiotics, HAPs attack multiple low-affinity targets. This makes it more challenging for target microbes to defend themselves through a single resistance mechanism.
B. Likelihood of Resistance: Experts believe that the development of widespread bacterial resistance toward HAPs is unlikely. Antimicrobial peptide resistance is less widespread than that of conventional antibiotics.
C. Susceptibility to Various Microorganisms: HAPs are effective against a variety of microorganisms, including bacteria, fungi, viruses, and some parasites.
D. Adaptability: HAPs can operate across a wide range of illnesses, making them adaptable.
E. Swift Action: HAPs often act swiftly, either eliminating or halting the development of bacteria.
F. Significance in Acute Infections: The quick intervention of HAPs is particularly important when treating acute infections.
G. Immunomodulatory Actions: HAPs facilitate bacterial clearance in the host. They have a wide spectrum of immunomodulatory actions. Actions include

stimulation of chemotaxis, modification of immune cell differentiation, and initiation of adaptive immunity.

H. Reduction of Proinflammatory Responses: Immunomodulatory actions of HAPs reduce proinflammatory cytokine production mediated by toll-like receptors (TLRs) and/or cytokines. This reduction helps prevent excessive and detrimental proinflammatory responses, such as sepsis.
I. Altering Immune Response: Certain HAPs can alter the immune response due to their immunomodulatory characteristics. This may strengthen the immune system's general capacity to combat infections.
J. Potential as a Treatment Strategy: The potential of HAPs as a treatment strategy against drug-resistant infections is significant.
K. Low-Molecular-Weight Oligopeptides: HAPs are low-molecular-weight oligopeptides. They possess broad-spectrum antibacterial properties against harmful microbes.
L. Nonspecific Action and Immune Response: HAPs, as the initial line of defense against invasive pathogenic bacteria, are nonspecific. They target components of microbes that aid in the immune response.
M. Potency and Diversity: Due to their potency and diversity, HAPs are excellent options for various applications.
N. Alternate Applications: HAPs can be used alone or in conjunction with different biomaterials for increased therapeutic action.
O. Targeting Drug-Resistant Bacteria: HAPs can be utilized to target drug-resistant bacteria in vaccine manufacturing.
P. Prospects and Developments: Ongoing prospects and developments in HAP discovery and development are focused on targeting antimicrobial resistance (AMR) bacteria.
Q. Reasons for Limited Resistance: HAPs target several microbial components simultaneously. This hinders the development of disease resistance.

11 Challenges in Clinical Advancement

Synthetic AMPs face challenges in progressing effectively in the clinic as antibiotics. Challenges include an unclear toxicological profile when delivered systemically.

A. Factors Limiting Progress: High cost is a significant factor limiting the clinical advancement of synthetic HAPs. Lability to proteolytic breakdown is another limitation, attributed to quick breakdown by blood plasma proteolytic enzymes. Quick excretion from the bloodstream by kidneys and liver (hepatic clearance) restricts systemic delivery, especially through intravenous injection.
B. Short Half-Life: Systemic delivery of peptides is limited by their short half-life. This limitation is particularly relevant when delivered through intravenous injection.

C. Cost of Large-Scale Production: Large-scale production of antimicrobial peptides can be costly. Cost considerations may restrict their broad usage, especially in resource-limited environments.
D. Metabolic Stability Issues: Low metabolic stability of HAPs is a significant issue, common in the medicinal peptides-systemic enzymatic breakdown and inadequate intestinal mucosal penetration contribute to minimal oral bioavailability.
E. Regulatory Challenges: Regulatory licensing of HAPs as medicinal agents may be challenging. Concerns include production standards, long-term effects, and safety issues.
F. Potential Delay in Clinical Development: Regulatory obstacles may potentially significantly delay the clinical development of HAPs. Omiganan and Pexiganan, two well-characterized HAPs, have encountered regulatory obstacles during clinical testing.
G. Despite challenges, HAPs are considered viable substitutes for established treatments. They are immunomodulatory and anti-infective drugs with a lower likelihood of causing resistance compared to traditional antibiotics.

12 Conclusion

The current topic covers a broad spectrum of multidisciplinary research that helped us achieve our objective of learning more about AMPs, including their novel sources, identification, isolation, and characterization as well as their bioactivity, molecular mechanism, and structure–activity characteristics. This topic was effective in compiling extensive data on a few AMPs from natural sources that are connected to the field of antibiotic discovery. There are several methods for separating antimicrobial peptides from plants, and the parameters of the experiment can be modified to extract specific peptides or members of peptide families. It does, however, remove the option of screening plants for a broad range of AMPs, particularly those that may yet be unknown. The majority of the protein-peptide extract's components that are not intended for a specific use go unclaimed. Nevertheless, some AMPs require alternative isolation approaches. There are two common methods, though the isolation of different AMP families and the isolation of cyclic peptides. The whole diversity of the composition of plant AMPs can be obtained by optimizing these methods. There has been a resurgence of interest in AMPs as a possible substitute for traditional antibiotics due to the rise of antibiotic-resistant bacteria. When it comes to treating different infectious diseases, AMPs outperform traditional antibiotics with obvious advantages. The threat posed by multidrug-resistant (MDR) bacteria may be eliminated by their (i) potential immunomodulatory properties, (ii) rapid action mode, non-specific, multi-hit limiting the emergence of resistance, (iii) broad spectrum of activity, and (iv) synergistic interactions with conventional antibiotics. Cytotoxic effects, production costs, and issues with peptide bioavailability and efficacy are obstacles to the therapeutic application of HAPs.

Many approaches have been developed to address these issues, including the creation of ultra-short or truncated HAPs, delivery mechanisms and chemical alterations, and the meticulous choice of a counter-ion of HAP synthesis. While not every HAP in the clinical pipeline will make it to market, these tactics may increase HAPs' clinical trial success rates. Although there has been a tendency toward better clinical results, certain HAPs have failed clinical trials because there was insufficient evidence of their efficacy or superiority over traditional antibiotics. To develop novel antibiotics using HAPs, we believe that this special collection will assist and motivate experts from various study fields to make appropriate use of the knowledge gained. In the near future, it is anticipated that novel treatments based on AMPs through engineering techniques will replace antibiotics that cause resistance.

References

Afroz M, Akter S, Ahmed A, Rouf R, Shilpi JA, Tiralongo E, Sarker SD, Göransson U, Uddin SJ (2020) Ethnobotany and antimicrobial peptides from plants of the solanaceae family: an update and future prospects. Front Pharmacol 11:565

Ageitos JM, Sánchez-Pérez A, Calo-Mata P, Villa TG (2017) Antimicrobial peptides (AMPs): ancient compounds that represent novel weapons in the fight against bacteria. Biochem Pharmacol 133:117–138

Alam MZ, Wu X, Mascio C, Chesnel L, Hurdle JG (2015) Mode of action and bactericidal properties of surotomycin against growing and nongrowing Clostridium difficile. Antimicrob Agents Chemother 59(9):5165–5170

Almasia NI, Molinari MP, Maroniche GA, Nahirñak V, Barrios Barón MP, Taboga OA, Vazquez RC (2017) Successful production of the potato antimicrobial peptide Snakin-1 in baculovirus-infected insect cells and development of specific antibodies. BMC Biotechnol 17(1):75

Anaya K, Podszun M, Franco OL, de Almeida Gadelha CA, Frank J (2020) The coconut water antimicrobial peptide CnAMP1 is taken up into intestinal cells but does not alter p-glycoprotein expression and activity. Plant Foods Hum Nutr 75(3):396–403

Astafieva AA, Enyenihi AA, Rogozhin EA, Kozlov SA, Grishin EV, Odintsova TI (2015) Novel proline-hydroxyproline glycopeptides from the dandelion (Taraxacum officinale Wigg.) flowers: de novo sequencing and biological activity. Plant Sci Int J Exp Plant Biol 238:323–329

Astafieva AA, Rogozhin EA, Andreev YA, Odintsova TI, Kozlov SA, Grishin EV (2013) A novel cysteine-rich antifungal peptide ToAMP4 from Taraxacum officinale Wigg. flowers. Plant Physiol Biochem 70:93–99

Aweya JJ, Zheng Z, Zheng X, Yao D, Zhang Y (2021) The expanding repertoire of immune-related molecules with antimicrobial activity in penaeid shrimps: a review. Rev Aquac 13(4):1907–1937

Bakare OO, Gokul A, Fadaka AO, Wu R, Niekerk LA, Barker AM, Keyster M, Klein A (2022) Plant antimicrobial peptides (PAMPs): features, applications, production, expression, and challenges. Molecules (Basel) 27(12):3703

Bakare OO, Gokul A, Wu R, Niekerk LA, Klein A, Keyster M (2021) Biomedical relevance of novel anticancer peptides in the sensitive treatment of cancer. Biomol Ther 11(8):1120

Barber D, Limas GG, Gavilanes JG, Méndez E (1988) Isolation and characterization of thirteen new salt-soluble proteins from barley by reversed-phase high-performance liquid chromatography. Planta 176:221–229

Beisswenger C, Bals R (2005) Functions of antimicrobial peptides in host defense and immunity. Curr Protein Peptide Sci 6(3):255–264

Bhandari D, Rafiq S, Gat Y, Gat P, Waghmare R, Kumar V (2020) A review on bioactive peptides: physiological functions, bioavailability and safety. Int J Pept Res Ther 26(1):139–150

Bondaryk M, Staniszewska M, Zielińska P, Urbańczyk-Lipkowska Z (2017) Natural antimicrobial peptides as inspiration for design of a new generation antifungal compounds. J Fungi (Basel) 3(3):46

Borrajo P, Pateiro M, Gagaoua M, Franco D, Zhang W, Lorenzo JM (2020) Evaluation of the antioxidant and antimicrobial activities of porcine liver protein hydrolysates obtained using alcalase, bromelain, and papain. Appl Sci 10(7):2290

Brogi S, Ramalho TC, Kuca K, Medina-Franco JL, Valko M (2020) In silico methods for drug design and discovery. Front Chem 8:612

Bulger EM, Maier RV, Sperry J, Joshi M, Henry S, Moore FA (2014) A novel drug for treatment of necrotizing soft-tissue infections: a randomized clinical trial. JAMA Surg 149(6):528–536

Caceres-Delpiano J, Ibañez R, Alegre P, Sanhueza C, Paz-Fiblas R, Correa S, Retamal P, Cristóbal Jiménez J, Álvarez L (2020) Deep learning enables the design of functional de novo antimicrobial proteins. CSH bioRxiv:1–22. https://doi.org/10.1101/2020.08.26.266940

Cai T, Verze P, Brugnolli A, Tiscione D, Luciani LG, Eccher C, Lanzafame P, Malossini G, Wagenlehner FME, Mirone V, Bjerklund Johansen TE, Pickard R, Bartoletti R (2016) Adherence to european association of urology guidelines on prophylactic antibiotics: an important step in antimicrobial stewardship. Eur Urol 69(2):276–283

Campos ML, de Souza CM, de Oliveira KBS, Dias SC, Franco OL (2018) The role of antimicrobial peptides in plant immunity. J Exp Bot 69(21):4997–5011

Chan YS, Ng TB (2013) Northeast red beans produce a thermostable and pH stable defensin-like peptide with potent antifungal activity. Cell Biochem Biophys 66(3):637–648

Chauhan S, Dhawan DK, Saini A, Preet S (2021) Antimicrobial peptides against colorectal cancer-a focused review. Pharmacol Res 167:105529

Corrêa JAF, Evangelista AG, de Melo NT, Luciano FB (2019) Fundamentals on the molecular mechanism of action of antimicrobial peptides. Materialia 8:100494

Crabbé A, Jensen P, Bjarnsholt T, Coenye T (2019) Antimicrobial tolerance and metabolic adaptations in microbial biofilms. Trends Microbiol 27(10):850–863

Crowther GS, Baines SD, Todhunter SL, Freeman J, Chilton CH, Wilcox MH (2013) Evaluation of NVB302 versus vancomycin activity in an in vitro human gut model of Clostridium difficile infection. J Antimicrob Chemother 68(1):168–176

Das K, Datta K, Karmakar S, Datta SK (2019) Antimicrobial peptides - small but mighty weapons for plants to fight phytopathogens. Protein Pept Lett 26(10):720–742

David JM, Rajasekaran AK (2015) Gramicidin a: a new mission for an old antibiotic. J Kidney Cancer VHL 2(1):15–24

de Barros E, Gonçalves RM, Cardoso MH, Santos NC, Franco OL, Cândido E (2019) Snake venom cathelicidins as natural antimicrobial peptides. Front Pharmacol 10:1415

De Souza CE, Silva Cardoso MH, Sousa DA, Viana JC, De Oliveira- Júnior NG, Miranda V (2014) The use of versatile plant antimicrobial peptides in agribusiness and human health. Peptides 55:65–78

de Zélicourt A, Letousey P, Thoiron S, Campion C, Simoneau P, Elmorjani K, Marion D, Simier P, Delavault P (2007) Ha-DEF1, a sunflower defensin, induces cell death in Orobanche parasitic plants. Planta 226(3):591–600

Deng M, Peng J, Zhang J, Ran S, Cai C, Yu L, Ni S, Huang X, Li L, Wang X (2021) The Cysteine-rich peptide Snakin-2 negatively regulates tubers sprouting through modulating lignin biosynthesis and H2O2 accumulation in potato. Int J Mol Sci 22(5):2287

Dhama K, Karthik K, Khandia R, Munjal A, Tiwari R, Rana R, Khurana SK, Sana Ullah Khan RU, Alagawany M, Farag MR, Dadar M, Joshi SK (2018) Medicinal and therapeutic potential of herbs and plant metabolites / extracts countering viral pathogens - current knowledge and future prospects. Curr Drug Metab 19(3):236–263

Diamond G, Beckloff N, Weinberg A, Kisich K (2009) The roles of antimicrobial peptides in innate host defense. Curr Pharm Des 15(21):2377–2392

Dijksteel GS, Ulrich WMM, Middelkoop E, Boekema LHK, Bouke. (2021) Review: lesson learned from clinical trials using antimicrobial peptides (AMPs). Frontier. Microbiology 12:616979

Divyashree M, Mani MK, Reddy D, Kumavath R, Ghosh P, Azevedo V, Barh D (2020) Clinical applications of antimicrobial peptides (AMPs): where do we stand now? Protein Peptide Lett 27(2):120–134

Doi K, Hu X, Yuen PST, Leelahavanichkul A, Yasuda H, Kim SM (2008) AP214, an analogue of a-melanocyte-stimulating hormone, ameliorates sepsis-induced acute kidney injury and mortality. Kidney Int 73(1):1266–1274

dos Santos LA, Taveira GB, Ribeiro SFF, Pereira LS, Carvalho AO, Rodrigues R (2017) Purification and characterization of peptides from Capsicum annuum fruits which are α-amylase inhibitors and exhibit high antimicrobial activity against fungi of agronomic importance. Protein Expr Purif 132:97–107

Easton DM, Nijnik A, Mayer ML, Hancock REW (2009) Potential of immunomodulatory host defense peptides as novel anti-infectives. Trends Biotechnol 27(10):582–590

Edsfeldt S, Holm B, Mahlapuu M, Reno C, Hart DA, Wiig M (2017) PXL01 in sodium hyaluronate results in increased PRG4 expression: a potential mechanism for anti-adhesion. Upsala J Med Sci 122(1):28–34

Egorov TA, Odintsova TI, Pukhalsky VA, Grishin EV (2005) Diversity of wheat antimicrobial peptides. Peptides 26(1):2064–2073

Evaristus NA, Wan Abdullah WN, Gan CY (2018) Extraction and identification of α-amylase inhibitor peptides from Nephelium lappacheum and Nephelium mutabile seed protein using gastro-digestive enzymes. Peptides 102:61–67

Finkina EI, Melnikova DN, Bogdanov IV, Ovchinnikova TV (2019) Peptides of the innate immune system of plants. Part II. Biosynthesis, biological functions, and possible practical applications. Russian J Bioorganic Chem 45(2):55–65

Finkina EI, Melnikova DN, Bogdanov IV, Ovchinnikova TV (2018) Peptides of the innate immune system of plants. Part I. Structure, biological activity, and mechanisms of action. Russian J Bioorg Chem 44(6):573–585

Fulco P, Wenzel RP (2006) Ramoplanin: a topical lipoglycodepsipeptide antibacterial agent. Expert Rev Anti-Infect Ther 4(6):939–945

García-Olmedo F, Molina A, Alamillo JM, Rodriguez-Palenzuéla P (1998) Plant defense peptides. Biopolymers 47(6):479–491

Ghanbari R, Ebrahimpour A (2018) Separation and identification of bromelain-generated antibacterial peptides from Actinopyga lecanora. Food Sci Biotechnol 27(2):591–598

Gottler LM, Ramamoorthy A (2009) Structure, membrane orientation, mechanism, and function of pexiganan—a highly potent antimicrobial peptide designed from magainin. Biochim Biophys Acta 1788(8):1680–1686

Gründemann C, Stenberg KG, Gruber CW (2019) T20K: an immunomodulatory cyclotide on its way to the clinic. Int J Pept Res Ther 25(1):9–13

Håkansson J, Ringstad L, Umerska A, Johansson J, Andersson T, Boge L (2019) Characterization of the in vitro, ex vivo, and in vivo efficacy of the antimicrobial peptide DPK-060 used for topical treatment. Front Cell Infect Microbiol 9:174

Hammami R, Ben Hamida J, Vergoten G, Fliss I (2009) PhytAMP: a database dedicated to antimicrobial plant peptides. Nucleic Acids Res 37(Database issue):D963-8

Hancock RE (2001) Cationic peptides: effectors in innate immunity and novel antimicrobials. Lancet Infect Dis 1(3):156–164

Haney EF, Straus SK, Hancock REW (2019) Reassessing the host defense peptide landscape. Front Chem 7:43

Hao G, Bakker MG, Kim HS (2020) Enhanced resistance to Fusarium graminearum in transgenic arabidopsis plants expressing a modified plant thionin. Phytopathology 110(5):1056–1066

Havenga B, Ndlovu T, Clements T, Reyneke B, Waso M, Khan W (2019) Exploring the antimicrobial resistance profiles of WHO critical priority list bacterial strains. BMC Microbiol 19(1):303

Hughes P, Dennis E, Whitecross M, Llewellyn D, Gage P (2000) The cytotoxic plant protein, β-Purothionin, forms ion channels in lipid membranes. J Biol Chem 275(2):823–827

Iqbal A, Khan RS, Shehryar K, Imran A, Ali F, Attia S, Shah S, Mii M (2019) Antimicrobial peptides as effective tools for enhanced disease resistance in plants. Plant Cell Tissue Organ Cult 139(38):1–15

Isah T (2019) Stress and defense responses in plant secondary metabolites production. Biol Res 52(1):39–64

Itoh H, Tokumoto K, Kaji T, Paudel A, Panthee S, Hamamoto H (2018) Total synthesis and biological mode of action of WAP-8294A2: a menaquinone-targeting antibiotic. J Org Chem 83(13):6924–6935

Karunaratne AS, Azam-Ali SN, Al-Shareef I, Sesay A, Jørgensen ST, Crout NMJ (2010) Modelling the canopy development of bambara groundnut. Agric For Meteorol 150(7–8):1007–1015

Karwal P, Vats ID, Sinha N, Singhal A, Sehgal T, Kumari P (2020) Therapeutic applications of peptides against Zika virus: a review. Curr Med Chem 27(23):3906–3923

Kayser O (2018) Ethnobotany and medicinal plant biotechnology: from tradition to modern aspects of drug development. Planta Med 84(12–13):834–838

Kenny RG, Marmion CJ (2019) Toward multi-targeted platinum and ruthenium drugs—a new paradigm in cancer drug treatment regimens? Chem Rev 119(2):1058–1137

Kudrimoti M, Curtis A, Azawi S, Worden F, Katz S, Adkins D (2016) Dusquetide: a novel innate defense regulator demonstrating a significant and consistent reduction in the duration of oral mucositis in preclinical data and a randomized, placebo-controlled Phase 2a clinical study. J Biotechnol 239:115–125

Kul'Ko AB, Kisil OV, Sadykova VS, Mikhailov VF, Vasilieva IM, Shulenina LV (2016) Investigation of thionins from black cumin (Nigella sativa L.) seeds showing cytotoxic, regulatory and antifungal activity. Antibiot Khimioter 61(9–10):8–16

Kulaeva O, Kliukova M, Afonin A, Sulima A, Zhukov V, Tikhonovich I (2020) The role of plant antimicrobial peptides (AMPs) in response to biotic and abiotic environmental factors. Biol Commun 65(2):187–199

Kumarasamy KK, Toleman MA, Walsh TR, Bagaria J, Butt F, Balakrishnan R (2010) Emergence of a new antibiotic resistance mechanism in India, Pakistan, and the UK: a molecular, biological, and epidemiological study. Lancet Infect Dis 10(9):597–602

Le AV, Parks SE, Nguyen MH, Roach PD (2019) Optimised extraction of trypsin inhibitors from defatted gac (Momordica cochinchinensis Spreng) seeds for production of a trypsin inhibitor-enriched freeze-dried powder. Separations 6(1):8

Lee G, Bae H (2016) Anti-inflammatory applications of melittin, a major component of bee venom: detailed mechanism of action and adverse effects. Molecules 21(5):616

Li J, Hu S, Jian W, Xie C, Yang X (2021) Plant antimicrobial peptides: structures, functions, and applications. Bot Stud 62(1):5

Li Y, Rebuffat S (2020) The manifold roles of microbial ribosomal peptide–based natural products in physiology and ecology. J Biol Chem 295(1):34–54

Limas GG, Salinas M, Moneo I, Fischer S, Wittmann-Liebold B, Méndez E (1990) Purification and characterization of ten new rice NaCl-soluble proteins: identification of four protein-synthesis inhibitors and two immunoglobulin- binding proteins. Planta 181(1):1–9

López-Gutiérrez N, Romero-González R, Plaza-Bolaños P, Martínez Vidal JL, Garrido FA (2015) Identification and quantification of phytochemicals in nutraceutical products from green tea by UHPLC-Orbitrap-MS. Food Chem 173:607–618

Maan AM, Hofman AH, de Vos WM, Kamperman M (2020) Recent developments and practical feasibility of polymer-based antifouling coatings. Adv Funct Mater 30(32):2000936

Magana M, Pushpanathan M, Santos AL, Leanse L, Fernandez M, Ioannidis A, Giulianotti MA, Apidianakis Y, Bradfute S, Ferguson AL (2020) The value of antimicrobial peptides in the age of resistance. Lancet Infect Dis 20(9):e216–e230

Mahlapuu M, Håkansson J, Ringstad L, Björn C (2016) Antimicrobial peptides: an emerging category of therapeutic agents. Front Cell Infect Microbiol 6:194

Majid A, Priyadarshini CGP (2020) Millet derived bioactive peptides: a review on their functional properties and health benefits. Crit Rev Food Sci Nutr 60(19):3342–3351

Malanovic N, Leber R, Schmuck M, Kriechbaum M, Cordfunke RA, Drijfhout JW (2015) Phospholipid-driven differences determine the action of the synthetic antimicrobial peptide OP-145 on Gram-positive bacterial and mammalian membrane model systems. Biochim Biophys Acta 1848(10 Pt A):2437–2447

Mensa B, Howell G, Scott R, DeGrado W (2014) Comparative mechanistic studies of brilacidin, daptomycin, and the antimicrobial peptide LL16. Antimicrob Agents Chemother 58(9):5136–5145

Mercer DK, Robertson JC, Miller L, Stewart CS, O'Neil D (2020) A, NP213 (NovexatinR): a unique therapy candidate for onychomycosis with a differentiated safety and efficacy profile. Med Mycol 58(8):1064–1072

Muchintala D, Suresh V, Raju D, Sashidhar RB (2020) Synthesis and characterization of cecropin peptide-based silver nanocomposites: its antibacterial activity and mode of action. Mater Sci Eng C 110:110712

Mudgil P, Omar LS, Kamal H, Kilari BP, Maqsood S (2019) Multi-functional bioactive properties of intact and enzymatically hydrolysed quinoa and amaranth proteins. LWT Food Sci Technol 110:207–213

Mukherjee PK, Wiest A, Ruiz N, Keightley A, Moran-Diez ME, McCluskey K, Pouchus YF, Kenerley CM (2011) Two classes of new peptaibols are synthesized by a single non-ribosomal peptide synthetase of Trichoderma virens. J Biol Chem 286(6):4544–4554

Mwale SS, Azam-Ali SN, Massawe FJ (2007) Growth and development of bambara groundnut (Vigna subterranea) in response to soil moisture. Eur J Agron 26(4):345–353

Mwangi J, Hao X, Lai R, Zhang ZY (2019) Antimicrobial peptides: new hope in the war against multidrug resistance. Zool Res 40(6):488–505

Nganso YOD, Sidjui LS, Amang AGAN, Kahouo AD, Abah K, Fomena H, Hamadou M (2020) Identification of peptides in the leaves of Bauhinia rufescens Lam (Fabaceae) and evaluation of their antimicrobial activities against pathogens for aquaculture. Sci J Chem 8(4):81–91

Nolde SB, Vassilevski AA, Rogozhin EA, Barinov NA, Balashova TA, Samsonova OV (2011) Disulfide-stabilized helical hairpin structure and activity of a novel antifungal peptide EcAMP1 from seeds of barnyard grass (Echinochloa crus-galli). J Biol Chem 286(28):25145–25153

Odintsova TI, Korostyleva TV, Odintsova MS, Pukhalsky VA, Grishin EV, Egorov TA (2008a) Analysis of Triticum boeoticum and Triticum urartu seed defensins: to the problem of the origin of polyploid wheat genomes. Biochimie 90(6):939–946

Odintsova TI, Rogozhin EA, Baranov Y, Musolyamov AK, Yalpani N, Egorov TA (2008b) Seed defensins of barnyard grass Echinochloa crusgalli (L.) Beauv. Biochimie 90:1667–1673

Odintsova TI, Vassilevski AA, Slavokhotova AA, Musolyamov AK, Finkina EI, Khadeeva NV (2009) A novel antifungal hevein-type peptide from Triticum kiharae seeds with a unique 10-cysteine motif. FEBS J 276(15):4266–4275

Okada T, Yoshizumi H (1970) A lethal toxic substance for brewing yeast in wheat and barley: part II isolation and some properties of toxic principle. Agric Biol Chem 34:1089–1094

Okafor N, Okeke BC (2017) Modern industrial microbiology and biotechnology. CRC Press, Boca Raton

Oliveira JT, Souza PF, Vasconcelos IM, Dias LP, Martins TF, Van Tilburg MF, Guedes MI, Sousa DO (2019) Mo-CBP3-PepI, Mo-CBP3-PepII, and Mo-CBP3-PepIII are synthetic antimicrobial peptides active against human pathogens by stimulating ROS generation and increasing plasma membrane permeability. Biochimie 157:10–21

Oyama LB, Olleik H, Teixeira ACN, Guidini MM, Pickup JA, Cookson AR, Vallin H, Wilkinson T, Bazzolli D, Richards J (2019) In silico identification of novel peptides with antibacterial activity against multidrug resistant Staphylococcus aureus. BioRxiv:577221

Palmer N, Maasch JR, Torres MD, de la Fuente-Nunez C (2021) Molecular dynamics for antimicrobial peptide discovery. Infect Immun 8(4):e00703–e00720

Parisi K, Shafee TMA, Quimbar P, van der Weerden NL, Bleackley MR, Anderson MA (2019) The evolution, function and mechanisms of action for plant defensins. Semin Cell Dev Biol 88:107–118

Parsley NC, Kirkpatrick CL, Crittenden CM, Rad JG, Hoskin DW, Brodbelt JS, Hicks LM (2018) PepSAVI-MS reveals anticancer and antifungal cycloviolacins in Viola odorata. Phytochemistry 152:61–70

Paudel KR, Patel V, Vishwas S, Gupta S, Sharma S, Chan Y, Jha NK, Shrestha J, Imran M, Panth N, Shukla SD, Jha SK, Devkota HP, Warkiani ME, Singh SK, Ali MK, Gupta G, Chellappan DK, Hansbro PM, Dua K (2022) Nutraceuticals and COVID-19: a mechanistic approach toward attenuating the disease complications. J Food Biochem 46(12):e14445

Phazang P, Negi NP, Raina M, Kumar D (2020) Plant antimicrobial peptides: next-generation bioactive molecules for plant protection. In: Environmental and microbial biotechnology. Springer, Singapore, pp 281–293

Porto WF, Irazazabal L, Alves ES, Ribeiro SM, Matos CO, Pires ÁS, Fensterseifer IC, Miranda VJ, Haney EF, Humblot V (2018) In-silico optimization of a guava antimicrobial peptide enables combinatorial exploration for peptide design. Nat Commun 9(1):1490

Prasad BD, Sahni S, Ranjan T, Kumari D (2019) Antimicrobial proteins: key components of innate immunity. Curr J Appl Sci Technol 36(3):1–7

Prince A, Sandhu P, Kumar P, Dash E, Sharma S, Arakha M (2016) Lipid-II independent antimicrobial mechanism of nisin depends on its crowding and degree of oligomerization. Sci Rep 6:37908

Raghubeer EV, Phan BN, Onuoha E, Diggins S, Aguilar V, Swanson S, Lee A (2020) The use of High-Pressure Processing (HPP) to improve the safety and quality of raw coconut (Cocos nucifera L) water. Int J Food Microbiol 331:108697

Rauf S, Ortiz R, Shehzad M, Haider W, Ahmed I (2020) The exploitation of sunflower (Helianthus annuus L.) seed and other parts for human nutrition, medicine and the industry. Helia 43(73):167–184

Rempel S, Gati C, Nijland M, Thangaratnarajah C, Karyolaimos A, de Gier JW, Guskov A, Slotboom D (2020) A mycobacterial ABC transporter mediates the uptake of hydrophilic compounds. Nature 580(7803):409–412

Rogozhin E, Ryazantsev D, Smirnov A, Zavriev S (2018) Primary structure analysis of antifungal peptides from cultivated and wild cereals. Plants (Basel) 7(3):74

Rubinchik E, Dugourd D, Algara T, Pasetka C, Friedland HD (2009) Antimicrobial and antifungal activities of a novel cationic antimicrobial peptide, omiganan, in experimental skin colonisation models. Int J Antimicrob Agents 34(9):457–461

Salas CE, Badillo-Corona JA, Ramírez-Sotelo G, Oliver-Salvador C (2015) Biologically active and antimicrobial peptides from plants. Biomed Res Int 102129:1

Sánchez A, Vázquez A (2017) Bioactive peptides: a review. Food Quality Safety 1(1):29–46

Santini A, Novellino E (2018) Nutraceuticals- shedding light on the grey area between pharmaceuticals and food. Expert Rev Clin Pharmacol 11(6):545–547

Santos-Silva CAD, Zupin L, Oliveira-Lima M, Vilela LMB, Bezerra-Neto JP, Ferreira-Neto JR, Ferreira JDC, Oliveira- Silva RLD, Pires CDJ, Aburjaile FF (2020) Plant antimicrobial peptides: state of the art, in silico prediction and perspectives in the omics era. Bioinform Biol Insights 14:1177932220952739

Schmidt M, Arendt EK, Theory TLC (2019) Isolation and characterisation of the antifungal activity of the cowpea defensin Cp-thionin II. Food Microbiol 82:504–514

Schultz H, Weiss JP (2007) The bactericidal/permeability-increasing protein (BPI) in infection and inflammatory disease. Clin Chim Acta 384(1–2):12–23

Sharma S, Gupta N (2020) Defense signaling in plants against micro-creatures: do or die. Indian Phytopathol 73(1):605–613

Sharma S, Sanyal SK, Sushmita K, Chauhan M, Sharma A, Anirudhan G (2021) Modulation of phototropin signalosome with artificial illumination holds great potential in the development of climate-smart crops. Curr Genomics 22(3):181–213

Shen Y, Xu L, Huang J, Serra A, Yang H, Tam JP (2019) Potentides: new Cysteine- rich peptides with unusual disulfide connectivity from Potentilla anserine. Chembiochem 20(15):1995–2004

Shwaiki LN, Arendt EK, Lynch KM (2020) Study on the characterisation and application of synthetic peptide Snakin-1 derived from potato tubers – action against food spoilage yeast. Food Control 118:107362

Singh P, Kumar S (2019) Microbial enzyme in food biotechnology. In: Enzymes in food biotechnology. Academic Press, pp 19–28. https://doi.org/10.1016/B978-0-12-813280-7.00002-5

Sivertsen A, Isaksson J, Leiros HKS, Svenson J, Svendsen JS, Brandsdal BO (2014) Synthetic cationic antimicrobial peptides bind with their hydrophobic parts to drug site II of human serum albumin. BMC Struct Biol 14:4. https://doi.org/10.1186/1472-6807-14-4

Srinivas N, Jetter P, Ueberbacher BJ, Werneburg M, Zerbe K, Steinmann J (2010) Peptidomimetic antibiotics target outer-membrane biogenesis in Pseudomonas aeruginosa. Science 327(5968):1010–1013

Stec B (2006) Plant thionins--the structural perspective. Cell Mol Life Sci 63(12):1370–1385

Swedberg JE, Mahatmanto T, Abdul Ghani H, de Veer SJ, Schroeder CI, Harris JM, Craik DJ (2016) Substrate-guided design of selective FXIIa inhibitors based on the plant-derived Momordica cochinchinensis trypsin inhibitor-II (MCoTI-II) scaffold. J Med Chem 59(15):7287–7292

Tam J, Wang S, Wong K, Tan W (2015) Antimicrobial peptides from plants. Pharmaceuticals 8(4):711–757

Tang SS, Prodhan ZH, Biswas SK, Le CF, Sekaran SD (2018) Antimicrobial peptides from different plant sources: isolation, characterisation, and purification. Phytochemistry 154:94–105

Taylor SD, Palmer M (2016) The action mechanism of daptomycin. Bioorgan Med Chem 24(24):6253–6268

Toldrá F, Reig M, Aristoy MC, Mora L (2018) Generation of bioactive peptides during food processing. Food Chem 267:395–404

Travis S, Yap LM, Hawkey C, Warren B, Lazarov M, Fong T (2005) RDP58 is a novel and potentially effective oral therapy for ulcerative colitis. Inflamm Bowel Dis 11(8):713–719

Udeh EL, Nyila MA, Kanu SA (2020) Nutraceutical and antimicrobial potentials of Bambara groundnut (Vigna subterranean): a review. In Heliyon 6(10):e05205

Van Groenendael R, Kox M, Van Eijk LT, Pickkers P (2018) Immunomodulatory and kidney-protective effects of the human chorionic gonadotropin derivate EA-230. Nephron 140(2):148–151

Vasilchenko AS, Smirnov AN, Zavriev SK, Grishin EV, Vasilchenko AV, Rogozhin EA (2016) Novel Thionins from black seed (Nigella sativa L.) demonstrate antimicrobial. Int J Pept Res Ther 23:171–180

Vats SLW, Tiwari R, Alam A, Behera KK, Pareek R (2012) Evaluation of phytochemicals, antioxidant and antimicrobial activity of in vitro culture of Vigna unguiculata. Researcher 4(11):70

Wang J, Dou X, Song J, Lyu Y, Zhu X, Xu L, Li W, Shan A (2019) Antimicrobial peptides: promising alternatives in the post feeding antibiotic era. Med Res Rev 39(3):831–859

Wang M, Odom T, Cai J (2020) Challenges in the development of next-generation antibiotics: opportunities of small molecules mimicking mode of action of host-defense peptides. Expert Opin Ther Pat 30(5):303–305

Wang S, Zeng X, Yang Q, Qiao S (2016) Antimicrobial peptides as potential alternatives to antibiotics in food animal industry. Int J Mol Sci 17(5):603

Wang Y, Hong J, Liu X, Yang H, Liu R, Wu J, Wang A, Lin D, Lai R (2008) Snake cathelicidin from Bungarus fasciatus is a potent peptide antibiotics. PLoS One 3(9):e3217

Welling M, Brouwer C, Roscini L, Cardinali G, Corte L, Casagrande Pierantoni D et al (2018) Structure-activity relationship study of synthetic variants derived from the highly potent human antimicrobial peptide hLF(1-11). Cohesive J Microbiol Infect Dis 3(1). https://doi.org/10.31031/CJMI.2018.01.000512

Wimley WC, Hristova K (2011) Antimicrobial peptides: successes, challenges, and unanswered questions. J Membr Biol 239(1–2):27–34

Wong JH, Ng TB, Wang H, Cheung RCF, CCW N, Ye X, Yang J, Liu F, Ling C, Chan K, Ye X, Chan WY (2019) Antifungal proteins with antiproliferative activity on cancer cells and hiv-1 enzyme inhibitory activity from medicinal plants and medicinal fungi. Curr Protein Pept Sci 20(3):265–276

Woong SJ, Xuewei SL, Sun JN, Edgerton M (2008) The P-113 fragment of histatin 5 requires a specific peptide sequence for intracellular translocation in Candida albicans, which is independent of cell wall binding. Antimicrobial Agents Chemother 52(2):497–504

Wu CL, Hsueh JY, Yip BS, Chih YH, Peng KL, Cheng JW (2020) Antimicrobial peptides display strong synergy with vancomycin against vancomycin-resistant E. faecium, S. aureus, and wild-type E. coli. Int J Mol Sci 21(13):4578

Xia X, Cheng L, Zhang S, Wang L, Hu J (2018) The role of natural antimicrobial peptides during infection and chronic inflammation. Antonie Van Leeuwenhoek 111(1):5–26

Yasir M, Dutta D, Hossain KR, Chen R, Ho KKK, Kuppusamy R (2020) Mechanism of action of surface immobilized antimicrobial peptides against Pseudomonas aeruginosa. Front Microbiol 10:3053

Yasir M, Dutta D, Willcox MDP (2019) Mode of action of the antimicrobial peptide Mel4 is independent of Staphylococcus aureus cell membrane permeability. PLoS One 14(7):e0215703

Yu Z, Qin W, Lin J, Fang S, Qiu J (2015) Antibacterial mechanisms of polymyxin and bacterial resistance. Biomed Res Int 2015:679109. https://doi.org/10.1155/2015/679109

Zahedifard F, Lee H, No JH, Salimi M, Seyed N, Asoodeh A, Rafati S (2020) Comparative study of different forms of Jellein antimicrobial peptide on Leishmania parasite. Exp Parasitol 209:107823

Zanetti M (2004) Cathelicidins- multifunctional peptides of the innate immunity. J Leukoc Biol 75(1):39–48

Zhou Y, Manghwar H, Hu W, Liu F (2022) Degradation mechanism of autophagy-related proteins and research progress. Int J Mol Sci 23(13):7301

Chapter 7
Bioprocessing and Market Aspects of Antimicrobial Peptides

Kanti Nandan Mihooliya and Alka Kumari

1 Introduction

Before the onset of the twentieth century, infectious disease was a significant contributor to global morbidity and mortality, with meager average life expectancy at birth being 47 years, even in industrialized nations (Center for Health Statistics 2020; Adedeji 2016). Infectious diseases such as tuberculosis, plaque, pneumonia, typhoid, cholera, syphilis, smallpox, diphtheria were pervasive with no options for treatment (Adedeji 2016). The discovery of penicillin in the early twentieth century revolutionized the field of medicine, followed by the discovery of several antibiotics that aid in treating deadly microbial infections (Center for Health Statistics 2020; Fleming 1929; Lobanovska and Pilla 2017). However, the increase in the emergence of resistance to antibiotics in the last few decades poses a grave threat to public health around the world (Aslam et al. 2018; Murray et al. 2022; Antibiotic resistance threats in the United States 2019; Uddin et al. 2021).

Resistance to antibiotics is the natural process possessed by several types of bacteria from different mechanisms such as genetic mutations, selection pressure (Munita and Arias 2016; Davies and Davies 2010; Martínez and Baquero 2014). Unfortunately, the suboptimal doses and misuse of antibiotics in the healthcare system of humans and animals, as well as in agriculture, have been instrumental in the emergence and spread of antibiotic resistance (Martínez and Baquero 2014; Ayukekbong et al. 2017; Almansour et al. 2023; Dijksteel et al. 2021). Currently, the various antibiotic-resistant strains such as methicillin-resistant *Staphylococcus aureus* (MRSA), penicillin-resistant *Streptococcus pneumonia* (PRSP), multidrug-resistant gram-negative bacilli (MDRGNB), vancomycin-resistant enterococci (VRE), and carbapenem-resistant enterococci became a cause of deadly bacterial infections in humans, especially in those who have compromised immune system

K. N. Mihooliya (✉) · A. Kumari (✉)
Department of Chemical Engineering, Indian Institute of Technology, Delhi, India

P. Baindara, S. M. Mandal (eds.), *Evolution of Antimicrobial Peptides*,
https://doi.org/10.1007/978-3-031-67515-7_7

(Adedeji 2016; Dijksteel et al. 2021; Lee Ventola 2015; Yoshikawa 2002; Assis et al. 2017; Laws et al. 2019; de Oliveira et al. 2022). As a result, addressing antibiotic resistance has become a critical global health priority. Scientists, healthcare, and policymakers are working diligently to develop new antibiotic alternatives and promote responsible antibiotic use to combat this growing public health crisis.

Antimicrobial peptides (AMPs), a category of small biologically active compounds referred to as host defense peptides, have emerged as a promising alternative to conventional antibiotics due to their wide-ranging antimicrobial efficacy against a diverse array of microorganisms, encompassing bacteria, viruses, protozoans, eukaryotic pathogens, and more (de Oliveira et al. 2022; Wang et al. 2019; Zhang et al. 2021a; Shanmugaraj et al. 2021). The significance of AMPs extends beyond their direct antimicrobial activity. They have gained attention for their immunomodulatory functions, which can escalate the body's defense mechanisms against various infections (Shanmugaraj et al. 2021; Ngambenjawong et al. 2022; Duarte-Mata and Salinas-Carmona 2023; Viana de Freitas et al. 2023). Additionally, AMPs have demonstrated the potential to influence wound healing and tissue repair and even exhibit anticancer properties (Jafari et al. 2022; Luong et al. 2020; Gera et al. 2022). As we scrabble around the understanding of AMPs, their distinctive attributes, and multifaceted roles make them a subject of enormous scientific research in therapeutics as well as for other beneficial applications.

In this context of achieving multiple applications concomitantly, these molecules possess highly significant values at commercial levels, especially in the sector of biopharmaceuticals and medicine. To achieve the desired outcomes, it is imperative to efficiently investigate the bioprocessing facets of these peptides, aiming for their pilot-scale production to facilitate more comprehensive pharmaceutical and other biotechnological evaluations. The bioprocessing of AMPs refers to the systematic production and purification of these naturally occurring molecules through biotechnological methods involving an array of steps to cultivate the source microbial source, extract, purify, and subsequently analyze the quality of the final molecule, which will be checked for various applications (Fig. 7.1). The bioprocessing aspects usually consist of the following criteria.

1.1 Selection, Identification, and Isolation of Source Organisms

The choice of source organism holds significant importance in producing any biomolecule. Natural biodiversity boasts a plethora of organisms that produce various biomolecules with applications spanning numerous fields (Pham et al. 2019). Nevertheless, within this diversity, certain sources exhibit a pronounced abundance of specific biomolecules akin to valuable repositories tailored for particular applications (Chin et al. 2006). AMPs have been documented as being isolated from diverse sources, ranging from terrestrial ecosystems, such as soil environments, to the rich

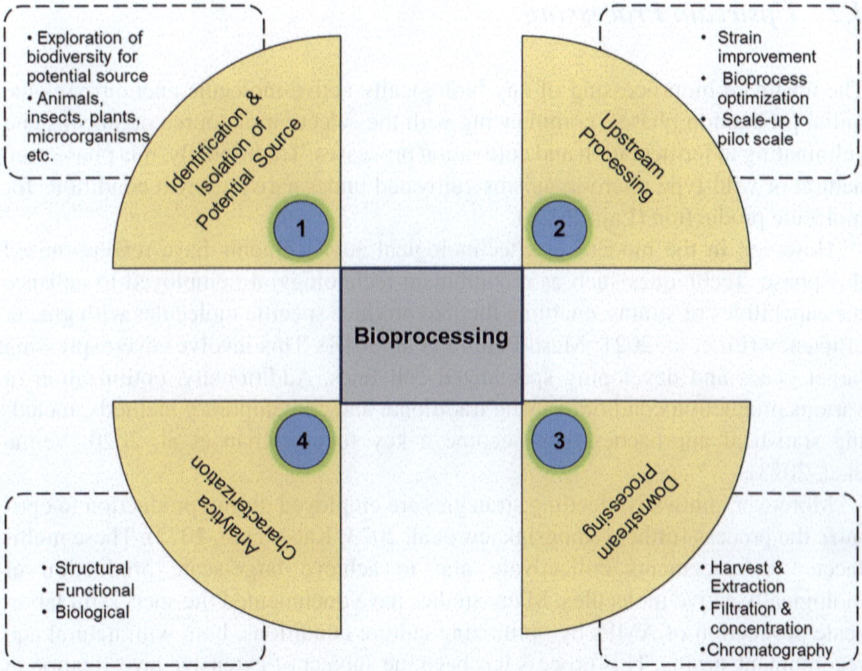

Fig. 7.1 Array of steps involved in bioprocessing; (1) Identification & isolation of potential source; (2) Upstream bioprocessing; (3) Downstream processing; and (4) Analytical characterization

and diverse biota inhabiting marine ecosystems (Semreen et al. 2018; Atipairin et al. 2022; Wu et al. 2021; Wang et al. 2017). AMPs have been extensively isolated from microorganisms in terrestrial soil ecosystems, illustrating their wide ecological prevalence. Specifically, two distinct types of AMPs have been isolated and characterized from *Bacillus subtilis* strain SK.DU.4, obtained from a rhizosphere soil sample (Baindara et al. 2013). Furthermore, investigations have provided compelling evidence of isolating and identifying microorganisms responsible for producing AMPs from onion (*Allium cepa L.*), underscoring the breadth of potential sources for these antimicrobial molecules (Taggar et al. 2021). Additionally, many potent AMPs have been notably sourced from marine environments, covering approximately 70% of the Earth's surface (Falanga et al. 2016). This vast marine realm boasts a rich diversity of microbial life, and research efforts have extensively explored this ecosystem to isolate and produce AMPs. Various marine sources, including bacteria, yeasts, and marine invertebrates, have been investigated for their AMP-producing potential (Wu et al. 2021; Guryanova et al. 2023; Macedo et al. 2021). This underscores the comprehensive exploration of the marine environment as a valuable reservoir for discovering and producing AMPs. The versatility and adaptability of AMPs in nature are evident in the wide spectrum of sources from which they are derived, signifying their potential in diverse ecological niches.

1.2 Upstream Processing

The upstream bioprocessing of any biologically active molecule encompasses the initial production phases, commencing with the selection of source organisms and culminating in fermentation and cultivation processes. Traditionally, this phase used natural or wild-type microorganisms cultivated under native growth conditions for molecule production (Lajis 2020).

However, in the modern era, technological advancements have revolutionized this phase. Techniques such as recombinant technology are employed to enhance the capabilities of strains, enabling them to produce specific molecules with greater efficiency (Lu et al. 2021; Mesa-Pereira et al. 2018). This involves overexpressing target genes and developing specialized cell lines. Additionally, optimization of various production conditions using traditional and contemporary methods, including statistical approaches, has become a key focus (Khan et al. 2020; Verma et al. 2023).

Moreover, innovative feeding strategies are employed during production to optimize the process further (Kongsinkaew et al. 2023; Käßer et al. 2022). These multifaceted enhancements collectively aim to achieve large-scale production of biologically active molecules. Many studies have documented the successful large-scale production of AMPs by optimizing culture conditions, both with natural and recombinant strains. This process has been the subject of extensive investigation in scientific research.

1.3 Downstream Processing

Downstream processing includes a series of critical procedures following the production of desired molecules, extending from harvesting microbial biomass or the culture medium containing the target product to attaining the final purified substance. In the context of conventional biological molecules, this often involves separating the target product from impurities through methods such as centrifugation, filtration, or other separation techniques (de Oliveira et al. 2022; Tripathi and Shrivastava 2019; Gupta and Shukla 2017).

AMPs are small molecules typically produced extracellularly by natural or wild-type strains. These AMP-containing fractions are ideally separated from impurities using organic solvents like ethyl acetate, methanol, petroleum ether, etc. (Verma et al. 2023; Zhang et al. 2018a). This choice of organic solvents is strategic because they can effectively extract compounds with lipophilic and hydrophilic characteristics in their structure. AMPs and several other low molecular weight compounds possess these dual characteristics, making organic solvent extraction methods conducive to partial purification of most AMPs from natural strains (Zhang et al. 2018a). In cases involving recombinant or genetically engineered strains, the downstream processing of AMPs relies on conventional chromatography methods for extraction and purification.

Downstream processing of AMPs is often regarded as challenging compared to other biological molecules due to their small size (Luo et al. 2023; da Cunha et al. 2017). Nevertheless, with technological advancements, robust techniques are now employed for AMPs downstream processing. These modern methods encompass utilizing specific extraction conditions, applying membrane-based filtration techniques such as ultrafiltration and microfiltration, resin-based adsorption, and culminating in final purification achieved through advanced chromatographic techniques (Schwarz et al. 2022; Sinha and Shukla 2018; Liang et al. 2023). The chromatographic techniques include interaction, capture, or size exclusion chromatography, etc., employing specially designed chromatography columns tailored to the unique characteristics of AMPs (Selsted 1997; Niu et al. 2022).

1.4 Analytical Characterization

The postproduction and purification analysis of biomolecules, including AMPs, is an integral component for ascertaining their purity, structural attributes, and functional characteristics. This analytical characterization is instrumental in discerning their potential applications within specific domains. A spectrum of advanced methodologies is available for effectively elucidating the unique features of AMPs.

Amino acid sequencing serves as the foundation for their classification, enabling a comprehensive understanding of their primary structure. Mass spectrometry–based techniques, such as matrix-assisted laser desorption/ionization (MALDI) and electrospray ionization (ESI), are pivotal for determining molecular size and weight, aiding in evaluating purity and identifying post-translational modifications (Stegemann and Hoffmann 2008; Kumari et al. 2021).

Circular dichroism (CD) and light scattering techniques, notably dynamic light scattering (DLS), provide insights into the secondary structure and diffusion behavior of AMPs (Avitabile et al. 2014; Ilyas et al. 2019; Freire et al. 2021). Nuclear magnetic resonance (NMR) spectroscopy is indispensable for visualizing the intricate three-dimensional structure of AMPs, allowing a detailed examination of atomic spatial arrangements, disulfide bond positions, and side-chain orientations (Porcelli et al. 2013). Fourier-transform infrared (FTIR) spectroscopy and fluorescence spectroscopy are deployed to ascertain the chemical and functional groups present in AMPs, delivering valuable information regarding their chemical composition and functional attributes (Verma et al. 2023; Kumari et al. 2021; Castellano et al. 2007).

The biotechnology field has witnessed substantial progress in developing peptide-specific chromatographic columns tailored for the precise analysis of AMPs, aiding in determining signature profiles and the amino acid composition of various AMPs (Grant and Crankshaw 2003; Rose et al. 1983). Notably, many AMPs resist proteolytic degradation structural stability, even at high temperatures exceeding 100 °C (Verma et al. 2023; Baindara et al. 2016; Heymich et al. 2021). This resilience underscores the indispensable role of the aforementioned analytical techniques in deciphering amino acid sequences and identifying the specific amino acids responsible for their functional properties. In this context, the characterization of AMPs holds paramount significance in unraveling their intricacies and refining their design for diverse applications, encompassing the development of antimicrobial drugs, food preservation, and a myriad of other biotechnological endeavors.

The demand and scale of bioprocessing for any product are intrinsically linked to its significance and global market requirements. In this regard, the global demand for AMPs is anticipated to increase substantially in the coming years (Mazurkiewicz-Pisarek et al. 2023). This projection is underpinned by the persistent and alarming escalation of drug-resistant infections observed over the last few decades. The proliferation of antibiotic-resistant bacteria has emerged as a global concern of paramount importance. Within this context, AMPs emerge as a beacon of hope for researchers and healthcare professionals, offering the potential to serve as effective antibiotics against various life-threatening infectious diseases. Conditions such as pneumonia and tuberculosis, which have exhibited resistance to conventional antibiotics, stand to benefit from the unique properties of AMPs. Beyond their applications in therapeutics, AMPs find utility in various sectors, including food preservation, cosmetic and personal care products, feed additives, transgenic crop resistance, agriculture, and veterinary medicine (Erdem Büyükkiraz and Kesmen 2022). These multifaceted applications render AMPs exceptionally attractive for commercial development. Therefore, several vital drivers contribute to the dynamism of the global AMPs market.

1.5 Technological Advancements

Owing to their significance, AMPs have gained considerable attention and enthusiasm among diverse scientific communities worldwide. Ongoing technological advances have propelled research and development efforts, promising novel applications and enhanced production methods.

1.6 Global Surge in Infectious Diseases

The heightened prevalence of infectious diseases globally has heightened the urgency for innovative solutions. The adaptability and effectiveness of AMPs in addressing these diseases make them increasingly indispensable (Mazurkiewicz-Pisarek et al. 2023).

1.7 Pressing Need for AMPs

The indispensable need for AMPs across various sectors is instrumental in shaping market dynamics. The current chapter will delve into avant-garde methodologies for the bioprocessing of AMPs at both laboratory and pilot scales, encompassing the entire spectrum from upstream processing to the attainment of finished products. This comprehensive exploration aims to shed light on the global market demand for AMPs, contingent upon their extensive array of applications at the commercial level.

2 Overview of AMPs

AMPs represent a critical component of the innate immune system, serving as host defense peptides present in various life forms, including mammals, insects, animals, birds, fish, plants, crustaceans, and microbes. These AMPs are short bioactive molecules typically composed of 5–100 amino acids, exhibiting a broad spectrum of antimicrobial properties against highly pathogenic strains of both gram-positive and gram-negative bacteria, enveloped and non-enveloped viruses, fungi, parasites, and other microorganisms. Particularly noteworthy is their effectiveness in combatting drug-resistant microbial infections (Bahar and Ren 2013). Despite their antimicrobial potential, only a limited number of AMPs, such as polymyxin B and gramicidin S, have found application in clinical medicine. This restricted use can be attributed to their broad-spectrum activity and tendency to lose efficacy under physiological conditions (Drayton et al. 2020).

AMPs, in addition to their antagonistic antimicrobial functions, possess immune-activating properties and may contribute to wound healing (Nasseri and Sharifi 2022). These peptides are primarily localized in regions of organisms that are exposed to environmental challenges, such as the lungs, epithelial surfaces, and mucosal secretions, and are notably abundant in neutrophils (Moretta et al. 2021; Mahlapuu et al. 2016). The prevalence of AMPs across diverse life forms and the range of environmental stressors encountered by the organisms producing them have led to the development of distinct classes of AMPs.

The amino acid composition of AMPs greatly influences their hydrophobic and amphiphilic properties, including their anionic and cationic charges, which exert specific effects on different microorganisms. However, AMPs are primarily amphiphilic due to the presence of both hydrophobic and hydrophilic charged residues (Lei et al. 2019). These charged residues are crucial for the initial interaction between AMPs and susceptible organisms. In the case of gram-negative bacteria, this interaction occurs between the positive charge residues of AMPs and the anionic lipopolysaccharides, while for gram-positive bacteria, it involves the anionic teichoic acids (Malanovic and Lohner 2016). Subsequently, the hydrophobic residues of AMPs come into play, disrupting or solubilizing the cell membrane and forming pores. This disrupts the membrane's barrier function and alters the membrane potential, which are critical mechanisms for the survival of the microorganisms (Dijksteel et al. 2021).

2.1 AMPs Origin

AMPs are ancient components of the immune system, dating back to approximately 2.6 billion years ago, before the emergence of adaptive immunity (Gordon et al. 2005). These peptides are a crucial aspect of innate immunity, representing the first line of defense against various pathogens. Their origins can be traced to the early stages of life on Earth, making them ancient defenders in the realm of immune defense mechanisms (Gordon et al. 2005).

AMPs are remarkably conserved among different species, including mammals, insects, fish, plants, and a myriad of other organisms (Mahlapuu et al. 2016; Hancock 2000; Kościuczuk et al. 2012). Their evolutionary conservation and rapid response capabilities imply that they have persisted throughout the evolution of various species, providing essential defense even in the absence of more advanced immune systems. The diverse functions AMPs serve, including roles in immune response, wound healing, and inflammation, further support their ancient origins. This cumulative evidence points to the existence of AMPs in early life forms on Earth, with their persistence and conservation playing a pivotal role in the evolution of various species.

In terms of the discovery of AMPs, the journey began with the pioneering work of scientists. The first known AMP was identified in 1939 by Dubos from a soil bacterial strain, *Bacillus*, which protected mice from pneumococcal infections (Dubos 1939). This AMP was subsequently identified the following year and named gramicidin by Hotchkiss and Dubos (Bahar and Ren 2013; Hotchkiss and Dubos 1940). In 1941, two additional AMPs, tyrocidine and purothionin, were isolated by different research, demonstrating their antimicrobial capability against gram-positive and gram-negative bacteria and fungi (Bahar and Ren 2013). The first animal-derived AMP, defensin, was reported in 1956, followed by the discovery of two more animal-derived AMPs in 1957, known as bombinin and lactoferrin (Pärn et al. 2015). According to the data repository of antimicrobial peptides (DRAMP), there are currently around 6100 entries for various AMPs, encompassing both natural and synthetic variants, reflecting the ongoing exploration of these essential components of host defense (Shi et al. 2022).

2.2 Classification of AMPs

AMPs constitute a complex and diverse group of molecules that pose challenges in categorization due to their multifaceted characteristics. However, classifying AMPs is essential to establish a framework for comprehending their diversity and functionality. This classification is crucial for unlocking their potential applications in different fields such as microbiology, immunology, and biotechnology. AMPs are typically classified based on various factors.

2.2.1 Origin Source

AMPs can be categorized according to their source of production, including host defense peptides (e.g., defensins and cathelicidins), microbial peptides (such as bacteriocins), plant-derived peptides (e.g., thionins and cyclotides), animal peptides (e.g., indolicidin and lactoferrecin), and synthetic or engineered peptides (e.g., Epi-1) (Rima et al. 2021; Huan et al. 2020).

2.2.2 Structural Features

The predominant structural elements within AMPs dictate their classification. Common structural categories include (a) α-helical AMPs, characterized by α-helical secondary structures that often form pores in the cell membranes of the target, (b) β-sheets AMPs, composed of β structures, which can effectively disrupt membranes or interact with intracellular targets, and (c) loop structures AMPs, characterized by the presence of disulfide bridges or cyclic structures (Saeed et al. 2022).

2.2.3 Based on Mechanism

AMPs employ various mechanisms to combat pathogens, leading to their classification as (a) membrane-disrupting AMPs, which destabilize the integrity of microbial cell membranes, ultimately causing cell lysis (Rima et al. 2021); (b) nonmembrane-disrupting AMPs, targeting intracellular components of microbes without necessarily affecting the cell membrane (Huan et al. 2020); (c) immunomodulatory AMPs, capable of modulating the host's immune response to enhance its ability to combat infections (Saeed et al. 2022); and (d) dual-action AMPs, which exhibit a multifaceted defense mechanism by combining both membrane-disrupting and intracellular-targeting mechanisms (Huan et al. 2020; Saeed et al. 2022).

2.2.4 Sequence and Amino Acid Composition

AMPs are characterized by diverse sequences and amino acid compositions. This forms the basis for categorization into (a) cationic AMPs, rich in amino acids having positive charges like arginine and lysine, that interact with microbe's membrane, which is negatively charged; (b) anionic AMPs, featuring an abundance of negatively charged amino acids that interact with positively charged components in microbial membranes, and (c) amphipathic AMPs, containing both hydrophobic and hydrophilic amino acid residues, allowing them to insert into and disrupt microbial membranes (Huan et al. 2020).

2.2.5 The Spectrum of Target

AMPs vary in their effectiveness against a range of pathogens, leading to classification as (a) broad-spectrum AMPs, with the capability to combat a wide array of pathogenic microbes, including bacteria, fungi, viruses, etc., and (b) narrow-spectrum AMPs, which specifically target particular groups of pathogenic microbes.

It is important to note that an individual AMP may fit into multiple categories simultaneously, with its classification depending on the specific criteria applied. This classification system is indispensable in organizing and comprehending AMPs, enabling their efficient utilization in various applications.

2.3 AMPs Mode of Action

AMPs can exert their antimicrobial effects in two main ways: extracellular mechanism and intracellular mechanism.

2.3.1 Extracellular Targeting

Extracellularly, AMPs primarily act on the plasma membrane of microorganisms by forming pores. This process involves several key steps.

Attraction
For the initial AMPs mechanism, the AMPs attach to the bacteria surface by the electrostatic association between the positively charged AMPs and the negatively charged components of the bacteria membrane, such as lipopolysaccharides (LPS) in gram-negative and teichoic acids in gram-positive bacteria (Talapko et al. 2022). However, this initial attraction is not sufficient for antimicrobial activity.

Attachment
After initial attraction, AMPs reach the cell membrane and interact with the lipid bilayer. Research has shown that the interaction between AMPs and lipid membranes can create tension in the membrane, with the extent of tension being proportional to the concentration of AMPs (Huan et al. 2020).

Insertion and Membrane Permeability
The way in which AMPs bind to the membrane depends on the ratio of AMPs to lipids. At a low ratio, AMPs bind in a parallel orientation to the membrane. In contrast, they insert perpendicularly into the membrane at a high ratio, forming pores in

the cell membrane. There are four models describing AMP interaction with cell membranes.

- *Barrel-stave model:* This model involves the perpendicular insertion and aggregation of AMPs into the membrane, forming transmembrane pores (Lohner and Prossnigg 2009).
- *Carpet model:* It involves the accumulation of peptides on the membrane's surface, leading to pore formation as their concentration increases (Talapko et al. 2022; Oren and Shai 1998).
- *Toroidal pore model:* This model involves helical peptides binding to the lipid bilayer and inducing pore formation by bending the lipid monolayer (Matsuzaki et al. 1995).
- *Aggregate model:* This model describes the interaction of AMPs with the bacterial membrane, forming complex micelles (Bechinger and Lohner 2006).

2.3.2 Intracellular Targeting

Certain AMPs exert their effects on bacteria's internal components. They activate autolysins, which, in turn, lead to cell lysis. Additionally, these AMPs can trigger the activation of phospholipases in the bacterial membrane, resulting in the degradation of phospholipids. To access these intracellular targets, AMPs need to cross the bacterial membrane (Le et al. 2017).

One such AMP, buforin II, a linear α-helical peptide with a proline hinge, facilitates penetration of the plasma membrane (Park et al. 2000). Once inside, it accumulates in the cytosol, effectively combating bacteria. Lantibiotics, derived from gram-positive bacteria, contain thioether amino acids like lanthionine (McAuliffe et al. 2001). They target lipid II and inhibit cell wall synthesis, ultimately forming pores in the bacterial membrane (McAuliffe et al. 2001). These intricate mechanisms underscore the multifaceted ways AMPs combat bacterial infections, both externally and internally.

All these mechanisms demonstrate the diverse strategies employed by AMPs to combat pathogenic microorganisms, both by disrupting the cell membrane from the outside and targeting intracellular components.

3 Bioprocessing of AMPs

Bioprocessing is a branch of biotechnology that harnesses the potential of living organisms or biological systems to transform emerging technologies into products for society's benefit and plays a pivotal role in various industries (Esmonde-White et al. 2022; Jozala et al. 2016). It is instrumental in producing a wide array of products, including biopharmaceuticals, organic acids, and other industrial goods (Rathore et al. 2021). Bioprocessing encompasses a series of critical steps to achieve the desired end products, ranging from isolating biological sources to identifying promising molecules and producing purified products at the required scale (Esmonde-White et al. 2022; Jozala et al. 2016).

In the specific realm of AMPs bioprocessing, certain translational steps can be particularly challenging due to the necessity of using organic solvents in their extraction, low biomass levels in some instances because of self-inhibitory mechanisms, or their relatively small molecular size. Nonetheless, some AMPs, like polymyxin B and gramicidin S, are commercially available and employed for medicinal purposes.

Therefore, in this section of the chapter, we delve into the various facets of AMPs bioprocessing. We examine the intricacies and challenges associated with producing these AMPs, exploring methods for production, extraction, purification, analysis, and scale-up processes. Additionally, we consider the broader scientific and commercial implications of AMPs in medicine and related industries that influence its market demands.

However, these antimicrobial peptides exhibit several benefits over traditional antibiotics, including broad-range activity, sensitivity to drug-resistant bacteria, and a method of action devoid of specific binding sites. However, using these peptides poses several difficulties due to their expensive manufacture, potential toxicity, vulnerability to proteases (also present in wound fluid), and uncertain pharmacokinetics (Poojitha et al. 2023).

3.1 Upstream Processing of AMPs

The first step in translating AMPs production from laboratory to pilot scale is upstream processing, which encompasses recognition of potential sources, that is, natural or engineered, optimization of production parameters, harvest, etc.

3.1.1 Process Development to Produce AMPs from Wild-Type or Natural Strains

In process development, the main goal is the efficient manufacturing of the required product, and for fermentation-based goods, it becomes crucial to optimize the composition of the medium and conditions for growth to increase product yield (Chopra et al. 2015). The effects of pH, temperature, and medium components on bacteriocins' synthesis and antimicrobial properties have been studied (De Vuyst 1995; Krier et al. 1998). Since case-specific tailoring of numerous factors for bioproduct creation is the norm, generating bacteriocins frequently does not require a complex medium's components (Han et al. 2019). The traditional one-factor-at-a-time (OFAT) method of medium composition optimization involves changing one independent variable while fixing the values of the other constants. This is exceedingly time-consuming, costly for numerous variables and may result in incorrect findings

(Adinarayana et al. 2003). This requirement is met by statistical experimental strategies like one factor at a time (factorial design) and response surface methodology (RSM). When comparing Plackett–Burman designs (PBDs) to equivalent fractional designs, PBDs require fewer runs than comparable fractional designs because they use a two-level experimental design. RSM is a group of statistical methods for creating models by using the critical factors obtained from PBD, designing experiments, assessing the effects of factors, and looking for the best combinations of factors to produce desired results (Li et al. 2002). A significant benefit of RSM is that fewer experiments are required to test several parameters simultaneously while accounting for their interactions (Chopra et al. 2015). Numerous biotechnology fields have successfully employed it, including improving enzyme synthesis conditions and media compositions. Using these statistical experimental designs, the author has reported a 2.5-to-15-fold increase in AMPs production (Verma et al. 2023; Chopra et al. 2015).

3.1.2 Biological Expression Systems to Produce AMPs

It is challenging to isolate and purify natural AMPs originating from plants and animals. A synthetic AMP has a lower yield and costs more than those produced by animals and microbes. In recent years, it has become clear that the commercial manufacturing of AMPs can significantly benefit from the recombinant expression of AMPs using genetic engineering technologies. Bacteria and yeast expression systems comprise most of the standard expression systems for AMPs. Bacterial expression has the benefits of fast expression times, high expression levels, and low expression costs; however, difficulties with lack of modification and AMP purification may occur. The fact that yeast expression is nontoxic to host cells and facilitates facile purification is one of its benefits, but it has a high expression cost and prolonged expression times as drawbacks (Lyu et al. 2023). The use of transgenic plants for AMP generation has recently gained popularity because the desired peptide can be directly expressed in the crop without having to be purified. For these reasons, bacteria, yeasts, and plants have been selected as promising possibilities for producing peptides (Parachin et al. 2012).

Bacterial Expression Systems

The most common microorganism used to manufacture AMP is the bacterium *Escherichia coli* (Ingham and Moore 2007). Multiple factors contribute to this, one of which is the cost-effectiveness of utilizing *E. coli* for recombinant protein production. This is attributed to its fast growth rate, abundant expression vectors, established DNA engineering techniques, and comprehensive understanding of its physiology, genetics, and biochemistry (Sørensen and Mortensen 2005). In conclusion, *E. coli* remains the preferred host for AMP synthesis, with the highest yields observed thus far. In addition to *E. coli*, *Bacillus subtilis* was employed to express the hybrid peptide cecropin AD (Feng et al. 2012). It has also been noted that *Propionibacterium freudenreichii* serves as a host for the heterologous production of AMPs (Brede et al. 2005). In terms of the number of genetic tools available, the majorly exploited expression vectors for AMP generation are the pET series created by Novagen. The most popular expression hosts are Origami and Rosetta, C41 (DE3), and BL21 (DE3) (Novagen).

However, some obstacles must be surmounted to establish efficient AMP production in bacteria, like protease degradation of AMPs, lethality of AMPs toward host strain, and solubility of AMPs (Li 2011). To overcome these problems, AMPs are fused to a carrier protein (harboring the net anionic charge), which prevents degradation by neutralizing the AMPs charge and toxicity toward the host (Li 2011). To increase the solubility of AMPs, glutathione S-transferase (GST), F4 fragment of PurF, green fluorescent protein (GFP), and small ubiquitin-like modifier (SUMO) are a few of the fusion proteins that have been employed for the AMPs expression and purification, with thioredoxin being the most popular one (Parachin et al. 2012).

Modifying or engineering the natural sequences of AMPs is another method for increasing their expression, solubility rates, and antibacterial spectrum (Cao et al. 2005). Furthermore, codon optimization makes the recombinant expression of multiple AMPs from different sources easier (Huang et al. 2008). Furthermore, post-translational modifications are impossible in prokaryotes, and in these circumstances, other expressions to synthesize AMP must be addressed (Parachin et al. 2012).

Yeast-Derived Expression Systems

Pichia pastoris and *Saccharomyces cerevisiae* are often employed in practice to express AMPs. Yeast cells offer distinct advantages compared to prokaryotic cells in expressing AMPs, mainly when the expression involves post-transcriptional and post-translational modifications, such as glycosylation. Both types of yeast enable the secretion of recombinant proteins, leading to reduced purification steps, high protein recovery, and a more straightforward scale-up process (Cregg et al. 2009). *S. cerevisiae*, a commercially accessible yeast, has been exploited as a host for the bulk of recombinant molecules developed in recent years (Mattanovich et al. 2012). However, *S. cerevisiae*, a fermentative yeast, channels a few of its carbon resources into ethanol production, leading to decreased biomass and power protein yields (Mattanovich et al. 2012). Consequently, *S. cerevisiae* might not be the optimal host for the large-scale synthesis of heterologous proteins (Parachin et al. 2012).

In recent times, there has been a substantial elevation in the use of *P. pastoris* as a heterologous expression host, with a growing number of studies. *P. pastoris* is not in the category of fermentative microbes like *S. cerevisiae*, allowing it to be grown at a greater cell density with no harmful metabolites or metabolism rerouting toward ethanol synthesis. Additional benefits of utilizing *P. pastoris* include cost-effective culture conditions, elevated heterologous protein expression, easy scale-up, simple purification, and the ability to perform post-translational modifications essential for functionality (Cereghino et al. 2002).

Most research on generating recombinant or engineered AMPs utilizing *P. pastoris* is still done in shaking flasks. A new method for the production of porcine interferon has been recently developed. Once a high cell density of 110–130 g/L was reached, the methanol feed was fine-tuned to achieve optimal recombinant protein production without inhibiting cell growth (Yu et al. 2010). While *P. pastoris* is undoubtedly a top choice for yeast-based host expression, it requires concurrently developing a more suitable promoter (inducible or constitutive) along with process optimization to accomplish the elevated yield of recombinant AMPs (Parachin et al. 2012).

Plant-Based Expression Systems

Developing genetically modified plants for recombinant production of proteins, peptides, and other molecules is mainly used to improve crops. Consequently, there have been limited initiatives to measure the quantity of recombinant AMP generated (Delaunois et al. 2009). The antibacterial effect is characteristic in AMPs, which inhibits their expression in microbial systems. Plant-based expression platforms have been established to generate such peptides without needing controlled expression, as they can express high amounts and contain complex folding and processing processes (Chen et al. 2005). Plants, superior eukaryotes, are well-adapted for developing AMPs with intricate disulfide link structures, including cyclotides with six cysteines. However, managing the yield of AMPs produced by plant expression might be challenging. Additionally, specialized glycosylation routes exist in plants compared to animal glycosylation processes. Animal AMP expression may change their original structure and reduce their antibacterial action (Lyu et al. 2023). Past studies have unveiled the production of heterologous AMPs in plants, aiming to enhance crop resistance against fungal and bacterial infections. Plant and animal defensins are also categorized as a typical class of AMPs expressed in plants although microbial AMPs are rarely found (Pelegrini et al. 2008). Resistance to *Rhizoctonia solani*, *Erwinia carotovora*, *Phytophthora infestans*, *Streptomyces scabies*, and *Fusarium solani* was recently demonstrated in potato plants reinforced by a combination of three antimicrobial-expressing genes (Rivero et al. 2012). The findings suggested that combining AMPs could be a promising strategy in crop improvement for achieving resistance to bacterial and fungal diseases (Rahnamaeian and Vilcinskas 2012). The use of plants as AMP expression carriers is still in the early stages and requires more research.

3.1.3 Scale-Up of AMP Production

After selecting an efficient production source, the primary objective is to develop an optimized production process in shake flasks, which can be scaled up to the pilot level. This involves refining various physiological and nutritional parameters at the laboratory scale. Following the successful optimization of bioprocess conditions in shake flasks, it is standard practice to validate these optimized conditions in small-scale bioreactors, typically ranging from 2 L to 20 L in volume (Mihooliya et al. 2020). Once validated, the objective is to smoothly transition the process to a larger scale while maintaining key factors at constant volumetric levels (Fig. 7.2).

For a successful scale-up, two critical factors come into play. First, it is essential to maintain a consistent agitation rate on a volumetric basis. This is achieved by keeping the tip speed of the agitation system constant across bioreactors of different scales (Karimi Alavijeh et al. 2022). Second, it is imperative to maintain a constant volumetric rate of dissolved oxygen transfer throughout the scaling process (Shin et al. 2013; Garcia-Ochoa and Gomez 2009). This ensures that the biological processes, including microbial growth or cell culture, remain under optimal conditions as the production volume increases (Garcia-Ochoa and Gomez 2009). These meticulous steps are crucial for the seamless transition of bioprocesses from laboratory-scale validation to larger production scales.

One notable achievement is the enhanced production of the AMP called peptide-MT110, derived from the new source *Vibrio proteolyticus* MT100 (Verma et al. 2023). Through the optimization of nutritional and physiological factors in shake flask experiments, production increased by 2.5-fold compared to the initial unoptimized state. This success facilitated the scale-up of peptide-MT110 production from shake flasks to a 2 L bioreactor, with appropriate control of dissolved oxygen and agitation rate (Verma et al. 2023). In another study, the production of the AMP sonorensin was improved by 15-fold through the optimization of various parameters and scaling up to a 5 L bioreactor, with particular attention to volumetric oxygen transfer (Chopra et al. 2015).

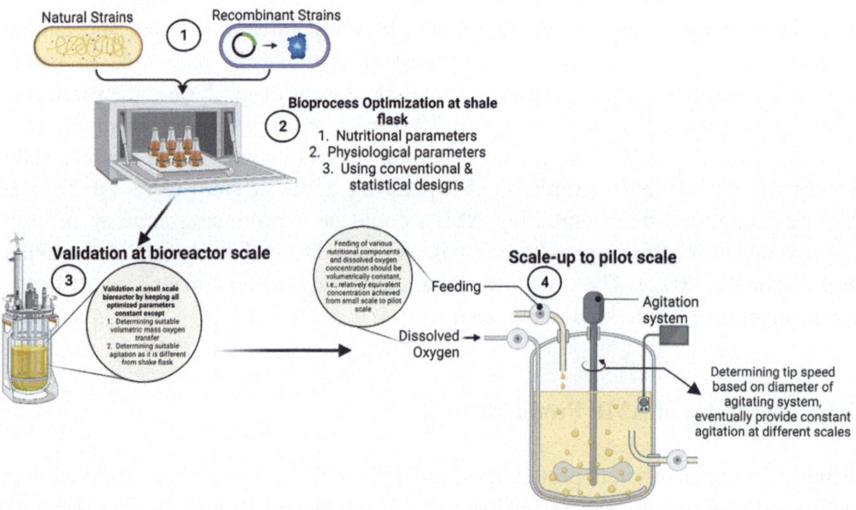

Fig. 7.2 Upstream bioprocessing of different biologically active molecules involving scale-up from lab to pilot level. (Note: This image is created by BioRender)

Different fermentation modes are available for large-scale production of AMPs, including batch, fed-batch, and continuous processes. However, fed-batch processes are predominantly employed for submerged fermentation in AMP production. This approach was demonstrated in producing clavanin-MO peptide on a 5 L scale. Furthermore, a separate study successfully produced the peptide piscidin 4 at a 500 L scale using fed-batch processes in a bioreactor (Tai et al. 2021).

3.2 *Downstream Processing of AMPs*

Downstream processing constitutes the vital postproduction phase dedicated to recovering and purifying specific molecules. It is pivotal in upholding bioactive substances' safety, purity, and efficacy for diverse applications, including pharmaceuticals, food, cosmetics, and industrial processes. In the realm of biotechnology, downstream processing stands out as the most intricate manufacturing phase, characterized by a series of complex, time-intensive steps that require a high level of expertise and incur substantial process costs (Matte 2022). Remarkably, downstream processing alone typically accounts for approximately 50–80% of the total process cost in the biotechnology and biopharmaceutical industry (Matte 2022; Eidenberger et al. 2023; Kruschitz and Nidetzky 2020). Consequently, developing a robust and cost-effective downstream processing system is a prerequisite for ensuring the viability of producing biological molecules at a commercial level.

AMPs are small bioactive compounds characterized by their relatively modest molecular size and a diverse array of peptides with distinct properties and structures. These unique characteristics present both challenges and opportunities in the downstream processing of AMPs (Luo et al. 2023; da Cunha et al. 2017). The downstream processing of AMPs entails a series of intricate steps, starting with the initial recovery process involving the isolating and extracting fractions containing the AMPs (Verma et al. 2023). Subsequently, the process includes concentration methods such as precipitation or filtration, with techniques like ultrafiltration and diafiltration employed (Pieracci et al. 2018). Finally, the purification phase relies on various chromatography methods to yield high-quality, purified bioactive AMPs (Fig. 7.3). This intricate sequence of steps is vital for obtaining AMPs with the required levels of purity and safety for their intended applications.

3.2.1 Isolation or Extraction and Concentration of AMPs

Following the production of AMPs, the initial step in downstream processing aims to recover the fraction containing these bioactive molecules. The choice of isolation and extraction methods varies depending on the sources of the AMPs. In the case of natural sources, AMPs are often secreted into the culture medium due to their small size. To extract these AMPs, various organic solvents, such as acetone, ethanol, ethyl acetate, chloroform, methanol, acetonitrile, and mixtures like chloroform

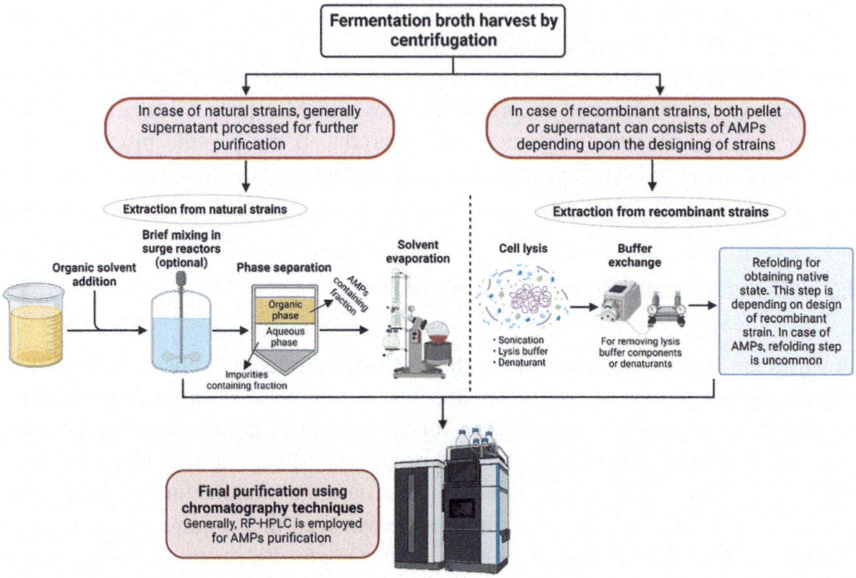

Fig. 7.3 Various steps involved in the downstream processing of AMPs. (Note: This image is created by BioRender)

methanol, are employed. Organic solvents are advantageous due to the amphipathic nature of AMPs, which have both hydrophilic and hydrophobic regions. Organic solvents can effectively dissolve the hydrophobic components of AMPs, making it easier to extract them from the surrounding biological compounds (Pieracci et al. 2018). Numerous studies have demonstrated the successful extraction of AMPs from crude samples using these organic solvents. For example, the peptide-MT110 was effectively extracted from the culture medium of *Vibrio proteolyticus* MT100 using ethyl acetate (Verma et al. 2023). In another study involving tissue samples of marine invertebrates, multiple organic solvents were employed for peptide extraction (Musthafa Poyil et al. 2021).

Given that AMPs themselves are hydrophobic, various reverse-phase matrices such as Diaion HP-20, Amberlite XAD-4, and Amberlite XAD-16 are utilized for initial extraction (Pereira-Dias et al. 2023). For example, peptides like sonorensin from *Bacillus sonorensis* MT93 and peptide-Ba49 from *Bacillus subtilis* subsp. *spizizenii* strain Ba49 were successfully adsorbed and extracted from the culture medium Diaion HP-20 resins (Taggar et al. 2021; Chopra et al. 2014). These matrices work based on the hydrophobicity of AMPs, effectively separating them from other components in the culture medium and yielding a partially purified fraction of the desired peptides (Pereira-Dias et al. 2023). Besides organic solvents and hydrophobic resins, salt-based precipitation methods, such as ammonium sulfate precipitation, are also used for AMP extraction (Rai et al. 2016). However, it is important to note that this method often results in higher impurities because many proteins and other biological components also precipitate in the presence of salts.

In cases involving recombinant or engineered sources, AMPs can be intentionally produced intracellularly, in the periplasmic space, or extracellularly. Intracellular or periplasmic AMPs are first harvested through centrifugation and then extracted by disrupting the cells using techniques like sonication or the use of chaotropic or denaturing agents. For extracellular AMPs, they are already present in the culture medium. In both scenarios, the final extraction and concentration of the AMP-containing fraction are often achieved using ultrafiltration or diafiltration techniques with appropriate membranes and hollow filters (Liang et al. 2023; Pieracci et al. 2018). In a large-scale study, a two-step filtration process was employed, involving ultrafiltration with hollow fibers followed by reverse osmosis concentration using a reverse osmosis membrane with a small pore size (0.001–0.0001 μm). This process effectively concentrated the desired fraction while removing water molecules, inorganic salts, and other impurities (Letian et al. 2015).

3.2.2 Final Purification Using Chromatographic Techniques

Chromatographic techniques represent a cornerstone in the final purification of a wide array of biological molecules at the industrial scale. These techniques are chosen based on the unique attributes of the target molecule, with the selection of chromatographic resins or matrices tailored to these attributes. The applications of chromatographic resins can be categorized according to the type of ligand involved:

- *Gel permeation or size-exclusion chromatography (GPC or SEC):* This method separates molecules based on their molecular size, allowing larger molecules to elute first.
- *Ion-exchange chromatography (IEC):* It utilizes the property of charges the molecules carry for separation.
- *Hydrophobic interaction chromatography (HIC):* It exploits the hydrophobic nature of molecules, especially in the presence of high salt concentrations.
- *Reverse-phase chromatography (RPC):* It separates molecules by their strength of hydrophobic interactions with the stationary phase.
- *Affinity chromatography:* Leverages the specific binding affinity of molecules to the resin, ensuring high selectivity.

In the context of purifying AMPs derived from natural sources, the most commonly employed chromatographic technique is reverse-phase high-pressure liquid chromatography (RP-HPLC) (Verma et al. 2023; Chen et al. 2007). This is due to its compatibility with various organic solvents, which is essential as many natural peptides, including AMPs, are only soluble in organic solvents. In RP-HPLC, AMPs bind to a carbon-based stationary phase attached to a silica matrix. Subsequently, these bound molecules are eluted using acetonitrile as the eluent. The molecules are separated and eluted at different time intervals, referred to as "retention time," based on their varying hydrophobicity toward the carbon molecules. The most frequently used RP-HPLC columns for AMP purification feature C18 (18 carbons), C8 (8 carbons), and C4 (4 carbons) as the stationary phase. Notably, a study achieved highly homogenous AMPs through a three-step purification process employing a C18 RP-HPLC column.

IEC is another popular choice for AMP purification. For instance, the strong cationic resin SP Sepharose is utilized to initially purify peptides like sonorensin and peptide-Ba49, followed by further refinement through Sephadex G-25 and G-10 (SEC) (Taggar et al. 2021; Chopra et al. 2014). In another large-scale investigation, the initial purification of AMP was accomplished using CM Sepharose cationic resins, followed by a refinement step involving a CM Sephadex A-25 column (Letian et al. 2015).

In the case of recombinant or engineered strains, the purification process for biological molecules, including AMPs, is considered relatively straightforward. In these cases, affinity or purification tags are typically affixed to the AMP molecules during production, providing a handle for their purification (Zhang et al. 2021b; Wu et al. 2022). These tags can be engineered into the sequence of interest or naturally present in the host organism used for peptide expression. However, it is essential to note that such tags may not be deemed suitable for the final commercial product and can present regulatory challenges. Consequently, these tags are often connected to molecules featuring cleavage sites. For example, in a study involving the recombinant production of the host defense peptide IDR1 (one type of innate defense regulator) in the *E. coli* BL21 strain, successful purification was achieved using a Ni-NTA Sepharose column due to its affinity for the His-tag. Subsequently, the tag was cleaved from the peptide, and a final purification step was conducted using RPC (Bommarius et al. 2010). In another study, multiple recombinant strains expressing AMPs were constructed and purified through Ni-NTA column chromatography (Zhang et al. 2021b). Similarly, a His affinity tag was employed to purify cecropin-like AMPs recombinantly expressed in *E. coli* (Wu et al. 2022).

Despite the efficiency and well-established nature of chromatographic techniques for purification, it is essential to recognize that they constitute a significant portion of the overall manufacturing costs. Therefore, optimizing and evaluating each step of the process is critical, taking into account the specific characteristics of the molecule and its intended application.

3.3 Analytical Characterization of AMPs

Analytical characterization is essential to scientific inquiry, enabling a profound understanding of compounds' composition, structure, and functional attributes. Its primary objective is to provide precise and comprehensive insights into biologically active molecules, thereby elucidating their potential commercial applications. Following the final purification of AMPs, a plethora of analytical techniques are deployed to unravel their amino acid sequence, structural conformation, and functional properties. Some fundamental methods include:

- *Peptide sequencing:* The foremost objective is ascertaining the amino acid composition and sequence. Mass spectrometry and Edman degradation, recognized as gold standards, are pivotal in elucidating the amino acid composition of AMPs (Stegemann and Hoffmann 2008; Kumari et al. 2021). For instance, mass spectrometry has been proficiently used to determine the structure of AMPs from *Odorrana jingdongensis* (Asian frogs). Additionally, it aids in molecular weight determination, often in conjunction with liquid chromatography (Verma et al. 2023; Kumari et al. 2021). The Pico-Tag method is another valuable tool for determining the amino acid composition of shorter peptides, as demonstrated with peptide-MT110 produced from *Vibrio proteolyticus* MT110 (Verma et al. 2023).
- *Nuclear magnetic resonance (NMR) spectroscopy:* NMR spectroscopy is a powerful tool in the arsenal for AMP characterization. It facilitates the determination of the three-dimensional structure of peptides and their functional attributes. Researchers have effectively harnessed NMR to explore the structure–functional relationships of multiple AMPs (Porcelli et al. 2013).
- *Fourier transform infrared (FTIR) spectroscopy:* FTIR spectroscopy is employed to discern the types of chemical groups present and the alterations they undergo during molecular interactions (Mihooliya et al. 2023). This is invaluable for deducing the functional groups within AMPs. For instance, FTIR analysis has successfully identified functional groups within peptide-MT110 produced from *Vibrio proteolyticus* MT110 (Verma et al. 2023). In another study, FTIR proved indispensable in elucidating the functional relationship between lactocin 705 and liposomal membranes (Castellano et al. 2007).
- *Biological characterization:* It aims to gauge the efficacy of AMPs against various pathogenic strains. In the context of AMPs, it involves assessing their antimicrobial activity, thereby determining potential utility as antimicrobial agents.

In addition to these techniques, many other methods are available to delve deeper into the analytical characterization of AMPs. The comprehensive characterization of AMPs necessitates the judicious combination of these techniques to gain a profound understanding of their structural attributes, functional properties, and interactions. The selection of specific methodologies is contingent on the unique research objectives and the nature of the AMPs under investigation.

4 Challenges in Bioprocessing of AMPs from Lab Scale to Manufacturing: From the Author's Perspectives

Compared to traditional proteins and other biological molecules, the scalability of AMPs production from laboratory-scale to industrial manufacturing poses several formidable challenges due to the following factors:

- *Self-inhibitory effect:* AMPs are known for inhibiting the host cells that produce them. Higher cell concentrations are required when aiming to manufacture AMPs at a larger scale. However, this often results in a decreased cell biomass due to the inhibitory nature of AMPs (Aoki and Ueda 2013). Consequently, producing AMPs necessitates using significantly larger bioreactors, which has substantial economic implications. This issue is significant enough that it discourages their production by many smaller or semi-industrial enterprises.
- *Solubility of AMPs:* AMPs are amphiphilic molecules, but many have a substantial hydrophobic component. Organic solvents are often employed to isolate AMPs from other biological components due to their propensity to precipitate in aqueous buffers (Zhang et al. 2018b; Ratrey et al. 2020). This requires the use of specialized surge tanks or reservoirs for the extraction process. Furthermore, the downstream processing of AMPs demands high-pressure-tolerant and specialized chromatographic instruments (such as HPLC) or their accessories, specifically designed for organic solvent compatibility and at a relatively higher cost than aqueous buffer-compatible systems. Using organic solvents poses safety concerns, necessitating additional steps to meet regulatory requirements (Rosinha Grundtvig et al. 2018). However, various recombinant technologies and synthetic engineering efforts have been explored to enhance the solubility of AMPs in aqueous buffers, rendering the manufacturing of AMPs at a pilot scale a daunting task.
- *Low bioavailability and short half-life:* AMPs face challenges related to their bioavailability and stability (Drayton et al. 2020; Ratrey et al. 2020). They are susceptible to degradation by digestive enzymes when administered orally, and some AMPs exhibit short half-lives in the bloodstream (Drayton et al. 2020). Moreover, systemic toxicity can arise due to their nonspecific mechanisms against pathogens, leading to rapid clearance from the body without efficient action (Drayton et al. 2020). Although efforts have been made to improve the bioavailability of AMPs and mitigate these challenges, their applicability remains a significant concern. These issues affect the feasibility of manufacturing AMPs at a pilot scale due to the complexities of their use.

5 Market Demands of AMPs

The market of AMPs has exhibited a complex and evolving landscape characterized by promising applications but challenges in practical implementation. The intricate structural and functional nature of AMPs has impeded their widespread success across various markets. Nonetheless, the global AMPs market has experienced steady growth, primarily driven by heightened awareness of antimicrobial resistance, the demand for innovative antimicrobial agents, and advances in AMP production. In 2022, the worldwide AMPs market was approximately USD 4.6 billion, with a projected market size of USD 7.8 billion by 2029, signifying a remarkable 1.70-fold increase (Peptide antibiotics market size 2023). This growth is expected to maintain a 4.6% compound annual growth rate (CAGR) from 2022 to 2030. Several factors contribute to the expansion of the AMPs market (Fig. 7.4).

- *Type of AMPs:* Non-ribosomally synthesized AMPs accounted for the largest revenue share in 2021, at 72.2%, and are expected to maintain their dominance in the market from 2022 to 2030. An illustrative example is BLEOMYCIN, manufactured by Pfizer Inc. and employed in treating lymphoma, testicular carcinoma, and squamous carcinoma (Peptide antibiotics market size 2023).
- *Type of diseases:* In 2021, skin infections held the highest revenue share in the AMPs market at 30%. This sector is anticipated to continue leading the market in the coming decade, driven by the increasing incidence of skin infections such as furuncles, impetigo, and cellulitis. Lipo peptide AMPs like CUBICIN RF (daptomycin), produced by Merck & Co., Inc., are used for treating skin infections caused by *Staphylococcus pyrogenes*, *Staphylococcus aureus*, and *Staphylococcus agalactiae*. Additionally, bloodstream infections are poised to become a significant market segment due to the growing prevalence of such infections and the availability of effective AMPs (Mazurkiewicz-Pisarek et al. 2023). For example, polymyxin B, manufactured by Xellia Pharmaceuticals, is used to treat bloodstream infections caused by *Pseudomonas aeruginosa*, *E. coli*, and *Haemophilus influenzae* (Velkov et al. 2013).
- *Type of bioavailability:* Bioavailability concerns have limited the market demand for AMPs. Oral administration of AMPs has faced mechanisms, short half-lives, and cytotoxicity challenges. In contrast, injectable AMPs have the largest market share at 42.7%, driven by their efficiency and diverse mechanisms. Injectable AMPs, exemplified by DALVANCE and licensed by AbbVie Inc., are used to treat various skin infections (Global antimicrobial peptides market 2022).

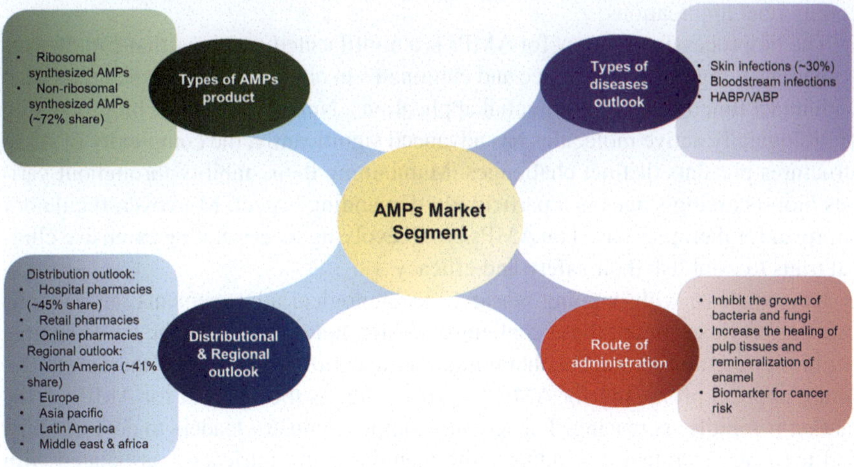

Fig. 7.4 Different factors responsible for AMPs market segmentation

- *Distribution and regional factors:* Hospital pharmacies have demonstrated the AMPs market, capturing 45.4% of the market share in 2021. Local pharmacies are expected to witness rapid growth due to increasing AMP demand. Regarding regional prevalence, North America accounted for the largest share of the AMPs market in 2021 at approximately 41%, followed by Europe (Peptide antibiotics market size 2023; Global antimicrobial peptides market 2022).

While AMPs hold diverse and promising applications, their market growth has been constrained by challenges such as limitations of efficient mechanisms and complexities associated with pilot-scale manufacturing. However, technological advancements in biopharmaceutical manufacturing and the involvement of key market players, like Pfizer Inc.; Merck & Co., Inc.; AuroMedics Pharma LLC; Theravance Biopharma; AbbVie Inc.; Xellia Pharmaceuticals; Sandoz International GmbH; and Boehringer Ingelheim, among others, are expected to facilitate the development of AMPs at a commercial scale, thereby enhancing their market demand significantly.

6 Conclusions and Future Perspectives

AMPs represent a promising avenue for combating various infectious diseases and addressing concerns related to antibiotic resistance. These peptides can be derived from a wide range of sources, such as microorganisms, plants, animals, and insects, leading to a diverse array of AMP types. Among these sources, microorganisms stand out as particularly appealing due to their ease of isolation, cultivation, and the ability to optimize bioprocessing conditions to achieve higher microbial biomass, ultimately resulting in increased AMP production to desired levels. Consequently, the bioprocessing of AMPs assumes critical importance when considering their commercial applications.

The bioprocessing journey for AMPs is a multifaceted endeavor that commences with identifying a suitable source and culminates in obtaining a purified end product with intact functionality for potential applications. Notably, while the bioprocessing of biologically active molecules has advanced significantly, the complexity of AMP structures presents distinct challenges. Maintaining their stability throughout various bioprocessing stages is a particularly demanding aspect. Moreover, regulatory approval for therapies based on AMPs is still evolving, necessitating extensive clinical trials to establish their safety and efficacy.

Nevertheless, with ongoing research, technological advancements, and a dedicated effort to surmount these challenges, AMPs hold promise in the battle against antibiotic-resistant pathogens, heralding a new era of antimicrobial therapies. The future appears promising for AMP bioprocessing, as the demand for AMPs in the market is rapidly increasing. This growth compels industry leaders to devise robust and innovative strategies to address the manufacturing intricacies associated with these molecules. Leveraging progress in genetic engineering, manufacturing technology, and a deepening understanding of the considerable potential of these natural peptides, large-scale AMP bioprocessing is poised to become more cost-effective, efficient, and resilient in the near future.

Conflicts of Interest The authors do not have any conflict of interest related to this book chapter.

References

Adedeji W (2016) The treasure called antibiotics. Ann Ib Postgrad Med 14:56–57

Adinarayana K, Ellaiah P, Srinivasulu B, Bhavani Devi R, Adinarayana G (2003) Response surface methodological approach to optimize the nutritional parameters for neomycin production by Streptomyces marinensis under solid-state fermentation. Process Biochem 38:1565–1572

Almansour AM et al (2023) The silent threat: antimicrobial-resistant pathogens in food-producing animals and their impact on public health. Microorganisms 11:2127

Antibiotic resistance threats in the United States, 2019. https://stacks.cdc.gov/view/cdc/82532 (2019) https://doi.org/10.15620/cdc:82532

Aoki W, Ueda M (2013) Characterization of antimicrobial peptides toward the development of novel antibiotics. Pharmaceuticals 6:1055–1081

Aslam B et al (2018) Antibiotic resistance: a rundown of a global crisis. Infect Drug Resist 11:1658

Assis LM, Nedeljković M, Dessen A (2017) New strategies for targeting and treatment of multi-drug resistant Staphylococcus aureus. Drug Resist Updat 31:1–14

Atipairin A et al (2022) Identification and characterization of a potential antimicrobial peptide isolated from soil Brevibacillus sp. WUL10 and its activity against MRSA pathogens. Trop Med Infect Dis 7

Avitabile C, D'Andrea LD, Romanelli A (2014) Circular dichroism studies on the interactions of antimicrobial peptides with bacterial cells. Sci Rep 4:4293

Ayukekbong JA, Ntemgwa M, Atabe AN (2017) The threat of antimicrobial resistance in developing countries: causes and control strategies. Antimicrob Resist Infect Control 6:1–8

Bahar AA, Ren D (2013) Antimicrobial peptides. Pharmaceuticals 6:1543–1575

Baindara P et al (2013) Characterization of two antimicrobial peptides produced by a halotolerant Bacillus subtilis strain SK.DU.4 isolated from a rhizosphere soil sample. AMB Express 3:1–11

Baindara P et al (2016) Characterization of the antimicrobial peptide penisin, a class Ia novel lantibiotic from Paenibacillus sp. strain A3. Antimicrob Agents Chemother 60:580–591

Bechinger B, Lohner K (2006) Detergent-like actions of linear amphipathic cationic antimicrobial peptides. Biochim Biophys Acta Biomembr 1758:1529–1539

Bommarius B et al (2010) Cost-effective expression and purification of antimicrobial and host defense peptides in Escherichia coli. Peptides (NY) 31:1957–1965

Brede DA et al (2005) Heterologous production of antimicrobial peptides in Propionibacterium freudenreichii. Appl Environ Microbiol 71:8077–8084

Cao W, Zhou Y, Ma Y, Luo Q, Wei D (2005) Expression and purification of antimicrobial peptide adenoregulin with C-amidated terminus in Escherichia coli. Protein Expr Purif 40:404–410

Castellano P, Vignolo G, Farías RN, Arrondo JL, Chehín R (2007) Molecular view by Fourier transform infrared spectroscopy of the relationship between lactocin 705 and membranes: speculations on antimicrobial mechanism. Appl Environ Microbiol 73:415–420

Center for Health Statistics, N. Health, United States 2020–2021 (2020). https://www.cdc.gov/nchs/hus/contents2020-2021.htm#Table-LExpMort

Cereghino GPL, Cereghino JL, Ilgen C, Cregg JM (2002) Production of recombinant proteins in fermenter cultures of the yeast Pichia pastoris. Curr Opin Biotechnol 13:329–332

Chen B et al (2005) Isolation and characterization of novel cyclotides from Viola hederaceae: solution structure and anti-HIV activity of vhl-1, a leaf-specific expressed cyclotide. J Biol Chem 280:22395–22405
Chen Y, Mant CT, Hodges RS (2007) Preparative reversed-phase high-performance liquid chromatography collection efficiency for an antimicrobial peptide on columns of varying diameters (1mm to 9.4mm I.D.). J Chromatogr A **1140**:112–120
Chin Y-W, Balunas MJ, Chai HB, Kinghorn AD (2006) Drug discovery from natural sources. AAPS J 8:E239–E253
Chopra L, Singh G, Choudhary V, Sahoo DK (2014) Sonorensin: an antimicrobial peptide, belonging to the heterocycloanthracin subfamily of bacteriocins, from a new marine isolate, Bacillus sonorensis MT93. Appl Environ Microbiol 80:2981–2990
Chopra L, Singh G, Jena KK, Verma H, Sahoo DK (2015) Bioprocess development for the production of sonorensin by Bacillus sonorensis MT93 and its application as a food preservative. Bioresour Technol 175:358–366
Cregg JM et al (2009) Expression in the yeast Pichia pastoris. In: Burgess RR, Deutscher MP (eds) Methods in enzymology, vol 463. Academic, pp 169–189
da Cunha NB et al (2017) The next generation of antimicrobial peptides (AMPs) as molecular therapeutic tools for the treatment of diseases with social and economic impacts. Drug Discov Today 22:234–248
Davies J, Davies D (2010) Origins and evolution of antibiotic resistance. Microbiol Mol Biol Rev 74:417–433
de Oliveira KBS et al (2022) Current trends and prospects in antimicrobial peptide bioprocessing. In: Rehm B, Wibowo D (eds) Microbial production of high-value products, vol vol. 37. Springer, Cham, pp 109–141
De Vuyst L (1995) Nutritional factors affecting nisin production by Lactococcus lactis subsp. actis NIZO 22186 in a synthetic medium. J Appl Bacteriol 78:28–33
Delaunois B, Cordelier S, Conreux A, Clément C, Jeandet P (2009) Molecular engineering of resveratrol in plants. Plant Biotechnol J 7:2–12
Dijksteel GS, Ulrich MMW, Middelkoop E, Boekema BKHL (2021) Review: lessons learned from clinical trials using antimicrobial peptides (AMPs). Front Microbiol 12:616979
Drayton M, Kizhakkedath JN, Strau SK (2020) Towards robust delivery of antimicrobial peptides to combat bacterial resistance. Molecules 25:3048
Duarte-Mata DI, Salinas-Carmona MC (2023) Antimicrobial peptides' immune modulation role in intracellular bacterial infection. Front Immunol 14:1119574
Dubos RJ (1939) Studies on a bactericidal agent extracted from a soil Bacillus: I. Preparation of the agent. Its activity in vitro. J Exp Med 70:1–10
Eidenberger L, Kogelmann B, Steinkellner H (2023) Plant-based biopharmaceutical engineering. Nat Rev Bioengin 1:426–439
Erdem Büyükkiraz M, Kesmen Z (2022) Antimicrobial peptides (AMPs): a promising class of antimicrobial compounds. J Appl Microbiol 132:1573–1596
Esmonde-White KA, Cuellar M, Lewis IR (2022) The role of Raman spectroscopy in biopharmaceuticals from development to manufacturing. Anal Bioanal Chem 414:969–991
Falanga A et al (2016) Marine antimicrobial peptides: nature provides templates for the design of novel compounds against pathogenic bacteria. Int J Mol Sci 17:1–18
Feng X et al (2012) Recombinant expression, purification, and antimicrobial activity of a novel hybrid antimicrobial peptide LFT33. Appl Microbiol Biotechnol 95:1191–1198
Fleming A (1929) On the antibacterial action of cultures of a Penicillium, with special reference to their use in the isolation of B. influenzae. Br J Exp Pathol 10:226–236
Freire RVM et al (2021) Antimicrobial peptide induced colloidal transformations in bacteria-mimetic vesicles: combining in silico tools and experimental methods. J Colloid Interface Sci 596:352–363
Garcia-Ochoa F, Gomez E (2009) Bioreactor scale-up and oxygen transfer rate in microbial processes: an overview. Biotechnol Adv 27:153–176

Gera S, Kankuri E, Kogermann K (2022) Antimicrobial peptides—unleashing their therapeutic potential using nanotechnology. Pharmacol Ther 232:1–23

Global antimicrobial peptides market—Industry trends and forecast to 2029 (2022). https://www.databridgemarketresearch.com/reports/global-antimicrobial-peptides-market

Gordon YJ, Romanowski EG, McDermott AM (2005) A review of antimicrobial peptides and their therapeutic potential as anti-infective drugs. Curr Eye Res 30:505–515

Grant GA, Crankshaw MW (2003) Identification of PTH-amino acids by HPLC. In: Smith BJ (ed) Protein sequencing protocols. Humana Press, pp 247–268. https://doi.org/10.1385/1-59259-342-9:247

Gupta SK, Shukla P (2017) Sophisticated cloning, fermentation, and purification technologies for an enhanced therapeutic protein production: A review. Front Pharmacol 8:1–17

Guryanova SV, Balandin SV, Belogurova-Ovchinnikova OY, Ovchinnikova TV (2023) Marine invertebrate antimicrobial peptides and their potential as novel peptide antibiotics. Mar Drugs 21:1–37

Han B, Yu Z, Liu B, Ma Q, Zhang R (2019) Optimization of bacteriocin production by Lactobacillus plantarum YJG, isolated from the mucosa of the gut of healthy chickens. Adv J Microbiol Res 13:001–009

Hancock REW (2000) Cationic antimicrobial peptides: towards clinical applications. Expert Opin Investig Drugs 9:1723–1729

Heymich ML, Srirangan S, Pischetsrieder M (2021) Stability and activity of the antimicrobial peptide leg1 in solution and on meat and its optimized generation from chickpea storage protein. Food Secur 10:1192

Hotchkiss RD, Dubos RJ (1940) Fractionation of the bactericidal agent from cultures of a soil bacillus. J Biol Chem 132:791–792

Huan Y, Kong Q, Mou H, Yi H (2020) Antimicrobial peptides: classification, design, application and research progress in multiple fields. Front Microbiol 11:582779

Huang L, Ching CB, Jiang R, Leong SSJ (2008) Production of bioactive human beta-defensin 5 and 6 in Escherichia coli by soluble fusion expression. Protein Expr Purif 61:168–174

Ilyas H, Kim JW, Lee DK, Malmsten M, Bhunia A (2019) Structural insights into the combinatorial effects of antimicrobial peptides reveal a role of aromatic-aromatic interactions in antibacterial synergism. J Biol Chem 294:14615–14633

Ingham AB, Moore RJ (2007) Recombinant production of antimicrobial peptides in heterologous microbial systems. Biotechnol Appl Biochem 47:1–9

Jafari A, Babajani A, Sarrami Forooshani R, Yazdani M, Rezaei-Tavirani M (2022) Clinical applications and anticancer effects of antimicrobial peptides: from bench to bedside. Front Oncol 12:1–19

Jozala AF et al (2016) Biopharmaceuticals from microorganisms: from production to purification. Braz J Microbiol 47:51–63

Karimi Alavijeh M, Baker I, Lee YY, Gras SL (2022) Digitally enabled approaches for the scale up of mammalian cell bioreactors. Digital Chem Eng 4:100040

Käßer L, Rotter M, Coletta L, Salzig D, Czermak P (2022) Process intensification for the continuous production of an antimicrobial peptide in stably-transformed Sf-9 insect cells. Sci Rep 12:1–10

Khan MM et al (2020) Response surface optimization of culture conditions for cyclic lipopeptide MS07 from Bacillus siamensis reveals diverse insights targeting antimicrobial and antibiofilm activity. PRO 8:1–17

Kongsinkaew C et al (2023) Enhancing antimicrobial peptide productivity in Pichia pastoris (Muts strain) by improving the fermentation process based on increasing the volumetric methanol consumption rate. Fermentation 9:1–13

Kościuczuk EM et al (2012) Cathelicidins: family of antimicrobial peptides. A review. Mol Biol Rep 39:10957–10970

Krier F, Revol-Junelles AM, Germain P (1998) Influence of temperature and pH on production of two bacteriocins by Leuconostoc mesenteroides subsp. mesenteroides FR52 during batch fermentation. Appl Microbiol Biotechnol 50:359–363

Kruschitz A, Nidetzky B (2020) Downstream processing technologies in the biocatalytic production of oligosaccharides. Biotechnol Adv 43:107568

Kumari A, Kumari S, Prasad GS, Pinnaka AK (2021) Production of sophorolipid biosurfactant by insect derived novel yeast Metschnikowia churdharensis f.a., sp. nov., and its antifungal activity against plant and human pathogens. Front Microbiol 12:1–13

Lajis AFB (2020) Biomanufacturing process for the production of bacteriocins from Bacillaceae family. Bioresour Bioprocess 7:1–26

Laws M, Shaaban A, Rahman KM (2019) Antibiotic resistance breakers: current approaches and future directions. FEMS Microbiol Rev 43:490–516

Le CF, Fang CM, Sekaran SD (2017) Intracellular targeting mechanisms by antimicrobial peptides. Antimicrob Agents Chemother 61:e02340–e02316

Lee Ventola C (2015) The antibiotic resistance crisis part 1: causes and threats. Pharm Therap 40:277–283

Lei J et al (2019) The antimicrobial peptides and their potential clinical applications. Am J Transl Res 11:3919–3931

Letian H, Ding C, Fenggeng X, Kuiying H, Feiping M (2015) Industrial separation and purification method of bacterial source antibacterial peptide

Li Y (2011) Recombinant production of antimicrobial peptides in Escherichia coli: a review. Protein Expr Purif 80:260–267

Li C, Bai J, Cai Z, Ouyang F (2002) Optimization of a cultural medium for bacteriocin production by Lactococcus lactis using response surface methodology. J Biotechnol 93:27–34

Liang T, Lu H, Ma J, Sun L, Wang J (2023) Progress on membrane technology for separating bioactive peptides. J Food Eng 340:111321

Lobanovska M, Pilla G (2017) Penicillin's discovery and antibiotic resistance: lessons for the future? Yale J Biol Med 90:135–145

Lohner K, Prossnigg F (2009) Biological activity and structural aspects of PGLa interaction with membrane mimetic systems. Biochim Biophys Acta Biomembr 1788:1656–1666

Lu SY, Skory CD, El Enshasy HA, Liu S (2021) Fermentative production of alternative antimicrobial peptides and enzymes. Biocatal Agric Biotechnol 37:1–8

Luo X et al (2023) Advancements, challenges and future perspectives on peptide-based drugs: focus on antimicrobial peptides. Eur J Pharm Sci 181:1–15

Luong HX, Thanh TT, Tran TH (2020) Antimicrobial peptides—advances in development of therapeutic applications. Life Sci 260:118407

Lyu Z, Yang P, Lei J, Zhao J (2023) Biological function of antimicrobial peptides on suppressing pathogens and improving host immunity. Antibiotics 12:1037

Macedo MWFS et al (2021) Marine organisms as a rich source of biologically active peptides. Front Mar Sci 8:1–23

Mahlapuu M, Håkansson J, Ringstad L, Björn C (2016) Antimicrobial peptides: an emerging category of therapeutic agents. Front Cell Infect Microbiol 6:194

Malanovic N, Lohner K (2016) Antimicrobial peptides targeting Gram-positive bacteria. Pharmaceuticals 9:59

Martínez JL, Baquero F (2014) Emergence and spread of antibiotic resistance: setting a parameter space. Ups J Med Sci 119:68–77

Matsuzaki K, Murase O, Fujii N, Miyajima K (1995) Translocation of a channel-forming antimicrobial peptide, magainin 2, across lipid bilayers by forming a pore. Biochemistry 34:6521–6526

Mattanovich D et al (2012) Recombinant protein production in yeasts. In: Lorence A (ed) Recombinant gene expression, vol 824. Humana Press, pp 329–358

Matte A (2022) Recent advances and future directions in downstream processing of therapeutic antibodies. Int J Mol Sci 23:8663

Mazurkiewicz-Pisarek A, Baran J, Ciach T (2023) Antimicrobial peptides: challenging journey to the pharmaceutical, biomedical, and cosmeceutical use. Int J Mol Sci 24:9031

McAuliffe O, Ross RP, Hill C (2001) Lantibiotics: structure, biosynthesis and mode of action. FEMS Microbiol Rev 25:285–308

Mesa-Pereira B, Rea MC, Cotter PD, Hill C, Ross RP (2018) Heterologous expression of biopreservative bacteriocins with a view to low cost production. Front Microbiol 9:1–15

Mihooliya KN et al (2020) Studies on efficient production of a novel L-asparaginase by a newly isolated Pseudomonas resinovorans IGS-131 and its heterologous expression in Escherichia coli. 3 Biotech 10:148

Mihooliya KN et al (2023) Assessment of structural behaviour of a new L-asparaginase and SAXS data-based evidence for catalytic activity in its monomeric form. Int J Biol Macromol 253:126803

Moretta A et al (2021) Antimicrobial peptides: a new hope in biomedical and pharmaceutical fields. Front Cell Infect Microbiol 11:668632

Munita JM, Arias CA (2016) Mechanisms of antibiotic resistance. Microbiol Spectr 4

Murray CJ et al (2022) Global burden of bacterial antimicrobial resistance in 2019: a systematic analysis. Lancet 399:629–655

Musthafa Poyil M, Bari N, Prasad SS, Alfaki MA (2021) Extraction of antimicrobial peptides (AMPs) from Portunus sanguinolentus herbst, Perna viridis Linneaus and Octopus indicus Orbigny: identifying the best solvent for AMP recovery and determining its anti-ESKAPE activity. J Res Med Dent Sci 9:20–26

Nasseri S, Sharifi M (2022) Therapeutic potential of antimicrobial peptides for wound healing. Int J Pept Res Ther 28:38

Ngambenjawong C, Chan LW, Fleming HE, Bhatia SN (2022) Conditional antimicrobial peptide therapeutics. ACS Nano 16:15779–15791

Niu C et al (2022) A rapid one-step process for the isolation of antibacterial peptides by silica-decorated Fe3O4nanoparticles. LWT 155:1–8

Oren Z, Shai Y (1998) Mode of action of linear amphipathic α-helical antimicrobial peptides. Pept Sci 47:451–463

Parachin NS, Mulder KC, Viana AAB, Dias SC, Franco OL (2012) Expression systems for heterologous production of antimicrobial peptides. Peptides (NY) 38:446–456

Park CB, Yi K-S, Matsuzaki K, Kim MS, Kim SC (2000) Structure-activity analysis of buforin II, a histone H2A-derived antimicrobial peptide: the proline hinge is responsible for the cell-penetrating ability of buforin II. Proc Natl Acad Sci USA 97:8245–8250

Pärn K, Eriste E, Langel Ü (2015) The antimicrobial and antiviral applications of cell-penetrating peptides. In: Cell-penetrating peptides: methods and protocols. Springer, New York, pp 223–245. https://doi.org/10.1007/978-1-4939-2806-4_15

Pelegrini PB, Lay FT, Murad AM, Anderson MA, Franco OL (2008) Novel insights on the mechanism of action of α-amylase inhibitors from the plant defensin family. Proteins Struct Funct Bioinformatics 73:719–729

Peptide antibiotics market size, share & trends analysis report by product type, by disease, by route of administration (oral, injectable, topical), by distribution dhannel, by region, and segment forecasts, 2022–2030 (2023). https://www.grandviewresearch.com/industry-analysis/peptide-antibiotics-market-report/request/rs2

Pereira-Dias L et al (2023) Peptaibiotics: harnessing the potential of microbial secondary metabolites for mitigation of plant pathogens. Biotechnol Adv 68:108223

Pham JV et al (2019) A review of the microbial production of bioactive natural products and biologics. Front Microbiol 10:1404

Pieracci JP, Armando JW, Westoby M, Thommes J (2018) Industry review of cell separation and product harvesting methods. In: Jagschies G, Lindskog E, Łącki K, Galliher P (eds) Biopharmaceutical processing. Elsevier, pp 165–206. https://doi.org/10.1016/B978-0-08-100623-8.00009-8

Poojitha R et al (2023) Antimicrobial peptides and their therapeutic aspect. Int J Livestock Res 13:1–7

Porcelli F, Ramamoorthy A, Barany G, Veglia G (2013) On the role of NMR spectroscopy for characterization of antimicrobial peptides. In: Ghirlanda G, Senes A (eds) Membrane proteins: folding, association, and design. Humana Press, pp 159–180. https://doi.org/10.1007/978-1-62703-583-5_9

Rahnamaeian M, Vilcinskas A (2012) Defense gene expression is potentiated in transgenic barley expressing antifungal peptide metchnikowin throughout powdery mildew challenge. J Plant Res 125:115–124

Rai M, Pandit R, Gaikwad S, Kövics G (2016) Antimicrobial peptides as natural bio-preservative to enhance the shelf-life of food. J Food Sci Technol 53:3381–3394

Rathore AS, Mishra S, Nikita S, Priyanka P (2021) Bioprocess control: current progress and future perspectives. Life 11:557

Ratrey P, Dalvi SV, Mishra A (2020) Enhancing aqueous solubility and antibacterial activity of curcumin by complexing with cell-penetrating octaarginine. ACS Omega 5:19004–19013

Rima M et al (2021) Antimicrobial peptides: a potent alternative to antibiotics. Antibiotics 10:1095

Rivero M et al (2012) Stacking of antimicrobial genes in potato transgenic plants confers increased resistance to bacterial and fungal pathogens. J Biotechnol 157:334–343

Rose K, Bairoch A, Offord RE (1983) Amino acid sequence determination by gas chromatography—mass spectrometry of permethylated peptides: the application of capillary columns. J Chromatogr A 268:197–206

Rosinha Grundtvig IP et al (2018) Screening of organic solvents for bioprocesses using aqueous-organic two-phase systems. Biotechnol Adv 36:1801–1814

Saeed SI, Mergani AE, Aklilu E, Kamaruzzaman NF (2022) Antimicrobial peptides: bringing solution to the rising threats of antimicrobial resistance in livestock. Front Vet Sci 9:851052

Schwarz S, Gerlach D, Fan R, Czermak P (2022) GbpA as a secretion and affinity purification tag for an antimicrobial peptide produced in Vibrio natriegens. Electron J Biotechnol 56:75–83

Selsted ME (1997) HPLC methods for purification of antimicrobial peptides. In: Shafer WM (ed) Antibacterial peptide protocols. Humana Press, pp 17–33. https://doi.org/10.1385/0-89603-408-9:17

Semreen MH et al (2018) Recent updates of marine antimicrobial peptides. Saudi Pharm J 26:396–409

Shanmugaraj B, Bulaon CJI, Malla A, Phoolcharoen W (2021) Biotechnological insights on the expression and production of antimicrobial peptides in plants. Molecules 26:1–23

Shi G et al (2022) DRAMP 3.0: an enhanced comprehensive data repository of antimicrobial peptides. Nucleic Acids Res 50:D488–D496

Shin WS, Lee D, Kim S, Jeong YS, Chun GT (2013) Application of scale-up criterion of constant oxygen mass transfer coefficient (Kla) for production of itaconic acid in a 50 l pilot-scale fermentor by fungal cells of aspergillus terreus. J Microbiol Biotechnol 23:1445–1453

Sinha R, Shukla P (2018) Antimicrobial peptides: recent insights on biotechnological interventions and future perspectives. Protein Pept Lett 26:79–87

Sørensen HP, Mortensen KK (2005) Advanced genetic strategies for recombinant protein expression in Escherichia coli. J Biotechnol 115:113–128

Stegemann C, Hoffmann R (2008) Sequence analysis of antimicrobial peptides by tandem mass spectrometry. In: Otvos L (ed) Peptide-based drug design. Humana Press, pp 31–46. https://doi.org/10.1007/978-1-59745-419-3_3

Taggar R et al (2021) Bacteriocin isolated from the natural inhabitant of Allium cepa against Staphylococcus aureus. World J Microbiol Biotechnol 37:1–16

Tai HM et al (2021) Scale-up production of and dietary supplementation with the recombinant antimicrobial peptide tilapia piscidin 4 to improve growth performance in Gallus gallus domesticus. PLoS One 16:e0253661

Talapko J et al (2022) Antimicrobial peptides—mechanisms of action, antimicrobial effects and clinical applications. Antibiotics 11:1417

Tripathi NK, Shrivastava A (2019) Recent developments in bioprocessing of recombinant proteins: expression hosts and process development. Front Bioeng Biotechnol 7:1–35
Uddin TM et al (2021) Antibiotic resistance in microbes: history, mechanisms, therapeutic strategies and future prospects. J Infect Public Health 14:1750–1766
Velkov T, Roberts KD, Nation RL, Thompson PE, Li J (2013) Pharmacology of polymyxins: new insights into an 'old' class of antibiotics. Future Microbiol 8:711–724
Verma H, Mihooliya KN, Nandal J, Sahoo DK (2023) Studies on a new antimicrobial peptide from Vibrio proteolyticus MT110. Prep Biochem Biotechnol 1–14:193. https://doi.org/10.1080/10826068.2023.2209892
Viana de Freitas T et al (2023) Release of immunomodulatory peptides at bacterial membrane interfaces as a novel strategy to fight microorganisms. J Biol Chem 299:1–15
Wang L, Dong C, Li X, Han W, Su X (2017) Anticancer potential of bioactive peptides from animal sources (review). Oncol Rep 38:637–651
Wang J et al (2019) Antimicrobial peptides: promising alternatives in the post feeding antibiotic era. Med Res Rev 39:831–859
Wu R et al (2021) Marine invertebrate peptides: antimicrobial peptides. Front Microbiol 12:1–12
Wu CL et al (2022) High level expression and purification of Cecropin-like antimicrobial peptides in Escherichia coli. Biomedicine 10:1351
Yoshikawa TT (2002) Antimicrobial resistance and aging: beginning of the end of the antibiotic era? J Am Geriatr Soc 50:S226–S229
Yu R et al (2010) Effective and stable porcine interferon-α production by pichia pastoris fed-batch cultivation with multi-variables clustering and analysis. Bioprocess Biosyst Eng 33:473–483
Zhang QW, Lin LG, Ye WC (2018a) Techniques for extraction and isolation of natural products: a comprehensive review. Chin Med (UK) 13:1–26
Zhang M-K et al (2018b) Enhanced solubility and antimicrobial activity of alamethicin in aqueous solution by complexation with γ-cyclodextrin. J Funct Foods 40:700–706
Zhang QY et al (2021a) Antimicrobial peptides: mechanism of action, activity and clinical potential. Mil Med Res 8:48
Zhang Y et al (2021b) A method for rapid screening, expression, and purification of antimicrobial peptides. Microorganisms 9:1858

Chapter 8
Relationship Between Antimicrobial Peptides, Probiotics, Postbiotics, and Parabiotics in the Maintenance of a Well-Balanced Gut Ecosystem

Priyanka Parhi

1 Introduction

The gut ecosystem, also known as the gut microbiota or gut microbiome, comprises a heterogeneous collection of microorganisms, including bacteria, viruses, fungi, and other microbes, which reside within the gastrointestinal tract of humans and other animal species (Zong et al. 2020). The ecosystem plays a vital role in numerous physiological processes, including digestion, nutrient absorption, immune system regulation, and synthesis of certain vitamins and metabolites (Martín et al. 2014).

The significance of gut health holds considerable importance within the domain of human welfare. The gastrointestinal system, commonly known as the gut, is crucial for preserving general health and vigor (Bengmark 2013). The gut possesses an importance that surpasses its primary function of digestion, as it harbors a sophisticated ecology that includes billions of microorganisms, collectively referred to as the gut microbiota (Martín et al. 2014). Microbes influence a diverse range of physiological processes within the human body, thereby affecting not only the digestive system but also immunological responses, metabolic homeostasis, and even neurological well-being (Wang et al. 2017).

Recent scientific research has elucidated the significant ramifications of gastrointestinal well-being on multiple facets of human health. The maintenance of healthy gut microbiota has been reported to be crucial for various aspects of human health, including the avoidance of gastrointestinal problems, regulation of immunological responses, and treatment of chronic diseases (Bengmark 2013). As a result,

P. Parhi (✉)
Monash University Malaysia, Selagnor, Malaysia

Mr Biologist LLP, Pune/Guwahati, India
e-mail: priyankaparhi@askmrbiologist.com

P. Baindara, S. M. Mandal (eds.), *Evolution of Antimicrobial Peptides*,
https://doi.org/10.1007/978-3-031-67515-7_8

there has been a significant surge of interest in the scientific community and the general population regarding the comprehension and enhancement of gut health.

Dysbiosis explores the complex and captive realms of the gut microbiome. The term "gut dysbiosis" denotes perturbation of the intricate equilibrium of bacteria residing within the gastrointestinal tract, resulting in an imbalanced and detrimental microbial consortium (Aziz et al. 2013). Maintaining this equilibrium is crucial for our holistic welfare, as these minuscule organisms play a major role in digestion, food assimilation, and even our immunological response. Dysbiosis can lead to the proliferation of pathogenic bacteria and a fall in helpful bacteria, establishing an environment that is favorable for the development of various health complications. The potential outcome encompasses a spectrum of physiological issues, including, but not limited to, gastrointestinal disorders such as irritable bowel syndrome and broader systemic implications such as obesity, immunological conditions, and even psychiatric disorders (Aziz et al. 2013; Bengmark 2013; Wang et al. 2017). Comprehension of dysbiosis has provided novel opportunities in the realm of medicine, emphasizing the significance of cultivating a varied and balanced gut microbiome for our overall well-being in the long run.

Maintaining a well-balanced gut ecosystem is crucial for overall health, and several key elements such as antimicrobial peptides (AMPs), probiotics, postbiotics, and parabiotics play pivotal roles in achieving this equilibrium (Salminen et al. 2021; Singh et al. 2018). AMPs are endogenous proteins synthesized inside the gastrointestinal tract, serving the purpose of regulating the proliferation of pathogenic bacteria, hence maintaining a state of equilibrium within the microbial community for optimal health (Ageitos et al. 2017). Probiotics refer to living bacteria that offer advantageous effects when ingested since they contribute to the enhancement of the gut microbiota by replenishing the population of beneficial microorganisms (Piqué et al. 2019). Postbiotics are metabolic byproducts of probiotics, encompassing various organic acids and enzymes (Jastrząb et al. 2021). Parabiotics refer to probiotic compounds that lack viability but demonstrate beneficial effects on gut health, including the facilitation of a harmonized immunological response (Taverniti and Guglielmetti 2011). The potential synergistic effects of AMPs, when combined with probiotics, postbiotics, or parabiotics, offer a promising direction in the realm of health and immunity. Antimicrobial peptides (AMPs) are integral constituents of the innate immune system that exhibit antimicrobial capabilities that can synergistically enhance the effects of probiotics or their derivatives. Within the gastrointestinal tract, this cooperative interaction may potentially enhance the immune response against detrimental pathogens, thus promoting a harmonious microbial community composition. Furthermore, the coalescence of AMPs with probiotics and their derivatives may have implications for immunomodulation, potentially augmenting the overall efficacy of the immune response. Understanding the interplay between these components is crucial for fostering and maintaining a balanced gastrointestinal environment, which holds significant ramifications for general health and welfare.

The aim of this chapter is to explore the interrelationships between AMPs, probiotics, postbiotics, and parabiotics in maintaining a balanced gut environment.

Within this discussion, this chapter will comprehensively investigate each component, beginning with AMPs that regulate the activity of pathogenic bacteria. Subsequently, the examination will be extended to the domains of probiotics, postbiotics, and parabiotics. This systematic approach was designed to provide readers with a thorough understanding of how these constituents synergistically contribute to the sustenance of gut health and overall well-being.

2 Probiotics, Postbiotics, and Parabiotics and Their Potential Effect on Gut Microbiome

The human gastrointestinal tract contains more than ten trillion microbial cells, including a variety of mutually beneficial microbes (Senchukova 2023). The microbiota produces metabolites and bioactive chemicals that activate the host's immune and metabolic systems (Senchukova 2023). The host organism provides nourishment and a consistent environment in exchange for dietary assistance, gastrointestinal tract maturation, early immune system priming, human immune system development, inhibition of harmful microbial growth, and intestinal barrier preservation (Aziz et al. 2013). These interactions establish microbial symbiosis and maintain the human intestinal microbiota, which is necessary for lifelong health and resilience (Aziz et al. 2013). Consuming fermented foods, vegetables, whole grains, polyphenol-rich meals, a plant-based diet, and prebiotics or probiotics can increase and modify gut microbiota (Bhadekar and Parhi 2016; Plamada and Vodnar 2021). Prebiotics are specific dietary components used by microorganisms to confer positive effects on host health (Gibson et al. 2017). Probiotics, which are live microorganisms that offer health benefits to animals and humans when consumed in appropriate quantities, have recently been accompanied by postbiotics and parabiotics as further members of the biotics category (Cuevas-González et al. 2020.). These newcomers have been reported to confer valuable benefits to the host's body, either directly or indirectly.

2.1 *Probiotics*

Probiotics are living microorganisms that are nonpathogenic and are supplied to improve the balance of microbes, particularly inside the gastrointestinal system (Fig. 8.1). According to the Expert Consultation by the Food and Agriculture Organization (FAO) and the World Health Organization (WHO), probiotics are defined as "live microorganisms that, when administered in sufficient quantities, provide health benefits to the host" (Hotel and Cordoba 2001). Numerous definitions of probiotics exist, but the current predominant definition was established by the International Scientific Association for Probiotics and Prebiotics (ISAPP) in

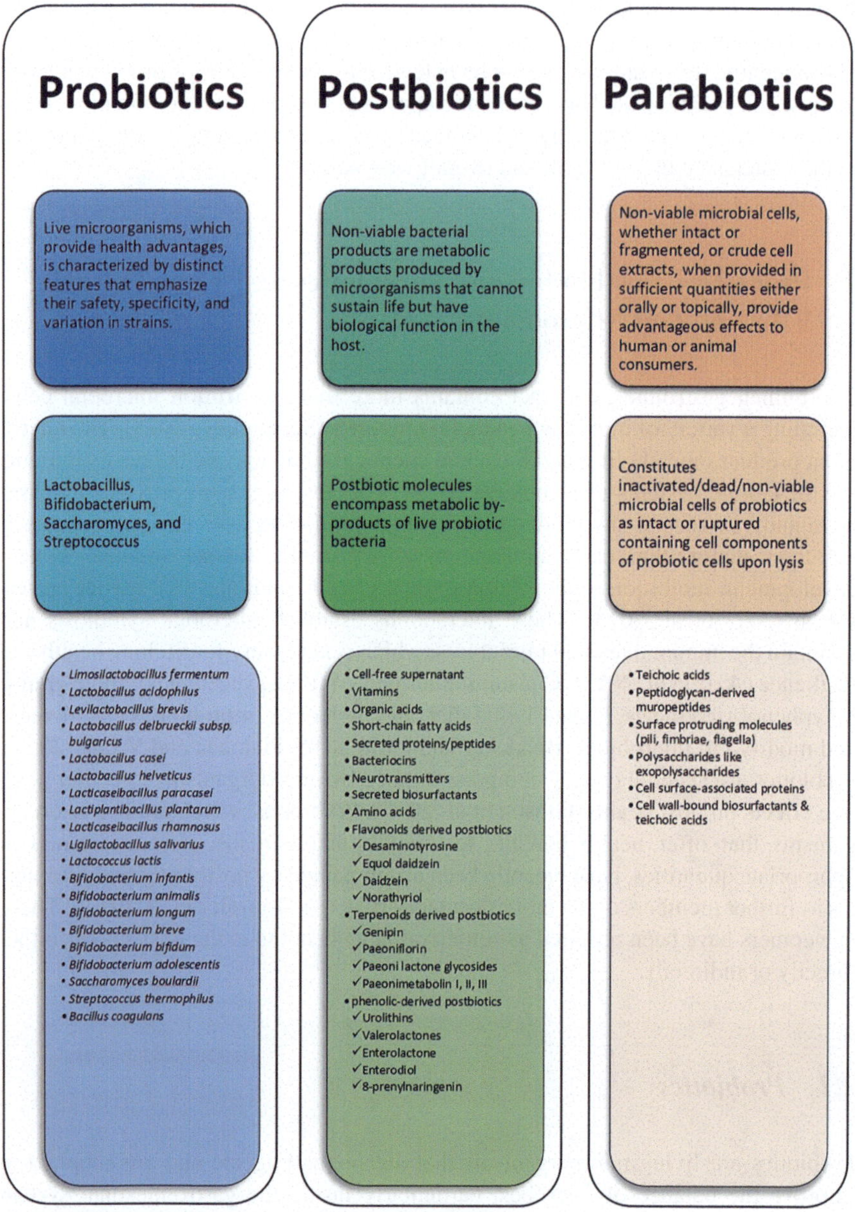

Fig. 8.1 Comparison of probiotics, postbiotics, and parabiotics: a visual representation illustrating the distinct categories of beneficial microbial products. Probiotics are live microorganisms, postbiotics are their fermentation byproducts, and parabiotics consist of inactivated microbial components. Examples include live cultures, bioactive compounds, and inactivated cells, respectively

2014. ISAPP defined probiotics as "live microorganisms that, when administered in sufficient quantities, confer a health benefit on the host" (Hill et al. 2014). These advantageous microorganisms are introduced into the human body through various means, including both dairy and nondairy fermented products.

The selection of probiotics is influenced by biological factors, including considerations, such as food safety and functional attributes, as well as technical features. Key elements in the choice and production of probiotics include pathogenicity, antibiotic resistance, positive effects on the immune system, antagonistic characteristics, viability, and stability within the gastrointestinal tract as well as physiological and genetic stability (Rafique et al. 2023). The microorganisms involved in this context include various species from the genera *Lactobacillus* and *Bifidobacterium*, as well as *Streptococcus thermophilus*, *Bacillus coagulans*, *Saccharomyces boulardii*, and yeast (Fig. 8.1). These microorganisms are subject to regulatory oversight as dietary supplements and food products. Probiotics elicit their beneficial effects by many methods, such as decreasing intestinal pH, preventing colonization and invasion by harmful organisms, and modulating the human immunological response.

Probiotics typically do not establish permanent residence in the gastrointestinal tract (GIT), necessitating frequent consumption to replenish those eliminated through feces for a sustained exogenous probiotic effect. Achieving the desired effect requires the use of an appropriate quantity of probiotic microorganisms at the correct dosage. It is recommended to maintain a minimum population of probiotics, typically ranging from 10^8 to 10^9 colony-forming units (CFU) per serving portion (Parhi et al. 2021; Senchukova 2023). Probiotics must meet several criteria, including survival through the upper digestive tract, adhesion to the human intestinal epithelium, and the ability to endure and proliferate in the intestine with acid and bile tolerance. Probiotic characteristics are not inherently tied to a microorganism's genus or species, emphasizing the importance of marketed products featuring strains consistent with documented health effects (Parhi et al. 2021). Studies highlighting the probiotic attributes of a specific strain at a tested dose do not necessarily validate these qualities at different doses of the same strain. Additionally, the choice of carrier or matrix is crucial, as it can affect product characteristics by influencing the viability of a specific strain (Parhi et al. 2023).

Probiotic bacteria produce a range of anti-pathogenic compounds that exert direct antagonistic effects on pathogenic microorganisms (Ageitos et al. 2017). Antimicrobial compounds such as organic acids, hydrogen peroxide, and bacteriocins are involved in the inhibition of pathogen adherence to the intestinal lining (Ageitos et al. 2017). The clinical efficacy of probiotics is strongly supported by substantial evidence in the treatment of acute diarrhea, particularly when linked to rotavirus, as well as in the management of pouchitis. Probiotics are widely acknowledged to be safe and well-tolerated, with bloating and gas being the most frequently documented adverse effects (Hill et al. 2014). Additional studies have clarified the significance of probiotics in the prevention of several illnesses, including antibiotic-associated diarrhea, *Clostridium difficile* infection, traveler's diarrhea, irritable bowel syndrome, ulcerative colitis, Crohn's disease, and vulvovaginal candidiasis (Aziz et al. 2013; Bengmark 2013; Wang et al. 2017).

2.2 Postbiotics

The significance of cell viability in conferring health benefits through probiotics has been a long-standing perspective. Nevertheless, there has been a prolonged acknowledgment that even nonviable microbes, along with their cellular components and metabolites, can exert an impact on health (Vinderola et al. 2022). Within the field of probiotics, there is increasing recognition of the substantial influence of nonviable microorganisms, as well as their cellular constituents and metabolites, on human health (Durack et al. 2018). As a result, postbiotics have become a central focus in the field of functional foods (Fig. 8.1).

Postbiotics are bioactive compounds produced by probiotic microbes through fermentation or metabolic processes, which play a vital role in promoting health. According to ISAPP, postbiotics are defined as the "preparation of inanimate microorganisms and/or their components that confers a health benefit on the host" (Salminen et al. 2021). Antimicrobial compounds are classified into low-molecular-weight forms (such as carbon dioxide, diacetylene, and hydrogen peroxide) and higher-molecular-weight compounds, often referred to as postbiotics, specifically bacteriocins. Additionally, postbiotics encompass soluble substances that bacteria secrete or release during lysis, including enzymes, peptides, teichoic acids, peptidoglycan-derived micropeptides, polysaccharides, cell surface proteins, and organic acids (Wegh et al. 2019). The gastrointestinal tract, characterized by distinct parameters such as pH, temperature, and substrate accessibility, creates a favorable milieu for the synthesis of postbiotics (Aguilar-Toalá et al. 2021). Upon release, these chemicals have the potential to exert an influence on gastrointestinal processes in the immediate vicinity or undergo absorption into circulation, thereby affecting several organs and systems.

Researchers have been captivated by the term "postbiotic" owing to its extended shelf life, established safety dose limits, distinctive chemical structural properties, and diverse array of end products, including various signaling molecules. These molecules play a regulatory role in immunomodulatory, antihypertensive, anti-inflammatory, antiproliferative, hypocholesterolemic, antiobesogenic, and antioxidant activities (Rafique et al. 2023). Postbiotics, which function as immunomodulators, aid in the mitigation of inflammation induced by infections and promote resilience of intestinal epithelial cells. Postbiotics, which originate from probiotics, encompass compounds like teichoic acid, indole, lipopolysaccharide, muramyl dipeptide, and lactospin (Rad et al. 2022). Among these, short-chain fatty acids, including butyric acid, acetic acid, and propionic acid, have been particularly scrutinized. These fatty acids play roles in both the digestive and physiological functions of the intestine to maintain intestinal health and specific metabolic pathways are involved in the digestion and absorption of food (Holanda et al. 2020). These compounds contribute significantly to the maintenance of gut barrier integrity and support a balanced gut microbiome. Moreover, the synergy between probiotics and postbiotics, in which the consumption of probiotics leads to increased postbiotic levels through fermentation, enhances the overall impact on gut health, providing a comprehensive approach to digestive health and immune function support.

2.3 Parabiotics

Parabiotics, paraprobiotics, ghost probiotics, or inactivated probiotics refer to non-viable microbial cells or their components that provide health benefits when consumed (Fig. 8.1). Paraprobiotics are inactivated, deceased, or nonviable microbial cells derived from probiotics. These cells, whether intact or ruptured, contain components from probiotic cells, such as teichoic acids, surface-protruding molecules (pili, fimbriae, and flagella), peptidoglycan-derived muropeptides, polysaccharides like exopolysaccharides, cell surface–associated proteins, cell wall–bound biosurfactants, and teichoic acids, especially upon lysis (Singh et al. 2018). In contrast to active probiotics, parabiotics offer the benefits of microbial-derived substances without the considerations linked to introducing live microorganisms into the body. The concept of parabiotics/ghost probiotics was proposed by Taverniti and Guglielmetti (2011) as, "the inactivated/dead/non-viable microbial cells of probiotics (intact or ruptured containing probiotic cell components upon lysis) or crude cell extracts (i.e. with complex chemical composition)". Recently, the currently parabiotics have been defined as "non-viable microbial cells (either intact or broken) or crude cell extracts which when administered (either orally or topically) in adequate amounts, confer a benefit on the human or animal consumer" (Nataraj et al. 2020). Similar to postbiotics, parabiotics have advantages over probiotics such as the elimination of bacterial translocation risk, absence of antibiotic resistance gene transfer, natural extraction processes, beneficial effects from cell lysis, and enhanced interaction with epithelial cells (Piqué et al. 2019).

Parabiotics can be sourced from cell lysates, extracellular vesicles, inactivated or deceased cells of probiotic, or microbial cell components. The preferred method for achieving the inactivation of microbial cells has traditionally been heat treatment. However, alternative methods such as the use of chemicals (e.g., formalin), gamma or ultraviolet irradiation, and sonication can also be employed for the inactivation of live bacteria. The stability and standardization of parabiotics make them appealing, ensuring consistent therapeutic components and reliable products (Taverniti and Guglielmetti 2011). Heat-killed (HK) probiotics demonstrated anti-adhesion potential, suggesting their ability to combat diarrheal and food-borne pathogens by outcompeting them for adhesion sites on gut enterocytes in the Caco-2 experimental model (Singh et al. 2017). Like its live counterpart *Lactobacillus reuteri* GMNL-263, heat-killed (HK) Lr263 demonstrated notable effects in mitigating obesity-induced metabolic abnormalities (Hsieh et al. 2016). Similarly, both live and heat-killed *Bifidobacterium breve* M-16V cells suppressed pro-inflammatory cytokines, influencing intestinal metabolism and host gene expression (Sugahara et al. 2017). Furthermore, heat-killed strains of *Lactobacillus* demonstrated anti-inflammatory and anti-oxidative properties. In both in vitro and in vivo models, these strains were observed to scavenge free radicals, reduce levels of inflammatory markers like IL-6 and TNF-α, and increase the production of anti-inflammatory cytokines like IL-10 (Jang et al. 2018; Chung et al. 2019). These findings suggest that not only live bacteria, but also their heat-killed counterparts (parabiotics) have the potential to modulate host physiology. Furthermore, current evidence indicates a similar mode of action between parabiotics and viable cells.

The health impacts associated with probiotics, postbiotics, and parabiotics are closely interconnected with the metabolites that are either produced or present in their compositions. In essence, the richness and variety of metabolites present or generated by probiotics, postbiotics, and parabiotics contribute significantly to their respective health impacts. Understanding the specific metabolic profiles of these categories becomes crucial in unraveling the complex mechanisms underlying their beneficial effects on health. One such common metabolite is antimicrobial peptides that are either produced or present in the formulation of probiotics, postbiotics, and parabiotics.

3 Antimicrobial Peptides (AMPs)

Antimicrobial Peptides (AMPs) are peptides synthesized by bacterial or fungal ribosomes (ribosomally-synthesized and post-translationally-modified peptides (RiPPs)) to kill or inhibit pathogenic bacteria, viruses, and fungi (Arnison et al. 2013; Erdem Büyükkiraz and Kesmen 2022). AMPs are alternatively referred to as bioactive peptides, bacteriocins, or peptides with antimicrobial activity. AMPs synthesized by lactic acid bacteria (LAB) and other bacterial taxa exhibit distinct characteristics such as cationic charge, thermal stability, and amphiphilic properties (Tajbakhsh et al. 2017). The peptides under consideration are synthesized using a high concentration of lysyl and arginyl residues and exhibit features that enable them to permeate cell membranes (Malanovic and Lohner 2016.). AMPs have a notable array of structural variations, and they typically display amphipathic properties, wherein they possess regions that are both hydrophilic, displaying an affinity toward water, and hydrophobic, exhibiting a repulsion toward water. Despite fitting the AMPs or bacteriocin definition, newly discovered RiPP subgroups with antimicrobial activity do not align with existing classifications (Arnison et al. 2013). Alvarez-Sieiro et al. (2016) propose a slightly adjusted LAB classification based on biosynthesis mechanisms and biological activity, consistent with prior suggestions. Class I RiPPs, including Ia (lantibiotics), Ib (labyrinthopeptins), and Ic (sanctibiotics), exhibit diverse peptide characteristics; class II, consisting of peptides without modified amino acids, is further categorized into IIa (pediocin-like), IIb (two-peptide), IIc (circular), and IId (linear non-pediocin-like); class III encompasses large and heat-labile bacteriocins like colicin; and class IV includes complex proteins with associations to lipid or carbohydrate moieties (Alvarez-Sieiro et al. 2016; Ouertani et al. 2022). The distinctive architecture of AMPs facilitates their interaction with and subsequent disruption of microbial cell membranes, resulting in the demise of the invading pathogens (Pasupuleti et al. 2012).

AMPs exhibit high potency and low toxicity, and they can be either naturally produced by probiotics in the host's environment or engineered through bioengineering techniques (Table 8.1). LABs are the primary contributors to the production of AMPs, which serve as a competitive adaptation mechanism. The synthesis of AMPs by LAB is dependent on the physical and biological parameters of the microbial culture. The production of AMPs is significantly influenced by environmental variables. The peptides play a crucial role in hindering the growth of closely related strains of the producer organism and other bacterial genera, thereby exerting a bactericidal or bacteriostatic effect. The scope of their uses continues to increase as novel AMPs are continuously being discovered (Table 8.1).

These AMPs demonstrate a wide range of antibacterial activity and antifungal activity via diverse mechanisms (Table 8.2) while causing minimal harm to cells. AMPs constitute a diverse set of peptides that manifest a spectrum of antimicrobial activities by sequestering essential growth nutrients, disrupting bacterial membranes, and employing other related mechanisms (Table 8.2). Nisin A, a peptide composed of 34 amino acids produced by *Lactococcus lactis*, belongs to the type A (I) lantibiotics and is recognized for its robust antibacterial properties (Negash and Tsehai 2020). Among the antimicrobial peptides (AMPs) identified within LAB, namely, lactococcin A, lactococcin B, lactococcin M, lactococcin 972, LsbA, and LsbB, it has been observed that they exhibit a limited and specific antibacterial range akin to that of lactococcin Q (Zendo et al. 2006).

Lactic acid bacteria like *Lactobacillus acidophilus* generate additional inhibitory compounds, including lactocidin, acidolin, acidophilin, and lactacium-B. Similarly, *Bifidobacterium* species produce bifidolin and bifilong, both exhibiting inhibitory effects on various pathogenic bacteria. AMPs from *L. acidophilus* (LA-5) and *B. lactis* (BB-12) were found in cheese whey demonstrating remarkable stability under elevated temperatures and varying pH levels. These peptides exhibit broad-spectrum antimicrobial efficacy against pathogenic bacteria, such as *Staphylococcus aureus* PTCC 1431, *Listeria monocytogenes* PTCC 1298, *Escherichia coli* PTCC 1763, *Pseudomonas aeruginosa* PTCC 1707, and *Salmonella enterica* PTCC 1709 (Amiri et al. 2022). In vitro studies revealed that the AMP derived from *L. acidophilus* ATCC-4356 displays antagonistic activity against *Aeromonas hydrophila* (Akter et al. 2020). AMPs LR14, produced by *Lactobacillus plantarum* strain LR/14, demonstrate antifungal efficacy against four spoilage fungi: *Aspergillus niger*, *Rhizopus stolonifer*, *Mucor racemosus*, and *Penicillium chrysogenum* (Gupta and Srivastava 2014). Proline-rich antimicrobial peptide PR39 (4.7 kDa) integrated into the secretory expression vector plasmid, pPG:612, in *Lactobacillus casei* resulted in the recombinant strain pPG:612-PR39/*L. casei* 393. Administration of this recombinant strain to BALB/c mice showed improved growth, immunological parameters, and resistance against *E. coli* compared to the control group (Zhang et al. 2016). AMPs produced by a novel strain of *L. fermentum* HF-D1, isolated from the human gut, exhibit inhibitory effects against the growth of *P. aeruginosa* and *Serratia marcescens* (Pavlova et al. 2020).

It is important to note that despite their ability to inhibit bacterial growth, antimicrobial peptides (AMPs) are distinct from antibiotics (Table 8.1). Unlike antibiotics, which are secondary metabolites enzymatically synthesized, bacteriocins, or AMPs, are primary metabolites synthesized by the ribosomes of the producing organism. Given the growing concern surrounding antimicrobial resistance (AMR), AMPs present a potential alternative. Several AMPS have shown in vitro efficacy against AMR like *Streptococcus pneumoniae*, vancomycin-resistant enterococci (VRE),

Table 8.1 A concise overview of the key advantages and disadvantages associated with the use of AMPs

Advantages associated with the use of antimicrobial peptides	
Natural defense	AMPs are part of the body's innate immune system, providing a natural and immediate defense against pathogens.
Broad-spectrum activity	AMPs exhibit activity against a wide range of microorganisms, including bacteria, viruses, fungi, and parasites.
Low resistance development	Compared to traditional antibiotics, there is a lower likelihood of pathogens developing resistance to AMPs. This is due to their diverse mechanisms of action.
Target specificity	Some AMPs demonstrate selective targeting of pathogens while sparing beneficial microbiota, minimizing disruptions to the host's normal flora.
Potential for synergy	AMPs may work synergistically with conventional antibiotics, enhancing the overall antimicrobial effect.
Rapid action	AMPs often exhibit a rapid onset of action, quickly neutralizing pathogens, which can be crucial in acute infections.
Potential for biofilm disruption	Some AMPs can disrupt biofilms, which are protective layers formed by bacteria, making them more susceptible to antimicrobial agents.
Immunomodulatory effects	AMPs can modulate the host's immune response, promoting the recruitment and activation of immune cells to enhance overall defense.
Application in wound healing	AMPs play a role in wound healing by promoting tissue repair and reducing the risk of infection in damaged areas.
Diversity in sources	AMPs can be derived from various sources, including animals, plants, and microorganisms, providing a diverse pool for potential therapeutic development.
Disadvantages associated with the use of antimicrobial peptides	
Limited stability	Some AMPs may have limited stability in certain environments, affecting their effectiveness and shelf life.
Potential cytotoxicity	At high concentrations, certain AMPs may exhibit cytotoxic effects on host cells, limiting their therapeutic application.
Immunogenicity	The immune system may recognize AMPs as foreign, leading to an immune response that could impact their efficacy or safety.
Cost of production	Large-scale production of AMPs can be expensive, hindering their widespread use and accessibility.
Variable efficacy	The effectiveness of AMPs can vary depending on factors such as the specific peptide sequence, concentration, and the target pathogen.
Route of administration challenges	Administering AMPs can be challenging, as they may be degraded in the gastrointestinal tract or face other absorption barriers, impacting their bioavailability.
Off-target effects	Some AMPs may exhibit off-target effects, affecting nonpathogenic microorganisms or host cells, potentially leading to unintended consequences.
Selectivity challenges	Achieving high selectivity for pathogens while avoiding harm to host cells can be challenging, especially in complex biological environments.
Limited clinical data	Despite promising preclinical studies, there might be limited clinical data for certain AMPs, hindering their widespread adoption in medical practice.
Regulatory approval hurdles	Regulatory approval for AMP-based therapies may face challenges due to concerns about safety, efficacy, and manufacturing processes.

methicillin-resistant *S. aureus* (MRSA), various mycobacteria, *Propionibacterium acnes*, and *Clostridium difficile* (Manoharan and Balasubramaniam 2022). Their effectiveness in treating enteric infections stems from their potent nature, lower likelihood of eliciting resistance, and the ability to be precisely tuned to target specific pathogens. This makes AMPs a promising option for preventing dysbiosis and addressing the challenges posed by AMR. AMPs have also been widely utilized in several domains like food processing, biomedical research, and personal care.

4 Synergy of AMPs with Probiotic, Parabiotic, and Postbiotic in Maintaining Gut Microbial Balance

The AMPs in the gastrointestinal tract employ a variety of mechanisms to intricately balance the defense against pathogens and the preservation of symbiotic relationships with commensal bacteria. The AMPs collectively function within the gastrointestinal tract to form a resilient protective barrier against pathogenic microorganisms, which is essential for maintaining the homeostasis of the gut. The significance of this resides in the antibacterial activity and immunological regulatory activities exhibited by these entities, as demonstrated by Zong et al. (2020), which have profound consequences for the overall well-being of individuals. The direct antibacterial capabilities of AMPs involve disrupting microbial cell membranes and inhibiting cell wall synthesis (Table 8.2). AMPs also exhibit antiviral properties by hindering replication or modulating host cell responses. Additional roles include inhibiting biofilm formation, enhancing wound healing, and selectively targeting pathogenic microbes while sparing commensal ones (Table 8.2).

LAB-produced AMPs demonstrate a wide-ranging antibacterial impact against foodborne pathogens. The results of in vitro antibacterial testing revealed that the presence of the PR39 peptide in recombinant *L. casei* led to the manifestation of antibacterial properties against *E. coli* and *Salmonella*. However, the impact on *S. aureus* was found to be limited. According to Lüders et al. (2003), AMPs such as nisin and lacticin 3147, which are commonly employed as food preservatives, have exhibited specific efficacy against gram-positive *Listeria ivanovii*. Certain AMPs utilize a dual mechanism by interacting with the lipid II receptor, inducing pore formation that disrupts membrane potential and efflux of metabolites, halting biosynthetic processes, and inhibiting peptidoglycan synthesis, leading to bacterial cell death such as nisin and PlnC (Table 8.2) (Wiedemann et al. 2006). In addition, bacteriocin Abp118 produced by *L. salivarius* UCC118 protects mice against *L. monocytogenes*, confirming that the antimicrobial effect arises from direct antagonism between probiotic strain and the pathogen, mediated by Abp118 (Corr et al. 2007). Antifungal activities against *Aspergillus parasiticus* and *Penicillium expansum* were observed by peptides present in the cell-free supernatants derived from *L. rhamnosus*, *L. johnsoni*, and *L. plantarum* (Luz et al. 2017).

Peptide F1 from *L. paracasei* FX-6 AMS exhibits promise as an antimicrobial agent against colistin-resistant *E. coli* SHP45 by co-upregulating key genes and

Table 8.2 The diverse mechanisms of action utilized by antimicrobial peptides (AMPs) produced by lactic acid bacteria (LAB)

Mechanism	Description	Specific examples and details
Cell membrane disruption 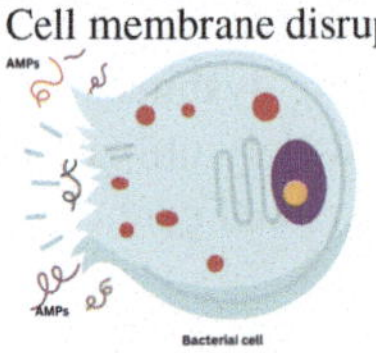	AMPs interact with bacterial membranes, disrupting lipid bilayers, forming transient pores, and causing cellular content leakage, ultimately leading to bacterial cell death.	PlnC, classified among lipid II-interacting antimicrobial peptides, strongly inhibits cell wall synthesis (Wiedemann et al. 2006).
Pore formation 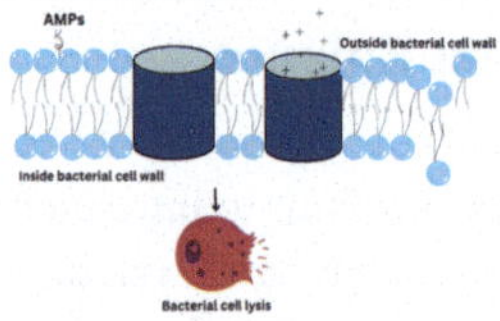	AMPs can insert into bacterial membranes, forming stable pores that create ion imbalances, disrupt osmotic equilibrium, and induce cell lysis, ultimately killing the bacterium.	PlnC dissipates proton motive force and triggers the immediate release of preaccumulated solutes in *Lactobacillus fermentum* and *Lactobacillus sake* cells (Wiedemann et al. 2006).
Inhibition of cell wall synthesis 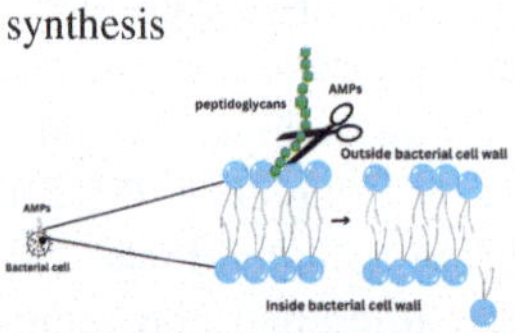	AMPs disrupt bacterial cell wall synthesis by targeting key components like lipids and peptidoglycans, weakening the structural integrity of the bacterial cell.	PlnA possesses cationic properties and demonstrates antimicrobial membrane-permeabilizing activity through a nonchiral interaction (Kristiansen et al. 2005)
Binding to intracellular targets 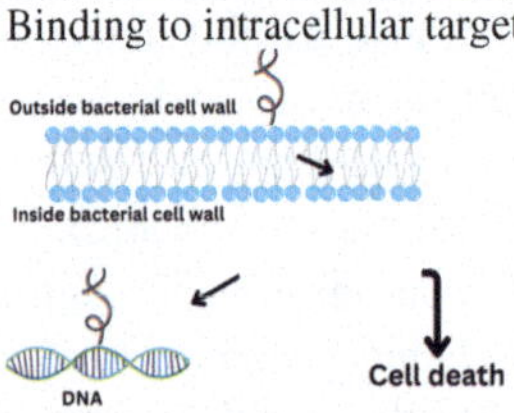	AMPs can enter bacterial cells, binding to intracellular targets like DNA, RNA, or proteins; disrupting essential cellular processes; and causing bacterial dysfunction and death.	Peptide F1 from *Lactobacillus paracasei* subsp. *tolerans* FX-6 shows a bactericidal effect on *Streptomyces aureus* by damaging cell membranes and binding to cytoplasmic genomic DNA (Miao et al. 2016).
Immunomodulation 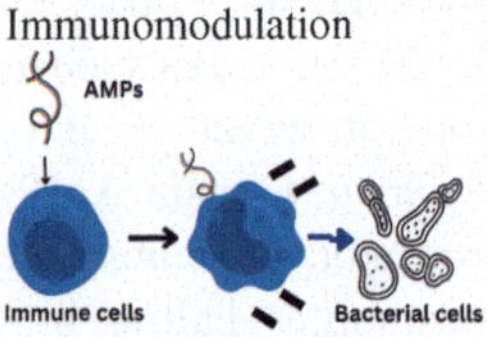	AMPs modulate the gut's host immune response by enhancing the activity of immune cells, such as macrophages and neutrophils, to recognize and eliminate bacterial pathogens more effectively.	Nisin A increased intracellular neutrophil superoxide activating neutrophils, promoting the formation of NETs known for neutralizing virulence factors and eliminating bacterial pathogens (Guryanova 2023).
Anti-biofilm activity 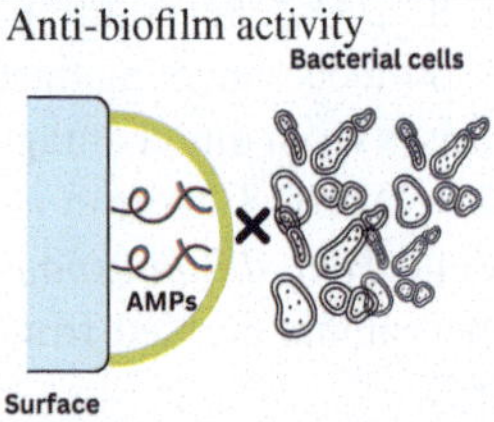	Certain AMPs show anti-biofilm properties, preventing bacterial biofilm formation or disrupting existing biofilms, thereby reducing bacterial adherence and persistence in the gut.	Nisin inhibited oral bacteria growth at low concentrations and dose-dependently reduced biofilm development, biomass, and thickness (Shin et al. 2015).

(continued)

Table 8.2 (continued)

Mechanism	Description	Specific examples and details
Synergy with antibiotics 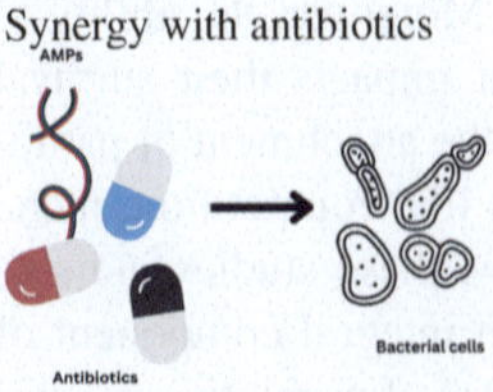 	AMPs can synergize with conventional antibiotics, enhancing the overall antimicrobial effect and potentially reducing the required antibiotic dosage, thereby minimizing the risk of antibiotic resistance.	Lacticin 3147 synergizes with polymyxin B and polymyxin E, achieving eight-fold and 16-fold reductions in their required levels to inhibit *Escherichia coli* and *Cronobacter* (Draper et al. 2013).

proteins involved in membrane damage and protein hydrolysis, leading to DNA damage, metabolic disorders, and inhibition of nutrient transport and energy metabolism (Zhang et al. 2023). According to Hernández-Granados and Franco-Robles (2020), the utilization of genetically modified *L. lactis* that synthesizes antimicrobial peptides has been found to effectively impede the growth of Enterococci and multidrug-resistant *Enterococcus faecium* strains within the gastrointestinal system. However, these AMPs have not demonstrated any antimicrobial activity against gram-negative *E. coli*. Additionally, other AMPs like pediocin PA-1, sakacin P, and curvacin A have also been found to possess individual properties against *L. ivanovii*. The potential of boosting specific action and expanding target-cell range is demonstrated by the synergistic combination of various AMPs.

AMPs also contribute to immunomodulation by influencing inflammatory responses, recruiting immune cells, and acting as chemoattractants (Table 8.2). Studies on the immunomodulatory properties of AMPs have shed light on their ability to affect immune responses at different points along cellular signaling pathways. The group of mice that were administered the recombinant pPG:612-PR39/*L. casei* 393 had enhanced growth and elevated percentages of peripheral blood lymphocytes, as well as higher quantities of white blood cells and spleen indices, in comparison to the control group (Zhang et al. 2016). According to a study conducted by Jastrząb et al. (2021), the application of nisin in a clinical endometritis animal model resulted in modifications to the cytokine expression profile and effectively mitigated the changes in uterine tissue caused by *S. aureus* infection. In vivo studies show that AMPs such as mersacidin from *Bacillus* sp. HIL-Y85/54728 and bacteriocin KT11 from *E. faecalis* KT11 inhibit antibiotic-resistant bacteria, including methicillin-resistant *S. aureus* and various gram-positive and gram-negative strains (Kruszewska et al. 2004; Abanoz and Kunduhoglu 2018).

AMPs produced by LAB demonstrate a protective influence in the GIT either by excluding pathogens or by fostering colonization in the gut (Alvarez-Sieiro et al. 2016; Ouertani et al. 2022). Kimchi, prepared using the bacteriocin-producing strain *Leuconostoc citreum* GJ7 as a starter culture, may serve as a protective strategy for the fermented product, preventing the introduction of pathogens upon consumption (Chang and Chang 2011). Similarly, the presence of the AMPs-producing strain *L. plantarum* MCS1 in a naturally contaminated salami sausage led to a reduction in the survival of *L. monocytogenes* (Ndlovu et al. 2015). AMPs such as nisin and pediocin PA-1 have found commercial applications as food additives, as has enterocin AS-48 or lacticin 3147 (Zhang et al. 2023). Moreover, the ability to generate AMPs is a desirable characteristic in LAB as it impacts their survival within the gastrointestinal system of humans by hindering the attachment of pathogens to intestinal epithelial cells by engaging mechanisms like competition, niche clearance, and spatial segregation (Table 8.3). Collectively, these studies indicate that antimicrobial peptides (AMPs) play a vital role as an integral component of probiotics, postbiotics, and parabiotics. The beneficial effects of probiotics, postbiotics, and parabiotics span various mechanisms, including competitive exclusion, antagonism against pathogens, modulation of the intestinal microbiota, generation of AMPs, stimulation of the immune system, and improvement of nutrient digestibility and intestinal morphology development.

5 Future Challenges and Perspectives

Future research in functional foods and biotherapeutics holds promising prospects. Understanding the synergistic effects of AMPs with probiotics, postbiotics, and parabiotics presents opportunities for enhancing health and immunity. Investigating the specific metabolic profiles of these components is crucial for unraveling the complex mechanisms underlying their beneficial effects. Additionally, exploring the potential therapeutic applications of these entities, including their role in preventing dysbiosis and addressing antimicrobial resistance, is an avenue worth exploring.

As scientific knowledge advances, the development of novel strategies for maintaining gut health and preventing associated disorders will likely emerge. The exploration of these interrelationships among AMPs, probiotics, postbiotics, and parabiotics provides a comprehensive foundation for further research and potential therapeutic interventions to optimize gut health and overall well-being.

The complex interaction between AMPs and the gut flora is a fundamental aspect of comprehending and improving overall well-being. As the exploration of probiotics, postbiotics, and parabiotics deepens, it becomes apparent that a complete strategy, which incorporates several elements, is needed to fully harness the capabilities of gut health formulations.

Future research should prioritize efforts to elucidate the precise mechanisms that underlie the observed synergies in these complex formulations. Undoubtedly, the investigation of the intricate interactions and influences between various substances will significantly assist in the advancement of more precise and efficacious therapies aimed at enhancing gastrointestinal well-being.

Table 8.3 A detailed compilation presenting the names of AMPs, their respective microbial producers, and the specific antimicrobial activities they manifest. This table sheds light on the intricate relationships between AMPs and microorganisms, emphasizing the multifaceted nature of these peptides in maintaining microbial balance and contributing to host defense mechanisms

MP	Produced by	Activity	References
Doderlein	*L. acidophilus*	Antimicrobial activity against *Streptococcus agalactiae* ATCC12386, *Micrococcus luteus* A270, *S. aureus* ATCC 29213, *P. aeruginosa* ATCC 27853, and *Candida albicans*	Da Silva et al. (2023)
LGH2	*L. casei* HZ1	Antimicrobial activity against *S. aureus*.	He et al. (2018)
Lactolisterin BU	*L. lactis subsp. lactis bv. diacetylactis BGBU1-4*	Antimicrobial activity against *L. monocytogenes*, *S. aureus*, *Bacillus* spp., and streptococci	Lozo et al. (2017)
Plantaricin DL3	*L. plantarum* DL3	Strongly inhibiting *L. monocytogenes, P. aeruginosa, Shewanella putrefaciens,* and *Psychrobacter* sp.	Lv et al. (2017)
Paracin 1.7	*L. paracasei* HD1-7	Antimicrobial activity against *Staphylococcus, Micrococcus, Bacillus, Lactobacillus, Proteus, Escherichia, Enterobacter, Pseudomonas,* and *Salmonella*	Ge et al. (2016)
Peptide F1	*L. paracasei* subsp. *tolerans* FX-6	Bactericidal effect on *Bacillus thuringiensis, S. aureus, Pseudomonas putida, and E. coli*	Miao et al. (2016), Zhang et al. (2023)
Plantaricin Q7	*L. plantarum* Q7	Exhibited sensitivity against *Pseudomonas putida* AS1.1819, *P. aeruginosa* CICC21636, *L. monocytogenes* ATCC19111, *E. coli* ATCC25922, *Salmonella typhimurium* ATCC14028, *S. aureus*, *Shigella flexneri* ATCC12022, and *Shigella sonnei* ATCC25931.	Liu et al. (2016)
PR39	Recombinant strain pPG:612-PR39/*L. casei 393*	Antibacterial activity against *E. coli* and *Salmonella*	Zhang et al. (2016)
Plantaricin K25	*L. plantarum* K25	Inhibited the growth of *B. cereus*	Wen et al. (2016)
Plantaricin KL-1Y	*L. plantarum* KL-1	Inhibitory activity against *S. enterica* serovar, *Enteritidis* DMST 17368, *P. aeruginosa* ATCC 15442, *P. aeruginosa* ATCC 9027, *E. coli* O157:H7, and *E. coli* ATCC 8739	Rumjuankiat et al. (2015)
Plantaricin ZJ008	*L. plantarum* ZJ008	Inhibitory activity against *Streptomyces citreus, Micrococcus luteus, S. aureus, E. coli, Shigella flexineri, Vibrio parahaemolyticus, L. monocytogenes,* and *P. aeruginosa*	Zhu et al. (2014)
AMPs LR14	*L. plantarum* strain LR/14	Antifungal efficacy against *A. niger, M. racemosus, R. stolonifer,* and *P. chrysogenum*	Gupta and Srivastava (2014)

(continued)

Table 8.3 (continued)

MP	Produced by	Activity	References
Plantaricin 163	*L. plantarum* 163	Antibacterial activity against *S. aureus, L. monocytogenes, B. pumilus, B. cereus, Micrococcus luteus, L. thermophilus, L. rhamnosus, E. coli, P. aeruginosa,* and *P. fluorescens*	Hu et al. (2013)
Bisin	*B. longum* DJO10A	Inhibitory activity against *S. thermophilus ST403, Clostridium perfringens, Staphylococcus epidermidis, B. subtilis, Serratia marcescens, E. coli DH5a*	Lee et al. (2011)
Plantaricin MG	*L. plantarum* KLDS1	Inhibitory activity against *L. monocytogenes, S. aureus, S. typhimurium*, and *E. coli*	Gong et al. (2010)
Bifidin I	*B. infantis* BCRC 14602	Inhibitory activity against LAB strains, *Staphylococcus, Bacillus, Streptococcus, Salmonella, Shigella, E. coli*	Cheikhyoussef et al. (2009, 2010)
AP48-MapA	*L. reuteri* LA92	Inhibitory activity against *B. cereus, S. aureus*, and *L. monocytogenes*	Bøhle et al. (2010)
Plantaricin W	*L. plantarum* PMU 33	Inhibitory activity against *L. monocytogenes, B. cereus, S. aureus, Enterococcus faecium,* and *Enterococcus faecalis*	Noonpakdee et al. (2009)
Plantaricin UG1	*L. plantarum* UG1	Inhibition activity against *L. monocytogenes* LMG10470	Altalhi (2008)
Lacticin Q	*L. lactis* QU 5	Antimicrobial activity against *Bacillus* sp., *Lactobacillus* sp., *Enterococcus* sp., *Lactococcus* sp., *and S. aureus*	Fujita et al. (2007)
Lactococcin Q	*L. lactis* QU 4	Antibacterial activity only against *L. lactis* strains	Zendo et al. (2006)
Plantaricin C (PlnC)	*L. plantarum* strain LL441	*L. fermentum, L. sake* whole cells, *Micrococcus flavus* DSM 1790, and *L. lactis* subsp. *cremoris* HP	Wiedemann et al. (2006)
Thermophilicin B67	*B. thermophilum* RBL67	Inhibitory activity against *Listeria* sp., *L. acidophilus*	Von Ah (2006)
Plantaricin OL15	*L. plantarum* OL15	Inhibitory action against *Lactobacillus, Lactococcus, and Propionibacterium*	Mourad et al. (2005)
Bifidocin B	*B. bifidum* NCFB	Antimicrobial activity against *B. cereus, E.s faecalis, Pediococcus acidolactici, S. faecalis, L. monocytogenes, Bochothrix thermosphacta, L. fermentum,* and *Leuconostoc* spp	Collado et al. (2005)
Nisin	*L. lactis*	Antimicrobial activity against several Gram-positive bacteria	Kramer et al. (2004)
Bifilact Bb-46	*B. longum* Bb-46	Inhibitory activity against *S. aureus, S. typhimurium, B. cereus, E. coli*	Saleh and El-Sayed (2004)
Bifilact Bb-12	*B. lactis* Bb-12	Inhibitory activity against *S. aureus, S. typhimurium, B. cereus, E. coli*	Saleh and El-Sayed (2004)
LsbA	*L. lactis* subsp. *lactis* BGMN1-5	Very narrow antibacterial spectrum	Gajic et al. (2003)

6 Conclusion

In conclusion, the gut ecosystem, comprising a diverse array of microorganisms, plays a pivotal role in maintaining overall health. Dysbiosis, the imbalance within the gut microbiome, can lead to various health complications, emphasizing the importance of cultivating a diverse and balanced gut microbiome. The interconnected roles of AMPs, probiotics, postbiotics, and parabiotics are crucial in achieving and sustaining this balance.

AMPs, synthesized within the gastrointestinal tract, contribute to microbial community equilibrium by regulating the proliferation of pathogenic bacteria. Probiotics, living microorganisms introduced into the body, enhance the gut microbiota by replenishing beneficial microorganisms. Postbiotics, metabolic byproducts of probiotics, serve as immunity boosters and contribute to intestinal resilience. Parabiotics, nonviable microbial compounds, offer microbial-derived benefits without the concerns associated with live microorganisms.

The chapter highlights the essential requirement for embracing an integrated approach to understanding gut health, by incorporating the concepts of probiotics, postbiotics, and parabiotics. The importance of promoting a harmonious interplay among these varied constituents is emphasized. As researchers dive into the relatively unexplored domains of the gut microbiota, significant opportunities arise for the creation of formulations that use the collective effectiveness of AMPs, probiotics, postbiotics, parabiotics, and prebiotics. These formulations not only offer optimistic possibilities but also serve as evidence of the changing nature of gastrointestinal health. The inherent potential for synergy between antimicrobial peptides AMPs and different elements of gut health offers a new aspect to our comprehension, opening possibilities for innovative approaches that show considerable potential in improving gastrointestinal well-being in the coming years.

References

Abanoz HS, Kunduhoglu B (2018) Antimicrobial activity of a bacteriocin produced by Enterococcus faecalis KT11 against some pathogens and antibiotic-resistant bacteria. Korean J Food Sci Anim Resour 38(5):1064

Ageitos JM, Sánchez-Pérez A, Calo-Mata P, Villa TG (2017) Antimicrobial peptides (AMPs): ancient compounds that represent novel weapons in the fight against bacteria. Biochem Pharmacol 133:117–138

Aguilar-Toalá JE, Arioli S, Behare P, Belzer C, Berni Canani R, Chatel JM, D'Auria E, de Freitas MQ, Elinav E, Esmerino EA, García HS (2021) Postbiotics—when simplification fails to clarify. Nat Rev Gastroenterol Hepatol 18(11):825–826

Akter N, Hashim R, Pham HQ, Choi SD, Lee DW, Shin JH, Rajagopal K (2020) Lactobacillus acidophilus antimicrobial peptide is antagonistic to Aeromonas hydrophila. Front Microbiol 11:570851

Altalhi AA (2008) Antilisterial activity of plantaricin UG1 during manufacture of Zabady and Kareesh cheese: two Arabian dairy products. Int J Biomed Sci 4(4):319

Alvarez-Sieiro P, Montalbán-López M, Mu D, Kuipers OP (2016) Bacteriocins of lactic acid bacteria: extending the family. Appl Microbiol Biotechnol 100:2939–2951
Amiri S, Mokarram RR, Khiabani MS, Bari MR, Khaledabad MA (2022) Characterization of antimicrobial peptides produced by lactobacillus acidophilus LA-5 and Bifidobacterium lactis BB-12 and their inhibitory effect against foodborne pathogens. LWT 153:112449
Arnison PG, Bibb MJ, Bierbaum G, Bowers AA, Bugni TS, Bulaj G, Camarero JA, Campopiano DJ, Challis GL, Clardy J, Cotter PD (2013) Ribosomally synthesized and post-translationally modified peptide natural products: overview and recommendations for a universal nomenclature. Nat Prod Rep 30(1):108–160
Aziz Q, Doré J, Emmanuel A, Guarner F, Quigley EMM (2013) Gut microbiota and gastrointestinal health: current concepts and future directions. Neurogastroenterol Motil 25(1):4–15
Bengmark S (2013) Gut microbiota, immune development and function. Pharmacol Res 69(1):87–113
Bhadekar R, Parhi P (2016) Non-dairy functional foods: potential of probiotics. Microbes Food Health:1–27
Bøhle LA, Brede DA, Diep DB, Holo H, Nes IF (2010) Specific degradation of the mucus adhesion-promoting protein (MapA) of Lactobacillus reuteri to an antimicrobial peptide. Appl Environ Microbiol 76(21):7306–7309
Chang JY, Chang HC (2011) Growth inhibition of foodborne pathogens by kimchi prepared with bacteriocin-producing starter culture. J Food Sci 76(1):M72–M78
Cheikhyoussef A, Pogori N, Chen H, Tian F, Chen W, Tang J, Zhang H (2009) Antimicrobial activity and partial characterization of bacteriocin-like inhibitory substances (BLIS) produced by Bifidobacterium infantis BCRC 14602. Food Control 20(6):553–559
Cheikhyoussef A, Cheikhyoussef N, Chen H, Zhao J, Tang J, Zhang H, Chen W (2010) Bifidin I–a new bacteriocin produced by Bifidobacterium infantis BCRC 14602: purification and partial amino acid sequence. Food Control 21(5):746–753
Chung IC, OuYang CN, Yuan SN, Lin HC, Huang KY, Wu PS, Liu CY, Tsai KJ, Loi LK, Chen YJ, Chung AK (2019) Pretreatment with a heat-killed probiotic modulates the NLRP3 inflammasome and attenuates colitis-associated colorectal cancer in mice. Nutrients 11(3):516
Collado MC, Hernandez M, Sanz Y (2005) Production of bacteriocin-like inhibitory compounds by human fecal Bifidobacterium strains. J Food Prot 68(5):1034–1040
Corr SC, Li Y, Riedel CU, O'Toole PW, Hill C, Gahan CG (2007) Bacteriocin production as a mechanism for the antiinfective activity of Lactobacillus salivarius UCC118. Proc Natl Acad Sci 104(18):7617–7621
Cuevas-González PF, Liceaga AM, Aguilar-Toalá JE (2020) Postbiotics and paraprobiotics: from concepts to applications. Food Res Int 136:109502
da Silva BS, Díaz-Roa A, Yamane ES, Hayashi MA, Junior PIS (2023) Doderlin: isolation and characterization of a broad-spectrum antimicrobial peptide from Lactobacillus acidophilus. Res Microbiol 174(3):103995
Draper LA, Cotter PD, Hill C, Ross RP (2013) The two peptide lantibiotic lacticin 3147 acts synergistically with polymyxin to inhibit Gram negative bacteria. BMC Microbiol 13(1):1–8
Durack J, Kimes NE, Lin DL, Rauch M, McKean M, McCauley K, Panzer AR, Mar JS, Cabana MD, Lynch SV (2018) Delayed gut microbiota development in high-risk for asthma infants is temporarily modifiable by Lactobacillus supplementation. Nat Commun 9(1):707
Erdem Büyükkiraz M, Kesmen Z (2022) Antimicrobial peptides (AMPs): a promising class of antimicrobial compounds. J Appl Microbiol 132(3):1573–1596
Fujita K, Ichimasa S, Zendo T, Koga S, Yoneyama F, Nakayama J, Sonomoto K (2007) Structural analysis and characterization of lacticin Q, a novel bacteriocin belonging to a new family of unmodified bacteriocins of gram-positive bacteria. Appl Environ Microbiol 73(9):2871–2877
Gajic O, Buist G, Kojic M, Topisirovic L, Kuipers OP, Kok J (2003) Novel mechanism of bacteriocin secretion and immunity carried out by lactococcal multidrug resistance proteins. J Biol Chem 278(36):34291–34298
Ge J, Sun Y, Xin X, Wang Y, Ping W (2016) Purification and partial characterization of a novel bacteriocin synthesized by Lactobacillus paracasei HD1-7 isolated from Chinese sauerkraut juice. Sci Rep 6(1):19366

Gibson GR, Hutkins R, Sanders ME, Prescott SL, Reimer RA, Salminen SJ, Scott K et al (2017) Expert consensus document: the International Scientific Association for Probiotics and Prebiotics (ISAPP) consensus statement on the definition and scope of prebiotics. Nat Rev Gastroenterol Hepatol 14(8):491–502

Gong HS, Meng XC, Wang H (2010) Plantaricin MG active against Gram-negative bacteria produced by Lactobacillus plantarum KLDS1. 0391 isolated from "Jiaoke", a traditional fermented cream from China. Food Control 21(1):89–96

Gupta R, Srivastava S (2014) Antifungal effect of antimicrobial peptides (AMPs LR14) derived from Lactobacillus plantarum strain LR/14 and their applications in prevention of grain spoilage. Food Microbiol 42:1–7

Guryanova SV (2023) Immunomodulation, bioavailability and safety of bacteriocins. Life 13(7):1521

He J, Luo X, Jin D, Wang Y, Zhang T (2018) Identification, recombinant expression, and characterization of LGH2, a novel antimicrobial peptide of Lactobacillus casei HZ1. Molecules 23(9):2246

Hernández-Granados MJ, Franco-Robles E (2020) Postbiotics in human health: possible new functional ingredients? Food Res Int 137:109660

Hill C, Guarner F, Reid G, Gibson GR, Merenstein DJ, Pot B, Morelli L, Canani RB, Flint HJ, Salminen S, Calder PC (2014) Expert consensus document: the International Scientific Association for Probiotics and Prebiotics consensus statement on the scope and appropriate use of the term probiotic. *Nat Rev Gastroenterol Hepatol* 11:506

Holanda DM, Yiannikouris A, Kim SW (2020) Investigation of the efficacy of a postbiotic yeast cell wall-based blend on newly-weaned pigs under a dietary challenge of multiple mycotoxins with emphasis on deoxynivalenol. Toxins 12(8):504

Hotel ACP, Cordoba A (2001) Health and nutritional properties of probiotics in food including powder milk with live lactic acid bacteria. Prevention 5(1):1–10

Hsieh FC, Lan CCE, Huang TY, Chen KW, Chai CY, Chen WT, Fang AH, Chen YH, Wu CS (2016) Heat-killed and live Lactobacillus reuteri GMNL-263 exhibit similar effects on improving metabolic functions in high-fat diet-induced obese rats. Food Funct 7(5):2374–2388

Hu M, Zhao H, Zhang C, Yu J, Lu Z (2013) Purification and characterization of plantaricin 163, a novel bacteriocin produced by Lactobacillus plantarum 163 isolated from traditional Chinese fermented vegetables. J Agric Food Chem 61(47):11676–11682

Jang HJ, Song MW, Lee NK, Paik HD (2018) Antioxidant effects of live and heat-killed probiotic Lactobacillus plantarum Ln1 isolated from kimchi. J Food Sci Technol 55:3174–3180

Jastrząb R, Graczyk D, Siedlecki P (2021) Molecular and cellular mechanisms influenced by postbiotics. Int J Mol Sci 22(24):13475

Kramer NE, Smid EJ, Kok J, de Kruijff B, Kuipers OP, Breukink E (2004) Resistance of Gram-positive bacteria to nisin is not determined by lipid II levels. FEMS Microbiol Lett 239(1):157–161

Kristiansen PE, Fimland G, Mantzilas D, Nissen-Meyer J (2005) Structure and mode of action of the membrane-permeabilizing antimicrobial peptide pheromone plantaricin A. J Biol Chem 280(24):22945–22950

Kruszewska D, Sahl HG, Bierbaum G, Pag U, Hynes SO, Ljungh Å (2004) Mersacidin eradicates methicillin-resistant Staphylococcus aureus (MRSA) in a mouse rhinitis model. J Antimicrob Chemother 54(3):648–653

Lee JH, Li X, O'Sullivan DJ (2011) Transcription analysis of a lantibiotic gene cluster from Bifidobacterium longum DJO10A. Appl Environ Microbiol 77(17):5879–5887

Liu H, Zhang L, Yi H, Han X, Chi C (2016) Identification and characterization of plantaricin Q7, a novel plantaricin produced by Lactobacillus plantarum Q7. LWT-Food Sci Technol 71:386–390

Lozo J, Mirkovic N, O'Connor PM, Malesevic M, Miljkovic M, Polovic N, Jovcic B, Cotter PD, Kojic M (2017) Lactolisterin BU, a novel class II broad-spectrum bacteriocin from Lactococcus lactis subsp. lactis bv. diacetylactis BGBU1-4. Appl Environ Microbiol 83(21):e01519–e01517

Lüders T, Birkemo GA, Fimland G, Nissen-Meyer J, Nes IF (2003) Strong synergy between a eukaryotic antimicrobial peptide and bacteriocins from lactic acid bacteria. Appl Environ Microbiol 69(3):1797–1799

Luz C, Saladino F, Luciano FB, Mañes J, Meca G (2017) In vitro antifungal activity of bioactive peptides produced by Lactobacillus plantarum against Aspergillus parasiticus and Penicillium expansum. LWT-Food Sci Technol 81:128–135

Lv X, Lin Y, Jie Y, Sun M, Zhang B, Bai F et al (2017) Purification, characterization, and action mechanism of plantaricin DL3, a novel bacteriocin against Pseudomonas aeruginosa produced by lactobacillus plantarum DL3 from Chinese Suan-Tsai. Eur Food Res Technol 244(2):323–331

Malanovic N, Lohner K (2016) Antimicrobial peptides targeting gram-positive bacteria. Pharmaceuticals 9(3):59

Manoharan M, Balasubramanian TS (2022) An extensive review on rroduction, purification, and bioactive application of different classes of bacteriocin. Journal of Tropical Biodiversity & Biotechnology 7(3).

Martín R, Miquel S, Ulmer J, Langella P, Bermudez-Humaran LG (2014) Gut ecosystem: how microbes help us. Benefic Microbes 5(3):219–233

Miao J, Zhou J, Liu G, Chen F, Chen Y, Gao X, Dixon W, Song M, Xiao H, Cao Y (2016) Membrane disruption and DNA binding of Staphylococcus aureus cell induced by a novel antimicrobial peptide produced by Lactobacillus paracasei subsp. tolerans FX-6. Food Control 59:609–613

Mourad K, Halima ZK, Nour-Eddine K (2005) Detectionn and activity of plantaricin OL15 a bacteriocin produced by Lactobacillus plantarum OL15 isolated from Algerian fermented olives. Grasas Aceites 56(3):191–197

Nataraj BH, Ali SA, Behare PV, Yadav H (2020) Postbiotics-parabiotics: the new horizons in microbial biotherapy and functional foods. Microb Cell Factories 19(1):1–22

Ndlovu B, Schoeman H, Franz CMAP, Du Toit M (2015) Screening, identification and characterization of bacteriocins produced by wine-isolated LAB strains. J Appl Microbiol 118(4):1007–1022

Negash AW, Tsehai BA (2020) Current applications of bacteriocin. Int J Microbiol 2020:4374891

Noonpakdee W, Jumriangrit P, Wittayakom K, Zendo J, Nakayama J, Sonomoto K, Panyim S (2009) Two-peptide bacteriocin from Lactobacillus plantarum PMU 33 strain isolated from som-fak, a Thai low salt fermented fish product. Asia Pacific J Mol Biol Biotechnol 17(1):19–25

Ouertani A, Mosbah A, Cherif A (2022) Anti-microbial peptides: the importance of structure-function analysis in the design of new AMPs. In: Insights on antimicrobial peptides. p 7

Parhi P, Song KP, Choo WS (2021) Effect of inulin and fructooligosaccharide supplementation on the growth and survival of Lactobacillus casei in model sugar systems. J Food Process Preserv 45(3):e15228

Parhi P, Song KP, Choo WS (2023) Effect of prebiotic supplementation on the low-temperature storage stability and in vitro gastrointestinal tolerance of Bifidobacterium breve and Bifidobacterium longum in lactose and sucrose systems. Bioact Carbohydr Diet Fibre 30:100369

Pasupuleti M, Schmidtchen A, Malmsten M (2012) Antimicrobial peptides: key components of the innate immune system. Crit Rev Biotechnol 32(2):143–171

Pavlova AS, Ozhegov GD, Arapidi GP, Butenko IO, Fomin ES, Alemasov NA, Afonnikov DA, Yarullina DR, Ivanov VT, Govorun VM, Kayumov AR (2020) Identification of antimicrobial peptides from novel Lactobacillus fermentum strain. Protein J 39:73–84

Piqué N, Berlanga M, Miñana-Galbis D (2019) Health benefits of heat-killed (Tyndallized) probiotics: an overview. Int J Mol Sci 20(10):2534

Plamada D, Vodnar DC (2021) Polyphenols—gut microbiota interrelationship: a transition to a new generation of prebiotics. Nutrients 14(1):137

Rad AH, Hosseini S, Pourjafar H (2022) Postbiotics as dynamic biological molecules for antimicrobial activity: a mini-review. Biointerface Res Appl Chem 12(5):6543–6556

Rafique N, Jan SY, Dar AH, Dash KK, Sarkar A, Shams R, Pandey VK, Khan SA, Amin QA, Hussain SZ (2023) Promising bioactivities of postbiotics: a comprehensive review. J Agri Food Res 14:100708

Rumjuankiat K, Perez RH, Pilasombut K, Keawsompong S, Zendo T, Sonomoto K, Nitisinprasert S (2015) Purification and characterization of a novel plantaricin, KL-1Y, from Lactobacillus plantarum KL-1. World J Microbiol Biotechnol 31:983–994

Saleh FA, El-Sayed EM (2004) Isolation and characterization of bacteriocins produced by Bifidobacterium lactis BB-12 and Bifidobacterium longum BB-46. 9th Egyptian Conference for Dairy Science and Technology, Research Papers, Cairo, pp 323–37
Salminen S, Collado MC, Endo A, Hill C, Lebeer S, Quigley EM, Sanders ME, Shamir R, Swann JR, Szajewska H, Vinderola G (2021) The International Scientific Association of Probiotics and Prebiotics (ISAPP) consensus statement on the definition and scope of postbiotics. Nat Rev Gastroenterol Hepatol 18(9):649–667
Senchukova MA (2023) Microbiota of the gastrointestinal tract: friend or foe? World J Gastroenterol 29(1):19
Shin JM, Ateia I, Paulus JR, Liu H, Fenno JC, Rickard AH, Kapila YL (2015) Antimicrobial nisin acts against saliva derived multi-species biofilms without cytotoxicity to human oral cells. Front Microbiol 6:617
Singh TP, Kaur G, Kapila S, Malik RK (2017) Antagonistic activity of Lactobacillus reuteri strains on the adhesion characteristics of selected pathogens. Front Microbiol 8:486
Singh A, Vishwakarma V, Singhal B (2018) Metabiotics: the functional metabolic signatures of probiotics: current state-of-art and future research priorities—metabiotics: probiotics effector molecules. Adv Biosci Biotechnol 9(04):147–189
Sugahara H, Yao R, Odamaki T, Xiao JZ (2017) Differences between live and heat-killed bifidobacteria in the regulation of immune function and the intestinal environment. Benefic Microbes 8(3):463–472
Tajbakhsh M, Karimi A, Fallah F, Akhavan MM (2017) Overview of ribosomal and non-ribosomal antimicrobial peptides produced by Gram positive bacteria. Cell Mol Biol 63(10):20–32
Taverniti V, Guglielmetti S (2011) The immunomodulatory properties of probiotic microorganisms beyond their viability (ghost probiotics: proposal of paraprobiotic concept). Genes Nutr 6(3):261–274
Vinderola G, Sanders ME, Salminen S (2022) The concept of postbiotics. Food Secur 11(8):1077
von Ah U (2006) Identification of Bifidobacterium thermophilum RBL67 isolated from baby faeces and partial purification of its bacteriocin. Culture. Swiss Federal Institute of Technology. PhD thesis, Zurich, pp 1–192
Wang B, Yao M, Lv L, Ling Z, Li L (2017) The human microbiota in health and disease. Engineering 3(1):71–82.
Wegh CA, Geerlings SY, Knol J, Roeselers G, Belzer C (2019) Postbiotics and their potential applications in early life nutrition and beyond. Int J Mol Sci 20(19):4673
Wen LS, Philip K, Ajam N (2016) Purification, characterization and mode of action of plantaricin K25 produced by Lactobacillus plantarum. Food Control 60:430–439
Wiedemann I, Böttiger T, Bonelli RR, Schneider T, Sahl HG, Martínez B (2006) Lipid II-based antimicrobial activity of the lantibiotic plantaricin C. Appl Environ Microbiol 72(4):2809–2814
Zendo T, Koga S, Shigeri Y, Nakayama J, Sonomoto K (2006) Lactococcin Q, a novel two-peptide bacteriocin produced by Lactococcus lactis QU 4. Appl Environ Microbiol 72(5):3383–3389
Zhang L, Guo D, Liu Y, Shao Y, Wang Y, Xu Y, Jiang Y, Cui W, Li Y, Tang L (2016) Probiotic Lactobacillus casei expressing porcine antimicrobial peptide PR39 elevates antibacterial activity in the gastrointestinal tract. Can J Microbiol 62(11):961–969
Zhang H, Chen C, Yang Z, Ye L, Miao J, Lan Y, Wang Q, Ye Z, Cao Y, Liu G (2023) Combined transcriptomic and proteomic analysis of the antibacterial mechanisms of an antimicrobial substance produced by Lactobacillus paracasei FX-6 against colistin-resistant Escherichia coli. LWT 173:114394
Zhu X, Zhao Y, Sun Y, Gu Q (2014) Purification and characterisation of plantaricin ZJ008, a novel bacteriocin against Staphylococcus spp. from Lactobacillus plantarum ZJ008. Food Chem 165:216–223
Zong X, Fu J, Xu B, Wang Y, Jin M (2020) Interplay between gut microbiota and antimicrobial peptides. Anim Nutr 6(4):389–396

Chapter 9
Strategic Defenders: Antimicrobial Peptides and Their Impact on the Innate Immunity

Gurpreet Kaur and Pratyushaa Trivedi

1 Introduction

Antimicrobial peptides are small molecular peptides that play a pivotal role as essential constituents within the inherent defense mechanisms against a spectrum of pathogens, encompassing viruses, bacteria, fungi, parasites, and cancer cells. These short, positively charged peptides are not only crucial components of the innate immune response, serving as formidable guardians against a myriad of pathogens, but also possess the capacity to influence the very core of our immunological identity. A total of 3791 AMPs from six kingdoms have been recorded in the AMP database to date, including 431 from bacteria, 4 from archaea, 7 from protozoa, 6 from fungi, 824 from plants, and 2519 from animals. For a very long time, antibiotics have served as the mainstay of treatment for pathogenic bacteria. Although many antibiotics are employed to treat intracellular bacterial pathogens, the growing concern surrounding antibiotic resistance and the increasing emergence of multidrug resistance clinical strains represent a significant global issue. This phenomenon markedly diminishes or, in some cases, entirely nullifies the efficacy and success of these antibiotics.

Hence, there is a pressing demand for innovative and more efficacious therapies to control intracellular infections. AMPs are recognized for their diverse biological roles, encompassing antibacterial, antifungal, immunomodulation, antiviral, wound healing, stimulating angiogenesis, promoting cell migration, and also antitumor properties. With their small yet mighty structure, AMPs pack a remarkable punch in their ability to combat pathogens. AMPs exhibit extensive activity, not only targeting bacteria, yeasts, fungi, and viruses but also demonstrating efficacy against cancer cells. Alternatively known as "host defense peptides," AMPs emphasize their

G. Kaur (✉) · P. Trivedi
Department of Biotechnology, Chandigarh College of Technology (CCT), Chandigarh Group of Colleges, Landran, Mohali, India

P. Baindara, S. M. Mandal (eds.), *Evolution of Antimicrobial Peptides*,
https://doi.org/10.1007/978-3-031-67515-7_9

multifaceted immunomodulatory roles, actively contributing to the preservation of immune homeostasis (Huan et al. 2020; Zhang and Gallo 2016). Beyond their antimicrobial properties, these peptides possess the capacity to finely regulate both the innate and adaptive immune systems and their resultant responses, thereby shaping the fundamental framework of our immunological defenses. Antimicrobial peptides (AMPs) emerge as a promising and compelling alternative. These remarkable molecules can manage diverse infections, whether by directly eradicating pathogens or by modulating specific immune responses.

The key features that make AMPs the earliest and most important host defense mechanism against pathogen attack include: (1) AMPs produced by host cells as the first line of defense mounts rapid immune response to numerous kinds of pathogens. Due to the fact that mechanisms relying on specific antigen recognition hinge on the clonal proliferation of immunological cells, a process that takes days to weeks to fully develop, our initial defense mechanisms recognize microbe-specific class characteristics. These mechanisms also need to employ strategies that are either constitutive or rapidly inducible. This adaptability is crucial for mounting swift and effective responses, particularly in the early stages of an immune reaction, where timely recognition and action are paramount. In organisms like unicellular invertebrates and plants that lack adaptive immunity, AMPs play an important role in immune defense responses (Fritig et al. 1998). (2) Additionally, the smaller size of these peptides allows for quicker targeting than antibodies. (3) Furthermore, certain eukaryotes, such as insects, which lack lymphocyte-based immune systems, rely primarily on antimicrobial peptides and proteins to attack invaders and protect themselves. (4) AMPs induce lower levels of resistance compared to conventional antibiotics, (5) They possess broad-spectrum activity, (6) AMPs act quickly through diverse mechanisms, and (7) AMPS exhibit additional immunomodulatory properties (Fig. 9.1).

The body's first defense mechanism against pathogens is the innate immune response, which relies on a complex interplay of biological elements such as epithelial barriers and a diverse array of leukocytes. Among these cellular defenders are neutrophils, macrophages, NK cells, lymphocytes, and mast cells. To understand the pivotal role of AMPs, one must first understand the response induced by the innate immune system—the host's rapid and nonspecific defense mechanism

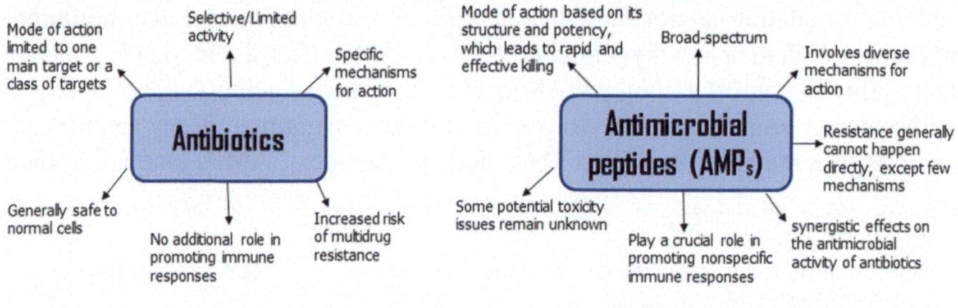

Fig. 9.1 Comparative analysis illustrating the efficacy and versatility of AMPs over traditional antibiotics

against pathogens. The innate immune system acts as our biological border control, maintaining constant vigilance and swiftly reacting to potential threats. It deploys a range of mechanisms, from physical barriers like the skin to the rapid responses of white blood cells and, indeed, antimicrobial peptides. These molecules are poised as nature's multi-tool, combating a broad spectrum of invaders without the need for prior exposure or immunological memory.

Natural antimicrobial peptides (AMPs) exhibit robust and wide-ranging effectiveness against various categories of pathogens, including bacteria, viruses, fungi, yeasts, protozoa, and parasites (Huan et al. 2020; Luong et al. 2020). They manifest multifaceted properties such as inhibiting bacterial growth, directly exterminating microorganisms, and inducing cell lysis (Pasupuleti et al. 2012). Furthermore, the appeal and exploration of AMPs have intensified in the wake of the recent global health crisis caused by the viral SARS-CoV-2 outbreak (Kurpe et al. 2020; Saini et al. 2022). Microbiologist René Dubos made the groundbreaking discovery of antimicrobial peptides (AMPs) in 1939. Pioneering the field, Dubos isolated the antimicrobial agent known as gramicidin from a soil *Bacillus* strain, thereby not only making a pivotal contribution to the understanding of AMPs but also laying the foundation for subsequent explorations into the remarkable world of these natural defense compounds. This breakthrough not only marked the birth of AMP research but also showcased gramicidin's capacity to protect animals against pneumococcal infections (Van Epps 2006). Subsequently, a number of AMPs have been discovered spanning across the domains of both prokaryotic and eukaryotic life and have been registered in the AMP database (Boparai and Sharma 2020; Huan et al. 2020). Antimicrobial peptides exhibit significant variations in their structure, sequence, secondary structural preferences, and mechanism of action. Numerous antimicrobial peptides (AMPs) combat microbial pathogens through a membrane-disrupting mechanism. The positively charged nature of antimicrobial peptides (AMPs) serves a dual purpose, attracting them toward the negatively charged surfaces of microbial membranes. Additionally, their hydrophobic characteristics aid in the seamless insertion into these membranes. This integration, in turn, leads to the disruption of membrane integrity, causing the release of bacterial cell contents and eventually cell death. The apparent selectivity of AMPs toward bacteria is attributed to the prevalence of anionic lipids, lipopolysaccharides (in gram-negative bacteria), or teichoic acids (in gram-positive bacteria) on the surfaces of bacterial membranes. These components carry a significantly more negative charge compared to mammalian cell membranes. The interaction with the membrane serves as a pivotal determinant of AMPs' antimicrobial activity, whether targeting the bacterial membrane itself or intracellular components. The intrinsic ability of AMPs to selectively target specific microbial components highlights their sophisticated and discerning nature as natural defenders against bacterial threats.

Additionally, AMPs can also exhibit another facet of their efficacy by traversing the bacterial membrane and directly engaging with intracellular biological macromolecules, including nucleic acids or proteins. This indirect approach influences critical processes of a host-like transcription, replication, or protein synthesis (Zhang et al. 2021). Furthermore, research emphasizes the substantial therapeutic potential of AMPs, particularly in relation to their anticancer properties (Tornesello et al. 2020). With cancer posing a substantial global health burden and its incidence doubling over the last decade, research efforts are intensifying to discover biologically active molecules. These molecules, when paired with conventional treatment therapies, hold the promise of serving as efficient contributors to anticancer treatment efficacy. AMPs, characterized by unique features such as anticancer activity, demonstrate the ability to selectively impact cancer cells, making them a valuable focus of social research. With a wide spectrum of activity, minimal toxicity to host cells, and a limited induction of resistance in target cells, specific antimicrobial peptides (AMPs) emerge as compelling candidates for novel therapeutic applications. Their potential spans across antibacterial, antiviral, antifungal, and anticancer realms, offering promising prospects when strategically integrated into diverse treatment approaches. Moreover, recent investigations reveal the remarkable capacity of AMPs to serve as crucial links, bridging the divide between the innate and adaptive immune systems, thus underscoring the potential of AMPs not only as direct defenders against pathogens but also as orchestrators of a dynamic interplay between different arms of the immune response. They are capable of efficiently modulating immune responses, serving as pivotal players in safeguarding against infections, regulating excessive inflammation, and influencing normal homeostasis. The extensive array of immunomodulatory activities linked with antimicrobial peptides spans a diverse range of functions, including leukocyte recruitment, stimulation of chemotaxis, induction of pro- and anti-inflammatory cytokines, neutralization of endotoxins, and activation of immune cell lines. Defensins and cathelicidins from various species are among the most extensively studied peptides exhibiting these immunomodulatory capacities. Despite this progress, the mechanisms underlying AMPs are not fully comprehended, necessitating further investigation. Understanding the interplay between various physicochemical properties is crucial for developing cost-effective and highly safe AMPs with potent antimicrobial effects. Additionally, there is a need for further exploration into enhancing the specificity and synergistic capabilities of AMPs. By exploring further their multifaceted mechanisms and expanding our understanding of their interactions with the immune system, ongoing research promises to unlock new dimensions in the application of AMPs, potentially revolutionizing treatment strategies against a spectrum of infectious and pathological challenges. With this backdrop, this review seeks to unravel the multifaceted roles of antimicrobial peptides. In the present work, we describe the abilities of various AMPS to modulate the broader immune response, the mechanisms they use to protect us from pathogens, and the intricate balance they maintain between innate immunity and host defense. By comprehensively understanding the dual roles of AMPs, our objective is to unravel the intricate interplay between innate immunity and these peptides. Exploration of the complex interplay between the microscopic world of antimicrobial peptides and the vast landscape of immunology promises a deeper understanding of the immune system's complexities and the possibility of unlocking new avenues for treating and managing various diseases.

2 Antimicrobial Peptides: Structure and Classification of AMPs

Antimicrobial peptides (AMPs), also known as host defense peptides (HDPs), are naturally occurring compounds present across diverse life forms, including bacteria, protozoa, fungi, plants, insects, and other animals. These peptides, spanning prokaryotes and eukaryotes, serve as the initial barrier in innate responses against infectious organisms. Rapidly inhibiting microbial proliferation, AMPs play a crucial role in innate host defense, disrupting pathogen lipid bilayers and extending their impact to modulate adaptive and cellular immunity. As multifunctional therapeutic agents effective against a broad spectrum of microorganisms, AMPs offer a promising solution for the increasing demand for new antimicrobial compounds (Radek and Gallo 2007). Typically, AMPs are described as peptides containing fewer than 100-amino-acid residues. They display a net positive charge spanning from +2 to +9, primarily attributed to the presence of positively charged amino acids such as arginine and lysine, coupled with a significant proportion of hydrophobic residues (Malmsten 2014). The unique structural and physicochemical attributes of AMPs are critical factors that shape their selectivity and affinity for target cells. They show remarkable variability and diversity in their structure and sequence (Fig. 9.1). Nevertheless, they tend to share several common characteristics. Currently, various methods for classification exist for AMPs, with criteria, including their origin, function, or structure (Bin Hafeez et al. 2021). Typically, AMPs can be grouped into four principal structural types, primarily dictated by their secondary configuration (Fig. 9.1): (1) amphipathic α-helical peptides, exemplified by cathelicidin and LL-37; (2) β-strands stabilized by disulfide bonds, represented by defensins; (3) peptides notably abundant in one or two specific amino acids, such as tryptophan-rich peptides like indolicidin; and (4) loop peptides, for example, Bactenecin contains an intramolecular covalent disulfide bond at its C-terminal end, giving rise to a distinctive loop formation.

Further, AMPs are categorized based on a number of factors, including their structure, place of origin, and method of action (Li et al. 2012).

The following are examples of some typical categories.

2.1 *Based on Structure*

- *Linear AMPs*: These peptides have a straight, unbranched chain of amino acids.
- *Cyclic AMPs*: Their structure forms a closed loop, which can enhance stability.
- *Alpha-Helical AMPs*: These molecules possess helical structure and often disrupt bacterial membranes.
- *Beta-Sheet AMPs*: They acquire a sheet-like structure and have the ability to form pores in membranes.

2.2 Based on Origin

- *Host Defence Peptides*: Many organisms produce naturally occurring antimicrobial peptides as part of their immune defense.
- *Non-host Defence Peptides*: AMPs sourced from entities other than the host organism, such as plants or bacteria.

2.3 Based on Amino Acid Composition

- *Cationic AMPs*: The antibacterial activity of AMPs is directly influenced by their cationic properties, which is linked to the abundance of lysine and arginine residues that bind with negatively charged microbial membranes.
- *Anionic AMPs*: Have a net negative charge and may have a different mechanism of action.

2.4 Based on Target Specificity

- *Broad-Spectrum AMPs*: Effective against a wide range of microorganisms.
- *Narrow-Spectrum AMPs*: Target-specific types of pathogens.

2.5 Based on Mechanism of Action

- *Membrane-Disrupting AMPs*: These peptides disrupt the cell membranes of the pathogen, causing lysis of the cell.
- *Non-membrane-Disrupting AMPs*: These peptides have the capacity to interfere with normal microbial processes inside the cell, such as DNA replication or protein translation, without disrupting the integrity of the cell membrane.

2.6 Based on Secondary Functions

- Some AMPs may have additional roles, such as modulating the immune response or promoting wound healing (Fig. 9.2).

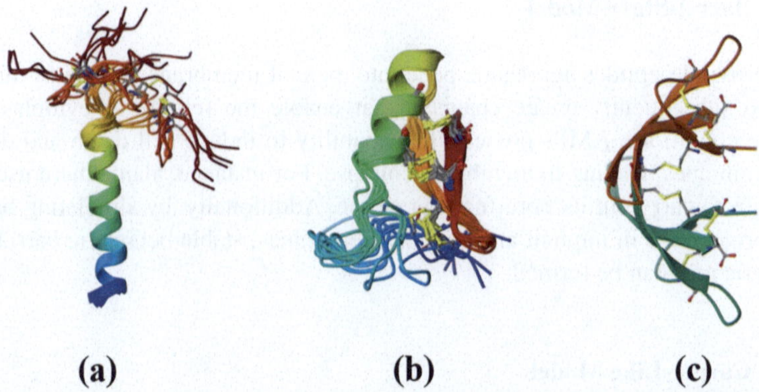

Fig. 9.2 PDB structures of some antimicrobial peptides (AMPs) that show great versatility in their structure and sequence. (**a**) Nicomicin-1, novel antimicrobial peptide (AMP) from the Arctic polychaeta Nicomache minor in the Maldanidae family (PDB ID 6HN9); (**b**) three-dimensional structure of insect defensin A (PDB ID 1ICA); (**c**) human alpha-defensin 1 (PDB ID 2PM4) (www.rcsb.org)

3 Mechanism of AMPs

Multiple models have been suggested to explain the mechanism through which antimicrobial peptides induce membrane permeabilization although it is important to note that there is no universal view among researchers on this matter (Pasupuleti et al. 2012). Some of the widely accepted models are described below.

3.1 Membrane-Targeting Mechanism

The pore and carpet models, alongside the toroidal pore and barrel-stave variations of the pore model, contribute to understanding the membrane-targeting mechanisms employed by AMPs.

3.1.1 Toroidal Pore Model

The toroidal pore model, alternatively referred to as the wormhole model, illustrates the mechanism, wherein vertically integrated antimicrobial peptides (AMPs) accumulate within the cellular membrane, subsequently bending to create a ring-shaped orifice with a diameter ranging from one to two nanometers (Matsuzaki et al. 1995). Important AMPs like arenicin, lacticin Q, melittins, and magainin 2 are included in this model. Furthermore, cationic peptides such as BP2, TC84, and TC19 disrupt the microbial membrane barrier by inducing the formation of fluid domains.

3.1.2 Barrel-Stave Model

Antimicrobial peptides aggregate, penetrate the cell membrane bilayer as multimers, and subsequently create channels that enable the release of cytoplasm. In extreme conditions, AMPs possess the capability to induce cell death and disrupt cell membranes, leading to membrane collapse. For instance, alamethicin uses this paradigm to carry out its pore-forming action. Additionally, by simulating hairpin AMP protegrin-1 in implicit and explicit membranes, stable octameric barrels and tetrameric arcs can be formed.

3.1.3 Carpet-Like Model

Antimicrobial peptides (AMPs) align parallel to the cell membrane, positioning their hydrophilic ends toward the solution and their hydrophobic ends against the phospholipid bilayer. This arrangement resembles a carpet covering the membrane surface, leading to a "detergent-like destruction" of the cell membrane (Oren and Shai 1998). Human cathelicidin LL-37, cecropin, indolicidin, and AMPs with β-sheet structures operate through this model (Corrêa et al. 2019). Similar to the action of detergents, AMPs form a carpet-like covering on the membrane surface, resulting in the breakdown of the cell membrane. Achieving a substantial concentration of AMPs is critical for this pore-forming mechanism, requiring a specific concentration threshold. The carpet model does not necessitate specific peptide–peptide interactions among membrane-bound peptide monomers. Moreover, it does not require the insertion of peptides into the hydrophobic core to create transmembrane channels or adopt specific peptide structures (Yeaman and Yount 2003). Numerous peptides exhibit antimicrobial activity regardless of their precise amino acid composition or sequence length. AMPs that operate under the carpet model function effectively at elevated concentrations due to their inherently amphiphilic nature.

Cell membranes predominantly consist of lysolipids, sphingolipids, glycerophospholipids, and sterols. Among bacterial cell membranes, the principal anionic lipids include phosphatidylethanolamine, phosphatidylglycerol, and cardiolipin. In contrast, fungal cell membranes are characterized by the prevalence of phosphatidylcholine, phosphatidylinositol, phosphatidylethanolamine, and phosphatidic acid as the major glycerophospholipids (GPLs). Fungal cell membranes exhibit higher levels of phosphatidylcholine (PC) and greater anionic characteristics compared to mammalian cell membranes. Ergosterol, in contrast, serves as the sterol in the plasma membrane of lesser eukaryotes like fungi, while cholesterol is predominant in animal cell membranes. Many antimicrobial peptides (AMPs) employ the diverse composition of cell membranes to carry out their actions effectively. These peptides leverage the variability in membrane constituents to disrupt and neutralize microbial threats, demonstrating their adaptability and versatility in combating various pathogens.

Initially, it was believed that AMPs exclusively targeted the cell membrane, precluding the possibility of intracellular targets. Investigations using early α-helical peptides like cecropin, and melittin, where the all-D versions demonstrated comparable efficacy to the natural all-L peptides, seemed to support the notion that intracellular cell targets such as proteins/nucleic acids were unnecessary for the antibacterial activity of AMPs, reinforcing the membrane-centric targeting concept (Wade et al. 1990; Epand and Vogel 1999). However, further research suggests that certain AMPs indeed have intracellular targets, as these peptides induce bacterial death without causing membrane permeabilization at the minimal effective concentration (Brogden 2005).

3.2 Non-membrane-Targeting Mechanism

AMPs can enter host cells through either direct cell penetration or endocytosis. Once the AMPs are within the cytoplasm, they recognize and engage with their specific targets. The categorization of antimicrobial peptides (AMPs) according to their specific targets can be determined as follows.

3.2.1 Inhibition/Disruption of Cellular Protein Synthesis

Antimicrobial peptides disrupt the essential processes such as transcription and translation by disrupting the mode of action of relevant enzymes and other essential effector molecules. Additionally, they impede the assembly of functional peptides, a process facilitated by molecular chaperone folding. PR-39, an antimicrobial peptide isolated from pigs' intestine, is abundant in amino acids proline and arginine that exhibited swift penetration of the outer cell membrane of *Escherichia coli*. It accomplishes this by employing a mechanism that degrades the proteins required for DNA synthesis ultimately leading to bacterial killing (Boman et al. 1993). Peptides like Tur1A affect the translation process in both *E. coli* and *Thermus thermophilus* by inhibiting the transition from the initiation phase to the elongation phase. In contrast, Bac7 1–35 selectively targets ribosomes, hindering the process of protein translation (Mardirossian et al. 2014). Another study demonstrated the antibacterial activity of Bac5, a nonlytic proline-rich antimicrobial peptide (PrAMP). The N-terminal fragments (1–25) and (1–31) of Bac5 were shown to interfere with protein synthesis by binding to the ribosome tunnel in the bacteria. This action prevents the transition from the initiation stage to the elongation stage of translation (Mardirossian et al. 2018). Api137, a peptide derived from apidaecin, was found to interact with ribosomes in *E. coli*, capturing release factor 1 (RF1) or release factor 2 (RF2), thereby facilitating the release of nascent polypeptide chains and termination of translation (Florin et al. 2017). They, however, exhibit notable differences that influence their interactions with ribosomes and their attachment to the ribosomal peptide exit tunnel. Proline-enriched antimicrobial peptides (AMPs) inhibit the synthesis by binding to ribosomes. For example, oncocin-type peptides hinder mRNA translation by binding to the 70S ribosome export, while apidaecin-type peptides impede the assembly of the ribosome's 50S large subunit (Graf et al. 2017; Roy et al. 2015).

Chaperones play a crucial role in ensuring the proper folding, assembly, and attainment of stereoisomerism in newly synthesized proteins. Recent studies have highlighted the inhibitory impact of both pyrhocoricin and drosocinin on DnaK, a chaperone protein essential for the reconstitution of misfolded proteins. These peptides induce a sustained closure of the DnaK peptide-binding cavity, thereby significantly impeding the chaperone's ability to facilitate the proper refolding of misfolded proteins (Wrońska and Boguś 2020; Kragol et al. 2001).

3.2.2 Inhibition of Biosynthesis of Nucleic Acid

Antimicrobial peptides employ diverse and potent mechanisms to combat microbial threats. They can exert their effects by influencing critical enzymes or initiating the degradation of nucleic acid molecules, thereby inhibiting nucleic acid biosynthesis. For example, indolicidin, a C-terminal-amidated cationic Trp-rich antimicrobial peptide with 13 amino acids, selectively targets the abasic region of DNA, cross-linking single- or double-stranded DNA and inhibiting DNA topoisomerase (Subbalakshmi and Sitaram 1998). Another noteworthy example is tissue factor pathway inhibitor 1 TC24 (TFP1-TC24), an antimicrobial peptide derived from tongues. Upon breaching the cell membrane, TFP1-TC24 gains access to the cytoplasm of target cells, initiating the degradation of DNA and RNA within the cell and contributing to its antimicrobial properties. Similarly, buforin II, an AMP with 21 amino acids, demonstrates antibacterial activity against a wide range of bacteria. Sharing the same sequence as part of histone H2A, a protein that directly interacts with nucleic acids, buforin II penetrates lipid vesicles in vitro without affecting membrane permeability and binds to DNA and RNA (Cardoso et al. 2019). Peptide-P2, an anionic antimicrobial peptide isolated from *Xenopus laevis* skin, disrupts bacterial growth by disrupting the bacterial cell membrane and interacting with microbial genomic DNA (Zhang et al. 2017). Besides directly binding to DNA and inducing DNA damage, AMPs also indirectly inhibit DNA replication or transcription. These actions underline the diverse and potent mechanisms employed by antimicrobial peptides in combating microbial threats (He et al. 2017).

3.2.3 Inhibition of Protease and Other Enzymes

Numerous antimicrobial peptides (AMPs) target diverse metabolic processes by focusing on protease activity. Histatin 5, for example, significantly hinders the secretion of proteasesenzyme. Microbial elastase, chymotrypsin, and serine proteases are all inhibited by the AMPs eNAP-2 and indolicidin (Le et al. 2017). Further, a peptide derived from the venom of *Bungarus fasciatus* Cathelicidin-BF showcases diverse biological functions encompassing antimicrobial, anticancer, and anti-inflammatory properties. Further, it effectively inhibited thrombus formation and additionally hindered the activity of protease-activated receptor 4. This unique peptide's properties suggest its potential utility in modulating thrombotic processes and platelet activation (Shu et al. 2019). Additionally, studies suggest that antimicrobial peptides (AMPs) can impede the activity of various bacterial intracellular enzymes. Pyrrhocoricin, an antimicrobial peptide rich in amino acid proline (PrAMP), demonstrated specific binding affinity to the bacterial HSP DnaK within protein lysates extracted from *Escherichia coli* (Otvos et al. 2000). Further studies showed pyrrhocoricin was able to reduce ATPase activity of recombinant DnaK and thus inhibited the refolding of misfolded proteins (Kragol et al. 2001).

Yang et al. reported the remarkable antibacterial effects of the AMP LL-37 against *E. coli*. Yang et al. observed a pronounced antibacterial effect of LL-37 on *E. coli*, attributing it to the inhibition of palmitoyl transferase PagP a protein located in the outer membrane of gram-negative bacterial cells. PagP activates lipid A acylation and thus plays a crucial role in repairing membrane permeability (Yang et al. 2017). Microcin J25, a ribosomally synthesized and post-translationally modified peptide, was shown to bind to the secondary channel of RNA polymerase, obstructing the trigger loop folding crucial for efficient catalysis. This interaction leads to the inhibition of RNA polymerase activity by blocking the substrate entry (Braffman et al. 2019).

Another antimicrobial peptide, NP-6, sourced from Sichuan pepper seeds, was found to strongly inhibit *E. coli*'s β-galactosidase activity in a dose-dependent manner (Hou et al. 2019). These findings underscore the diverse and effective mechanisms by which antimicrobial peptides can combat bacterial infections.

3.2.4 AMPs Inhibitory Effect on Cell Division

Numerous antimicrobial peptides exert their antimicrobial effects via the inhibition of crucial cellular processes, including the synthesis of DNA and the DNA SOS response disrupting the normal progression of the cell cycle and preventing chromosome segregation to prevent the division of cells (Lutkenhaus 1990). GLARALTRLLRQLTRQLTRA, a 20-amino-acid antimicrobial peptide, has a high affinity for DNA and demonstrates remarkable efficacy in eradicating *C. albicans* due to its ability to penetrate cell membranes effectively, its strong affinity for binding to DNA, and its capacity to initiate S-phase arrest (Li et al. 2016).

Additionally, an antimicrobial peptide with a 40-amino-acid residue sequence, known as MciZ, has demonstrated potent inhibition of bacterial multiplication, by specifically targeting the formation and localization of the Z-ring and demonstrates remarkable antibacterial activity (Cruz et al. 2020).

Several antifungal peptides (AFPs) exhibit inhibitory effects on fungal organelles. Histatin 5, a human salivary antifungal peptide, selectively targets *Candida albicans* mitochondria, inhibiting respiration, inducing reactive oxygen species (ROS) formation, and leading to dose- and time-dependent cell death (Helmerhorst et al. 2001). These diverse mechanisms underscore the versatility of antimicrobial peptides in combating microbial threats.

Furthermore, AMPs may target various components, including lipid A, mannoproteins, lipopolysaccharides, and other intracellular entities. Lipopolysaccharides (LPS) in gram-negative bacteria, rich in negatively charged phosphate groups, form an electrostatic network through salt bridges with cations like Ca^{2+} contributing to the bacterial defense mechanism against hydrophobic drugs (Nikaido 2003). This serves as the primary defense against hydrophobic drugs. The fungal cell wall, on the other hand, primarily comprises β-glucans, chitin, mannoprotein, and glycoproteins. Notably, mutations in key genes related to LPS pathway and phospholipid trafficking confer resistance to antimicrobial peptides.

Mannoproteins, integral to fungal cell walls, constitute a diverse class of proteins with multifaceted functions. Encompassing various structural proteins, adhesion proteins, hydrolytic enzymes, and transglycosylase, mannoproteins play pivotal roles in both the formation and dynamic remodeling of the fungal cell wall (Spohn et al. 2019). These proteins serve as potential targets for antifungal peptides (AFPs). Their distinctive characteristics set them apart from human cell membrane proteins, making them specific candidates for targeted interventions against fungal cell walls. Additionally, teichoic and lipoteichoic acids found in cell walls are also thought to be possible targets for AMPs, and these hypotheses may help assist the development of AMPs with reduced cytotoxicity (Rautenbach et al. 2016).

3.2.5 Inhibitory Effects on Cell Wall

Nisin, a polypeptide consisting of 34-amino-acid residues, incorporates unusual compounds like lanthionine and β-methyllanthionine. It has the ability to induce bacterial cell death through mechanisms such as creating membrane pores or inhibiting peptidoglycan synthesis via interactions with lipid intermediates. However, the primary mechanism through which it operates is not fully understood (Brumfitt et al. 2002). In another study, the antimicrobial peptide KLKLLLLLKLK-NH2, developed from sapesin B using D-amino acids, exhibited improved antimicrobial activity compared to its L-form counterpart, particularly against bacteria such as *Staphylococcus aureus* and *E. coli*. This heightened membrane permeability was attributed to the peptide's specific interaction with cell wall components, including peptidoglycan. As a result of this interaction, the D-form displayed greater antibacterial activity compared to the L-form, highlighting the significance of peptide structure in influencing antimicrobial efficacy (Manabe and Kawasaki 2017).

Defensins, characterized by their small size and cysteine-rich composition, are prominent antimicrobial peptides highly present in human, rabbit, and guinea pig neutrophils. Notably, three defensins—human neutrophil peptide defensin 1 (HNP-1), HNP-2, and HNP-3—constitute a significant portion of the total protein content in azurophil granules of human neutrophils. In the study done to show the antibacterial efficacy of HNPs, HNP-1 sequentially permeabilized the outer membrane (OM) and inner membrane (IM) of *E. coli*. Concurrently, the bacterial synthesis of DNA, RNA, and protein ceased, leading to a decline in colony count (Lehrer et al. 1989).

Cycloserine, a recognized antibacterial agent, functions by inhibiting the enzymatic activity of both alanine racemase and D-Ala-D-Ala ligase. This dual inhibition disrupts the synthesis of the essential D-Ala-D-Ala dipeptide, a key constituent of lipid II in the peptidoglycan precursor (de Leeuw et al. 2010). This implies that HNP1 may utilize a comparable mechanism, interacting with lipid II, to eliminate bacteria and other microbes.

Further, teixobactin isolated from uncultured bacteria exerts its antibacterial effect by impeding cell wall synthesis. This is accomplished by binding to a highly conserved motif found in lipids II and III, both of which are essential precursors for the formation of cell wall teichoic acid (Ling et al. 2015). These insights underscore the diverse mechanisms through which antimicrobial agents act to combat bacterial infections. Also, its resistance-free profile against *Staphylococcus aureus* and *Mycobacterium tuberculosis* suggests a promising avenue for developing antibiotics resistant to resistance development.

Crucially, regardless of the specific metabolic targets involved, antimicrobial peptides (AMPs) must first interact with the membrane(s) before they can access the interior of the pathogens. It is worth mentioning that the manner in which individual peptides operate may differ depending on the type of microorganism they are targeting, the physical characteristics of the interacting membranes, and the concentrations of peptides (Jenssen et al. 2006). This variability underscores the complex and adaptable nature of AMPs in their antimicrobial actions.

4 AMPs and the Innate Immune System

4.1 Role of AMPs in Host Defence Against Pathogens

Antimicrobial peptides (AMPs) are integral components of the innate nonspecific immune defense, responding rapidly to microbial threats and contributing significantly to the host's protection against infections. Their multifaceted role involves the direct confrontation of pathogens and the modulation of immune responses, encompassing both innate and adaptive mechanisms. It is noteworthy that a single antimicrobial peptide can exhibit either or both of these properties, highlighting their versatile contributions to host defense. The key points that define the crucial role of AMPs in host defense and protection include the following:

- AMPs demonstrate the versatility to directly combat a spectrum of pathogenic organisms, spanning bacteria, viruses, fungi, and parasites. They accomplish this by interfering with crucial cellular functions or by compromising the integrity of microbial cell membranes.
- AMPs often have broad-spectrum antimicrobial action, which means they can target a number of different microbes rather than just one particular kind. This enables them adaptable defenses against different diseases kinds.

- In reaction to an illness or damage, AMPs can be produced and released quickly. This makes it possible for the innate immune system to use these peptides to fight pathogens that are infiltrating the body swiftly, acting as a first line of defense.
- To more efficiently destroy infections, AMPs cooperate with other immunological elements, including phagocytic cells like neutrophils and macrophages and the complement system. They can also improve phagocytic cells' capacity to take in and break down microorganisms.
- AMPs possess the capability to influence the host's immune system response to infections. They could contribute to maintaining the proper ratio of pro-inflammatory to anti-inflammatory signals, which is crucial for a well-controlled immune response.
- AMPs not only contribute to protection against infections but also play a crucial role in shaping and nurturing the body's microbiome. They aid in regulating the microbial populations that live on or on the host's surface.
- Some AMPs contribute to tissue repair and the growth of new cells, facilitating the healing of wounds and infections.
- AMPs have garnered significant attention for their promising therapeutic applications attributed to their antibacterial activities in the face of challenging microbial infections, researchers are actively investigating their applicability in pioneering the development of innovative antibiotics, novel pharmaceuticals, antiviral medications, and therapeutic interventions aimed at revolutionizing medical treatments.

4.2 *Significance of Innate Immune System*

The components of the innate immune system represent the primary defense mechanism, initiating rapid and nonspecific protective measures in the initial stages of an infection, often preceding the full maturation of adaptive immune responses. This intricate defense network comprises a multitude of components, including physical barriers like skin and mucous membranes, as well as a diverse array of cellular and molecular elements, such as phagocytes, leukocytes, complement proteins, and antimicrobial peptides. This prompt and vigilant response is crucial for the timely detection and elimination of pathogens, thereby bolstering the overall resilience of the host organism in its continuous battle against potential threats. This early immune defense not only serves as a crucial guardian against infections but also serves a fundamental role in the host's overall immune surveillance (Akira et al. 2006; Marshall et al. 2018). The innate immune system comprises functionally distinct pattern recognition receptors (PRRs), initiating the activation of antimicrobial defenses and stimulating the adaptive immune response. Upon the engagement of these pattern recognition receptors, a cascade of events is set in motion, triggering the activation of transcription factors such as NF-κB, AP-1, and IRF. These transcription factors play pivotal roles in orchestrating inflammatory responses and protective immune responses forming a sophisticated network of molecular signaling pathways that contribute to the host's ability to combat infections (Medzhitov 2007). The significance of the innate immune system lies in its rapid and nonspecific response, acting as the first barrier against potential threats. Upon the recognition of microbe-associated molecular patterns (MAMPs) or pathogen-associated molecular patterns (PAMPs) through pattern recognition receptors, both local and systemic responses are initiated, with variations observed across different taxa (Mogensen 2009).

One highly conserved aspect of this arm of immunity observed among various classes of organisms is the production of antimicrobial peptides (AMPs) in response to microbial infections (Zasloff 2002). Moreover, the multifaceted actions of AMPs extend beyond their immediate antimicrobial functions. These peptides play a crucial role in promoting phagocytosis, facilitating the engulfment and degradation of microbial invaders by immune cells. Their ability to stimulate the release of prostaglandins contributes to the regulation of inflammatory responses, helping to maintain a balanced immune environment. In addition to their direct antimicrobial effects, AMPs exhibit remarkable immunomodulatory properties. For instance, they induce the proliferation of fibroblast and epithelial cells in the airway, actively participating in tissue repair and remodeling processes. The impact on dendritic cells goes beyond surface-level changes, influencing both phenotypic and functional aspects. This comprehensive modulation extends to the inhibition of pro-inflammatory cytokines and molecules, including TNF-α, nitrite-oxide, and tissue factors, underscoring the intricate role of AMPs in orchestrating immune responses and maintaining immune homeostasis (Guaní-Guerra et al. 2010; Lei et al. 2019).

In animals, AMPs emerge as frontline guardians, strategically produced by epithelial cells in direct interface with the external environment. Their versatile nature allows them to play pivotal roles in safeguarding the host against microbial threats. These peptides, once generated, travel through the circulatory system, traversing the bloodstream or lymphoid drainage to reach infection sites. This systemic dissemination enables a swift and targeted response, contributing to the host's ability to combat infections efficiently. Contrastingly, in the intricate tapestry of plant defense mechanisms, AMPs exhibit distinctive patterns of expression and distribution. Unlike their mobile counterparts in animals, plant AMPs are not typically circulated. Instead, they display constitutive expression in specific sensitive organs or undergo induction in response to local and systemic microbial infections. This tailored deployment underscores the adaptability of AMPs in plants, ensuring a localized and rapid response to microbial challenges (Sels et al. 2008).

Recent progress in genomic research has uncovered a myriad of AMP-like genes across various eukaryotic organisms. This revelation underscores the ubiquitous and indispensable role of AMPs in the broader eukaryotic immune system. Notably, their significance becomes particularly pronounced in plants, where the absence of acquired immunity necessitates a heightened reliance on innate defense mechanisms. In essence, the evolutionary conservation of AMP-like genes highlights their fundamental contribution to the intricate tapestry of eukaryotic immunity, showcasing their indispensability across diverse organisms and underscoring their importance as key players in the defense against microbial threats (Silverstein et al. 2007).

Antimicrobial peptides orchestrate a multifaceted response against microbial threats. Mammalian cells contribute to the production of several AMPs. Notable among these defenders are defensins and cathelicidins, serving as stalwart guardians in the frontline defense mechanisms. The specific peptides encoded by AMP genes emerge as key players in the host defense, shaping the resilience and fortitude of tissues in the face of diverse microbial threats. An illustrative example of the tissue-specific effects of AMPs is found in cathelicidins. Cathelicidins stand as guardians and have shown protective effects against *Streptococcus* infections and *Pseudomonas aeruginosa* in mice. This underscores the nuanced and correlated impact of AMPs on different tissues in mammals, revealing the intricate dynamics of their antimicrobial defense mechanisms (Silverstein et al. 2007). While the exact role of AMPs is still under investigation, it is widely agreed that their chief function is to limit and control microbial threats. Antibiotics can indiscriminately kill various microbes, whereas natural AMPs have the potential to spare normal flora and selectively target pathogenic bacteria. Moreover, antimicrobial peptides (AMPs) are finely tuned to function within specific fluid compositions and environments, minimizing the potential for cross-reactions. Their overarching role is to prevent potential pathogens from colonizing and invading host epithelia. In recent years, there has been a renewed interest in AMPs, acknowledging their diverse roles in immune modulation and infection prevention.

4.3 Role of AMPs in the Innate Immune System

As essential elements of the innate immune system, AMPs are pivotal in the host's defense against infections. Serving as the frontline guardians of the body, AMPs act as first responders combating various microbes, encompassing bacteria, fungi, viruses, fungi, and parasites (Fig. 9.3).

4.3.1 Antibacterial Activity

AMPs employ diverse and specialized mechanisms to inhibit a wide range of microorganisms, primarily targeting cell membranes., membrane disruption, or targeting specific intracellular components. Cationic AMPs demonstrate increased binding affinity to microbial pathogens, due to the presence of specific anionic components in the cell membranes of pathogens like bacteria and fungi. Critical components comprise LPS in gram-negative bacteria, lipoteichoic acid in gram-positive bacteria, and mannan in fungi. These components play pivotal roles in the structure and function of microbial cell membranes. The interaction between cationic antimicrobial peptides (AMPs) and these specific anionic components influences the effectiveness of AMPs in targeting and disrupting microbial membranes. Certain antimicrobial peptides (AMPs) possess the capability to eliminate bacteria at low concentrations without compromising membrane integrity. Rather than directly

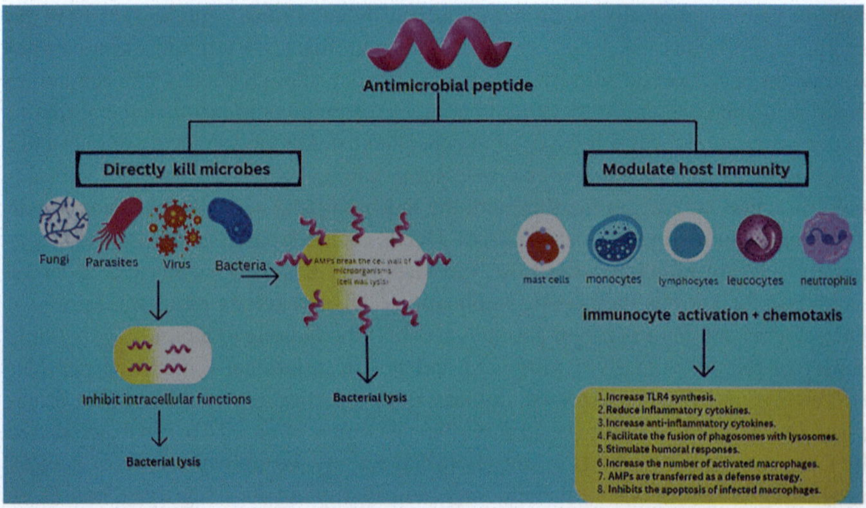

Fig. 9.3 The intricate immunomodulatory mechanisms of AMPs. These peptides exhibit dual functionality by directly eradicating pathogens and indirectly bolstering immune defenses through the activation of activating the immune system. AMPs activate various immune cells resulting in the production of cytokines and chemokines, orchestrating a coordinated response to engulf and eliminate pathogens

engagement with the membrane, these peptides focus on bacteria by impeding vital intracellular pathways, including DNA replication, cell division, and protein translation. The rapid bactericidal action and broad intracellular and membrane effects of antimicrobial peptides (AMPs), which do not specifically target particular molecules or pathways, serve as a deterrent against the development of increasing bacterial resistance. The distinctive ability of AMPs to target a broad spectrum of microbes, coupled with their unique mechanisms of action, makes them promising candidates for combating resistant strains of bacteria and fungi. Moreover, the inherent diversity in the amino acid sequences of AMPs contributes to their low likelihood of inducing resistance in microorganisms. As existing antibiotics encounter diminishing effectiveness, the exploration and application of AMPs present a hopeful avenue in the ongoing struggle against antimicrobial resistance. Their multifaceted modes of action, coupled with the potential for synthetic modifications to enhance stability and efficacy, make AMPs a valuable asset in the quest for novel and effective antimicrobial agents. The growing importance of AMPs in the pursuit of alternative strategies underscores their potential to shape the future landscape of antimicrobial therapeutics (Duarte-Mata and Salinas-Carmona 2023).

A newly identified short non-ribosomal antimicrobial peptide named albopeptide 6 has been identified in the culture broth of *Streptomyces*. This AMP exhibits a limited spectrum of antibacterial effects specifically targeting vancomycin-resistant *Enterococcus faecium* bacterial strain (Wang et al. 2021). Similarly, a bacteriocin-like inhibitory substance Mutacin-1140, a 22-amino-acid-long antimicrobial peptide (AMP), was isolated from *Streptococcus mutans* that displayed a broad-spectrum activity against clinically significant bacteria by inhibiting peptidoglycan (PGN) synthesis (Hillman et al. 1998). Lysostaphin, an antibacterial peptide, demonstrates significant bacteriolytic activity specifically against staphylococci while remaining inactive against bacteria from other genera. Lysostaphin demonstrates the ability to freely penetrate the capsule layer of encapsulated staphylococci (King et al. 1980), functioning as a cell wall lytic enzyme. Remarkably, lysostaphin provides substantial protection in mice, mitigating both the morbidity and mortality associated with *S. aureus* pneumonia (Chen et al. 2014). Due to its low toxicity and minimal potential for adverse effects on the host, lysostaphin stands as a viable and accepted candidate for various therapeutic applications. Nisins, belonging to the family of bacteriocins, are antimicrobial peptides (AMPs) that belong to a rapidly expanding subset of the ribosomally synthesized and post-translationally modified peptides (RiPPs) family. They have robust antibacterial properties against a variety of gram-positive and gram-negative bacteria (Galván Márquez et al. 2020). Nisin, produced by *Lactococcus lactis*, exerts its antimicrobial activity through lanthionine residues, interacting with the bacterial membrane and disrupting its integrity (Chan et al. 1996). Nisin has been noted for its anti-biofilm properties and its potential synergy with traditional therapeutic drugs. Moreover, akin to host defense peptides, nisin may stimulate the adaptive immune response, showcasing an immunomodulatory role. Notably, nisin has demonstrated the ability to impact tumor growth and display selective cytotoxicity against cancer cells (Shin et al. 2016). Additionally, a study by Tong et al. demonstrated the potential of nisin to significantly enhance the antibacterial and anti-biofilm activities of various antibiotics against *Enterococcus faecalis*, suggesting its promising role as an inhibitor against drug-resistant pathogens (Tong et al. 2014). In addition to its antimicrobial applications, Nisin finds utility in the food industry by extending shelf life through the inhibition of pathogenic gram-positive bacteria, thus preventing food spoilage. However, as discussed above, the utilization of nisin has transcended its traditional function as a food preservative. Another important AMP is laterosporulin10, a class IId bacteriocin isolated from *Brevibacillus strain* displayed distinct antimicrobial activity against *Staphylococcus aureus* and *Mycobacterium tuberculosis*, showcasing potential as a specific and effective antimicrobial agent. Crucially, the bacteriocin exhibits no toxicity toward mammalian cells, effectively eliminating intracellular *Mtb* and indicating its potential as a treatment for drug-resistant tuberculosis (Baindara et al. 2016). Further, studies by the same group also proved the anticancer properties of laterosporulin10 (LS10). LS10 was shown to induce apoptotic and necrotic death in cancer cells at lower and higher concentrations, respectively, suggesting its potential for therapeutic development (Baindara et al. 2017). These peptides, characterized by enhanced antimicrobial and anticancer efficacy and minimal cytotoxicity to eukaryotic cells, hold promise for therapeutic development. Another noteworthy example of novel AMP is penisin, a class Ia novel lantibiotic isolated from *Paenibacillus* sp. that belongs to the class of bacteriocins. It was identified as a unique lantibiotic exhibiting a wide-ranging antimicrobial spectrum and nonhemolytic characteristics. The antimicrobial effectiveness of penisin was associated with its capability to disrupt bacterial membranes. Molecular analysis indicates a lantibiotic gene cluster, and in vivo, experiments demonstrate penisin's promising potential in reducing the bacterial burden and protecting against *Staphylococcus aureus*-induced bacteremia (Baindara et al. 2015).

An important point to consider is that extended exposure to low concentrations of these peptides may also result in the emergence of resistance, highlighting the need for careful consideration in their therapeutic application. Consequently, to ensure optimal bactericidal activity, it is advised to maintain a high concentration of AMPs (Rodríguez-Rojas et al. 2021). This emphasizes the importance of understanding the intricate regulatory mechanisms involved in AMP expression, offering insights into effective strategies for bolstering the host's antimicrobial defenses against potential pathogens.

4.3.2 Antiviral Activity

Antimicrobial peptides (AMPs) also exhibit antiviral activity, contributing to their multifaceted roles in host defense predominantly through three key mechanisms: (i) inhibition of virus entry by inhibiting attachment and adherence to the cell membrane of the host; (ii) direct viral inactivation that often involves binding to viral envelope proteins or the viral membrane, destabilizing the integrity of the virus, and rendering it noninfectious; and (iii) inhibition of virus replication within host cells. This could involve inhibiting viral protein synthesis, interfering with nucleic acid replication, or disrupting other essential steps in the viral life cycle (Skalickova et al. 2015). FluPep family, a novel family of peptides derived from the Tkip peptide, represents a significant breakthrough in the pursuit of influenza A virus inhibitors. These peptides demonstrate the capability to inhibit the infection caused by influenza A virus, showcasing their potential as antiviral agents. The FluPep family's inhibitory action against influenza A virus infection suggests a specific and targeted approach to combatting this viral threat. This discovery opens avenues for further research and development, offering potential insights into the design of novel antiviral therapeutics to address influenza infections. The unique properties of the FluPep family peptides may contribute to the ongoing efforts to enhance our antiviral arsenal and improve the management of influenza-related illnesses. The peptide FP1 serves as a hemagglutinin (HA) blocking antiviral peptide (AVP) and prevents virus from binding to cell membranes. It mimicks the action of suppression of cytokine signaling protein, inhibiting viral attachment (Nicol et al. 2012). Notably, FluPep has been used as a viable antiviral drug effective against poxviruses, attributed to its ability to modulate inflammatory cytokine responses. N-acetylneuraminic acid is a pivotal component of the influenza virus receptor. Anti-lipopolysaccharide peptides (ALPs) are a class of broad-spectrum peptides that bind to N-acetylneuraminic acid and play a crucial role in inhibiting the attachment of the virus to host cells both in vitro and in vivo. Additionally, synthetic ALPs (SALPs) have been identified for their ability to effectively hinder the replication of various influenza virus subtypes (H7N7, H3N2, and 2009 pandemic H1N1) and protected mice from lethal infection of H7N7 influenza virus infection (Hoffmann et al. 2014). Moreover, synthetic ALPs (SALPs) emerge as promising antiviral agents, showcasing their potential to impede the replication of diverse influenza virus subtypes. These findings contribute valuable insights to the development of antiviral strategies targeting influenza infections and provide potential avenues for therapeutic interventions.

Bovine antimicrobial peptide-13 exhibits its antiviral efficacy by disrupting the synthesis of viral protein and interfering with the expression of viral genes, providing a multifaceted approach to inhibiting viral replication. This interference at the molecular level underscores its potential as a promising antiviral agent. This unique mode of action highlights its effectiveness in countering the transmissible gastroenteritis virus, providing a promising avenue for antiviral interventions. The ability of this peptide to impede crucial viral processes underscores its potential as a therapeutic agent against viral infections, particularly those involving the transmissible gastroenteritis virus (Liang et al. 2020). Cathelicidins and LL-37 (prominent human antimicrobial peptides) exhibit the ability to disrupt viral envelopes. LL-37, in particular, has demonstrated inhibitory effects across a spectrum of enveloped viruses. These include human immunodeficiency virus (HIV), herpes simplex virus (HSV), vaccinia virus (VV), Influenza A virus (IAV), dengue virus (DENV), and Zika virus (ZIKV). The broad-spectrum antiviral activity of LL-37 highlights its potential as a versatile therapeutic candidate against various viral infections (Barlow et al. 2011; Bergman et al. 2007; Tripathi et al. 2013; Howell et al. 2004, 2006; Ahmed et al. 2019).

Cathelicidins and LL-37 showcase their potent antiviral capabilities through a multifaceted approach. Their mechanism involves not only the disruption of the viral membrane but also the inhibition of critical processes like DNA replication, collectively contributing to a robust defense against viral infections. Furthermore, certain defensins, such as lactoferrin, contribute to the neutralization of herpes simplex viruses (HSVs) by impeding their attachment to host cell surfaces. This inhibitory action is achieved via the binding of defensins to viral glycoproteins, including heparan sulfate. These interactions contribute to the overall antiviral defense mediated by antimicrobial peptides (Jenssen et al. 2004). Le Messurier et al.'s research highlights the immune-modulating properties of antimicrobial peptides (AMPs) in response to influenza A virus (IAV), enhancing the host's defense against the virus (Le Messurier et al. 2016). Notably, peptides like pa-MAP and temporin B offer protection against herpes simplex virus 1 (HSV1) by preventing virus attachment. Temporin B goes further to disrupt the viral envelope and impact postinfection stages, while temporin G, an analog of temporin B, interferes with the influenza virus hemagglutinin protein, impeding crucial conformational changes needed for viral envelope fusion and entry into cells (Vilas Boas et al. 2017; Marcocci et al. 2018). Within the realm of antiviral peptides, there exists a diverse subset focused on combating HIV, including notable examples such as LL-37, alpha and beta defensins (gramicidin D, caerin 1, maximin 3, magainin 2, dermaseptin-S1, dermaseptin-S4, siamycin-I, and siamycin-II) (Madanchi et al. 2019). Noteworthy among these is the commercial success of Fuzeon (enfuvirtide) as an established anti-HIV drug (Ashkenazi et al. 2011). The peptide HD5 (1–9), derived from human α-defensin, has been identified as a potent inhibitor of viral infection, exerting its effects by preventing virus adherence and subsequent entry into cells (Böffert et al. 2020). Cathelicidin-derived AMPs, including GF-17 and BMAP-18, exhibit inhibitory effects against Zika virus (ZIKV) by directly inactivating the virus and interfering with the interferon (IFN) pathway (He et al. 2018). Additionally, AMPs demonstrate antiviral activities against non-enveloped viruses such as adenovirus, rhinovirus, and Aichi virus (Monteiro et al. 2018; Hu et al. 2019).

In vitro experiments in Vero and CaSki cell lines showed that NP-1, a pivotal antiviral peptide (AVP) sourced from rabbit neutrophils, inhibited the integration of the HSV-2 virus into cells (Skalickova et al. 2015). The transcriptional regulator VP16, responsible for stimulating the expression of immediate early viral genes crucial for the virus's initial interaction with the cellular response, encounters obstruction from NP-1, impeding its transport into the nucleus (Liu et al. 1999). In addition to their direct impact on viral particles and replication cycles, AMPs play a crucial role in controlling virus growth by modulating host immune responses, as demonstrated in recent studies (Chessa et al. 2020). Notably, recent research underscores the role of vitamin D in stimulating the synthesis of cathelicidins and defensins, thereby lowering virus replication rates and reducing the susceptibility to infections, including influenza and corona infection (COVID-19) (Grant et al. 2020).

4.3.3 Antifungal Activity

Antifungal peptides (AFPs) exhibit their effectiveness by targeting intracellular components or the cell wall, resulting in the disruption of membrane integrity in fungal cells. This disruption leads to altered permeability through the formation of pores in the membrane structure. A notable example of an AFP is the echinocandin family, which inhibits 1,3-β-glucan synthase, a crucial enzyme for maintaining cell wall integrity in fungi. Beta-glucan synthase is widely distributed among various fungal species, including *Candida*, *Aspergillus*, *Cryptococcus*, and *Pneumocystis* (Matejuk et al. 2010). Peptidyl nucleoside inhibitors, including nikkomycin and polyoxins derived from *Streptomyces tendae*, *Streptomyces ansochromogenes*, and *Streptomyces cacaoi*, constitute a class of antifungal agents targeting chitin synthase in fungal species, notably *Candida albicans* (Chen et al. 2000). Rs-ARF2, a plant defensin from radish seeds, shares structural features with other plant defensins like HsAFP1 and DmAMP1. It exerts its antifungal effects by inducing K^+ efflux, Ca^{2+} uptake, and alkalization of the medium, targeting glucosylceramide in fungal cell membranes and inducing membrane permeability. Additionally, Rs-ARF2 impairs fungi by introducing toxic reactive oxygen species into their cells (Thevissen et al. 2004). A 51-amino-acid antifungal peptide (AFP) produced by *Aspergillus giganteus* demonstrates antifungal activity by neutralizing the charge and condensing DNA (Martinez Del Pozo et al. 2002). While categorized as an antifungal peptide (AFP), the penicillium antifungal (PAF) proteins, obtained from *Penicillium chrysogenum*, are small, basic, and cysteine-rich antifungal proteins

(Marx 2004). Their mode of action includes apoptosis-like responses, such as the production of reactive oxygen species, the translocation of phosphatidylserine from the inner to the outer side of the plasma membrane, mitochondrial disintegration, metabolic stress, and membrane perturbation (Leiter et al. 2005). AMPs stand out as potent and versatile antifungal agents due to their ability to combat a broad spectrum of fungal infections. Expanding our understanding of AFPs, recent research has unveiled the diverse mechanisms employed by these peptides to combat fungal infections. These findings underscore the evolving landscape of antifungal peptides and their potential applications in addressing emerging challenges in fungal infections. The unique properties of AMPs position them as promising candidates for the development of innovative and effective antifungal therapeutics.

4.3.4 Antiparasitic Activity

The rising prevalence of parasitic diseases on a global scale has become a significant challenge for public health. Consequently, there is a growing interest among researchers to investigate the potential of both natural and synthetic peptides as a promising avenue to address these diseases. Insects and arthropods, serving as hosts, are emerging as promising reservoirs for obtaining antiparasitic peptides (APPs). This has spurred efforts to harness these peptides for the development of effective strategies against parasitic infections. A wide variety of parasites exist, from protozoa to worms. The World Health Organization (WHO) has classified 11 parasitic diseases as neglected tropical diseases (NTDs) due to their potential risk to millions of individuals, particularly affecting those in impoverished regions (Theel and Pritt 2016). The most significant parasite infections are schistosomiasis, leishmaniasis, trypanosomiasis, and malaria. Antiparasitic treatment approaches based on AMPs have lately attracted a lot of attention against parasitic infections. These peptides with chemotherapeutic potential primarily originate from insects and exert their effects by disrupting protozoan homeostasis, either by interfering with their cell membranes or by disrupting the host metabolism. Peptide drugs targeted for neglected parasitic infections are in the nascent stages of development, lagging behind the progress made in the design and development of drugs intended for chronic and non-infectious diseases (Lax 2010; Lacerda et al. 2016). Leishmanicidal AMPs have been identified across various animal species including:

- The eusocial honeybee's venom demonstrated significant antileishmanial activity when tested on mouse macrophages and human erythrocytes (Pitale et al. 2020).
- *Lutzomyia longipalpis*, a sand fly, produces attacin, cecropin, and defensin 2, which have shown leishmanicidal activity (Tinoco-Nunes et al. 2016).
- Dragomide E, a linear lipopeptide, demonstrated inhibition of adenosine triphosphate (ATP) generation in parasite-infected erythrocytes.

Furthermore, a peptide derived from snake cathelicidin, LZ1 notably reduced blood-stage *Plasmodium falciparum* infection (Fang et al. 2019). Epi-1, a synthetic marine antimicrobial peptide (AMP), exerts notable inhibitory effects on *Trichomonas vaginalis* by disrupting its membrane (Neshani et al. 2019). Furthermore, the peptide Jellein, extracted from bee royal jelly, and the 4-amino acid AMP KDEL (lysine, aspartic acid, glutamic acid, and leucine) have shown significant effectiveness in inhibiting Leishmania (Cao et al. 2019a, b; Zahedifard et al. 2020).

4.3.5 Immunomodulatory Potential of AMPs

Antimicrobial peptides (AMPs), commonly referred to as host defense peptides, play a pivotal role in immune defense by showcasing antibacterial and immunomodulatory properties (Hancock et al. 2016; Mack and Kim 2016). This class of peptides includes cathelicidins and defensins, playing a pivotal role in maintaining immune homeostasis. Neutrophils are the primary producers of these peptides, which, in turn, regulate various aspects of the immune system. AMPs play a crucial role in maintaining immune homeostasis by orchestrating various aspects of the immune system. They not only modulate the production and activity of key cytokines—such as interleukins, tumor necrosis factors (TNFs), and interferons (IFNs)—but also exert a profound influence on the behavior of diverse immune cells, including dendritic cells (DCs), monocytes, macrophages, mast cells, granulocytes, and lymphocytes. Furthermore, these peptides engage with cell surface receptors, such as cytokine receptors, chemokine receptors, and toll-like receptors (TLRs). Through interactions with intracellular signaling pathways, including NF-κB, ERK1/2, p38, JNK, and PI3K, AMPs actively contribute to the finely tuned regulation of immune responses. Overall, the multifaceted roles of AMPs extend beyond their antimicrobial functions, encompassing the intricate modulation of various physiological processes to maintain a balanced immune microenvironment. AMPs actively engage with immune cells to safeguard the host from infections. One notable interaction involves LL-37, a peptide constitutively produced by various immune cells and epithelial cells. LL-37 engages with the peptide receptor-like 1 (FPR1), formerly known as FPR2 to recruit attracting immune cells as neutrophils, NK, and mast cells to sites of pathogen entry. This recruitment enhances the local immune response, facilitates a targeted response, and contributes to the rapid defense against pathogens. Moreover, LL-37 also binds to the integrin M2 (Mac-1) on the surface of bone marrow cells. This dual interaction facilitates the efficient movement of mononuclear/macrophage cells and significantly enhances macrophage phagocytosis, demonstrating its broad efficacy against a spectrum of bacteria, including both gram-positive and gram-negative strains.

Beyond these interactions, LL-37 binds to the G protein–coupled receptor (GPCR) Mas-related gene X2 (MrgX2) on mast cells (Lishko et al. 2016). This interaction triggers mast cell chemotaxis, degranulation, and the recruitment of mast cells to inflammatory lesions. In essence, LL-37 orchestrates a series of dynamic interactions with immune cells, amplifying the host's defense mechanisms and contributing to the effective response against microbial threats. Another notable function is its involvement in the formation of neutrophil extracellular traps (NETs), which serve as a defense mechanism against microbial threats. By stabilizing NETs, LL-37 enhances their resistance to bacterial nuclease degradation, providing an additional layer of protection (Cao et al. 2019a, b; Neumann et al. 2014). Furthermore, LL-37 exhibits a dynamic role in immune modulation. It interacts with the P2X7 receptor in lipopolysaccharide (LPS) primed macrophages, leading to caspase-1 activation, interleukin-1 (IL-1) processing, and release. Additionally, the generation of reactive oxygen species (ROS) by LL 37 in neutrophils contributes to the inflammatory response (Yoon et al. 2021; Schiffmann et al. 2021). LL-37's impact extends to mast cells, where it directly induces the production of various chemokines and cytokines, including CCL2, CCL3, IL-1, IL-6, and tumor necrosis factors (TNFs). In mast cells, LL-37 also influences the expression of TLRs, potentially enhancing their capacity to detect invasive infections and contribute to more immune response (Bąbolewska and Brzezińska-Błaszczyk 2015; Agier et al. 2018).

In monocytes derived from human blood, LL-37 initiates the activation of ERK1/2 and p38, resulting in the upregulation of chemokines, such as CXCL8/IL-8, monocyte chemoattractant protein-1 (MCP-1), and monocyte chemoattractant protein-3 (MCP-3). This activation not only induces chemotaxis but also promotes the proliferation and differentiation of monocytes, thereby enhancing the overall immune response and contributing to the immune response (Bowdish et al. 2004). Defensins, another group of AMPs, also play a significant pro-inflammatory role. Human α-defensins, primarily represented by human neutrophil peptides (HNPs) 1–4 and human defensins (HD) 5–6, along with β-defensins, predominantly composed of HBD1–4, are present in various immune cell types. These antimicrobial peptides (AMPs) are distributed among neutrophils, mononuclear phagocytes, macrophages, lymphocytes, natural killer (NK) cells, and small intestinal Paneth cells, collectively contributing to the robust immune defenses across different components of the immune system (Nakamura et al. 2016; Takakuwa et al. 2019; Xu and Lu 2020). HNP1–3 increases the expression of CD32 (FcRIIB) and CD64 (FcRI), which induce the release of tumour necrosis factor alpha (TNFα) and interferon gamma (IFNγ), and thus enhances and improves macrophage phagocytosis from macrophages (Soehnlein et al. 2008). The intricate interplay of AMPs like LL-37 and defensins showcases their diverse roles in immune modulation, contributing to the host's ability to mount effective responses against microbial challenges. In brief, defensins exhibit potent broad-spectrum microbicidal activity, particularly in environments with high local concentrations or under low-ionic strength conditions. Human neutrophil defensins also demonstrate chemotactic properties for human monocytes, T cells, and immature dendritic cells in vitro, suggesting that their release from neutrophils may serve as a signaling mechanism to mobilize immunocompetent mononuclear cells. Further, defensins are known to prompt the generation of interleukin 8 (IL-8) by human airway epithelial cells to draw additional neutrophils to the sites of inflammation (Lehrer et al. 2005). Human beta defensins (HBDs) are pivotal in immune regulation, and different members of the HBD family exert diverse effects on immune cells. According to Niyonsaba et al., HBD2–4, in contrast to HBD1, can induce the production of several key cytokines and chemokines in human keratinocytes (Niyonsaba et al. 2007). This includes IL-6, IL-10, IP-10, MCP-1, macrophage inflammatory protein-3 (MIP-3), and RANTES/CCL5, highlighting their immunomodulatory potential in skin immune responses. Porcine beta-defensin 2 (PBD2) emerges as a robust defender against bacterial infections, employing a dual mechanism. First, PBD2 directly eliminates bacteria, contributing to the host's immediate defense against microbial invaders. Second, PBD2 modulates inflammation by disrupting the TLR4/NF-κB pathway. The impact of this disruption on pro-inflammatory cytokines, such as IL-6, IL-12, TNF-α, and IL-1β, contributes to the modulation of immune responses, highlighting the intricate regulatory role of human beta defensins (HBDs) in immune function (Huang et al. 2019). Through these actions, PBD2 not only acts as a direct antimicrobial agent but also helps regulate the immune response, preventing excessive inflammation.

In the realm of medicinal plants, Sechet et al. reported on the immunomodulatory effects of three small compounds—andrographolide, oridonin, and isoliquiritigenin. These compounds, derived from medicinal plants, were found to increase the expression of human beta defensin 3 (HBD3) in colon epithelial cells. This effect was achieved via the epidermal growth factor receptor (EGFR)/mitogen-activated protein kinase (MAPK) pathway, indicating a promising approach to modulating HBD expression using natural compounds (Sechet et al. 2018). These findings collectively underscore the diverse immunomodulatory roles of beta-defensins, shedding light on their potential therapeutic applications in managing immune responses and combating microbial infections. The dynamic interaction between antimicrobial peptides (AMPs) and pathogens represents a multifaceted interplay, with dual implications in the immune response. While AMPs are pivotal for their antimicrobial actions, fostering protection against infections, it is noteworthy that, in specific circumstances, these interactions may paradoxically contribute to the progression of infections. This intricate relationship highlights the need for a better and complete understanding of AMP-pathogen dynamics in the broader context of immune regulation. In the case of *Shigella* infection targeting the intestinal epithelium, human α-defensin 5 (HD5), a key host defense peptide in intestinal homeostasis and innate immunity was shown to promote *Shigella* infection by enhancing its adhesion to and invasion of mucosal tissues. *Shigella*, known for lacking an efficient host-adhesion apparatus, selectively interacts with HD5, utilizing it to bolster its colonization of the intestinal epithelium (Xu et al. 2018). Interestingly, HD5 in macrophages can also enhance the phagocytosis of *Shigella*. However, this leads to bacterial multiplication, resulting in the death of macrophage cells and the release of bacteria that can subsequently infect intestinal epithelial cells (Xu et al. 2019). This highlights the intricate balance between the protective and potentially facilitative roles of AMPs in the context of infection.

The expression levels of AMPs at inflammatory sites determine whether they exhibit pro- or anti-inflammatory actions. For instance, LL-37 administration suppressed the increase of damage-associated molecular patterns (DAMPs) and pro-inflammatory cytokines while maintaining a balanced immune cell profile. Notably, LL-37 increased the release of neutrophil extracellular traps (NETs), contributing to bactericidal activity and reduced bacterial burden. These findings suggest that

LL-37 enhances survival in CLP sepsis by modulating inflammatory responses and promoting NETs with antibacterial properties (Hosoda et al. 2017). LL-37 reduced proinflammatory cytokine secretion in LPS-stimulated neutrophils, induced reactive oxygen species (ROS) production, and enhanced bacteria engulfment by neutrophils. In addition, LL-37 contributed in controlling inflammatory mediator release and increasing antimicrobial activity in neutrophils, highlighting its dual role as an effector molecule in the innate immune system (Alalwani et al. 2010). Additionally, LL-37 protects against *Aspergillus fumigatus* by disrupting the cell wall and inhibiting mycelial growth. Further, it inhibited proinflammatory cytokine release and protected mice against *A. fumigatus* infection (Luo et al. 2019). The antimicrobial peptide clavanin, derived from the marine tunicate Styela clava AMP, exerted an inhibitory effect on the elevation of the pro-inflammatory cytokines IL-12 and TNF-α and enhanced the synthesis of anti-inflammatory cytokine IL-10 in macrophages in vitro. The peptide showed the ability to modulate innate immunity by promoting leukocyte recruitment to the infection site and inducing the production of immune mediators such as GM-CSF, IFN-γ, and MCP-1 (Silva et al. 2016).

A novel defensin-like bacteriocin, laterosporulin25 (LS25), from *Brevibacillus laterosporus* DSM25 was identified and studied using immunoinformatic tools (Dinata and Baindara 2023). LS25 exhibits immunogenic and physicochemical properties, making it a potential candidate for subunit vaccine development. Hence, the immunomodulatory effects of AMPs are context-dependent, and understanding their intricate roles in infection and inflammation is crucial for developing targeted therapeutic strategies. Further, some antimicrobial peptides (AMPs) exhibit notable potential in addressing various dermatological or skin conditions (Moravvej et al. 2021). Since they are essential for maintaining the skin's barrier and function, as demonstrated by their ability to treat diverse skin issues such as melanoma, acne, diabetic foot ulcers, and psoriasis. Experimental investigations have highlighted the antifibrogenic attributes of AMPs, exemplified by studies demonstrating that LL-37 (Yoo et al. 2015) and S100 peptides (Gauglitz et al. 2015) in keloid tissues contribute to the reduction of collagen formation. The antimicrobial peptide YD, derived from *Bacillus amyloliquefaciens* CBSYD1, demonstrated potent antibacterial and antioxidant properties. In a carbon tetrachloride-induced mouse model of liver fibrosis, YD improves liver function, inhibits fibrosis progression, and reduces levels of miR-155. YD acts through the miR-155–Casp12–NF-κB axis, suppressing inflammatory cytokines and providing a promising therapeutic candidate for hepatic fibrosis (Yan et al. 2021). In addition to their impact on fibrosis, recombinant LL37, particularly the N-terminally proline-modified form (P-LL37), retained its immunophysiological properties. P-LL37 demonstrated anti-inflammatory effects on macrophages, induced endothelial cell proliferation and migration, and enhanced angiogenesis in wound healing, suggesting its potential as a therapeutic agent for promoting vascularization and reepithelialization in regenerative medicine (Ramos et al. 2011). AMPs are implicated in diabetes, particularly in type 2 diabetes. A study revealed a positive correlation between plasma LL-37 levels and inflammatory markers, coupled with a negative correlation with HDL cholesterol in individuals with significance to the observed associations with type 2 diabetes mellitus. The significance of these associations suggests that LL-37 may play a role in the inflammatory processes associated with type 2 diabetes. These findings highlight the role of AMPs like LL-37 in linking innate immunity to inflammation and atherosclerosis, underscoring their significance in chronic disease processes and cardiovascular risk (Meguro et al. 2014). The production of cathelicidin-related antimicrobial peptide (CRAMP) by insulin-secreting β cells is deficient in nonobese diabetic (NOD) mice. Administering CRAMP to prediabetic NOD mice-induced regulatory immune cells reduces autoimmune diabetes incidence, thereby highlighting the need to understand the role of AMPs in autoimmune diseases (Sun et al. 2015).

Apart from their implications in dermatology and metabolism, AMPs are crucial for sepsis survival as they neutralize lipopolysaccharide (LPS) effects and inhibit macrophage pyroptosis (Hu et al. 2016). Moreover, AMPs play a vital role in the maintenance of gut homeostasis, tissue repair, and preventing the risk of cancer/tumor by regulating the colon microbiota balance (Yoshimura et al. 2018; Zhang et al. 2019; Rathinam and Chan 2018). The diverse therapeutic potential of AMPs underscores their multifaceted roles in various physiological processes, positioning them as promising and novel candidates for the development of targeted treatments across different medical domains.

4.3.6 Anticancer Activity

AMPs are also known to have anticancer properties (Kang et al. 2017). AMPs exhibit selective cytotoxicity against cancer cells through their action on the membrane surface. The distinctive anionic composition of cancer cell membranes provides the targeting specificity of AMPs, setting them apart from normal cells. This distinctive feature makes AMPs particularly intriguing candidates for potential use as antitumor agents. The preferential targeting of tumor cell membranes by AMPs presents a unique avenue for developing novel therapeutic strategies that leverage the inherent differences in membrane composition between cancerous and healthy cells. This targeted approach holds promise for the advancement of more selective and efficient treatments in the ongoing quest for innovative cancer therapies (Utsugi et al. 1991). AMPs demonstrate remarkable selectivity by interacting with the distinct membrane characteristics of cancer cells, offering significant potential for precision in therapeutic interventions. This targeted destruction on the membrane surface represents a promising strategy in the development of anticancer therapies, emphasizing the role of AMPs in the ongoing exploration of innovative and effective treatments against cancer. In some cancers, AMPs paradoxically can help the tumor advance. As a result, each form of tumor has a different functional role for AMPs in cancer cells. Anticancer peptides (ACPs) employ various mechanisms to combat cancer. Presently, three primary sources contribute to the discovery of anticancer peptides: the natural world, peptide libraries, and de novo design and synthesis. These peptides exhibit potent antineoplastic activities by suppressing tumor angiogenesis, recruiting immune cells like dendritic cells, inducing tumor apoptosis/necrosis, inhibiting the activities of kinases and proteases, and activating specific regulatory proteins to interfere with the transcription and translation (Wu et al. 2014; Ma et al. 2020). Striking a delicate balance between the charge and hydrophobicity of these peptides is essential for enhancing the efficacy of these peptides, emphasizing that the functional role of ACPs varies based on the specific type of tumor. This nuanced understanding is crucial for tailoring effective and targeted therapeutic strategies against different cancer types.

The sole cathelicidin identified in humans, LL-37, emerges as a potent deterrent against the development of stomach and colon cancer tumors (Piktel et al. 2016). LL-37, exclusive to humans in the cathelicidin family, not only hinders tumor growth in colon and gastric cancer but also initiates a captivating cascade of events. LL-37 is implicated in carcinogenesis through its antibacterial and immunomodulatory functions. Evidence suggests that LL-37 exhibits dual effects—promoting or inhibiting tumor growth—depending on tissue-specific mechanisms and interactions with varying membrane receptors on cancer cells. Studies on cathelicidin-deficient mice further underscore the protective role of LL-37, revealing an increased susceptibility to colon cancer in the absence of this cathelicidin (Wang et al. 2019). The human cathelicidin LL-37 and its analog, FF/CAP18, have been observed to restrict the proliferation of colon cancer cell lines, indicating a potential inhibitory effect on the growth of these cancer cells. FF/CAP18 treatment increases exosome formation in colon cancer cells, leading to the release of exosomes containing growth-suppressing microRNAs (miRNAs). These miRNAs, transported via exosomes, extend their inhibitory effects on proliferation to other cancer cells, suggesting that FF/CAP18 regulates cancer growth by modulating both direct effects on cancer cells and indirect effects via exosomes. This intricate network of interactions highlights the multifaceted impact of cathelicidins, particularly LL-37, in orchestrating a defense against cancers (Hayashi et al. 2018). The human cathelicidin LL-37, a versatile host defense peptide, is found to be down-regulated in gastric adenocarcinomas. LL-37 reduces cell growth, delays cell cycle transition, and suppresses in vivo tumor growth by upregulating BMP4, phosphorylating Smad1/5, and inducing p21 (Waf1/Cip1), suggesting a potential therapeutic avenue for gastric cancer treatment (Wu et al. 2010). Furthermore, other antimicrobial peptides (AMPs) like KT2, BG-4, and KL15 have demonstrated the ability to induce apoptosis or necrosis in cancerous cells, emphasizing the versatile and promising roles of AMPs in combating various facets of cancer development and progression (Maraming et al. 2019; Maijaroen et al. 2018; Dia and Krishnan 2016; Chen et al. 2015). Interestingly, the impact of LL-37 is context-dependent, as it can also stimulate tumor growth in several cancers, including squamous cell carcinoma of the skin, lung cancer, ovarian cancer, breast cancer, malignant melanoma, prostate cancer, and pancreatic cancer (Chen et al. 2018). In the context of ovarian tumors, LL-37 contributes to promoting ovarian tumor growth by recruiting multipotent mesenchymal stromal cells. The activation of TLR2 and vitamin D3 signals in macrophages seems to be part of the signaling pathway leading to LL-37 production (Lu et al. 2015). Further, the pro-tumorigenic role of cathelicidin AMP hCAP18/LL-37 primarily stems from its activation of the EGFR/HER2/Akt signaling pathway within hepatocellular carcinoma (HCC) cells and xenograft tumors exhibiting endogenous overexpression. This signifies that hCAP18/LL-37 contributes significantly to tumor growth by triggering the cascading events associated with the EGFR/HER2/Akt pathway in both cellular and xenograft contexts with inherent upregulation (Zhang et al. 2022). Additionally, LL-37expressed by myeloid and non-tumorous cells acts as a crucial regulator in the development and growth of lung tumors. The peptide modulates immune responses and influences the recruitment of myeloid cells, with NF-κB signaling and Wnt/β-catenin activation playing crucial roles in its pro-tumorigenic effects, particularly in enhancing lung tumor cell proliferation through phosphorylation of key signaling proteins (Ji et al. 2019). These findings highlight the complex role of LL-37 in cancer progression, with its effects varying across different cancer types and microenvironments.

Defensin not only possesses antimicrobial properties but also stimulates the proliferation of normal fibroblast and epithelial cells in vitro. Furthermore, it plays a pivotal role in the in vivo wound healing process. Dermaseptin-PP, an antimicrobial peptide (AMP) derived from the skin of the less-explored frog species *Phyllomedusa palliata*. Dermaseptin-PP exhibits selective cytotoxicity against various cancer cells, showcasing a dual nature by demonstrating both antitumor and angiostatic activities. In a mice tumor model, Dermaseptin-PP demonstrated potent in vivo antitumor activity in a dose-dependent manner without noticeable hepatopulmonary side effects. The peptide disrupts cancer cell membranes and induces apoptosis (van Zoggel et al. 2012; Dong et al. 2020). Additionally, PFR (PFWRIRIRR-NH2), an antimicrobial peptide derived from lactoferrin derivatives, has been identified. A novel mechanism of PFR-induced necroptosis in acute myeloid leukemia (AML) cells involved endoplasmic reticulum stress, elevated cytoplasmic calcium, and mitochondrial ROS generation, thus inhibiting proliferation of human acute myeloid leukemia HL60 cells (Lv et al. 2019). These studies reveal that AMPs, exhibit not only antimicrobial efficacy but also contribute to wound healing and display anticancer activities with potential applications in therapeutic interventions.

Neutrophil-derived peptides, such as HNP1–3, exhibit dual roles in cancer pathogenesis. Increased levels of HNP1–3 are positively associated with cell necrosis in renal cell carcinoma. Moreover, these levels are notably elevated in both metastatic colorectal cancer tissues and serum (Albrethsen et al. 2005). On the contrary, HNP1 exhibits antitumor properties, restraining the growth of lung adenocarcinoma, colon, and breast tumors. This occurs through the induction of apoptosis, suppression of angiogenesis, and facilitation of antitumor immune responses (Xu et al. 2008). In tumor tissues, HNP1 mediated intratumoral microvessel normalization and caused significant in situ tumor cell apoptosis. Our study suggests that intratumoral expression of HNP1 can significantly improve the therapeutic efficacy of doxorubicin in breast cancer, abrogate the influence of multidrug resistance, and enhance medication safety. These findings may be important for the clinical therapeutic strategies of cancer chemotherapy (Li et al. 2014). Also, HNP1–3 has shown cytotoxic effects on oral squamous cell carcinoma cells (Lundy et al. 2004). Two modified peptides, epinecidin-8 and pardaxin-6, demonstrated broad-spectrum antibacterial and anticancer activities, exhibiting synergy with antibiotics against resistant bacteria, without inducing significant hemolysis (Lin et al. 2013).

Cell-penetrating peptides (CPPs) have emerged as highly efficient and extensively investigated carriers for facilitating intracellular entry. A cell-penetrating γ-peptide (PEG-1) combined with antimicrobial undecapeptides exhibited enhanced cytotoxicity against MDA-MB-231 breast cancer cells while demonstrating lower toxicity in non-malignant cells, highlighting the potential of this strategy for developing improved peptide-based anticancer drugs (Rosés et al. 2012). Studying the connections between the structure of AMPs and their antitumor activity is crucial for designing synthetic peptides with optimal antitumor efficacy, selectivity, and minimal side effects. A study investigated the cytotoxic activity of various antimicrobial peptides against the U937 human histiocytic lymphoma cell line (Koszałka et al. 2011). Pexiganan MSI-78 exhibited the strongest cytotoxic activity, followed by Citropin 1.1, Protegrin 1, and the synthetic lipopeptide N-α-palmitoyll-lysyl-l--lysine amide, attributed to their robust membranolytic activity, leading to necrosis. These findings highlight the complex and multifaceted roles of AMPs in cancer, with the potential for both promoting and inhibiting tumor growth depending on the context and specific characteristics like the chain length of the peptides involved. Various cancers, including colon cancer, oral squamous cell carcinoma (OSCC), prostate cancer, renal cell carcinoma (RCC), liver cancer, and oral tongue squamous cell carcinoma, exhibit low expression levels of human β-defensins (HBD1, HBD2, and HBD3) (Abiko et al. 2001; Joly et al. 2009; Sawaki et al. 2002). In colon cancer, the oncogenic EGFR-ERK-MYC signal axis suppresses HBD1 expression, while HBD1 overexpression induces apoptosis in renal and epidermoid carcinoma cells (Bonamy et al. 2018; Sun et al. 2006). HBD2's tumor-inhibitory effect on colon tumors is linked to antitumor immunity (Salem et al. 2019). Conversely, HBD3 hinders head and neck cancer cell migration and lung cancer growth (Wang et al. 2012; Hanaoka et al. 2016). However, in cervical cancer, HBD3 assumes an oncogenic role by activating NF-κB signaling, and in head and neck squamous cell carcinomas (HNSCC), it induces the expression of programmed death-ligand 1 (PD-L1) (Xu et al. 2016; Gomez Hernandez et al. 2019). These observations highlight the complex and context-dependent roles of human β-defensins in various cancer types, with their expression and functions subject to intricate regulatory mechanisms and specific tumor microenvironments.

5 AMPs: Unleashing Innate Immune Responses for Enhanced Defense

Concerning innate immunity, antimicrobial peptides (AMPs) can inhibit pathogen growth, enhance host immunological functions, and promote intestinal health. They play a pivotal role in innate immunity by eliminating various pathogens, and their genes are expressed in diverse host cells. Studies reveal diverse immunomodulatory functions of AMPs, showcasing their unique structural characteristics and positioning these peptides as crucial elements connecting innate and adaptive immunity,

Table 9.1 List of various antimicrobial peptides exhibiting a pivotal role in innate immunity, inhibiting pathogens, enhancing host functions, and offering promising avenues for therapeutic applications

Antimicrobial peptide	Therapeutic properties	Advantages	Activity	Mechanism of action	References
Arenicin	An AMP has antibacterial qualities, especially against bacteria and fungus. It is frequently present in aquatic life.	Broad-spectrum antibacterial action	Various bacterial and fungal infections	Rupturing of microbial cell membranes; disrupting fungal phospholipid membranes.	Andrä et al. (2008); Park and Lee (2009)
Lacticin	Typically produced by lactic acid bacteria, lacticin is a bacteriocin with antimicrobial activity.	Specificity against certain bacteria	Bacterial infections, particularly those caused by gram-positive bacteria	Breaks down bacterial cell membranes or obstructs vital cellular functions.	McAuliffe et al. (1998)
Laterosporulin10	A class IId bacteriocin produced from a *B. laterosporus* strain GI-9.	Structural homology and sequence similarity to various human defensins; antimicrobial activity and low cytotoxicity towards eukaryotic cells	Effectively inhibited the growth of *Staphylococcus aureus* and *Mycobacterium tuberculosis* and anticancer	Membrane permeabilization and disruption	Baindara et al. (2016); Baindara et al. (2017)

(continued)

Table 9.1 (continued)

Antimicrobial peptide	Therapeutic properties	Advantages	Activity	Mechanism of action	References
Magainins	Magainins are a class of peptides produced by amphibians.	Broad-spectrum antibacterial/antifungal/ antiviral activities	Fungal and bacterial diseases being examined	Causes microbial membranes to rupture, and causes membrane thinning. Pexiganan, a derivative of magainin showed efficacy in phase III clinical trials	Madanchi et al. (2020); Grage et al. (2016)
Alamethicin	Alamethicin is an antimicrobial peptide produced by the fungus *Trichoderma viride*.	Antibacterial activity	*Staphylococcus aureus;* inducing various physiological and defense responses in eukaryotic cells, including plants	Causes the integrity of microbial membranes to be disrupted by forming channels/pore formation in the inner membranes	Dong et al. (2016); Maischak et al. (2010)
Protegrin-1	Protegrin-1 is a peptide produced from porcine leukocytes that has antimicrobial capabilities.	Broad-spectrum antimicrobial action, stable in vivo	Bacterial infections like methicillin-resistant *Staphylococcus aureus* (MRSA) or *Pseudomonas aeruginosa;* immunomodulatory role during inflammation	Causes microbial membranes to rupture; granulosa proliferation	Steinberg et al. (1997); Pan et al. (2021)
Human cathelicidin LL-37	LL-37 is an antibacterial and immunomodulatory cathelicidin derived from humans.	Immunomodulatory and dual antibacterial properties	Antibacterial and antiviral activity and participating in immunomodulation	Alters immunological responses and breaks down microbial membranes.	Barlow et al. (2011)

Antimicrobial peptide	Therapeutic properties	Advantages	Activity	Mechanism of action	References
Cecropin	Cecropins are antibacterial peptides obtained from insects, bacteria *Helicobacter pylori*, tunicates, ascarid nematodes, and mammals	Broad-spectrum antibacterial action	Bacterial infections	Causes microbial membranes to rupture.	Gholizadeh and Moradi (2021)
Indolicidin	A novel antimicrobial peptide purified from the cytoplasmic granules of bovine neutrophils	Indolicidin being cationic, highly specific for bacteria	Antifungal and antibacterial	Interferes with cellular functions or ruptures microbial membranes. In Phase II trials Omiganan (MBI-226), an analogue of indolicidin exhibited anti-inflammatory activity against the noninfectious skin condition rosacea.	Selsted et al. (1992)
Nisin	Often used as a food preservative, nisin is a bacteriocin with antibacterial action.	Robust antibacterial properties; anticancer properties	Checked for infections by bacteria, particularly gram-positive infections	Causes damage to bacterial cell membranes or obstructs vital biological functions.	Galván Márquez et al. (2020); Shin et al. (2016)
Albopeptide 6	A novel short non-ribosomal AMP, discovered in the culture broth of *Streptomyces albofaciens*, possesses antibacterial activity.	Narrow-spectrum activity against vancomycin-resistant *Enterococcus faecium*	Various bacterial infections	Interaction with microbial membranes.	Wang et al. (2021)

(continued)

Table 9.1 (continued)

Antimicrobial peptide	Therapeutic properties	Advantages	Activity	Mechanism of action	References
Mutacin-1140	Mutacin-1140 is an antimicrobial bacteriocin, produced by the gram-positive bacterium *Streptococcus mutans*	Specificity against certain bacteria, low toxicity	Various bacterial infections	Causes damage to bacterial cell membranes or obstructs vital biological functions.	Kers et al. (2018)
Lysostaphin	Lysostaphin is an antibacterial peptide	Antibacterial effect against *Staphylococcus aureus* and also *S. epidermidis*; low toxicity and minimal potential for adverse effects	*Staphylococcus aureus* and *Staphylococcus epidermidis*	Breaks down the peptidoglycan in the bacterial cell wall and penetrates the cell wall	King et al. (1980); Wu et al. (2003)
Histatins	Present in multiple human tissues and fluids, including the oral mucosa and ocular surface	Antibacterial and antifungal properties; pro-migratory actions, promotion of wound healing, immunomodulation	Mouth wash for fungal/yeast infection (peptide derivative PAC 113 is in phase II clinical trial); histatin 5, a product of histatin 3 proteolytic cleavage, is the most active histatin against several yeasts, such as *Cryptococcus neoformans*, *Candida dubliniensis*, and *Candida albicans*	Histatins exert their activity by targeting the mitochondria, affecting cell respiration	da Costa et al. (2015); Helmerhorst et al. (2001)
Gramicidin D	A peptide having antibacterial qualities, gramicidin D is frequently used in conjunction with other antibiotics.	Broad-spectrum antibacterial action	Particularly effective against gram-positive bacteria; also used in treatment of infected surface wounds, and eye, nose, and throat infections	Gramicidin D binds to and inserts itself into bacterial membranes, resulting in membrane disruption and permeabilization	Liou et al. (2015); Hancock and Chapple (1999)

Caerin 1	Caerin 1 is an antimicrobial peptide obtained from frogs.	Broad-spectrum antibacterial action showing significant antibiotic activity, particularly against gram-positive organisms	Antibacterial activity	Interact with the lipid bilayer of the bacterial cell wall and causes membrane rupture.	Steinborner et al. (1998)
Maximin 3	Frog-derived peptide maximin 3 has antibacterial qualities.	Broad-spectrum antibacterial action	Fungal and bacterial infections	Causes microbial membranes to rupture.	Benetti et al. (2019)
Dermaseptin-S1 and Dermaseptin-S4	A family of cationic (lysine-rich), amphipathic antimicrobial peptides, derived from frog	Broad-spectrum antimicrobial action	Fungal and bacterial diseases	Causes microbial membranes to rupture, hemolytic activity	Mor et al. (1994)
Siamycin type Lasso peptides	Belong to a specific family of ribosomally synthesized and post-translationally modified peptides with an unusual lasso structure.	Broad-spectrum antibacterial action and high thermal and proteolytic stability being good carriers for other bioactive peptides	Antimicrobial activity, antifungal, enzyme inhibition, receptor blocking, anticancer properties	Causes damage to microbial membranes/ inhibits cell wall biosynthesis.	Tan et al. (2019); Cheng and Hua (2020)
Penisin	AMP produced from a bacterial strain *Paenibacillus ehimensis* A3	Antibacterial with other potential therapeutic applications	Effectively killed methicillin-resistant *Staphylococcus aureus* (MRSA) and gram-negative bacteria	Disruption of membrane integrity	Baindara et al. (2015)

(continued)

Table 9.1 (continued)

Antimicrobial peptide	Therapeutic properties	Advantages	Activity	Mechanism of action	References
Melittin	Principal component of venom from the honeybee *Apis mellifera*	Broad spectrum antimicrobial activity and anti-inflammatory	In vivo protection against MRSA infections has been demonstrated; the FDA approved its usage in clinical practice in pain management during tendinitis, arthritis, multiple sclerosis	Induction of pore formation following interaction with microbial membrane	Choi et al. (1997); Lee and Bae (2016)
Defensins	Isolated from human monocytes; includes hBD1, hBD2, hBD3	Broad-spectrum antibacterial activity against gram-positive and negative bacteria, antifungal activity	In vivo protection against various bacteria like *Staphylococcus*, *E. coli*, and *Salmonella*	Membrane disruption and inhibition of bacterial cell wall synthesis	Maiti et al. (2014); Xu and Lu (2020)
PAF	AMP produced from *Penicillium chrysogenum*	Antifungal activity	Wide range of antimycotic applications	Disruption of Ca^{2+} signaling pathway/ homeostasis	Palicz et al. (2016); Marx (2004)

with potential applications in therapeutic development (Table 9.1). Primarily dedicated to intracellular pathogen extermination, these peptides display a dual nature. Upon release into the extracellular environment, they transform into chemokines, demonstrating the capability to attract a variety of cell types—including monocytes, neutrophils, mast cells, eosinophils, and lymphocytes—toward sites of infection (Auvynet and Rosenstein 2009; Hiemstra 2006). This chemotactic prowess extends to synthetic AMPs, including the notable Innate Defence Regulator Peptides (IDR peptides), showcasing their proficiency in attracting specific cell types such as monocytes and neutrophils. This chemotactic recruitment is a crucial aspect of the immune response, orchestrating the arrival and stimulation of specific immune cells to effectively combat microbial threats or resolve tissue damage. Antimicrobial peptides (AMPs) activate processes resulting in the release of essential chemokines, including MCP-1, RANTES, and IL-8, concurrently upregulating the expression of specific receptors like CCR2 and CXCR4. Notably, membrane-associated G protein-coupled receptors (GPCRs), with a particular emphasis on formyl peptide receptors, play a substantial role in mediating these pathways (Choi and Mookherjee 2012; Yang et al. 2000). Chemokines, akin to defensins, also exhibit antimicrobial properties in vitro, demonstrating their ability to eliminate many pathogens such as Listeria monocytogenes, *E coli*, and other gram-positive and negative bacteria (Cole et al. 2001; Egesten et al. 2007). The intricate interplay among chemokines, receptors, and G protein-coupled receptors highlights a sophisticated and multifaceted immune response strategy to effectively counter microbial threats.

Peptides like HNP-1 (Human Neutrophil Peptide) not only demonstrate chemotactic activity but also serve a pivotal role in amplifying the macrophage-mediated elimination of intracellular pathogens. This intricate mechanism encompasses the phagocytosis of apoptotic neutrophils and peptides, contributing to a nuanced and effective immune response (Tan et al. 2006). Additionally, HNP1–3 peptides have shown the ability to activate macrophages, enhancing their phagocytic capabilities both in vitro and in murine models in vivo (Soehnlein et al. 2008). As mentioned earlier, Cathelicidin LL-37 activates phagocytosis in neutrophils by inducing CXCL8 synthesis via MAPK p38 and ERK. LL-37 also inhibits neutrophil apoptosis, prolonging cell lifespan through FPRL1 and P2X7 nucleotide receptor activation (Nagaoka et al. 2006). The diverse mechanism and versatile approach for immune modulation exerted by AMPs highlight the multifaceted roles of these peptides in shaping immune responses.

AMPs serve a dual purpose, displaying not only antimicrobial prowess but also significant anti-inflammatory effects. Their pivotal role in orchestrating inflammatory processes is underscored by compelling studies that emphasize the substantial impact of their anti-inflammatory properties. For example, Mice lacking cathelicidin demonstrate a heightened inflammatory phenotype compared to wild-type mice, underscoring the pivotal role of antimicrobial peptides (AMPs) in inflammation regulation (Alalwani et al. 2010). This multifaceted contribution of AMPs adds a layer of complexity to our understanding of their functions in the intricate web of immune responses. Moreover, the therapeutic potential of certain peptides in managing inflammation associated with various diseases has been demonstrated. An illustrative example is the peptide IDR-1002, known for its ability to reduce inflammation and alveolar macrophage infiltration in the context of infection when applied in vivo (Wuerth et al. 2017). This showcases the versatility of AMPs in modulating immune responses and suggests their potential therapeutic applications in inflammatory conditions. AMPs showcase not only the ability to regulate cytokine production but also a dual function in fostering both inflammatory and anti-inflammatory responses. For example, HNP1–3 demonstrates a dual impact on pro-inflammatory cytokines, inducing the production of TNF-α, IL-12, IFN-γ, and IL-1, while concurrently decreasing the anti-inflammatory cytokine IL-10 by mono nuclear cells. While the direct in vitro microbicidal activity of AMPs has long been recognized as a crucial innate immune defense, recent findings establish the in vivo relevance for specific defensins and cathelicidins. These proteins also exert receptor-mediated effects on immune cells; for instance, beta-defensins interact with CCR6, and cathelicidins act on FPRL1-expressing cells. Additionally, certain defensins display notable immunoenhancing activity, highlighting their contribution to both innate and adaptive antimicrobial immunity (Yang et al. 2004).

The regulatory influence of AMPs on inflammatory mediators extends to their interaction with Toll-like receptors (TLRs). LL-37, a notable AMP, protects against endotoxemia/sepsis by modulating inflammatory responses and inhibiting TNF-alpha release in LPS-stimulated cells. LL-37 acts on the TLR-to-NF-kappaB pathway. The peptide prevents cytokine release in response to various TLR agonists, highlighting its role in balancing inflammatory responses and combating sepsis (Mookherjee et al. 2006). This intricate modulation underscores the nuanced regulatory role of AMPs in shaping immune responses. LL-37 exhibits a dichotomous role by not only inhibiting TLR2, TLR4, and TLR9-induced proinflammatory cytokines but also enhancing TLR3-mediated responses, consequently promoting IL-1β secretion (Lai et al. 2011). Moreover, the immunomodulatory peptide LL-37 alters epithelial cell responses to proinflammatory stimuli, synergistically enhancing cytokine production in conjunction with various Toll-like receptor agonists, suggesting a potential role in modulating epithelial responses to infections in vivo (Filewod et al. 2009). These dynamic interactions illustrate the versatile and context-dependent nature of AMPs in modulating immune responses.

AMPs showcase their modulatory impact on the innate immune response through various mechanisms (Duarte-Mata and Salinas-Carmona 2023). AMPs exhibit diverse functions in immune modulation acting as opsonins, influencing the activation or suppression of the complement pathway, and initiating the activation of the NLRP3 inflammasome, which forms a critical part of host immune defenses. Their impact on the classical complement system pathway involves either augmentation or suppression, depending on the context, contributing to the overall regulation of immune responses. Additionally, the ability of AMPs to trigger the activation of the NLRP3 inflammasome highlights their involvement in the regulation of inflammatory processes. These diverse functions underscore the intricate role of AMPs in modulating immune responses and maintaining immune homeostasis (Prohászka et al. 1997; Kahlenberg et al. 2013). This multifaceted modulation underscores the intricate and dynamic role of AMPs in fine-tuning immune responses against

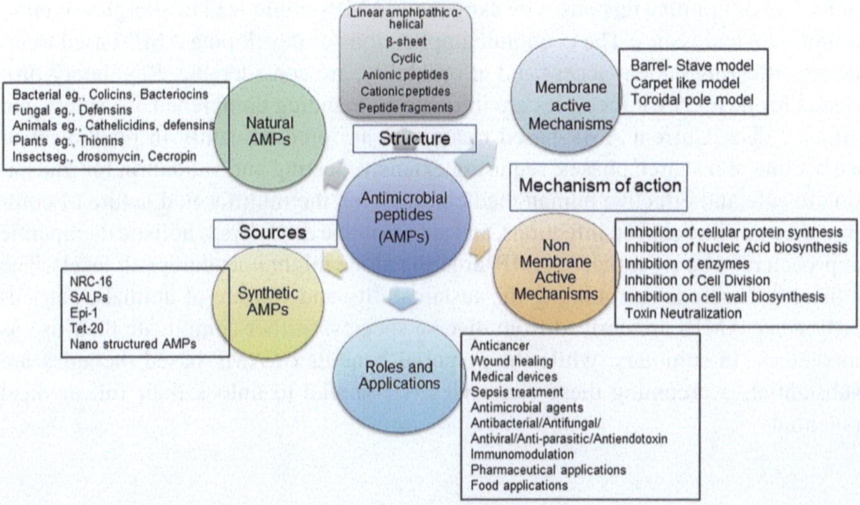

Fig. 9.4 Antimicrobial peptides' classification and diverse functions in host immune response. They play a pivotal role in inducing innate immune responses, safeguarding against infections, and modulating host-induced pro-inflammatory responses

various diseases. Their ability to engage in multiple facets of the immune system showcases the complexity and adaptability of AMPs in contributing actively to the host's defense mechanisms (Fig. 9.4).

6 Challenges and Limitations in Targeting AMPs for Therapy

While leveraging antimicrobial peptides (AMPs) for therapeutic purposes holds great promise across various medical scenarios, a myriad of challenges and constraints must be navigated. The intricate nature and diverse roles of AMPs within the innate immune system pose obstacles to precise therapeutic targeting. Careful examination of their functions in immunomodulation and antimicrobial defense is imperative. The broad-spectrum antimicrobial activity of many AMPs raises concerns about potential impacts on the host's cells and beneficial microbiota, alongside combating infections. Balancing antibacterial efficacy with host cell safety proves challenging, particularly given the potential for cytotoxicity at higher concentrations. The susceptibility of AMPs to enzymatic degradation, coupled with the technological difficulty of developing stable and long-lasting variants, adds another layer of complexity. Delivery to specific locations within the body poses a challenge, influenced by factors like administration mode, dosage, and penetration through biological barriers. Pathogen resistance, akin to bacterial resistance to conventional medicines, could diminish the efficacy of AMPs over time. Moreover, the induction of immune responses by exogenous AMPs might lead to allergies or other immunological issues. The economic implications of developing AMP-based treatments, affecting patient access and affordability, are considerable. Regulatory processes for novel AMP therapies are intricate, demanding comprehensive safety and efficacy data. Current AMP-based treatments are predominantly in preclinical or early clinical research phases, requiring extensive testing and validation for translation to safe and effective human medicines. Given the multifaceted nature of complex diseases like cancer, infections, and autoimmune disorders, a holistic therapeutic approach may be essential, as AMP targeting alone might not address all facets. The ethical dimensions surrounding the sustainability and welfare of animals, particularly when AMPs are derived from diverse species, further complicate their use as medicines. In summary, while the potential benefits of AMP-based therapies are substantial, overcoming these challenges is essential to unlock their full medical potential.

7 Emerging Research and Innovative Approaches

It has long been known that AMPs have antibacterial qualities. Some of these peptides are specifically engineered to target antibiotic-resistant bacteria, addressing a critical need in contemporary medicine. The development of novel AMPs or modifications to existing ones is underway to effectively treat various infections. Antimicrobial agents (AMPs) have demonstrated potential in preventing the growth of several viruses, such as influenza, herpes simplex virus (HSV), Zika virus, vaccinia virus, Influenza A, HIV, and autoimmune disorders. Furthermore, studies are underway to study how AMPs can potentially regulate the immune response, to alleviate symptoms associated with autoimmune disorders. Certain AMPs increase angiogenesis, tissue regeneration, and recruitment of immune cells that may aid in wound healing. Applications of these peptides in the management of burns, other tissue damage, and chronic wounds are now under investigation.

By rupturing the membranes surrounding cancer cells, preventing angiogenesis, or altering the immune system's reaction to tumors, several AMPs have demonstrated anticancer effects. Their potential as adjuvants in immunotherapy or as direct anticancer drugs is being investigated for anticancer treatment. The application of AMPs in ocular therapy is under investigation. To treat keratitis, eye infections, and other ocular disorders, they can be injected intravitreally or used topically. The potential of AMPs for treating periodontal disease and reducing dental caries is being investigated in relation to oral health. In relation to oral infections, several peptides have demonstrated antibacterial qualities. AMPs may be useful in treating neurodegenerative and neuroinflammatory disorders. It may be possible to use their immunomodulatory qualities to lessen neuroinflammation and save neuronal cells. Carriers for medication delivery can be made out of AMPs. Their ability to encapsulate medications and deliver them to precise locations within the body improves medication efficacy and minimizes adverse effects. Antibiotic-resistant microbes form intricate colonies known as biofilms. Efforts are underway to explore the potential of AMPs in breaking down biofilms, aiming to enhance the treatment of challenging infections that are resistant to complete elimination. Because of their antibacterial qualities, AMPs are also employed in biotechnology to avoid contamination in cell culture, fermentation, and other bioprocessing applications.

8 Conclusion and Future Perspectives

Antimicrobial peptides (AMPs) present a dynamic research field, yet key challenges persist in their design and application. Addressing these hurdles requires interdisciplinary collaboration involving chemistry, biology, computational, materials science, bioinformatics, and pharmacy. Modern computational methods may contribute to our complete and extended understanding of AMP mechanisms. Enhancing systematic and scientific validation necessitates comprehensive studies on the interaction between AMPs and diverse targets. Transitioning from cell-level experiments to complex physiological conditions in animal experiments is crucial for robust efficacy assessment.

Emphasizing applications in food, agriculture, and animal husbandry could unlock AMPs' potential, given their direct microbicidal effect and immune system modulation capabilities. These characteristics position AMPs as promising candidates for diverse infectious diseases. Notably, their immunoregulatory functions differentiate them from conventional antibiotics, diminishing the emergence of resistant strains. Antimicrobial and cell-penetrating peptides (CPPs) emerge as potent strategies against intracellular bacterial infections. Their capacity to traverse eukaryotic membranes, exert bactericidal effects on intracellular bacteria, and influence the immune system highlights the therapeutic potential of antimicrobial peptides. The enhancement of efficacy, particularly against intracellular bacteria and other pathogens, may be achieved by conjugating AMPs and CPPs with various molecules.

Despite research gaps, AMPs present significant promise as therapeutic agents for various infections. Their multifaceted mechanism, from direct bactericidal action to immune system modulation, positions them as valuable templates for drug development. Strategic approaches, including conjugation with diverse molecules, enhance the prospects of successfully combatting intracellular bacterial infections.

Antimicrobial peptides (AMPs) represent unconventional therapeutic molecules with significant potential as alternative treatments for infections caused by resistant microbes. Despite their potency, broad antimicrobial activity, and low accumulation in tissues, challenges like high cost of manufacturing, enzymatic degradation, instability, short shelf-life, and cytotoxicity hinder their widespread therapeutic application. Nevertheless, pharmaceutical companies are investing in developing AMPs as commercially available drugs. Identifying potential strategies to harness AMPs as drug candidates against multidrug-resistant pathogens may be a novel research area. While the efficacy of AMPs against bacterial and other infections has endured, the threat of resistance, including to cationic AMPs, necessitates ongoing efforts to combat resistance evolution. In conclusion, addressing AMP resistance in bacteria requires innovative approaches, alongside general precautions to manage conventional antibiotic resistance. Finally, the pressing demand for new antimicrobials has fuelled research on antimicrobial peptides (AMPs). While knowledge in the field is rapidly expanding and lead compounds are being identified, transitioning more antimicrobial peptides into clinical tests and treatments encounters obstacles. These challenges encompass concerns such as low specificity, high manufacturing costs, potential toxicity to animal cells, and the absence of a robust guideline for rational design. Overcoming these hurdles is essential for optimizing the effectiveness of AMPs in varied applications. The insights provided may pave the way for future therapeutic strategies, leveraging the inherent power of innate immunity for enhanced pathogen control and immune modulation.

References

Abiko Y, Suraweera AK, Nishimura M, Arakawa T, Takuma T, Mizoguchi I, Kaku T (2001) Differential expression of human beta-defensin 2 in keratinized and non-keratinized oral epithelial lesions; immunohistochemistry and in situ hybridization. Virchows Arch 438:248–253

Agier J, Brzezińska-Błaszczyk E, Żelechowska P, Wiktorska M, Pietrzak J, Różalska S (2018) Cathelicidin LL-37 affects surface and intracellular toll-like receptor expression in tissue mast cells. J Immunol Res 2018:7357162

Ahmed A, Siman-Tov G, Hall G, Bhalla N, Narayanan A (2019) Human antimicrobial peptides as therapeutics for viral infections. Viruses 11(8):704

Akira S, Uematsu S, Takeuchi O (2006) Pathogen recognition and innate immunity. Cell 124(4):783–801

Alalwani SM, Sierigk J, Herr C, Pinkenburg O, Gallo R, Vogelmeier C, Bals R (2010) The antimicrobial peptide LL-37 modulates the inflammatory and host defence response of human neutrophils. Eur J Immunol 40(4):1118–1126

Albrethsen J, Bøgebo R, Gammeltoft S, Olsen J, Winther B, Raskov H (2005) Upregulated expression of human neutrophil peptides 1, 2 and 3 (HNP 1-3) in colon cancer serum and tumours: a biomarker study. BMC Cancer 5(1):8

Andrä J, Jakovkin I, Grötzinger J, Hecht O, Krasnosdembskaya AD, Goldmann T, Gutsmann T, Leippe M (2008) Structure and mode of action of the antimicrobial peptide arenicin. Biochem J 410(1):113–122

Ashkenazi A, Wexler-Cohen Y, Shai Y (2011) Multifaceted action of Fuzeon as virus–cell membrane fusion inhibitor. Biochim Biophys Acta 1808(10):2352–2358

Auvynet C, Rosenstein Y (2009) Multifunctional host defence peptides: antimicrobial peptides, the small yet big players in innate and adaptive immunity. FEBS J 276(22):6497–6508

Bąbolewska E, Brzezińska-Błaszczyk E (2015) Human-derived cathelicidin LL-37 directly activates mast cells to proinflammatory mediator synthesis and migratory response. Cell Immunol 293(2):67–73

Baindara P, Chaudhry V, Mittal G, Liao LM, Matos CO, Khatri N, Franco OL, Patil PB, Korpole S (2015) Characterization of the antimicrobial peptide penisin, a class Ia novel lantibiotic from Paenibacillus sp. strain A3. Antimicrob Agents Chemother 60(1):580–591

Baindara P, Singh N, Ranjan M, Nallabelli N, Chaudhry V, Pathania GL, Sharma N, Kumar A, Patil PB, Korpole S (2016) Laterosporulin10: a novel defensin like class IId bacteriocin from Brevibacillus sp. strain SKDU10 with inhibitory activity against microbial pathogens. Microbiology (Reading) 162(8):1286–1299

Baindara P, Gautam A, Raghava G et al (2017) Anticancer properties of a defensin like class IId bacteriocin laterosporulin10. Sci Rep 7:46541

Barlow PG, Svoboda P, Mackellar A, Nash AA, York IA, Pohl J, Davidson DJ, Donis RO (2011) Antiviral activity and increased host defence against influenza infection elicited by the human cathelicidin LL-37. PLoS One 6(10):e25333

Benetti S, Timmons PB, Hewage CM (2019) NMR model structure of the antimicrobial peptide maximin 3. Eur Biophys J 48(2):203–212

Bergman P, Walter-Jallow L, Broliden K, Agerberth B, Soderlund J (2007) The antimicrobial peptide LL-37 inhibits HIV-1 replication. Curr HIV Res 5(4):410–415

Bin Hafeez A, Jiang X, Bergen PJ, Zhu Y (2021) Antimicrobial peptides: an update on classifications and databases. Int J Mol Sci 22(21):11691

Böffert R, Businger R, Preiß H, Ehmann D, Truffault V, Simon C, Ruetalo N, Hamprecht K, Müller P, Wehkamp J, Schindler M (2020) The human α-defensin-derived peptide HD5 (1–9) inhibits cellular attachment and entry of human cytomegalovirus. Antivir Res 177:104779

Boman HG, Agerberth B, Boman A (1993) Mechanisms of action on Escherichia coli of cecropin P1 and PR-39, two antibacterial peptides from pig intestine. Infect Immun 61(7):2978–2984

Bonamy C, Sechet E, Amiot A, Alam A, Mourez M, Fraisse L, Sansonetti PJ, Sperandio B (2018) Expression of the human antimicrobial peptide β-defensin-1 is repressed by the EGFR-ERK-MYC axis in colonic epithelial cells. Sci Rep 8(1):18043

Boparai JK, Sharma PK (2020) Mini review on antimicrobial peptides, sources, mechanism and recent applications. Protein Pept Lett 27(1):4–16

Bowdish DM, Davidson DJ, Speert DP, Hancock RE (2004) The human cationic peptide LL-37 induces activation of the extracellular signal-regulated kinase and p38 kinase pathways in primary human monocytes. J Immunol 172(6):3758–3765

Braffman NR, Piscotta FJ, Hauver J, Campbell EA, Link AJ, Darst SA (2019) Structural mechanism of transcription inhibition by lasso peptides microcin J25 and capistruin. Proc Natl Acad Sci USA 116(4):1273–1278

Brogden KA (2005) Antimicrobial peptides: pore formers or metabolic inhibitors in bacteria? Nat Rev Microbiol 3:238–250

Brumfitt W, Salton MR, Hamilton-Miller JM (2002) Nisin, alone and combined with peptidoglycan-modulating antibiotics: activity against methicillin-resistant Staphylococcus aureus and vancomycin-resistant enterococci. J Antimicrob Chemother 50(5):731–734

Cao Y, Chen F, Sun Y, Hong H, Wen Y, Lai Y, Xu Z, Luo X, Chen Y, Shi J, Li H (2019a) LL-37 promotes neutrophil extracellular trap formation in chronic rhinosinusitis with nasal polyps. Clin Exp Allergy 49(7):990–999

Cao L, Jiang W, Cao S, Zhao P, Liu J, Dong H, Guo Y, Liu Q, Gong P (2019b) In vitro leishmanicidal activity of antimicrobial peptide KDEL against Leishmania tarentolae. Acta Biochim Biophys Sin 51(12):1286–1292

Cardoso MH, Meneguetti BT, Costa BO, Buccini DF, Oshiro KGN, Preza SLE et al (2019) Non-lytic antibacterial peptides that translocate through bacterial membranes to act on intracellular targets. Int J Mol Sci 20(19):4877

Chan WC, Leyland M, Clark J, Dodd HM, Lian LY, Gasson MJ, Bycroft BW, Roberts GC (1996) Structure-activity relationships in the peptide antibiotic nisin: antibacterial activity of fragments of nisin. FEBS Lett 390(2):129–132

Chen W, Zeng H, Tan H (2000) Cloning, sequencing, and function of sanF: a gene involved in nikkomycin biosynthesis of Streptomyces ansochromogenes. Curr Microbiol 41:312–316

Chen C, Fan H, Huang Y, Peng F, Fan H, Yuan S, Tong Y (2014) Recombinant lysostaphin protects mice from methicillin-resistant Staphylococcus aureus pneumonia. Biomed Res Int 2014:602185

Chen YC, Tsai TL, Ye XH, Lin TH (2015) Anti-proliferative effect on a colon adenocarcinoma cell line exerted by a membrane disrupting antimicrobial peptide KL15. Cancer Biol Ther 16(8):1172–1183

Chen X, Zou X, Qi G, Tang Y, Guo Y, Si J, Liang L (2018) Roles and mechanisms of human cathelicidin LL-37 in cancer. Cell Physiol Biochem 47(3):1060–1073

Cheng C, Hua ZC (2020) Lasso peptides: heterologous production and potential medical application. Front Bioeng Biotechnol 8:571165

Chessa C, Bodet C, Jousselin C, Wehbe M, Lévêque N, Garcia M (2020) Antiviral and immunomodulatory properties of antimicrobial peptides produced by human keratinocytes. Front Microbiol 11:1155

Choi KY, Mookherjee N (2012) Multiple immune-modulatory functions of cathelicidin host defence peptides. Front Immunol 3:149

Choi CS, Yoe SM, Kim ES, Chae KS, Kim HR (1997) Purification and characterization of antibacterial peptides, spodopsin Ia and Ib induced in the larval haemolymph of the common cutworm, Spodoptera iitura. Korean J Biol Sci 1:457–462

Cole AM, Ganz T, Liese AM, Burdick MD, Liu L, Strieter RM (2001) Cutting edge: IFN-inducible ELR–CXC chemokines display defensin-like antimicrobial activity. J Immunol 167(2):623–627

Corrêa JAF, Evangelista AG, Nazareth TDM, Luciano FB (2019) Fundamentals on the molecular mechanism of action of antimicrobial peptides. Materialia 8:100494

Cruz GF, de Araujo I, Torres MD, de la Fuente-Nunez C, Oliveira VX, Ambrosio FN, Lombello CB, Almeida DV, Silva FD, Garcia W (2020) Photochemically-generated silver chloride nanoparticles stabilized by a peptide inhibitor of cell division and its antimicrobial properties. J Inorg Organomet Polym Mater 30:2464–2474

da Costa JP, Cova M, Ferreira R, Vitorino R (2015) Antimicrobial peptides: an alternative for innovative medicines? Appl Microbiol Biotechnol 99:2023–2040

de Leeuw E, Li C, Zeng P, Li C, Diepeveen-de Buin M, Lu WY, Breukink E, Lu W (2010) Functional interaction of human neutrophil peptide-1 with the cell wall precursor lipid II. FEBS Lett 584(8):1543–1548

Dia VP, Krishnan HB (2016) BG-4, a novel anticancer peptide from bitter gourd (Momordica charantia), promotes apoptosis in human colon cancer cells. Sci Rep 6(1):33532

Dinata R, Baindara P (2023) Laterosporulin25: a probiotically produced, novel defensin-like bacteriocin and its immunogenic properties. Int Immunopharmacol 121:110500

Dong P, Zhou Y, He W, Hua D (2016) A strategy for enhanced antibacterial activity against Staphylococcus aureus by the assembly of alamethicin with a thermo-sensitive polymeric carrier. Chem Commun (Camb) 52(5):896–899

Dong Z, Hu H, Yu X, Tan L, Ma C, Xi X, Li L, Wang L, Zhou M, Chen T, Du S (2020) Novel frog skin-derived peptide dermaseptin-PP for lung cancer treatment: in vitro/vivo evaluation and anti-tumor mechanisms study. Front Chem 8:476

Duarte-Mata DI, Salinas-Carmona MC (2023) Antimicrobial peptides immune modulation role in intracellular bacterial infection. Front Immunol 14:1119574

Egesten A, Eliasson M, Johansson HM, Olin AI, Mörgelin M, Mueller A, Pease JE, Frick IM, Björck L (2007) The CXC chemokine MIG/CXCL9 is important in innate immunity against Streptococcus pyogenes. J Infect Dis 195(5):684–693

Epand RM, Vogel HJ (1999) Diversity of antimicrobial peptides and their mechanisms of action. Biochim Biophys Acta Biomembr 1462:11–28

Fang Y, He X, Zhang P, Shen C, Mwangi J, Xu C, Mo G, Lai R, Zhang Z (2019) In vitro and in vivo antimalarial activity of LZ1, a peptide derived from snake cathelicidin. Toxins 11(7):379

Filewod NC, Pistolic J, Hancock RE (2009) Low concentrations of LL-37 alter IL-8 production by keratinocytes and bronchial epithelial cells in response to proinflammatory stimuli. FEMS Immunol Med Microbiol 56(3):233–240

Florin T, Maracci C, Graf M, Karki P, Klepacki D, Berninghausen O, Beckmann R, Vázquez-Laslop N, Wilson DN, Rodnina MV, Mankin AS (2017) An antimicrobial peptide that inhibits translation by trapping release factors on the ribosome. Nat Struct Mol Biol 24(9):752–757

Fritig B, Heitz T, Legrand M (1998) Antimicrobial proteins in induced plant defense. Curr Opin Immunol 10(1):16–22

Galván Márquez IJ, McKay B, Wong A, Cheetham JJ, Bean C, Golshani A, Smith ML (2020) Mode of action of nisin on Escherichia coli. Can J Microbiol 66(2):161–168

Gauglitz GG, Bureik D, Zwicker S, Ruzicka T, Wolf R (2015) The antimicrobial peptides psoriasin (S100A7) and koebnerisin (S100A15) suppress extracellular matrix production and proliferation of human fibroblasts. Skin Pharmacol Physiol 28(3):115–123

Gholizadeh A, Moradi B (2021) Cecropins activity against bacterial pathogens. Infect Dis Clin Pract 29(1):e6–e12

Gomez Hernandez MP, Bates AM, Starman EE, Lanzel EA, Comnick C, Xie XJ, Brogden KA (2019) HBD3 induces PD-L1 expression on head and neck squamous cell carcinoma cell lines. Antibiotics 8(4):161

Graf M, Mardirossian M, Nguyen F, Seefeldt AC, Guichard G, Scocchi M, Innis CA, Wilson DN (2017) Proline-rich antimicrobial peptides targeting protein synthesis. Nat Prod Rep 34(7):702–711

Grage SL, Afonin S, Kara S, Buth G, Ulrich AS (2016) Membrane thinning and thickening induced by membrane-active amphipathic peptides. Front Cell Dev Biol 4:65

Grant WB, Lahore H, McDonnell SL, Baggerly CA, French CB, Aliano JL, Bhattoa HP (2020) Evidence that vitamin D supplementation could reduce risk of influenza and COVID-19 infections and deaths. Nutrients 12(4):988

Guaní-Guerra E, Santos-Mendoza T, Lugo-Reyes SO, Terán LM (2010) Antimicrobial peptides: general overview and clinical implications in human health and disease. Clin Immunol 135(1):1–11

Hanaoka Y, Yamaguchi Y, Yamamoto H, Ishii M, Nagase T, Kurihara H, Akishita M, Ouchi Y (2016) In vitro and in vivo anticancer activity of human β-defensin-3 and its mouse homolog. Anticancer Res 36(11):5999–6004

Hancock RE, Chapple DS (1999) Peptide antibiotics. Antimicrob Agents Chemother 43(6):1317–1323

Hancock RE, Haney EF, Gill EE (2016) The immunology of host defence peptides: beyond antimicrobial activity. Nat Rev Immunol 16(5):321–334

Hayashi M, Kuroda K, Ihara K, Iwaya T, Isogai E (2018) Suppressive effect of an analog of the antimicrobial peptide of LL 37 on colon cancer cells via exosome encapsulated miRNAs. Int J Mol Med 42(6):3009–3016

He SW, Zhang J, Li NQ, Zhou S, Yue B, Zhang M (2017) A TFPI-1 peptide that induces degradation of bacterial nucleic acids, and inhibits bacterial and viral infection in half-smooth tongue sole, Cynoglossus semilaevis. Fish Shellfish Immunol 60:466–473

He M, Zhang H, Li Y, Wang G, Tang B, Zhao J, Huang Y, Zheng J (2018) Cathelicidin-derived antimicrobial peptides inhibit Zika virus through direct inactivation and interferon pathway. Front Immunol 9:722

Helmerhorst EJ, Troxler RF, Oppenheim FG (2001) The human salivary peptide histatin 5 exerts its antifungal activity through the formation of reactive oxygen species. Proc Natl Acad Sci 98(25):14637–14642

Hiemstra PS (2006) Defensins and cathelicidins in inflammatory lung disease: beyond antimicrobial activity. Biochem Soc Trans 34(2):276–278

Hillman JD, Novák J, Sagura E, Gutierrez JA, Brooks TA, Crowley PJ, Hess M, Azizi A, Leung KP, Cvitkovitch D, Bleiweis AS (1998) Genetic and biochemical analysis of mutacin 1140, a lantibiotic from Streptococcus mutans. Infect Immun 66(6):2743–2749

Hoffmann J, Schneider C, Heinbockel L, Brandenburg K, Reimer R, Gabriel G (2014) A new class of synthetic anti-lipopolysaccharide peptides inhibits influenza A virus replication by blocking cellular attachment. Antivir Res 104:23–33

Hosoda H, Nakamura K, Hu Z, Tamura H, Reich J, Kuwahara Arai K, Iba T, Tabe Y, Nagaoaka I (2017) Antimicrobial cathelicidin peptide LL 37 induces NET formation and suppresses the inflammatory response in a mouse septic model. Mol Med Rep 16(4):5618–5626

Hou X, Feng C, Li S, Luo Q, Shen G, Wu H, Li M, Liu X, Chen A, Ye M, Zhang Z (2019) Mechanism of antimicrobial peptide NP-6 from Sichuan pepper seeds against E. coli and effects of different environmental factors on its activity. Appl Microbiol Biotechnol 103:6593–6604

Howell MD, Jones JF, Kisich KO, Streib JE, Gallo RL, Leung DY (2004) Selective killing of vaccinia virus by LL-37: implications for eczema vaccinatum. J Immunol 172(3):1763–1767

Howell MD, Wollenberg A, Gallo RL, Flaig M, Streib JE, Wong C, Pavicic T, Boguniewicz M, Leung DY (2006) Cathelicidin deficiency predisposes to eczema herpeticum. J Allergy Clin Immunol 117(4):836–841

Hu Z, Murakami T, Suzuki K, Tamura H, Reich J, Kuwahara-Arai K, Iba T, Nagaoka I (2016) Antimicrobial cathelicidin peptide LL-37 inhibits the pyroptosis of macrophages and improves the survival of polybacterial septic mice. Int Immunol 28(5):245–253

Hu H, Guo N, Chen S, Guo X, Liu X, Ye S, Chai Q, Wang Y, Liu B, He Q (2019) Antiviral activity of Piscidin 1 against pseudorabies virus both in vitro and in vivo. Virol J 16(1):95

Huan Y, Kong Q, Mou H, Yi H (2020) Antimicrobial peptides: classification, design, application and research progress in multiple fields. Front Microbiol 11:2559

Huang C, Yang X, Huang J, Liu X, Yang X, Jin H, Huang Q, Li L, Zhou R (2019) Porcine beta-defensin 2 provides protection against bacterial infection by a direct bactericidal activity and alleviates inflammation via interference with the TLR4/NF-κB pathway. Front Immunol 10:1673

Jenssen H, Andersen JH, Uhlin-Hansen L, Gutteberg TJ, Rekdal Ø (2004) Anti-HSV activity of lactoferricin analogues is only partly related to their affinity for heparan sulfate. Antivir Res 61(2):101–109

Jenssen H, Hamill P, Hancock RE (2006) Peptide antimicrobial agents. Clin Microbiol Rev 19(3):491–511

Ji P, Zhou Y, Yang Y, Wu J, Zhou H, Quan W, Sun J, Yao Y, Shang A, Gu C, Zeng B, Firrman J, Xiao W, Bals R, Sun Z, Li D (2019) Myeloid cell-derived LL-37 promotes lung cancer growth by activating Wnt/β-catenin signaling. Theranostics 9(8):2209–2223

Joly S, Compton LM, Pujol C, Kurago ZB, Guthmiller JM (2009) Loss of human β-defensin 1, 2, and 3 expression in oral squamous cell carcinoma. Oral Microbiol Immunol 24(5):353–360

Kahlenberg JM, Carmona-Rivera C, Smith CK, Kaplan MJ (2013) Neutrophil extracellular trap–associated protein activation of the NLRP3 inflammasome is enhanced in lupus macrophages. J Immunol 190(3):1217–1226

Kang HK, Kim C, Seo CH, Park Y (2017) The therapeutic applications of antimicrobial peptides (AMPs): a patent review. J Microbiol 55:1–2

Kers JA, Sharp RE, Defusco AW, Park JH, Xu J, Pulse ME, Weiss WJ, Handfield M (2018) Mutacin 1140 lantibiotic variants are efficacious against Clostridium difficile infection. Front Microbiol 9:415

King BF, Biel ML, Wilkinson BJ (1980) Facile penetration of the Staphylococcus aureus capsule by lysostaphin. Infect Immun 29:892–896

Koszałka P, Kamysz E, Wejda M, Kamysz W, Bigda J (2011) Antitumor activity of antimicrobial peptides against U937 histiocytic cell line. Acta Biochim Pol 58(1):111–117

Kragol G, Lovas S, Varadi G, Condie BA, Hoffmann R, Otvos L (2001) The antibacterial peptide pyrrhocoricin inhibits the ATPase actions of DnaK and prevents chaperone-assisted protein folding. Biochemistry 40(10):3016–3026

Kurpe SR, Grishin SY, Surin AK, Panfilov AV, Slizen MV, Chowdhury SD, Galzitskaya OV (2020) Antimicrobial and amyloidogenic activity of peptides. Can antimicrobial peptides be used against SARS-CoV-2? Int J Mol Sci 21(24):9552

Lacerda AF, Pelegrini PB, de Oliveira DM, Vasconcelos ÉA, Grossi-de-Sá MF (2016) Anti-parasitic peptides from arthropods and their application in drug therapy. Front Microbiol 7:91

Lai Y, Adhikarakunnathu S, Bhardwaj K, Ranjith-Kumar CT, Wen Y, Jordan JL, Wu LH, Dragnea B, Mateo LS, Kao CC (2011) LL37 and cationic peptides enhance TLR3 signaling by viral double-stranded RNAs. PLoS One 6(10):e26632

Lax R (2010) The future of peptide development in the pharmaceutical industry. PharManufacturing 2:10–15

Le Messurier KS, Lin Y, McCullers JA, Samarasinghe AE (2016) Antimicrobial peptides alter early immune response to influenza A virus infection in C57BL/6 mice. Antivir Res 133:208–217
Le CF, Fang CM, Sekaran SD (2017) Intracellular targeting mechanisms by antimicrobial peptides. Antimicrob Agents Chemother 61(4):10–128
Lee G, Bae H (2016) Anti-inflammatory applications of melittin, a major component of bee venom: detailed mechanism of action and adverse effects. Molecules 21:616
Lehrer RI, Barton A, Daher KA, Harwig SS, Ganz T, Selsted ME (1989) Interaction of human defensins with Escherichia coli. Mechanism of bactericidal activity. J Clin Invest 84(2):553–561
Lehrer RI, Bevins CL, Ganz T (2005) Defensins and other antimicrobial peptides and proteins. Mucosal Immunol:95–110. https://doi.org/10.1016/B978-012491543-5/50010-3
Lei J, Sun L, Huang S, Zhu C, Li P, He J, Mackey V, Coy DH, He Q (2019) The antimicrobial peptides and their potential clinical applications. Am J Transl Res 11(7):3919
Leiter É, Szappanos H, Oberparleiter C, Kaiserer L, Csernoch L, Pusztahelyi T, Emri T, Pócsi I, Salvenmoser W, Marx F (2005) Antifungal protein PAF severely affects the integrity of the plasma membrane of Aspergillus nidulans and induces an apoptosis-like phenotype. Antimicrob Agents Chemother 49(6):2445–2453
Li Y, Xiang Q, Zhang Q, Su Z (2012) Overview on the recent study of antimicrobial peptides: origins, functions, relative mechanisms and application. Peptides 37:207–215
Li D, Qin Q, Wang XY, Shi HS, Luo M, Guo FC, Wang W, Wang YS (2014) Intratumoral expression of mature human neutrophil peptide-1 potentiates the therapeutic effect of doxorubicin in a mouse 4T1 breast cancer model. Oncol Rep 31(3):1287–1295
Li L, Sun J, Xia S, Tian X, Cheserek MJ, Le G (2016) Mechanism of antifungal activity of antimicrobial peptide APP, a cell-penetrating peptide derivative, against Candida albicans: intracellular DNA binding and cell cycle arrest. Appl Microbiol Biotechnol 100:3245–3253
Liang X, Zhang X, Lian K, Tian X, Zhang M, Wang S, Chen C, Nie C, Pan Y, Han F, Wei Z (2020) Antiviral effects of Bovine antimicrobial peptide against TGEV in vivo and in vitro. J Vet Sci 21(5):e80
Lin MC, Hui CF, Chen JY, Wu JL (2013) Truncated antimicrobial peptides from marine organisms retain anticancer activity and antibacterial activity against multidrug-resistant Staphylococcus aureus. Peptides 44:139–148
Ling LL, Schneider T, Peoples AJ, Spoering AL, Engels I, Conlon BP, Mueller A, Schäberle TF, Hughes DE, Epstein S, Jones M (2015) A new antibiotic kills pathogen without detectable resistance. Nature 517(7535):455–459
Liou JW, Hung YJ, Yang CH, Chen YC (2015) The antimicrobial activity of gramicidin A is associated with hydroxyl radical formation. PLoS One 10(1):e0117065
Lishko VK, Moreno B, Podolnikova NP, Ugarova TP (2016) Identification of human cathelicidin peptide LL-37 as a ligand for macrophage integrin αMβ2 (Mac-1, CD11b/CD18) that promotes phagocytosis by opsonizing bacteria. Res Rep Biochem 2016:39–55
Liu Y, Gong W, Huang CC, Herr W, Cheng X (1999) Crystal structure of the conserved core of the herpes simplex virus transcriptional regulatory protein VP16. Genes Dev 13(13):1692–1703
Lu Q, Quan W, Wu J, Zhang X, Ma W, Pang L, Li D (2015) Effect of antibacterial peptide hCAP18/LL-37 on ovarian cancer microenvironment and the regulatory mechanism of its expression. Zhonghua Zhong liu za zhi [Chin J Oncol] 37(10):725–730
Lundy FT, Orr DF, Gallagher JR, Maxwell P, Shaw C, Napier SS, Cowan CG, Lamey PJ, Marley JJ (2004) Identification and overexpression of human neutrophil α-defensins (human neutrophil peptides 1, 2 and 3) in squamous cell carcinomas of the human tongue. Oral Oncol 40(2):139–144
Luo XL, Li JX, Huang HR, Duan JL, Dai RX, Tao RJ, Yang L, Hou JY, Jia XM, Xu JF (2019) LL37 inhibits Aspergillus fumigatus infection via directly binding to the fungus and preventing excessive inflammation. Front Immunol 10:283
Luong HX, Thanh TT, Tran TH (2020) Antimicrobial peptides–advances in development of therapeutic applications. Life Sci 260:118407
Lutkenhaus J (1990) Regulation of cell division in E. coli. Trends Genet 6:22–55

Lv Y, Shao G, Zhang Q, Wang X, Meng Y, Wang L, Huang F, Yang T, Jin Y, Fu C (2019) The antimicrobial peptide PFR induces necroptosis mediated by ER stress and elevated cytoplasmic calcium and mitochondrial ROS levels: cooperation with Ara-C to act against acute myeloid leukemia. Signal Transduct Target Ther 4(1):38

Ma R, Wong SW, Ge L, Shaw C, Siu SW, Kwok HF (2020) In vitro and MD simulation study to explore physicochemical parameters for antibacterial peptide to become potent anticancer peptide. Mol Ther Oncolytics 16:7–19

Mack MR, Kim BS (2016) Superficial immunity: antimicrobial responses are more than skin deep. Immunity 45(1):6–8

Madanchi H, Shoushtari M, Kashani HH, Sardari S (2019) Antimicrobial peptides of the vaginal innate immunity and their role in the fight against sexually transmitted diseases. New Microbes New Infect 34:100627. https://doi.org/10.1016/j.nmni.2019.100627

Madanchi H, Shoushtari M, Kashani HH, Sardari S (2020) Antimicrobial peptides of the vaginal innate immunity and their role in the fight against sexually transmitted diseases. New Microbes New Infect 34:100627

Maijaroen S, Jangpromma N, Daduang J, Klaynongsruang S (2018) KT2 and RT2 modified antimicrobial peptides derived from Crocodylus siamensis Leucrocin I show activity against human colon cancer HCT-116 cells. Environ Toxicol Pharmacol 62:164–176

Maischak H, Zimmermann MR, Felle HH, Boland W, Mithöfer A (2010) Alamethicin-induced electrical long distance signaling in plants. Plant Signal Behav 5(8):988–990

Maiti S, Patro S, Purohit S, Jain S, Senapati S, Dey N (2014) Effective control of Salmonella infections by employing combinations of recombinant antimicrobial human beta-defensins hBD-1 and hBD-2. Antimicrob Agents Chemoter 58:6896–6903. https://doi.org/10.1128/AAC.03628-14

Malmsten M (2014) Antimicrobial peptides. Ups J Med Sci 119:199–204

Manabe T, Kawasaki K (2017) D-form KLKLLLLLKLK-NH2 peptide exerts higher antimicrobial properties than its L-form counterpart via an association with bacterial cell wall components. Sci Rep 7(1):43384

Maraming P, Klaynongsruang S, Boonsiri P, Peng SF, Daduang S, Leelayuwat C, Pientong C, Chung JG, Daduang J (2019) The cationic cell-penetrating KT2 peptide promotes cell membrane defects and apoptosis with autophagy inhibition in human HCT 116 colon cancer cells. J Cell Physiol 234(12):22116–22129

Marcocci ME, Amatore D, Villa S, Casciaro B, Aimola P, Franci G, Grieco P, Galdiero M, Palamara AT, Mangoni ML, Nencioni L (2018) The amphibian antimicrobial peptide temporin b inhibits in vitro herpes simplex virus 1 infection. Antimicrob Agents Chemother 62(5):10–128

Mardirossian M, Grzela R, Giglione C, Meinnel T, Gennaro R, Mergaert P, Scocchi M (2014) The host antimicrobial peptide Bac71-35 binds to bacterial ribosomal proteins and inhibits protein synthesis. Chem Biol 21(12):1639–1647

Mardirossian M, Barrière Q, Timchenko T, Müller C, Pacor S, Mergaert P, Scocchi M, Wilson DN (2018) Fragments of the nonlytic proline-rich antimicrobial peptide Bac5 kill Escherichia coli cells by inhibiting protein synthesis. Antimicrob Agents Chemother 62(8):10–128

Marshall JS, Warrington R, Watson W, Kim HL (2018) An introduction to immunology and immunopathology. Allergy Asthma Clin Immunol 14(2):49

Martinez Del Pozo A, Lacadena V, Mancheno JM, Olmo N, Onaderra M, Gavilanes JG (2002) The antifungal protein AFP of Aspergillus giganteus is an oligonucleotide/oligosaccharide binding (OB) fold-containing protein that produces condensation of DNA. J Biol Chem 277(48):46179–46183

Marx F (2004) Small, basic antifungal proteins secreted from filamentous ascomycetes: a comparative study regarding expression, structure, function and potential application. Appl Microbiol Biotechnol 65:133–142

Matejuk A, Leng Q, Begum MD, Woodle MC, Scaria P, Chou ST, Mixson AJ (2010) Peptide-based antifungal therapies against emerging infections. Drugs Future 35(3):197

Matsuzaki K, Murase O, Fujii N, Miyajima K (1995) Translocation of a channel-forming antimicrobial peptide, magainin 2, across lipid bilayers by forming a pore. Biochemistry 34:6521–6526
McAuliffe O, Ryan MP, Ross RP, Hill C, Breeuwer P, Abee T (1998) Lacticin 3147, a broad-spectrum bacteriocin which selectively dissipates the membrane potential. Appl Environ Microbiol 64(2):439–445
Medzhitov R (2007) Recognition of microorganisms and activation of the immune response. Nature 449:819–826
Meguro S, Tomita M, Katsuki T, Kato K, Oh H, Ainai A, Ito R, Kawai T, Itoh H, Hasegawa H (2014) Plasma antimicrobial peptide LL-37 level is inversely associated with HDL cholesterol level in patients with type 2 diabetes mellitus. Int J Endocrinol 2014:703696
Mogensen TH (2009) Pathogen recognition and inflammatory signaling in innate immune defences. Clin Microbiol Rev 22(2):240–273
Monteiro JM, Oliveira MD, Dias RS, Nacif-Marçal L, Feio RN, Ferreira SO, Oliveira LL, Silva CC, Paula SO (2018) The antimicrobial peptide HS-1 inhibits dengue virus infection. Virology 514:79–87
Mookherjee N, Brown KL, Bowdish DM, Doria S, Falsafi R, Hokamp K, Roche FM, Mu R, Doho GH, Pistolic J, Powers JP (2006) Modulation of the TLR-mediated inflammatory response by the endogenous human host defence peptide LL-37. J Immunol 176(4):2455–2464
Mor A, Hani K, Nicolas P (1994) The vertebrate peptide antibiotics dermaseptins have overlapping structural features but target specific microorganisms. J Biol Chem 269(50):31635–31641
Moravvej H, Memariani M, Memariani H, Robati RM, Gheisari M (2021) Can antimicrobial peptides be repurposed as a novel therapy for keloids? Dermatology 237(2):293–295
Nagaoka I, Tamura H, Hirata M (2006) An antimicrobial cathelicidin peptide, human CAP18/LL-37, suppresses neutrophil apoptosis via the activation of formyl-peptide receptor-like 1 and P2X7. J Immunol 176(5):3044–3052
Nakamura K, Sakuragi N, Takakuwa A, Ayabe T (2016) Paneth cell α-defensins and enteric microbiota in health and disease. Biosci Microbiota Food Health 35(2):57–67
Neshani A, Zare H, Akbari Eidgahi MR, Khaledi A, Ghazvini K (2019) Epinecidin-1, a highly potent marine antimicrobial peptide with anticancer and immunomodulatory activities. BMC Pharmacol Toxicol 20(1):1–11
Neumann A, Völlger L, Berends E, Molhoek EM, Stapels DA, Midon M, Friães A, Pingoud A, Rooijakkers SH, Gallo RL, Mörgelin M (2014) Novel role of the antimicrobial peptide LL-37 in the protection of neutrophil extracellular traps against degradation by bacterial nucleases. J Innate Immun 6(6):860–868
Nicol MQ, Ligertwood Y, Bacon MN, Dutia BM, Nash AA (2012) A novel family of peptides with potent activity against influenza A viruses. J Gen Virol 93(5):980–986
Nikaido H (2003) Molecular basis of bacterial outer membrane permeability revisited. Microbiol Mol Biol Rev 67(4):593–656
Niyonsaba F, Ushio H, Nakano N, Ng W, Sayama K, Hashimoto K, Nagaoka I, Okumura K, Ogawa H (2007) Antimicrobial peptides human β-defensins stimulate epidermal keratinocyte migration, proliferation and production of proinflammatory cytokines and chemokines. J Invest Dermatol 127(3):594–604
Oren Z, Shai Y (1998) Mode of action of linear amphipathic α-helical antimicrobial peptides. Pept Sci 47:451–463
Otvos L, O I, Rogers ME, Consolvo PJ, Condie BA, Lovas S, Bulet P, Blaszczyk-Thurin M (2000) Interaction between heat shock proteins and antimicrobial peptides. Biochemistry 39(46):14150–14159
Palicz Z, Gáll T, Leiter É, Kollár S, Kovács I, Miszti-Blasius K, Pócsi I, Csernoch L, Szentesi P (2016) Application of a low molecular weight antifungal protein from Penicillium chrysogenum (PAF) to treat pulmonary aspergillosis in mice. Emerg Microbes Infect 5(11):e114
Pan B, Liu C, Zhan X, Li J (2021) Protegrin-1 regulates porcine granulosa cell proliferation via the EGFR-ERK1/2/p38 signaling pathway in vitro. Front Physiol 12:673777

Park C, Lee DG (2009) Fungicidal effect of antimicrobial peptide arenicin-1. Biochim Biophys Acta 1788(9):1790–1796
Pasupuleti M, Schmidtchen A, Malmsten M (2012) Antimicrobial peptides: key components of the innate immune system. Crit Rev Biotechnol 32(2):143–171
Piktel E, Niemirowicz K, Wnorowska U, Wątek M, Wollny T, Głuszek K, Góźdź S, Levental I, Bucki R (2016) The role of cathelicidin LL-37 in cancer development. Arch Immunol Ther Exp 64:33–46
Pitale DM, Kaur G, Baghel M, Kaur KJ, Shaha C (2020) Halictine-2 antimicrobial peptide shows promising anti-parasitic activity against Leishmania spp. Exp Parasitol 218:107987
Prohászka Z, Német K, Csermely P, Hudecz F, Mezõ G, Füst G (1997) Defensins purified from human granulocytes bind C1q and activate the classical complement pathway like the transmembrane glycoprotein gp41 of HIV-1. Mol Immunol 34(11):809–816
Radek K, Gallo R (2007) Antimicrobial peptides: natural effectors of the innate immune system. In: Seminars in immunopathology, vol 29. Springer-Verlag, pp 27–43
Ramos R, Silva JP, Rodrigues AC, Costa R, Guardão L, Schmitt F, Soares R, Vilanova M, Domingues L, Gama M (2011) Wound healing activity of the human antimicrobial peptide LL37. Peptides 32(7):1469–1476
Rathinam VA, Chan FK (2018) Inflammasome, inflammation, and tissue homeostasis. Trends Mol Med 24(3):304–318
Rautenbach M, Troskie AM, Vosloo JA (2016) Antifungal peptides: to be or not to be membrane active. Biochimie 130:132–145
Rodríguez-Rojas A, Baeder DY, Johnston P, Regoes RR, Rolff J (2021) Bacteria primed by antimicrobial peptides develop tolerance and persist. PLoS Pathog 17(3):e1009443
Rosés C, Carbajo D, Sanclimens G, Farrera-Sinfreu J, Blancafort A, Oliveras G, Cirac AD, Bardají E, Puig T, Planas M, Feliu L (2012) Cell-penetrating γ-peptide/antimicrobial undecapeptide conjugates with anticancer activity. Tetrahedron 68(23):4406–4412
Roy RN, Lomakin IB, Gagnon MG, Steitz TA (2015) The mechanism of inhibition of protein synthesis by the proline-rich peptide oncocin. Nat Struct Mol Biol 22(6):466–469
Saini J, Kaur P, Malik N, Lakhawat SS, Sharma PK (2022) Antimicrobial peptides: a promising tool to combat multidrug resistance in SARS CoV2 era. Microbiol Res 265:127206
Salem A, Almahmoudi R, Hagström J, Stark H, Nordström D, Salo T, Eklund KK (2019) Human β-defensin 2 expression in oral epithelium: potential therapeutic targets in oral lichen planus. Int J Mol Sci 20(7):1780
Sawaki K, Mizukawa N, Yamaai T, Yoshimoto T, Nakano M, Sugahara T (2002) High concentration of beta-defensin-2 in oral squamous cell carcinoma. Anticancer Res 22(4):2103–2107
Schiffmann S, Gunne S, Henke M, Ulshöfer T, Steinhilber D, Sethmann A, Parnham MJ (2021) Sodium bituminosulfonate used to treat rosacea modulates generation of inflammatory mediators by primary human neutrophils. J Inflamm Res 14:2569
Sechet E, Telford E, Bonamy C, Sansonetti PJ, Sperandio B (2018) Natural molecules induce and synergize to boost expression of the human antimicrobial peptide β-defensin-3. Proc Natl Acad Sci 115(42):E9869–E9878
Sels J, Mathys J, De Coninck BM, Cammue BP, De Bolle MF (2008) Plant pathogenesis-related (PR) proteins: a focus on PR peptides. Plant Physiol Biochem 46(11):941–950
Selsted ME, Novotny MJ, Morris WL, Tang YQ, Smith W, Cullor JS (1992) Indolicidin, a novel bactericidal tridecapeptide amide from neutrophils. J Biol Chem 267(7):4292–4295
Shin JM, Gwak JW, Kamarajan P, Fenno JC, Rickard AH, Kapila YL (2016) Biomedical applications of nisin. J Appl Microbiol 120(6):1449–1465
Shu G, Chen Y, Liu T, Ren S, Kong Y (2019) Antimicrobial peptide cathelicidin-BF inhibits platelet aggregation by blocking protease-activated receptor 4. Int J Pept Res Ther 25:349–358
Silva ON, de la Fuente-Nunez C, Haney EF, Fensterseifer ICM, Ribeiro SM, Porto WF, Brown P, Faria-Junior C, Rezende TMB, Moreno SE et al (2016) An anti-infective synthetic peptide with dual antimicrobial and immunomodulatory activities. Sci Rep 6:35465

Silverstein KA, Moskal WA Jr, Wu HC, Underwood BA, Graham MA, Town CD, VandenBosch KA (2007) Small cysteine-rich peptides resembling antimicrobial peptides have been underpredicted in plants. Plant J 51(2):262–280

Skalickova S, Heger Z, Krejcova L, Pekarik V, Bastl K, Janda J, Kostolansky F, Vareckova E, Zitka O, Adam V, Kizek R (2015) Perspective of use of antiviral peptides against influenza virus. Viruses 7(10):5428–5442

Soehnlein O, Kai-Larsen Y, Frithiof R, Sorensen OE, Kenne E, Scharffetter-Kochanek K, Eriksson EE, Herwald H, Agerberth B, Lindbom L (2008) Neutrophil primary granule proteins HBP and HNP1–3 boost bacterial phagocytosis by human and murine macrophages. J Clin Invest 118(10):3491–3502

Spohn R, Daruka L, Lázár V, Martins A, Vidovics F, Grézal G, Méhi O, Kintses B, Számel M, Jangir PK, Csörgő B (2019) Integrated evolutionary analysis reveals antimicrobial peptides with limited resistance. Nat Commun 10(1):4538

Steinberg DA, Hurst MA, Fujii CA, Kung AH, Ho JF, Cheng FC, Loury DJ, Fiddes JC (1997) Protegrin-1: a broad-spectrum, rapidly microbicidal peptide with in vivo activity. Antimicrob Agents Chemother 41(8):1738–1742

Steinborner ST, Currie GJ, Bowie JH, Wallace JC, Tyler MJ (1998) New antibiotic caerin 1 peptides from the skin secretion of the Australian tree frog Litoria chloris. Comparison of the activities of the caerin 1 peptides from the genus Litoria. J Pept Res 51(2):121–126

Subbalakshmi C, Sitaram N (1998) Mechanism of antimicrobial action of indolicidin. FEMS Microbiol Lett 160(1):91–96

Sun CQ, Arnold R, Fernandez-Golarz C, Parrish AB, Almekinder T, He J, Ho SM, Svoboda P, Pohl J, Marshall FF, Petros JA (2006) Human β-defensin-1, a potential chromosome 8p tumor suppressor: control of transcription and induction of apoptosis in renal cell carcinoma. Cancer Res 66(17):8542–8549

Sun J, Furio L, Mecheri R, van der Does AM, Lundeberg E, Saveanu L, Chen Y, van Endert P, Agerberth B, Diana J (2015) Pancreatic β-cells limit autoimmune diabetes via an immunoregulatory antimicrobial peptide expressed under the influence of the gut microbiota. Immunity 43(2):304–317

Takakuwa A, Nakamura K, Kikuchi M, Sugimoto R, Ohira S, Yokoi Y, Ayabe T (2019) Butyric acid and leucine induce α-defensin secretion from small intestinal Paneth cells. Nutrients 11(11):2817

Tan BH, Meinken C, Bastian M, Bruns H, Legaspi A, Ochoa MT, Krutzik SR, Bloom BR, Ganz T, Modlin RL, Stenger S (2006) Macrophages acquire neutrophil granules for antimicrobial activity against intracellular pathogens. J Immunol 177(3):1864–1871

Tan S, Ludwig KC, Müller A, Schneider T, Nodwell JR (2019) The lasso peptide siamycin-I targets lipid II at the gram-positive cell surface. ACS Chem Biol 14(5):966–974

Theel ES, Pritt BS (2016) Parasites. Microbiol Spectr 4(4):DMIH2-0013-2015. https://doi.org/10.1128/microbiolspec

Thevissen K, Warnecke DC, François IE, Leipelt M, Heinz E, Ott C, Zähringer U, Thomma BP, Ferket KK, Cammue BP (2004) Defensins from insects and plants interact with fungal glucosylceramides. J Biol Chem 279(6):3900–3905

Tinoco-Nunes B, Telleria EL, da Silva-Neves M, Marques C, Azevedo-Brito DA, Pitaluga AN, Traub-Csekö YM (2016) The sandfly Lutzomyia longipalpis LL5 embryonic cell line has active Toll and Imd pathways and shows immune responses to bacteria, yeast and Leishmania. Parasit Vectors 9(1):1–1

Tong Z, Zhang Y, Ling J, Ma J, Huang L, Zhang L (2014) An in vitro study on the effects of nisin on the antibacterial activities of 18 antibiotics against Enterococcus faecalis. PLoS One 9(2):e89209

Tornesello AL, Borrelli A, Buonaguro L, Buonaguro FM, Tornesello ML (2020) Antimicrobial peptides as anticancer agents: functional properties and biological activities. Molecules 25(12):2850

Tripathi S, Tecle T, Verma A, Crouch E, White M, Hartshorn KL (2013) The human cathelicidin LL-37 inhibits influenza A viruses through a mechanism distinct from that of surfactant protein D or defensins. J Gen Virol 94(Pt 1):40

Utsugi T, Schroit AJ, Connor J, Bucana CD, Fidler IJ (1991) Elevated expression of phosphatidylserine in the outer membrane leaflet of human tumor cells and recognition by activated human blood monocytes. Cancer Res 51(11):3062–3066

Van Epps HL (2006) René Dubos: unearthing antibiotics. J Exp Med 203(2):259

van Zoggel H, Carpentier G, Dos Santos C, Hamma-Kourbali Y, Courty J, Amiche M, Delbé J (2012) Antitumor and angiostatic activities of the antimicrobial peptide dermaseptin B2. PLoS One 7(9):e44351

Vilas Boas LC, de Lima LM, Migliolo L, Mendes GD, de Jesus MG, Franco OL, Silva PA (2017) Linear antimicrobial peptides with activity against herpes simplex virus 1 and Aichi virus. Pept Sci 108(2):e22871

Wade D, Boman A, Wåhlin B, Drain CM, Andreu D, Boman HG, Merrifield RB (1990) All-D amino acid-containing channel-forming antibiotic peptides. Proc Natl Acad Sci USA 87:4761–4765

Wang K, Wang JH, Baskaran H, Wang R, Jurevic R (2012) Effect of human beta-defensin-3 on head and neck cancer cell migration using micro-fabricated cell islands. Head Neck Oncol 4(1):1–8

Wang J, Cheng M, Law IK, Ortiz C, Sun M, Koon HW (2019) Cathelicidin suppresses colon cancer metastasis via a P2RX7-dependent mechanism. Mol Ther Oncolytics 12:195–203

Wang S, Fang Q, Lu Z, Gao Y, Trembleau L, Ebel R, Andersen JH, Philips C, Law S, Deng H (2021) Discovery and biosynthetic investigation of a new antibacterial dehydrated non-ribosomal tripeptide. Angew Chem Int Ed 60(6):3229–3237

Wrońska AK, Boguś MI (2020) Heat shock proteins (HSP 90, 70, 60, and 27) in Galleria mellonella (Lepidoptera) hemolymph are affected by infection with Conidiobolus coronatus (Entomophthorales). PLoS One 15(2):e0228556

Wu JA, Kusuma C, Mond JJ, Kokai-Kun JF (2003) Lysostaphin disrupts Staphylococcus aureus and Staphylococcus epidermidis biofilms on artificial surfaces. Antimicrob Agents Chemother 47(11):3407–3414

Wu WK, Sung JJ, To KF, Yu L, Li HT, Li ZJ, Chu KM, Yu J, Cho CH (2010) The host defence peptide LL-37 activates the tumor-suppressing bone morphogenetic protein signaling via inhibition of proteasome in gastric cancer cells. J Cell Physiol 223(1):178–186

Wu D, Gao Y, Qi Y, Chen L, Ma Y, Li Y (2014) Peptide-based cancer therapy: opportunity and challenge. Cancer Lett 351(1):13–22

Wuerth KC, Falsafi R, Hancock RE (2017) Synthetic host defence peptide IDR-1002 reduces inflammation in Pseudomonas aeruginosa lung infection. PLoS One 12(11):e0187565

Xu D, Lu W (2020) Defensins: a double-edged sword in host immunity. Front Immunol 11:764

Xu N, Wang YS, Pan WB, Xiao B, Wen YJ, Chen XC, Chen LJ, Deng HX, You J, Kan B, Fu AF (2008) Human α-defensin-1 inhibits growth of human lung adenocarcinoma xenograft in nude mice. Mol Cancer Ther 7(6):1588–1597

Xu D, Zhang B, Liao C, Zhang W, Wang W, Chang Y, Shao Y (2016) Human beta-defensin 3 contributes to the carcinogenesis of cervical cancer via activation of NF-κB signaling. Oncotarget 7(46):75902

Xu D, Liao C, Zhang B, Tolbert WD, He W, Dai Z, Zhang W, Yuan W, Pazgier M, Liu J, Yu J (2018) Human enteric α-defensin 5 promotes Shigella infection by enhancing bacterial adhesion and invasion. Immunity 48(6):1233–1244

Xu D, Liao C, Xiao J, Fang K, Zhang W, Yuan W, Lu W (2019) Human enteric defensin 5 promotes Shigella infection of macrophages. Infect Immun 88(1):10–128

Yan Z, Wang D, An C, Xu H, Zhao Q, Shi Y, Song N, Deng B, Guo X, Rao J, Cheng L (2021) The antimicrobial peptide YD attenuates inflammation via miR-155 targeting CASP12 during liver fibrosis. Acta Pharm Sin B 11(1):100–111

Yang D, Chen Q, Schmidt AP, Anderson GM, Wang JM, Wooters J, Oppenheim JJ, Chertov O (2000) LL-37, the neutrophil granule–and epithelial cell–derived cathelicidin, utilizes formyl peptide receptor–like 1 (FPRL1) as a receptor to chemoattract human peripheral blood neutrophils, monocytes, and T cells. J Exp Med 192(7):1069–1074

Yang D, Biragyn A, Hoover DM, Lubkowski J, Oppenheim JJ (2004) Multiple roles of antimicrobial defensins, cathelicidins, and eosinophil-derived neurotoxin in host defence. Annu Rev Immunol 22:181–215

Yang H, Fu J, Zhao Y, Shi H, Hu H, Wang H (2017) Escherichia coli PagP enzyme-based de novo design and in vitro activity of antibacterial peptide LL-37. Med Sci Monit 23:2558

Yeaman MR, Yount NY (2003) Mechanisms of antimicrobial peptide action and resistance. Pharmacol Rev 55:27–55

Yoo JH, Ho S, Tran DH, Cheng M, Bakirtzi K, Kubota Y, Ichikawa R, Su B, Tran DH, Hing TC, Chang I (2015) Antifibrogenic effects of the antimicrobial peptide cathelicidin in murine colitis-associated fibrosis. Cell Mol Gastroenterol Hepatol 1(1):55–74

Yoon SH, Hwang I, Lee E, Cho HJ, Ryu JH, Kim TG, Yu JW (2021) Antimicrobial peptide Ll-37 drives rosacea-like skin inflammation in an Nlrp3-dependent manner. J Invest Dermatol 141(12):2885–2894

Yoshimura T, McLean MH, Dzutsev AK, Yao X, Chen K, Huang J, Gong W, Zhou J, Xiang Y, Badger H, J, O'hUigin C. (2018) The antimicrobial peptide CRAMP is essential for colon homeostasis by maintaining microbiota balance. J Immunol 200(6):2174–2185

Zahedifard F, Lee H, No JH, Salimi M, Seyed N, Asoodeh A, Rafati S (2020) Comparative study of different forms of Jellein antimicrobial peptide on Leishmania parasite. Exp Parasitol 209:107823

Zasloff M (2002) Antimicrobial peptides of multicellular organisms. Nature 415(6870):389–395

Zhang LJ, Gallo RL (2016) Antimicrobial peptides. Curr Biol 26(1):R14-9

Zhang Y, Liu S, Li S, Cheng Y, Nie L, Wang G et al (2017) Novel short antimicrobial peptide isolated from Xenopus laevis skin. J Pept Sci 23(5):403–409

Zhang M, Liang W, Gong W, Yoshimura T, Chen K, Wang JM (2019) The critical role of the antimicrobial peptide LL-37/CRAMP in protection of colon microbiota balance, mucosal homeostasis, anti-inflammatory responses, and resistance to carcinogenesis. Crit Rev Immunol 39(2):83

Zhang QY, Yan ZB, Meng YM, Hong XY, Shao G, Ma JJ, Cheng XR, Liu J, Kang J, Fu CY (2021) Antimicrobial peptides: mechanism of action, activity and clinical potential. Mil Med Res 8:1–25

Zhang H, Zhen J, Zhang R et al (2022) Cathelicidin hCAP18/LL-37 promotes cell proliferation and suppresses antitumor activity of 1,25(OH)2D3 in hepatocellular carcinoma. Cell Death Discov 8:27

Chapter 10
Antimicrobial Peptides from Silkworm and Their Therapeutic Application

Aashis Dutta and Manas Das

1 Introduction

Antimicrobial resistance is regarded as a prime concern to public health worldwide specifically in developing countries. It is a phenomenon that occurs when pathogens like bacteria, fungi and viruses emerge resistant to previously susceptible drugs, thereby making infections more hard to cure. The progress of newer antimicrobial agent discovery has dwindled over the past three decades probably due to long-term conventional strategies involved in the quest for newer antibiotic discovery or maybe because most of the antimicrobial activities possessing natural structures have already been reported (Armin et al. 2023). The quest for the discovery of newer peptides has landed us in the exploration of new therapeutics as these nature-derived compounds demonstrate vast chemical heterogeneity in turn transforming them into significant and authentic sources of novel therapeutics. AMPs belong to one such class of novel therapeutics. AMPs are low molecular weight proteins present in all kinds of organisms from invertebrates to vertebrates effective against fungi, bacteria and viruses. The first AMP was purified from the inoculated pupae of a giant silk moth, *Hyalophora cecropia* (Hultmark et al. 1980; Steiner et al. 1981). The production of AMPs along with melanization and reactive oxygen species (ROS) generation belong to the humoral immune response while the cellular response involves hemocyte-mediated activities including encapsulation, nodulation and phagocytosis (Janeway Jr and Medzhitov 2002). Both these types of responses, that is, cellular as well as humoral, are correlated with the innate immunity of silkworms that serve as the second line of defense for protection against invasion by pathogens (Lemaitre and Hoffmann 2007). The fat bodies of insects are responsible for the synthesis of AMPs. They function through both bacteriocidal and bacteriostatic action and have

A. Dutta · M. Das (✉)
Department of Zoology, Gauhati University, Assam, India
e-mail: manasdasne@gauhati.ac.in

P. Baindara, S. M. Mandal (eds.), *Evolution of Antimicrobial Peptides*,
https://doi.org/10.1007/978-3-031-67515-7_10

the capability to become futuristic drugs as their minute size causes them to disperse rapidly through the target bacterial cell plasma membrane and stimulate the defense mechanism of the host against microbes without resistance development or toxic effects generation (Islam et al. 2016). Certain other necessary factors, including charge, hydrophobicity, amphipathic stereogeometry and self-attachment with biomembranes, are crucially necessary for their wide range of antimicrobial activities making them probable candidates for the therapy of tumors and viral or parasitic infections (Tanaka et al. 2008; Chernysh et al. 2002). The differences between the lipid composition of both the prokaryotic and eukaryotic cell membranes are the sole target for AMP action. The antimicrobial nature of certain peptides is in their capability to hinder the particular mechanism of the bacterial cell without any alteration in the similar mechanism in the nearby cells and tissues of the infected organism.

2 Classes of AMPs

The following are the important classes of AMPs demonstrating numerous biological activities reported from both mulberry as well as nonmulberry silkworms.

*2.1 AMPs Reported from Mulberry Silkworm (***Bombyx mori***)*

2.1.1 Cecropins

Cecropins are a family of cationic peptides that were first isolated from *H. cecropia* hemolymph infected with bacteria (Steiner et al. 1981; Hultmark et al. 1982).

Structure

Cecropins are α-helical linear cationic peptides with 37 amino acids with a strong basic N-terminal without the cysteine residue attached to a neutral C-terminal by a flexible glycine–proline connection (Nesa et al. 2020). Five different cecropin types have been reported in *Bombyx mori*, namely, cecropin A, cecropin B, cecropin C, cecropin D and cecropin E. However, a total of 11 *Bmcec* genes, namely, *BmcecA1*, *BmcecA2*, *BmcecB1*, *BmcecB2*, *BmcecB3*, *BmcecB4*, *BmcecB5*, *BmcecC*, *BmcecD1*, *BmcecD2* and *BmcecE* are responsible for encoding these cecropins in silkworm (Cheng et al. 2006). These 11 cecropin genes have again been classified into five subtypes, namely, *Bmcec* A1 (2 genes), B6 (6 genes), C (1 gene), D (1 gene), and E (1 gene) (Nesa et al. 2020). Cecropins demonstrate a randomly coiled structure in aqueous solution but get transformed into an α-helix in hydrophobic environments. Previous reports are indicative of the identification of 30 cecropins from *Papilionoidea* species while some of these AMPs are reported from lepidopterans and dipterans (Wang et al. 2021). Cecropins from insects are also called bactericidin, lepidopteran and sarcotoxin (Ouyang et al. 2015).

Mechanism of Action

Cecropins are generated in the host against gram-positive and gram-negative bacterial infections. Although the mechanism of action of cecropin is still under shade, it is hypothesized that they bind to the helical domain axes parallel to the membranous lipid bilayer. The polar residues get attached to the lipid phosphates while the non polar side chains delve deep into the membranous hydrophobic core. Constant assemblage of cecropins generates a carpet-like structure on the lipid bilayer and fragments the bacterial membrane as it has detergent-like properties (Sato and Feix 2006). However, the cecropins from *H. cecropia* at mild concentration interplay with the membranes forming channels or pores impacting the cellular ionic balance and causing lysis of the cell (Sato and Feix 2006; Efimova et al. 2014). Besides these, cecropins and their derivatives like sarcotoxin-1, papiliocin, stomoxyn, hinnavin, SB-37 and Shiva function as specific inhibitors of *Trypanosoma* and *Plasmodium* while the latter two block the tumor cell line proliferation and HIV-1 multiplication (Barr et al. 1995; Suttmann et al. 2008). Cecropins demonstrate selective cytotoxicity and antiproliferative effects against different types of cancer like bladder cancer, gastric carcinoma, hepatocellular carcinoma, leukemia, breast cancer, small cell lung cancer and colon cancer (Suttmann et al. 2008).

Subtypes of Cecropin

The following are the different subtypes of cecropins reported in silkworms.

Cecropin A

Cecropin A has a stable helical structure. The AMP demonstrates both antibacterial activity and also antifungal properties against *Beauveria bassiana* (Silvestro and Axelsen 2000; Lu et al. 2016). This peptide lowers NADPH and glutathione levels in the body (Yun and Lee 2016). Cecropin A is capable of lysing both gram-positive and gram-negative bacteria by first binding to membrane lipids (negatively charged) by means of its very positively charged N-terminus followed by induction of pore formation by its hydrophobic C-terminus (Gregory et al. 2008).

Cecropin B

Cecropin B is a 35 naturally occurring amino acid containing linear-cationic peptide demonstrating maximum antibacterial activity amongst the entire cecropin family (Srisailam et al. 2001). The AMP is capable of significantly lowering the endotoxin levels in plasma and *E. coli*–induced septic shock rat model (Giacometti et al. 2001). The adult female mortality of the nematode worm *Brugia pahongi* was lowered while in the case of *Aedes aegypti* adult female, the larval development was decreased (Chalk et al. 1995). Furthermore, the peptide exhibits antifungal properties against *Candida albicans* as well as porcine bacterial infections (Hu et al. 2013). The utilization of recombinant DNA technology (RDT) has made it possible for the expression of this AMP in the cocoon of silkworms where it exhibits potent antibacterial properties (Liu et al. 2015). Even cecropin B1 has been overexpressed in silkworms which in turn produces silk with universal antibacterial action (Saviane et al. 2018). Similarly, the spread of *Xanthomonas oryzae* infections was restricted by the recombinant cecropin B expression in rice (Li et al. 2019). An AMP named enbocin having a similar amino acid composition to that of cecropin is also reported to be present in *B. mori* (Kim et al. 1998). CecropinXJ, another cationic AMP belonging to the cecropin B family, demonstrates potent antiproliferative action and functions specifically on cancer cells while having no RBC hemolytic property and null toxicity against healthy cells but the mode of cancerous cell death induction by cecropinXJ is probably through stimulation of dismantling of cytoskeleton and modulating the cytoskeletal protein expression. Thus, cecropinXJ might be a probable curative agent for futuristic cancer treatment (Xia et al. 2013). The AMP can generate amphipathic α-helices forming ion-permeable channels, consequently causing cell depolarization, irreversible cytolysis, and finally cell lysis and as such is very effective against both gram-positive and gram-negative bacteria as well as fungi (Pokorny et al. 2008). Further modification at the C-terminal generating CecXJ-37C and CecXJ-37N demonstrated antibacterial action through pore generation and morphological changes in *E. coli* (Liu et al. 2017).

Cecropin C
The hemolymph of *H. cecropia* is found to possess minute levels of cecropin C generated by cec A degradation. The antimicrobial property of cec C is still not well studied (Hultmark et al. 1982).

Cecropin D
Cecropin D was initially reported from *H. cecropia* inoculated with bacteria that demonstrate similarities with cecropin A and cecropin B but was discovered later than them (Hultmark et al. 1982). The recombinant cecropin D expression in *Pichia pastoris* demonstrated antibacterial potential against both gram-positive and gram-negative bacteria (Guo et al. 2012). The antibacterial potential is more marked in the lysine residue at the C-terminal of cecropin D due to active phosphorylation (Park et al. 2013). Furthermore, the AMP blocks the in vitro multiplication and affliction of the porcine reproductive and respiratory syndrome virus (PRRSV) (Liu et al. 2015).

Cecropin E
Another AMP, cecropin E was isolated from the lepidopterans which possesses 35 amino acids and is capable of interacting with both gram-positive and gram-negative bacteria (Hultmark et al. 1982).

2.1.2 Defensins

Defensins is a peptide class identified from immunized flesh fly, *Sacophoga peregrine* hemolymph (Matsuyama and Natori 1998). Even this AMP is also reported from a dipteran, *Phromia terranovae* hemolymph (Lambert et al. 1989).

Structure
Defensins are 4 kDa weighing cationic peptides consisting of six conserved cysteine residues (Matsuyama and Natori 1998). Phormicins, royalcins, sapecins and spodoptericis are a few well-characterized lepidopteran antibacterial defensins

(Fujiwara et al. 1990). The hemocytes, silk gland, head, ovary of the larvae and the fat body of *B. mori* pupae undergo transcription producing *BmDefensinA* gene transcripts. This gene *BmDefensinA* modulates the insect immune responses and also plays a significant role in both the immune system and insect metamorphosis (Wen et al. 2009). When the *BmDefA* gene was characterized, it was found to contain a 5′ upstream regulatory region with three preserved regulatory sequences of cis-elements, that is, NF-kβ-binding site, IL-6 responsive element and GATA motif. Translation of the *BmDefA* gene generates large peptides with an N-terminal signal peptide with 22 amino acids, a propeptide with 34 amino acids and a mature peptide weighing 4 kDa and having 36 amino acids (Nesa et al. 2020). The gene *BmDefensinB* was reported from the fat body of *B. mori* functioning as insect defensin homolog with three NF-κB binding sites and other cis-elements (KAIKOBLAST). The study implies that both *BmDefensinA* and *BmDefensinB* are responsible for the innate immunity of *B. mori* (Kaneko et al. 2007). *BmDefensinB* has around 27% sequence similarity with *BmDefA* and gets induced by bacteria such as *E. coli* and *B. subtilis* and also by fungi like *Beauveria bassiana*. The rabbit neutrophil defensins demonstrate strong bactericidal action against *Pseudomonas aeruginosa* MDR strains (Zhao et al. 2005). The Toll and Imd pathway modulates *BmDefB* expression and is also responsible for immunity modulation in *B. mori* against both bacterial and fungal infection (Buhroo et al. 2018).

Mechanism of Action

When the mechanism of activity of defensin was put to the test using transgenic insect defensin with *Micrococcus luteus* as a test organism, it was demonstrated that defensin disorganizes the porous barrier of *M. luteus* cytoplasmic membrane that causes a loss of cytoplasmic K^+, partial inner membrane depolarization, lowering of ATP in the cytoplasm and restricts respiration. It was noticeable that the alteration of the membrane permeability is a reflection of the generation of cytoplasmic membrane channels by defensin oligomers. The outcome was substantiated by the patch clamp experiments, which demonstrate that the insect defensins are capable of generating channels. The defensins are found to be effective against gram-positive bacteria like *Staphylococcus aureus*, *Bacillus megaterium*, *Micrococcus luteus*, *Bacillus subtilis*, *Aerococcus viridians*, and *B. thuringiensis* (Yi et al. 2014). A few antifungal defensins are also found including drosomycin from *Drosophila*, gallerimycin from *Galleria mellonella*, heliomycin from *Heliothis virescens*, termicin from *Pseudacanthotermes spiniger*, and Alo13 from the harlequin beetle *Acrocinus longimanus* (Barbault et al. 2003). *Bombyx mori* defensin-like peptide (BmDp) was spotted by wdS20658 cDNA from the expressed sequence tag (EST) database. This peptide was uploaded to GenBank in 2005 with GenBank Accession no. DQ118523. The wdS20658 cDNA comprises six cysteine residues [C…CXXXC…C…CXC] and demonstrated a predominant proportion of sequence similarity (75%) with spodoptericin. A translational study revealed that the gene expression of *BmDp* was stimulated by the bacterial inoculation that was found to hit the peak level after 8 h of inoculation. This report in turn is indicative of the fact that *BmDp* is correlated with bacterial immune response. A purified transgenic GSTmDp fusion protein did not exhibit any significant antibacterial and antifungal activity (Song et al. 2010).

2.1.3 Moricin

Moricin is a highly basic AMP with 42 residues first isolated from *E. coli* inoculated *B. mori* and is capable of inhibiting the growth of *S. aureus* with no major similarity with other antibacterial peptides (Hara and Yamakawa 1995b). Moricin is found to be present only in lepidopterans and cDNAs coding for it have also been found in *Manduca sexta*, *Spodoptera litura*, *G. mellonella*, *Helicoverpa armigera* and other insects (Oizumi et al. 2005; Wang et al. 2010). The basicity is the reason for the electrostatic association of the peptides to the oppositely charged bacterial surface (Gabay 1994).

Structure
The genome of *B. mori* has 12 moricin-encoding genes that are classified into three subtypes: Bmmor (1 gene), moricin-like A (3 genes) and moricin-like B (8 genes). The first gene, namely, *Bmmor* gene, demonstrates the highest antibacterial activity against *B. bombysepticus*, *E. coli*, *B. subtilis*, *B. thuringiensis*, *Serratia marcescens*, *B. thuringiensis galleriae*, *S. aureus*, *Ralstonia solaanacearum* and *P. aeruginosa* at 0.625–1.25 mL/L (Rd) dosage as visualizable through an in vitro experiment (Yang et al. 2011). Moricin-like A subtype demonstrates the next-in-line defense capability while the moricin-like B subtype has null effectiveness on the examined pathogens (Yang et al. 2011). The genes coding for mature moricin possess a positively charged C-terminus and an amphipathic α-helical N-terminus (Islam et al. 2016). The α-helical moricin motif is significant in elevating the membrane permeability to degrade the pathogens (Yi et al. 2014). The charged amino acids appear at a gap of three or four residues in the moricin N-terminal indicative of a structure common to amphipathic α-helix containing antibacterial proteins (Cociancich et al. 1994). These AMPs do not have modifications in the amino acid residues. The isoelectric point of moricin was found to be 12.0 which is higher than that of cecropin. The presence of an amphipathic α-helical structure is probably responsible for antimicrobial action (Hara and Yamakawa 1995b).

Mechanism of Action
Although moricins block the thriving of both gram-positive and gram-negative bacteria, the peptide manifests higher antibacterial action against the former as compared to cecropin B_1. *G. mellonella*–derived moricins demonstrate maximum action against filamentous fungi and yeasts (Hara and Yamakawa 1995b; Dai et al. 2008). The AMP has a predominant role in the protection of *B. mori* against bacterial attack as it demonstrates phenomenal antibacterial activity (Hara and Yamakawa 1995b).

2.1.4 Attacin

Attacin is a glycine-rich 20–23 kDa weighing peptide isolated from *H. cecropia* pupae (Buhroo et al. 2018).

Structure
Both acidic and basic attacins exist with an isoelectric point varying between 5.7 and 8.3 (Nesa et al. 2020). Attacins A–D belong to the basic group, while attacins E and F belong to the basic group (Buonocore et al. 2021). In general, the attacins are rich in amino acids like glycine, aspartic acid, alanine, and abnormal phenylalanine and threonine levels, while the acidic attacins have elevated levels of aspartic acid, arginine, and isoleucine while the basic ones have lysine, glutamic acid, and tryptophan in predominant amount (Islam et al. 2016). The post translational modifications of different attacin isoforms resulted in the generation of two parental pro-attacin sequences (Buhroo et al. 2018). Attacins are produced as pre-pro-proteins consisting of a signal peptide, a pro-peptide (P-domain) and an N-terminal attacin domain followed by two glycine-rich domains (G1 and G2 domains) (Hedengren et al. 2000). The N-terminal attacin pro-peptide contains a preserved motif of amino acids RXXR perceived by furin-like enzymes. This observation is suggestive of the fact that the furin-like enzymes modify the pro-attacins to generate mature forms (Gunne et al. 1990; Devi 1991). The presence of two subdomains in the G domains of *B. mori*, *H. cecropia* attacins and *S. peregrine* sarcotoxin IIA, the usual amino acids in the subdomains are preserved in the evolutionary course and appears to mediate a significant role in antibacterial protein functioning (Sugiyama et al. 1995).

Mode of Action
Attacins are the third antibacterial protein after cecropin and lysozyme present in the *H. cecropia* humoral immune system and function on growing cells and result in chain formation (Kockum et al. 1984). The AMP demonstrates its action against *Stenotrophomonas maltophilia*, *E. coli* and *Acinetobacter calcoaceticus* (Nesa et al. 2020). The attacin-A1 exhibits antimicrobial action against *E. coli* and *Trypanosoma brucei* while the attacin-B exhibits antibacterial action against *E. coli* and *Citrobacter freundii* and also antifungal activity against *C. albicans*. The AMP inhibits the generation of outer membranous proteins like OmpC, OmpF, OmpA and LamB which hinder bacterial cell wall formation (Nesa et al. 2020). The outcome of inhibition is demonstrated through the inhibition of the occurrence of transcription of the analogous genes, that is, a decrease in the constant mRNA levels. The *ompC* and *ompF* genes transcription occurs according to the genetic control of *ompB* and the regulatory locus consists of ompR coding for a DNA-binding protein in the cytoplasm and envZ coding for an environmental sensor protein centralized in the cytoplasmic membrane (Carlsson et al. 1991).

2.1.5 Golverin

Gloverin is a basic antibacterial ~14 kDa weighing AMP isolated from *Hyalophora gloveri* pupal hemolymph (Nesa et al. 2020). Gloverin was reported to be isolated only from lepidopterans like *Helicoverpa armigera*, *T. ni*, *A. mylitta*, *M. sexta* and *Diatrea saccharialis* (Mackintosh et al. 1998; Lundstrom et al. 2002; Gandhe et al.

2006; Abdel-latief and Hilker 2008; Kawaoka et al. 2008; Mrinal and Nagaraju 2008; Xu et al. 2012).

Structure
The AMP gloverin consists of about 18.5% glycine-containing amino acid residue while lacking cysteine (Axén et al. 1997). The AMP demonstrates a slight analogy with attacin peptides and is active against both gram-positive and gram-negative bacteria (Kaneko et al. 2007). They are generated as pre-pro-proteins (pro-gloverin) and the molecule possesses a preserved N-terminal RXXR motif (Xu et al. 2012). The 22- to 26-amino-acid residues in the N-terminal pro-regions are mostly removed by furin-like enzymes before mature gloverin synthesis. The fact that the transgenic *Trichoplusia ni* pro-gloverin is as efficacious against *E. coli* as mature *H. gloveri* gloverin is indicative of the fact that few pro-gloverins may not require pro-regions removal as a requirement to become active (Yi et al. 2014). Furthermore, the in silico sequence analysis demonstrates that a motif homologous to the nuclear factor κB (NF-kB) binding site is present in each *BmGlov* gene upstream. A transgenic expression of *BmGlov* gene in Baculovirus reveals that the thriving of *E. coli* was strongly inhibited by all the *BmGlov1-4* genes. The Northern Blot and RT-PCR study analysis reported that *BmGlov* genes have been stimulated in the *E. coli* inoculated fat body but the stimulation was slighter with yeast *Candida albicans* (Yi et al. 2014). A detailed study of the genome of silkworms reveals the generation of four genes coding for Bmglv 1, 2, 3, and 4 (Yang et al. 2011). Among them, the ancestral gene is gloverin 1 while the remaining ones, namely gloverins 2–4, are generated by gloverin 1 gene duplication (Mrinal and Nagaraju 2008). Gloverin 1 was expressed in the gonads of *B. mori* larval pupae but not in the adults while gloverin 2–4 is only expressed in the adult silkworm gonads (Mrinal and Nagaraju 2008). This AMP is predominantly expressed in *M. sexta* larva testis (Xu et al. 2012). All these genes were stimulated by *B. subtilis*, *E. coli* and *Salmonella abortusequi* while the gloverin gene expression was lowered with *S. aureus* infection. The variations in bacterial cell wall structure and composition among the different strains of bacteria might be the reason for distinctive gloverins expression (Kaneko et al. 2007). *BmGloverin2* (*BmGlv2*) possesses synergistic antifungal action against *B. bassiana* (Lu et al. 2016). The gloverin is also capable of inhibiting the thriving of *E. coli* and *Pseudomonas putida* by elevating the membrane porosity causing an interruption in the ionic balance between cytoplasm and the external environment and finally causing lysis (Wang et al. 2018). Extensive study involving gloverin peptides reveals that these peptides have antimicrobial activity along with significant insect reproductive development activity (Yi et al. 2014).

Mode of Action
Gloverins from *B. mori* are effective against *E. coli* and contain mutant strains (Df21f2, D21, and D22) while gloverin from *T. ni* and *S. exigua* is effective against virus and *Flavobacterium* sp., respectively. Furthermore, it is ineffective against smooth LPS-containing *E. coli* strains (Yi et al. 2014). As per the experimental reports, the attachment of gloverins to the inner part and lipid A region of LPS is necessary for its action. Gloverins undergo a conformational alteration during

penetration into the hydrophilic LPS regions and interplay with the negatively charged lipid hydrophobic regions (Yi et al. 2013). *BmGlvs* bind to rough LPS causing a conformational shift from random coil to α-helix thought to be the primary reason for bacterial cell membrane pore generation (Yi et al. 2013). The attachment of gloverin to the microbial body is indispensable for its conformational alteration and antibacterial activity.

2.1.6 Lebocin

Lebocin was originally reported as proline-rich, O-glycosylated 32-residue-containing AMP present in *E. coli*-challenged *B. mori* hemolymph (Hara and Yamakawa 1995a, b).

Structure
The active peptide with 32 residues is located in proximity to the precursor protein C-terminus containing 179 residues according to a cDNA clone for *B. mori* lebocin (Chowdhury et al. 1995). The amino acid residue, namely, 15-Thr, is a critical constituent for antibacterial action (Islam et al. 2016). The basic structure and antibacterial features of this AMP predominantly resemble that of abaecin (41% similarity in the peptide chain) of honey bees which has 34 residues of the proline-rich peptide without glycosylation (Casteels et al. 1990). The family of lebocin comprises four protein-encoding genes (*Leb1*, *Leb2*, *Leb3*, and *Leb4*). Lebocins 1 and 2 are the two distinctive lebocin analogs having indistinguishable amino acid residues while differing in their sugar chain length on the threonine residues. Glycosylation involves the addition of various sugars like galactose, N-acetyl galactosamine on lebocin 1 and N-acetyl galactosamine on lebocin 2. The amino acid sequence of lebocin 3 demonstrated single amino acid substitution at 16-Leu and 15-Thr (Islam et al. 2016). cDNA encoding lebocin precursors have been isolated from lepidopterans like *M. sexta*, *Trichoplusiani*, *Pieris rapea*, *Pseudoplusia includens*, *H. viresecns* and *Antheraea peryni*. All the previously mentioned lebocin precursors along with *Bombyx mori* precursors are proline-rich peptides with around 4–6 prolines in the N-terminus of mature precursor protein. The *Bombyx mori* precursor proteins have additional 32-residue peptides along with 7 prolines close to the C-termini (Liu et al. 2000; Lavine et al. 2005; Bao et al. 2005; Rao et al. 2012). It has been very well-demonstrated that the upregulation of an erythroblast transformation-specific (Ets) protein family referred to as BmEts is necessary for the increase in the functioning of lebocin promoters. The action of BmEts involving lebocin promoter depends upon at least two kB elements and the proximal GGAA/T motif situated in the 5′-upstream. Furthermore, it broadly increases the *E. coli* or BmRelish1-d2-induced lebocin promoter initiation (Tanaka et al. 2012).

Mode of Action
In vitro analysis revealed that lebocin induces bacterial cell membrane disruption below low ionic concentrations (Hara and Yamakawa 1995a). Its particular mode of antibacterial action is still under study as it has reduced sensitivity and requires low

ionic conductivity. Lebocin 3 has an additional cecropin D as their association remarkably elevates the efficacy of the latter's antimicrobial action on *B. mori* (Ponnuvel and Yamakawa 2002).

2.1.7 Enbrocin

Enbocin is a 59-amino-acid residue containing an antimicrobial peptide isolated from *B. mori* (Kaneko et al. 2007). Cloning of cDNA from silkworms was done in 1998 to visualize its first gene expression.

Structure
The amino acid sequence of enbocin reveals similarity with the cecropin family although the antibacterial properties of enbocin differ from that of cecropin (Kim et al. 1998). A minimum of 2 enbocin genes are present in silkworms (Kaneko et al. 2007).

Mode of Action
The AMP demonstrates antibacterial activity against both the gram-positive and gram-negative bacteria. According to induction mechanism analysis, the expression of enbocin is stimulated by *E. coli* and happens predominantly in the fat body (Li et al. 2019) (Table 10.1).

2.2 AMPs from Non-mulberry Silkworm

Along with AMPs from mulberry silkworm, there are also AMPs isolated from the non-mulberry silkworms of the Saturniidae family like *Antheraea assamensis* (muga), *Antheraea pernyi* (temperate oak tasar), *Antheraea mylitta* (tropical oak tasar), *Antheraea yamamai* (Japanese oak tasar) and *Samia cynthia ricini* (eri). An antifungal peptide, gallerimycin is being identified from *S. cynthia ricini* fat body. A cDNA clone of the Scr-gallerimycin (AB366558) gene codes for gallerimycin protein with 74 amino acids weighing 6.21 kDa and having a pKa of 7.6 (Hashimoto et al. 2008). A lebocin-like gene gets stimulated in bacteria that infect the silkworm fat body. Its cDNA encodes for 162 amino acid residues having homology with the *B. mori* and *Trichoplusia ni* lebocin precursor proteins (Bao et al. 2005). The attacin A and B cDNA clones were identified from the bacteria-infected *S. cynthia ricini* fat body. Both these genes encoded 233 amino acid residues and had 98% similarity at the protein level while a 92% sequence similarity at the nucleotide level was visualized (Kishimoto et al. 2002). An attacin-like gene was also identified from *A. pernyi*, which was remarkably upregulated in the fat body upon *E. coli* attack (Li et al. 2020). A gloverin-like peptide reported from muga silkworm immunized with *C. albicans* with 9.052 kDa molecular mass is active against gram-negative bacteria (Nayak et al. 2018). A glycine-predominant peptide (GGGGGGHLVA) isolated from *A. mylitta* is found to be effective against MDR *E. coli* causing urinary tract

Table 10.1 Major families of AMPs from silkworms

AMP	Features	Structure	Mode of Action	Amino acid sequence	Microbes	MIC	LC	References
CecropinA1	Cationic	α-Helix	Forms continuous leaky pores on the plasma membrane	MNFVRILSFVFALVLAL GAVSAAPEP RWKLFKKIEKVGRNV RDGLIKAGPAI AVIGQAKSLGK	*B. subtilis*	2.5 μM		Yang et al. (2011)
					B. thuringiensis	2.5 μM		
					E. coli	2.5 μM		
					P. aeruginosa	2.5 μM		
					Ralstonia dolaanacearum	2.5 μM		
Cecropin B	Cationic	α-Helix	Generates pores, produces a carpet structure on the surface of the lipid bilayer and causes bacterial membrane dismantling	KWKVFKKIEKMGR NIRNGIVKAGPAIAVLG EAKAL-NH2	*B. megatherium*		1.7 μM	Gazit et al. (1994) and Steiner et al. (1981)
					E. coli		0.35 μM	
					M. luteus		>207 μM	
					P. aeruginosa		10 μM	
					S. marcescens		17.22 μM	

(continued)

Table 10.1 (continued)

AMP	Features	Structure	Mode of Action	Amino acid sequence	Microbes	MIC	LC	References
Cecropin B1	Cationic	α-Helix	Forms continuous leaky pores on the plasma membrane	KWKVFKKIEKMG RNIRNGIVKAGPK WKVFKKIEK-NH2	*P. fluorescens*		1.6 μM	Morishima et al. (1990) and Srisailam et al. (2001)
					Xanthomonas campestris		1.2 μM	
					Chromobacterium iodinum		0.85 μM	
					Agrobacterium tumifaciens		0.41 μM	
					Alcaligenes faecalis		0.49 μM	
					Achromobacter polymorph		1.8 μM	
					E. coli K12		0.38 μM	
					S. marcescens		0.67 μM	
					M. luteus		0.59 μM	
					S. aureus		>10 μM	
					Brevibacterium ammoniagenes		0.49 μM	
					Lactobacillus planetarum		0.62 μM	
					Arthrobacter simplex		0.46 μM	
					B. subtilis		3.6 μM	
					B. sphaerius		4.4 μM	

Cecropin B6	Cationic	α-Helix	Forms continuous leaky pores on the plasma membrane	MNFAKILSFVFALVLA LSMTSAAPEPRWKIFKKI EKMGRNIRDGIVKAGPAIE VLGSAKAIGK	*B. bombysepticus*	2.5 μM		Yang et al. (2011)
					B. subtilis	2.5 μM		
					B. thuringiensis	1.25 μM		
					B. thuringiensis subsp. galleriae	1.25 μM		
					E. coli, S. marcescens, P. aeruginosa	0.625 μM		
					R. dolaanacearum	1.25 μM		
Cecropin C	Cationic	α-Helix	Forms continuous leaky pores on the plasma membrane	RWKLFKKIEKVGRN VRDGLIKAGPAIAV IGQAKSL				Yang et al. (2011)
Cecropin D	Cationic	α-Helix	Forms continuous leaky pores on the plasma membrane	MKISKIFVFVFAIV FATASVSAAPGNFF KDLVSIVLDVSGS	*B. bombysepticus*	2.5 μM		Yang et al. (2011) and Tanaka et al. (2008)
					B. subtilis	2.5 μM		
					B. thuringiensis	1.25 μM		
					B. thuringiensis subsp. galleriae	2.5 μM		
					E. coli, P. aeruginosa, R. dolaanacearum	1.25 μM		
					S. marcescens	2.5 μM		

(continued)

Table 10.1 (continued)

AMP	Features	Structure	Mode of Action	Amino acid sequence	Microbes	MIC	LC	References
Cecropin E	Cationic	α-Helix	Forms continuous leaky pores on the plasma membrane	MNFSRALFYVFAVFLV CASVMAAPEPRWKIFKKI EKVGQNIRDGIIKA GPAVAVVGQAATIAHGK	*B. thuringiensis*	1.25 μM		Yang et al. (2011) and Tanaka et al. (2008)
					E. coli, S. marcescens, P. aeruginosa, R. dolaanacearum	2.5 μM		
CecXJ-37C	Cationic	α-Helix	Forms continuous leaky pores on the plasma membrane		*E. coli* ATCC25922	3.9 μM		Liu et al. (2017)
					P. aeruginosa ATCC 27853	3.9 μM		
					K. pneumoniae ATCC 700603	15.7 μM		
					S. aureus ATCC 25923	0.25 μM		
					S. aureus ATCC 29213	1 μg/mL		
					S. aureus ATCC 43300	1 μg/mL		
					B. subtilis ATCC 6633	1 μg/mL		
					S. epidermidis ATCC 12228	1 μg/mL		

CecXJ-37 N	Cationic	α-Helix	Forms continuous leaky pores on the plasma membrane		*E. coli* ATCC25922	1 μM		Liu et al. (2017)
					P. aeruginosa ATCC 27853	1 μM		
					K. pneumoniae ATCC 700603	7.8 μM		
					S. aureus ATCC 25923	0.25 μM		
					S. aureus ATCC 29213	1 μg/mL		
					S. aureus ATCC 43300	1 μg/mL		
					B. subtilis ATCC 6633	1 μg/mL		
					S. epidermidis ATCC 12228	1 μg/mL		
Defensins	Cationic hydrophilic peptide rich in cysteine	Cysteine equlibratedαβ motif	Disorganizesbacterial cell membrane; Generation of voltage-dependent anion-selective cell membrane	MAHQRKSLVIFIFLTV LVFVFALPRDATVFD NQHSEVAIEKSTSK IDSSDVKIPGRIWC EFEEATETAICQEHCL PKGYSYGICVSNTCSCI	*S. aureus*			Lehrer and Lu (2012), Kagan et al. (1990), and Lehrer et al. (1989)

(continued)

Table 10.1 (continued)

AMP	Features	Structure	Mode of Action	Amino acid sequence	Microbes	MIC	LC	References
Moricins	Basic amphipathic	α-Helix	Changes the permeability of the membrane as voltage-dependant pores are produced		*E. coli* JM 109;		0.31 μM	Hara and Yamakawa (1995a), Hemmi et al. (2002)
					Acinetobacter sp. *NISL* B-4653		0.27 μM	
					P. fluorescens IAM 1179		0.53 μM	
					P. aeruginosa IAM 15140		0.81 μM	
					B. subtilis IAM 1107		0.19 μM	
					B. megaterium IAM 1030		0.09 μM	
					B. cereus IFO 3457		0.38 μM	
					S. aureus ATCC 6538P		0.21 μM	
					S. aureus ATCC 6538Pa		0.22 μM	
					S. aureus IFO 3083		0.46 μM	
					S. xylosus IAM 1312		0.27 μM	
					S. epidermidis IFO 12993		0.18 μM	
					S. pyogenes ATCC 21547		0.25 μM	

Moricin like A	Basic amphipathic	α-Helix	Changes in the permeability of the membrane as voltage-dependent pores are produced	AKIPIKAIKTVGKAVG KGLRAINIAS TANDVFNFLKPKKRKA				Hara and Yamakawa (1995b)
Moricin like B	Basic amphipathic	α-Helix	Changes in the permeability of the membrane as voltage-dependent pores are produced	MKVFSFFCVVLAMLVLI MGGTSAAPEPKGIGKIIRK GGKVIKHGLTAIGVGAAG HEVYQDSKNSG				Hara and Yamakawa (1995b)
Bmmor	Basicamphipathic	α-Helix	Changes in the permeability of the membrane as voltage-dependent pores are produced	MNILKLFFVFIVA MSLVSCSTAAPAK IPIKAIKTVGK AVGKGLRAINIASTA NDVFNFLKPKKRKH	*S. aureus, B. subtilis*	0.625 μM		Hara and Yamakawa (1995b)
					B. bombysepticus, B. thuringiensis, B. thuringiensis subsp. galleriae, E. coli, P. aeruginosa, R. dolaanacearum	1.25 μM		
					S. marcescens	0.625 μM		

(continued)

Table 10.1 (continued)

AMP	Features	Structure	Mode of Action	Amino acid sequence	Microbes	MIC	LC	References
Attacin	Glycine-rich	Random coil structure in aqueous solution	Changes the cell membrane permeability; Impedes bacterial plasma membrane protein synthesis	MSKSVALLLLCACLASGR HVPTRARRQAGSFTVNS DG TSGAALKVPLTGNDK NVLSAIGSADFNDRHKLS AASAGLALDNVNGHG LSLTGTRIPGFGEQLGV AGKVNLFHNN NHDLSA KAFAIRNSPSAFPNAPN FNTLG GGLDYMFKQKV GASLSAAHSDVINRNDY SAGGKLNLFRSPSSS LD FNAGFKKFDTPFYRSS WEPNVGFSFSKFF	Gram-positive and gram-negative bacteria			Buonocore et al. (2021)
Gloverin 1	Glycine-richacidic to neutral (pI: 5.5–7.2)	Random coil	Pore formation capacity	MYSKVLLSAALLVCVN AQVSMPPGYAEKYPIT SQFSRSV RHPRDIHDF IVTWDREMGGGKVFG TLGESDQGLFGKG GYNREFFNDDRG KLTGQAYGTRVLGPG GDST SYGGRLDWANE NAKAAIDLNRQIGGSAGIE ASAS GVWDLGKNTHLS AGGVVSKEFG HRRPDVGLQAQITHEW	*B. thuringiensis*	1.4 µM		Yang et al. (2011)
					B. thuringiensis subsp. galleriae	1.6 µM		
					E. coli; P. aeruginosa; R. dolaanacearum	1.4 µM		
					S. marcescens	1.2 µM		
					X. campestris	1.6 µM		
Gloverin 2	Glycine-rich acidic to neutral (pI: 5.5–7.2)	Random coil	Pore formation capacity	MNTNLFYIFATTLVCVNA EVYGPSDYAEDYSISGQSS RRHPRDVTWDKQMGGGKV FGTLGQNDDGLFGKAGY NR EIFNDDRGKLTGQAYG TRVLGPGGDSTNYGGRL DWANKNA QATIDLNRQIG GRSGMTASGSGVWDL DKNTHFSAGGMVSKEF GHKRPDVGLQAEIRHDW	*B. thuringiensis; B. thuringiensis subsp. galleriae; E. coli; S. marcescens; R. dolaanacearum*	1.6 µM		Yang et al. (2011)
					X. campestris; P. aeruginosa	1.8 µM		

Gloverin 3	Glycine- rich acidic to neutral (pI: 5.5–7.2)	Random coil	Pore formation capacity	MNSKLLFFIATVLVCV NAEVYRSPDYEEEYPIRG LFSKRHPRDVTWDTKMGGGKV FGTLGQNDDGLFGKAGYNR EIFNDDRGQLTGQAYGTR VLGPGGDSTNYGGRLDWAN KNAQAAIDINRQIGGRS GMTASGSGVWDL DKNTHISAGGMVSKE FGHRRPDVGLQAEIRHEW	*B. thuringiensis; B. thuringiensis subsp.; Galleriae; S. marcescens; R. dolaanacearum*	1.6 μM		Yang et al. (2011)
					E. coli	1.4 μM		
					P. aeruginosa; X. campestris	1.8 μM		
Gloverin 4	Glycine-rich acidic to neutral (pI: 5.5–7.2)	Random coil	Pore formation capacity	MNSKLLYFFATVLVCVN AEVYWEDEEGYPVSG QFSKRHPRDVT WDKQVGGGKVFGTLGQND DGLFGKAGYNREI FNDDRGKLTGQAYGTRV LGPAGDSTNYGGRLDWAN KNAEAAIDINRQIGGR SGMTATGSGVWDLDK NTRLSAGGMVSKEFGHR RPDVGVQAEFRHDW	*B. thuringiensis; E. coli; S. marcescens; R. dolaanacearum*	1.4 μM		Yang et al. (2011)
					B. thuringiensis subsp. galleriae; P. Aeruginosa; X. campestris	1.6 μM		
Lebocin 1	Proline-rich		Form leakage in plasma membrane lipid bilayer and demonstrated synergistic action on cecropin D	DLRFLYPRGKLPVPTPPP FNPKPIYIDMGNRY	Bacteria			Tanaka et al. (2012)
Lebocin 2	Proline-rich			DLRFLYPRGKLPVPT PPPFNPKPIYIDMGNRY	Bacteria			
Lebocin 3	Proline-rich			DLRFLYPRGKLPVPTPPPF NPKPIYIDMGNRY	Bacteria			
Lebocin 4	Proline-rich			DLRFLYPRGKLPVPTPPPF NPKPIYIDMGNRY	Bacteria			

infections (Dutta et al. 2017). Acetylated and non-acetylated forms of a tri-peptide AMP, NH2-Gln-Ala-Lys-COOH (QAK) isolated from the hemolymph obtained from immunized *A. mylitta* manifest antibacterial action against *E. coli* and *S. aureus* (Kim et al. 2012). A cecropin-like peptide identified from the Japanese oak silkworm, *A. yamamai* demonstrated antimicrobial properties with MIC values ranging between 1 and 2 μg/mL against gram-negative bacteria like *E. coli*, *P. aeruginosa*, and *K. pneumoniae*, values between 64 and 128 μg/mL against gram-positive bacteria like *M. luteus*, *S. aureus*, and *Enterococcus faecalis*, and values like 64 μg/mL for fungi like *C. albicans* (Kim et al. 2012).

3 Therapeutic Applications of AMPs

The resistance developed by the pathogenic microbes against conventional antibiotics is a big blow to global health, and as such with the elapse of time, diverse AMPs with natural origin have become potent candidates against different infectious pathogens like bacteria, fungi, and viruses. These AMPs are capable of rupturing the microbial membrane causing bacterial cell lysis as well as inhibiting the progression of microbial drug resistance. Reports have proven the point that AMP demonstrates synergistic properties with conventional antibiotics, thereby generating an option for amalgamated therapy. Defensin, cecropins, drosocin, attacins, diptericins, ponericins, metchnikowins, and melittin are some of the currently isolated insect AMP classes, and there is a high possibility for the discovery of new AMPs. As of now, 33 peptides are going through clinical trials while 43% of the 77 AMPs are presently in the preclinical stage (Makwana et al. 2023). Daptomycin and ovitavancin are antibacterial lipopeptides that the FDA has currently approved that are used for bacterial skin infections (Sangma et al. 2023). The insect AMPs can block multiple drug-resistant bacteria like *Escherichia coli*, *Streptococcus sanguinus*, *Klebsiella poneumoniae* and *Staphylococcus aureus*. Many fungal strains like *Aspergillus* sp., *Bptrytis* sp., *Cryptococcus* sp., and *Fusarium* sp. are reported to be vulnerable to insect AMPs (Manniello et al. 2021). Some studies have demonstrated their therapeutic capabilities in vivo as a mono-intraperitoneal dose of *H. cecropia* Cecs A and B and *Danaus plexipibus* DAN2 lowered mortality in rodent models with *E. coli* infection (Giacometti et al. 2001). Furthermore, mice administered with two fold higher doses of DAN2 than their most effectual antibacterial concentration did not exhibit any abnormality in their behavior or morphology indicating that these peptides do not exhibit any toxic effects after acute treatments. AMPs are considered to be potent and better as compared to traditional antibiotics for the following reasons:

- Possess antibacterial, antifungal, antiviral and antitumor properties.
- Have varying modes of action.
- Neutralize endotoxin and have considerably low MIC.
- Exhibit activity against biofilms to which conventional antibiotics are resistant.

- Low in vitro toxicity.
- Capable of maintaining their antimicrobial property in high ionic serum concentration in in vitro experiments.

AMPs are currently generated in pharma industries as L and D amino acids that aid them to resist proteolysis (Nagarajan et al. 2018). Certain AMPs are also able to kill the malignant cells abruptly and have a higher absorption rate along with decreased side effects as compared to traditional antibiotics. However, it has a loophole in the form of toxicity, permeability, stability, and route of drug administration (Hutchings et al. 2017). Regarding AMPs as one of the accurate approaches as an alternative to conventional drugs, some AMPs have even shown the capability of immune induction by stimulating processes like chemokines, cell trafficking, and cell multiplication. Presently, there are many insect-derived AMPs active against pathogens causing human diseases, namely, *Citrobacter freundii*, *Euterobacter aerogenes*, *K. pneumonia*, and *Legionella dumoffei*. Few fungal species in these insect antimicrobial peptides that are effective are *Pichia pastoris*, *Trichoderma viridae* and *Aspergillus fumigatus* (Altincicek and Vilcinskas 2007; Ursic-Bedoya et al. 2011; Vonkavaara et al. 2013; Poppel et al. 2015). Reports have shown that there are bacterial resistance development cases reported against non-Cec AMPs (Jozefiak and Engberg 2017; Lewies et al. 2019). Furthermore, a novel study involving the comparison of bacterial mutation rate induction of conventional antibiotics and cationic AMPs including Cec A from *H. cecropia* revealed that none of the AMPs could increase the *E. coli* mutation rates that might be attributed to their inability to stimulate the bacterial stress pathways linked with new intrinsic resistance factors development, which is associated with mutation rate augmentation promoting DNA mutagenesis in *E. coli* (Rodríguez-Rojas et al. 2014). There have been reports of hybrid CAM peptide production that at sublethal concentrations did not affect the MIC of the peptide. Furthermore, the CAM peptide retained its efficacy against both the laboratory and MDR *P. aeruginosa* strains while comparable continuous exposure to sublethal doses of gentamicin or LL-37 elevated their effective MICs on the same bacterial strains (Geitani et al. 2019). This indicates that Cec and their analogs do not easily induce antimicrobial resistance. A novel outlook earning popularity is the application of AMPs as adjuvants in association with traditional antibiotics (Grassi et al. 2017). Treatment procedures involving simultaneous administration of AMPs and antibiotics can direct synergistic therapeutic antimicrobial efficacy and decrease the administration dosage, thereby lowering toxicity after effects as in vitro study reports performed on Cec-analogs like CAM or Cec-LL37 hybrids demonstrated varying synergistic action according to the antibiotic types and bacterial species (Dosler and Karaaslan 2014; Wei et al. 2016; Garbacz et al. 2017; Geitani et al. 2019). Furthermore, there are many reasons for the hindrance of the activity of AMPs in humans as peptides suffer from proteolytic degradation by natural proteases like trypsin present in the digestive tract that digest *B. mori* Cec B and Cec XJ variants (Liu et al. 2017; Romoli et al. 2019). These peptides might also be the target for human elastase and can also be deactivated by pathogen proteases like *Pseudomonas* elastase and *S. aureus* V8 protease (Ji et al.

2014). However, the responsiveness of AMPs to proteolysis can be limited by the replacement of specific residues like a four-tryptophan-substitution variant (CAM-W) did not exhibit responsiveness to the aforementioned enzymatic proteolysis. Furthermore, the resistance to degradation can be accomplished by the replacement of natural L-residues with their respective D-enantiomers keeping the biocidal activity intact. Even the application of nanoparticles to generate new AMP formulation is essential to increase peptide stability, bioavailability, and efficacy at the effector organ (Kumar et al. 2018). Another setback in the generation of these AMPs as antimicrobial drugs is the enormous cost of production as chemical manufacture, transgenic technologies for the heterologous cell systems generation, cell-free expression systems and transgenic organisms are the various techniques adopted for this purpose which are very expensive. The entire process involved in the therapeutic application of AMPs is demonstrated in Fig. 10.1 (Table 10.2).

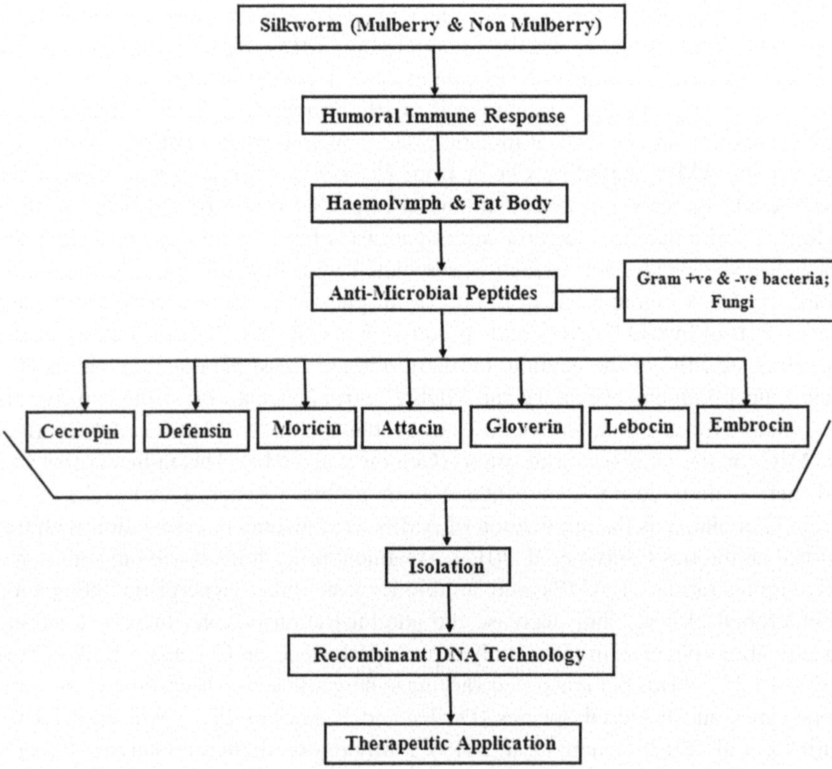

Fig. 10.1 Schematic steps showing the process of therapeutic application of silkworm AMPs

Table 10.2 Table depicting some insect-based AMPs with therapeutic properties

Sl. no.	Class of AMP	Source	Application	References
1	Cecropin A	*G. mellonella*	Destruction of planktonic and sessile biofilm generating Uropathogenic *Escherichia coli* cells	Kalsy et al. (2020)
2	Cec B	*H. cecropia*	Amelioration of the survivance of mice with malignant ascites	Moore et al. (1994)
3	Cec XJ	*B. mori*	Antitumor action in murine models	Xia et al. (2018)
4	Cec XJ	*B. mori*	Inhibition of the multiplication of human gastric cancer cells and stimulation of cell lysis both in vivo and in vitro	Xia et al. (2018)
5	Cec XJ	*B. mori*	Stimulation of cytoskeletal damage in oesophageal cancer cells	Xia et al. (2014)
6	Defensin	Diptera, Hymenoptera, Coleoptera, Lepidoptera, Hemiptera, Isoptera, Odonata	Effective against *M. luteus*, *A. viridans*, *B. megaterium*, *B. subtilis*, *B. thuringiensis* and *S. aureus*	Jozefiak and Engberg (2017)
7	CSαβ-DLP2 and DLP4	*H. illucens*	Antibacterial and immunomodulatory action of insect defensins-DLP2 and DLP4 against multidrug-resistant *S. aureus*	Li et al. (2017)
8	Tenecin 1 and 3	*T. molitor*	Inhibition of multiplication of leukemia cells	Jozefiak and Engberg (2017)
9	Hecate	*A mellifer*	Effective against Herplex simplex virus 1 (HSV-1)	Kruse and Kristensen (2008)
10	Alloferons	*C. vicina*	Effective against human influenza viruses A and B	Chernysh et al. (2002)

4 Significance of AMPs

The manner in which the MDR microbial strains are spreading is a cause of lethal infections. Antimicrobial drug resistance occurs due to the faster pace at which genetic mutations in microbes are occurring making them resistant to formerly effective antimicrobial agents. This property of the bacterial strains through which a rapid transfer of their genes occurs in turn implies that bacterial antibiotic resistance appears rapidly in the evolutionary course of bacterial development (Bulet and Stocklin 2005). Some common MDR bacteria like *Escherichia coli*, *Staphylococcus aureus*, *Klebsiella pneumonia*, *Streptococcus pneumonia*, *Neisseria gonorrhoeae*, and *Shigella* species have already acquired resistance against certain antibiotics like fluoroquinolones, cephalosporins, carbapenems, penicillin, and methicillin during evolution attributable to their genetic mutations. Even the rise of fungal infections is visualizable simultaneously in immunocompromised patients like HIV/AIDS patients, cancer patients with chemotherapy-stimulated neutropenia, and immunosuppressive therapy-receiving transplant recipients (Venkatesan et al. 2005). To fight the MDR menace, the involvement of insect AMPs has developed into a pivotal area of research for the innovation of naive biologically active, comparatively less toxic, and reduced-resistant antibiotics.

5 Conclusion

Although there are advantages of utilizing AMPs as a replacement to conventional antibiotics when it comes to the treatment involving the MDR strains of pathogens, still there are a lot of concerns regarding their production cost and proteolytic degradation inside the body. Nevertheless, the advancement of new technologies like nanotechnology will bring closure to these impediments. The application of these AMPs might allow the production of formulations effective against both gram-positive and gram-negative bacteria and also fungi. Although extrapolative research and studies are essential, scientific evidence substantiates the use of these AMPs as worthy tools to counter the worldwide outspread of MDR strains.

References

Abdel-latief M, Hilker M (2008) Innate immunity: eggs of Manduca sexta are able to respond to parasitism by Trichogramma evanescens. Insect Biochem Mol Biol 38:136–145

Altincicek B, Vilcinskas A (2007) Analysis of the immune-inducible transcriptome from microbial stress resistant, rat-tailed maggots of the drone fly Eristalis tenax. BMC Genomics 8:326

Armin S, Fallah F, Karimi A, Karbasiyan F, Alebouyeh M, Rafiei Tabatabaei S, Rajabnejad M, Ghanaie RM, Fahimzad SA, Abdollahi N et al (2023) Antibiotic susceptibility patterns for carbapenem-resistant Enterobacteriaceae. Int J Microbiol 2023:8920977

Axén A, Carlsson A, Engström A, Bennich H (1997) Gloverin an antibacterial protein from the immune hemolymph of Hyalophora pupae. Eur J Mol Biol Biochem 247:614–619

Bao Y, Yamano Y, Morishima I (2005) A novel lebocin-like gene from eri-silkworm, Samia Cynthia ricini, that does not encode the antibacterial peptide lebocin. Comp Biochem Mol Biol 140:127–131

Barbault F, Landon C, Guenneugues M, Meyer JP, Schott V, Dimarcq JL, Vovelle F (2003) Solution structure of Alo-3: a new knottintype antifungal peptide from the insect *Acrocinus longimanus*. Biochemistry 42:14434–14442

Barr SC, Rose D, Jaynes JM (1995) Activity of lytic peptides against intracellular Trypanosoma cruzi amastigotes in vitro and parasitemias in mice. J Parasitol 81:974–978

Buhroo ZI, Bhat MA, Bali GK, Kamili AS, Ganai NA (2018) Antimicrobial peptides from insects with special reference to silkworm Bombyx mori L.: a review. J Entomol Zool Stud 6:752–759

Bulet P, Stocklin R (2005) Insect antimicrobial peptides: structures, properties and gene regulation. Protein Pept Lett 12:3–11

Buonocore F, Fausto AM, Della Pelle G, Roncevic T, Gerdol M, Picchietti S (2021) Attacins: a promising class of insect antimicrobial peptides. Antibiotics 10:212

Carlsson A, Engstrom P, Palva ET, Bennich H (1991) Attacin, an antibacterial protein from Hyalophora cecropia, inhibits synthesis of outer membrane proteins in Escherichia coli by interfering with omp gene transcription. Infect Immun 59:3040–3045

Casteels P, Ampe C, Riviere L, Van Damme J, Elicone C, Fleming M, Jacobs F, Tempst P (1990) Isolation and characterization of abaecin, a major antibacterial response peptide in the honeybee (Apis mellifera). Eur J Biochem 187:381–386

Chalk R, Townson H, Ham PJ (1995) Brugia pahangi: the effects of cecropins on microfilariae in vitro and in Aedes aegypti. Exp Parasitol 80:401–406

Cheng T, Zhao P, Liu C, Xu P, Gao Z, Xia Q, Xiang Z (2006) Structures, regulatory regions, and inductive expression patterns of antimicrobial peptide genes in the silkworm Bombyx mori. Genomics 87:356–365

Chernysh S, Kim SI, Bekker G, Pleskach VA, Filatova NA, Anikin VB et al (2002) Antiviral and antitumor peptides from insects. Proc Natl Acad Sci USA 99:12628–12632

Chowdhury S, Taniai K, Hara S, Kadonookuda K, Kato Y, Yamamoto M, Xu J, Choi SK, Debnath NC, Choi HK, Miyanoshita A, Sugiyama M, Asaoka A, Yamakawa M (1995) cDNA cloning and gene expression of lebocin, a novel member of antibacterial peptides from the silkworm, Bombyx mori. Biochem Biophys Res Commun 214:271–278

Cociancich S, Bulet P, Hetru C, Hoffmann JA (1994) The inducible antibacterial peptides of insects. Parasitol Today 10:132–139

Dai H, Rayaprolu S, Gong Y, Huang R, Prakash O, Jiang H (2008) Solution structure, antibacterial activity, and expression profile of Manduca sexta moricin. J Pept Sci 14:855–863

Devi L (1991) Consensus sequence for processing of peptide precursors at monobasic sites. FEBS Lett 280:189–194

Dosler S, Karaaslan E (2014) Inhibition and destruction of Pseudomonas aeruginosa biofilms by antibiotics and antimicrobial peptides. Peptides 62:32–37

Dutta SR, Gauri SS, Ghosh T, Halder SK, DasMohapatra PK, Mondal KC, Ghosh AK (2017) Elucidation of structural and functional integration of a novel antimicrobial peptide from Antheraea mylitta. Bioorg Med Chem Lett 27:1686–1692

Efimova SS, Schagina LV, Ostroumova OS (2014) Channel-forming activity of cecropins in lipid bilayers: Effect of agents modifying the membrane dipole potential. Langmuir 30:7884–7892

Fujiwara S, Imai J, Fujiwara M, Yaeshima T, Kawashima T, Kobayashi K (1990) A potent antibacterial protein in royal jelly. Purification and determination of the primary structure of royalisin. J Biol Chem 265:11333–11337

Gabay JE (1994) Ubiquitous natural science. Science 264:373–374

Gandhe AS, Arunkumar KP, John SH, Nagaraju J (2006) Analysis of bacteria-challenged wild silkmoth, Antheraea mylitta (lepidoptera) transcriptome reveals potential immune genes. BMC Genomics 7:1–10

Garbacz K, Kamysz W, Piechowicz L (2017) Activity of antimicrobial peptides, alone or combined with conventional antibiotics, against Staphylococcus aureus isolated from the airways of cystic fibrosis patients. Virulence 8:94–100

Gazit E, Lee WJ, Brey PT, Shai Y (1994) Mode of action of the antibacterial cecropin B2: a spectrofluorometric study. Biochemistry 33:10681–10692

Geitani R, Moubareck CA, Touqui L, Sarkis DK (2019) Cationic antimicrobial peptides: alternatives and/or adjuvants to antibiotics active against methicillin-resistant Staphylococcus aureus and multidrug-resistant Pseudomonas aeruginosa. BMC Microbiol 19:54

Giacometti A, Cirioni O, Ghiselli R, Viticchi C, Mocchegiani F, Riva A, Saba V, Scalise G (2001) Effect of mono-dose intraperitoneal cecropins in experimental septic shock. Crit Care Med 29:1666–1669

Grassi L, Maisetta G, Esin S, Batoni G (2017) Combination strategies to enhance the efficacy of antimicrobial peptides against bacterial biofilms. Front Microbiol 8:2409

Gregory SM, Cavenaugh A, Journigan V, Pokorny A, Almeida PF (2008) A quantitative model for the all-or-none permeabilization of phospholipid vesicles by the antimicrobial peptide cecropin A. Biophys J 94:1667–1680. https://doi.org/10.1529/biophysj.107.118760
Gunne H, Hellers M, Steiner H (1990) Structure of preproattacin and its processing in insect cells infected with a recombinant baculovirus. Eur J Biochem 187:699–703
Guo C, Huang Y, Zheng H, Tang L, He J, Xiao L, Liu D, Jiang H (2012) Secretion and activity of antimicrobial peptide cecropin D expressed in Pichia pastoris. Exp Ther Med 4:1063–1068
Hara S, Yamakawa M (1995a) A novel antibacterial peptide family isolated from the silkworm, Bombyx mori. Biochem J 310:651–656
Hara S, Yamakawa M (1995b) Moricin, a novel type of antibacterial peptide isolated from the silkworm, Bombyx mori (*). J Biol Chem 270:29923–29927
Hashimoto K, Yamano Y, Morishima I (2008) Cloning and expression of a gene encoding gallerimycin, a cysteine-rich antifungal peptide, from eri-silkworm, Samia cynthia ricini. Comp Biochem Physiol B Biochem Mol Biol 150:229–232
Hedengren M, Borge K, Hultmark D (2000) Expression and evolution of the drosophila attacin/diptericin gene family. Biochem Biophys Res Commun 279:574–581
Hemmi H, Ishibashi J, Hara S, Yamakawa M (2002) Solution structure of moricin, an antibacterial peptide, isolated from the silkworm Bombyx mori. FEBS Lett 518:33–38
Hu H, Wang C, Guo X, Li W, Wang Y, He Q (2013) Broad activity against porcine bacterial pathogens displayed by two insect antimicrobial peptides moricin and cecropin B. Mol Cells 35:106–114
Hultmark D, Steiner H, Rasmuson T, Boman HG (1980) Insect immunity. Purification and properties of three inducible bactericidal proteins from hemolymph of immunized pupae of Hyalophora cecropia. Eur J Biochem 106:7–16
Hultmark D, Engström A, Bennich H, Kapur R, Boman HG (1982) Insect immunity. Isolation and structure of cecropin D and four minor antibacterial components from cecropia pupae. Eur J Mol Biol Biochem 127:207–217
Hutchings CJ, Koglin M, Olson WC, Marshall FH (2017) Opportunities for therapeutic antibodies directed at G-protein-coupled receptors. Nat Rev Drug Discov 16:787–810
Islam JS, Bezbaruah S, Kalita J (2016) A review on antimicrobial peptides from Bombyx mori L and their application in plant and animal disease control. J Adv Biol Biotechnol 9:1–15
Janeway CA Jr, Medzhitov R (2002) Innate immune recognition. Annu Rev Immunol 20:197–216
Ji S, Li W, Zhang L, Zhang Y, Cao B (2014) Cecropin A–melittin mutant with improved proteolytic stability and enhanced antimicrobial activity against bacteria and fungi associated with gastroenteritis in vitro. Biochem Biophys Res Commun 451:650–655
Jozefiak A, Engberg RM (2017) Insect proteins as a potential source of antimicrobial peptides in livestock production. A review. J Anim Feed Sci 26:87–99
Kagan BL, Selsted ME, Ganz T, Lehrer RI (1990) Antimicrobial defensin peptides form voltage-dependent ion-permeable channels in planar lipid bilayer membranes. Proc Natl Acad Sci USA 87:210–214
Kalsy M, Tonk M, Hardt M, Dobrindt U, Zdybicka-Barabas A, Cytrynska M et al (2020) The insect antimicrobial peptide cecropin A disrupts uropathogenic Escherichia coli biofilms. NPJ Biofilms Microbiomes 6:1–8
Kaneko Y, Furukawa S, Tanaka H, Yamakawa M (2007) Expression of antimicrobial peptide genes encoding Enbocin and Gloverin isoforms in the silkworm, Bombyx mori. Biosci Biotechnol Biochem 71:2233–2241
Kawaoka S, Katsuma S, Daimon T, Isono R, Omuro N, Mita K, Shimada T (2008) Functional analysis of four Gloverin-like genes in thesilkworm, Bombyx mori. Arch Insect Biochem Physiol 67:87–96
Kim SH, Park BS, Yun FY, Je YH, Woo SD, Kang SW, Kim KY, Kang SK (1998) Cloning and expression of a novel gene encoding a new antibacterial peptide from silkworm, Bombyx mori. Biochem Biophys Res Commun 246:388–392

Kim SR, Goo TW, Choi KH, Park SW, Kim SW, Hwang JS, Kang SW (2012) Isolation and purification of a cecropin-like antimicrobial peptide from the Japanese oak silkworm, Antheraea yamamai. J Sericultural Entomol Sci 50:145–149

Kishimoto K, Fujimoto S, Matsumoto K, Yamano Y, Morishima I (2002) Protein purification, cDNA cloning and gene expression of attacin, an antibacterial protein, from eri-silkworm, Samia cynthia ricini. Insect Biochem Mol Biol 32:881–887

Kockum K, Faye I, Hofsten PV, Lee JY, Xanthopoulos KG, Boman HG (1984) Insect immunity, isolation and sequence of two cDNA clones corresponding to acidic and basic attacins from Hyalophora cecropia. EMBO J 3:2071–2075

Kumar P, Kizhakkedathu JN, Straus SK (2018) Antimicrobial peptides: diversity, mechanism of action and strategies to improve the activity and biocompatibility in vivo. Biomol Ther 8:4

Kruse T, Kristensen HH (2008) Using antimicrobial host defense peptides as anti-infective and immunomodulatory agents. Expert Rev Anti Infect Ther 6:887–895.

Lambert J, Keppi E, Dimarcq JL, Wicker C, Reichhart JM, Dunbar B, Lepages P, Van Dorsselaer A, Hoffmann J, Forthergill J, Hoffmann D (1989) Insect immunity: isolation from immune blood of the dipteran Phormia terranovae of two insect antibacterial peptides with sequence homology to rabbit lung macrophage bactericidal peptides. Proc Natl Acad Sci USA 86:262–266

Lavine MD, Chen G, Strand MR (2005) Immune challenge differentially affects transcript abundance of three antimicrobial peptides in hemocytes from the moth Pseudoplusia includens. Insect Biochem Mol Biol 35:1335–1346

Lehrer RI, Lu W (2012) Alpha-Defensins in human innate immunity. Immunol Rev 245:84–112

Lehrer RI, Barton A, Daher KA, Harwig SS, Ganz T, Selsted ME (1989) Interaction of human defensins with Escherichia coli. Mechanism of bactericidal activity. J Clin Invest 84:553–561

Lemaitre B, Hoffmann J (2007) The host defense of Drosophila melanogaster. Annu Rev Immunol 25:697–743

Lewies A, Du Plessis LH, Wentzel JF (2019) Antimicrobial peptides: the Achilles' heel of antibiotic resistance? Probiotics Antimicrob Proteins 11:370–381

Li Z, Mao R, Teng D, Hao Y, Chen H, Wang X, Wang X et al (2017) Antibacterial and immunomodulatory activities of insect defensins-DLP2 and DLP4 against multidrug-resistant Staphylococcus aureus. Sci Rep 7:1–6

Li Z, Ma Y, Liu X, Li Y, Dai F (2019) Antimicrobial peptides in silkworm. Anim Biol 69:391–410

Li Q, Zhang J, Sun Y, Wang L, Qian C, Wei G, Zhu B, Liu C (2020) Immunological function of the antibacterial peptide Attacin-like in the Chinese oak silkworm, Antheraea pernyi. Protein Pept Lett 27:953–961

Liu G, Kang D, Steiner H (2000) Trichoplusianilebocin, an inducible immune gene with a downstream insertion element. Biochem Biophys Res Commun 269:803–807

Liu X, Guo C, Huang Y, Zhang X, Chen Y (2015) Inhibition of porcine reproductive and respiratory syndrome virus by Cecropin D in vitro. Infect Genet Evol 34:7–16

Liu D, Liu J, Li J, Xia L, Yang J, Sun S, Ma J, Zhang F (2017) A potential food biopreservative, CecXJ-37N, non-covalently intercalates into the nucleotides of bacterial genomic DNA beyond membrane attack. Food Chem 217:576–584

Lu D, Geng T, Hou C, Huang Y, Qin G, Guo X (2016) Bombyx mori cecropin A has a high antifungal activity to entomopathogenic fungus Beauveria bassiana. Gene 583:29–35

Lundstrom A, Liu G, Kang D, Berzins K, Steiner H (2002) Trichoplusia ni gloverin, an inducible immune gene encoding an antibacterial insect protein. Insect Biochem Mol Biol 32:795–801

Mackintosh JA, Gooley AA, Karuso PH, Beattie AJ, Jardine DR, Veal DA (1998) A gloverin-like antibacterial protein is synthesized in *Helicoverpa armigera* following bacterial challenge. Dev Comp Immunol 22:387–399

Makwana P, Kamisi R, Katsuhiko I, Bindu S (2023) Diversity of antimicrobial peptides in silkworm. Life 13:1161

Manniello MD, Moretta A, Salvia R, Scieuzo C, Lucchetti D, Vogel H, Sgambato A, Falabella P (2021) Insect antimicrobial peptides: potential weapons to counteract the antibiotic resistance. Cell Mol Life Sci 78:4259–4282

Matsuyama K, Natori S (1998) Purification of three antibacterial proteins from the culture medium of NIH-Sape-4, an embryonic cell line of Sarcophega peregrina. J Biol Chem 263:17112-17116

Moore AJ, Devine DA, Bibby MC (1994) Preliminary experimental anticancer activity of cecropins. Pept Res 7:265–269

Morishima I, Suginaka S, Ueno T, Hirano H (1990) Isolation and structure of cecropins, inducible antibacterial peptides, from the silkworm. Bombyx Mori Comp Biochem Physiol B 95:551–554

Mrinal N, Nagaraju J (2008) Intron loss is associated with gain of function in the evolution of the gloverin family of antibacterial genes in Bombyx mori. J Biol Chem 283:23376–23387

Nagarajan K, Marimuthu SK, Palanisamy S, Subbiah L (2018) Peptide therapeutics versus superbugs: highlight on current research and advancements. Int J Pept Res Ther 24:19–33

Nayak T, Mandal SM, Neog K, Ghosh AK (2018) Characterization of a gloverin-like antimicrobial peptide isolated from muga silkworm, Antheraea assamensis. Int J Pept Res Ther 24:337–346

Nesa J, Sadat A, Buccini DF, Kati A, Manda AK, Franco OL (2020) Antimicrobial peptides from Bombyx mori: a splendid immune defense response in silkworms. RSC Adv 10:512–523

Oizumi Y, Hemmi H, Minami M, Asaoka A, Yamakawa M (2005) Isolation, gene expression and solution structure of a novel moricin analogue, antibacterial peptide from a lepidopteran insect, Spodoptera litura. Biochim Biophys Acta-Proteins Proteom 1752:83–92

Ouyang L, Xu X, Freed S, Gao Y, Yu J, Wang S, Ju W, Zhang Y, Jin F (2015) Cecropins from Plutella xylostella and their interaction with Metarhizium anisopliae. PLoS One 10:e0142451

Park SI, An HS, Chang BS, Yoe SM (2013) Expression, cDNA cloning, and characterization of the antibacterial peptide cecropin D from Agrius convolvuli. Anim Cells Syst 17:23–30

Pokorny A, Kilelee EM, Wu D, Almeida PF (2008) The activity of the amphipathic peptide delta-lysin correlates with phospholipid acyl chain structure and bilayer elastic properties. Biophys J 95:4748–4755

Ponnuvel KM, Yamakawa M (2002) Immune responses against bacterial infection in Bombyx mori and regulation of host gene expression. Curr Sci 83:447–454

Poppel AK, Vogel H, Wiesner J, Vilcinskas A (2015) Antimicrobial peptides expressed in medicinal maggots of the blow fly Lucilia sericata show combinatorial activity against bacteria. Antimicrob Agents Chemother 59:2508–2514

Rao XJ, Xu XX, Yu XQ (2012) Functional analysis of two lebocin-related proteins from Manduca sexta. Insect Biochem Mol Biol 42:231–239

Rodríguez-Rojas A, Makarova O, Rol J (2014) Antimicrobials, stress and mutagenesis. PLoS Pathog 10:e1004445

Romoli O, Mukherjee S, Mohid SA, Dutta A, Montali A, Franzolin E, Brady D, Zito F, Bergantino E, Rampazzo C et al (2019) Enhanced silkworm Cecropin B antimicrobial activity against Pseudomonas aeruginosa from single amino acid variation. ACS Infect Dis 5:1200–1213

Sangma JA, Baroowa BP, Singh YR, Choudhury P (2023) Antimicrobial peptides: principal defense constituents of silk worms – a review 85:808–819. https://doi.org/10.55446/IJE.2023.951

Sato H, Feix JB (2006) Peptide–membrane interactions and mechanisms of membrane destruction by amphipathic-helical antimicrobial peptides. Biochim Biophys Acta 1758:1245–1256

Saviane A, Romoli O, Bozzato A, Freddi G, Cappelletti C, Rosini E, Cappellozza S, Tettamanti G, Sandrelli F (2018) Intrinsic antimicrobial properties of silk spun by genetically modified silkworm strains. Transgenic Res 27:87–101

Silvestro L, Axelsen PH (2000) Membrane-induced folding of cecropin A. Biophys J 79:1465–1477

Song KJ, Park BR, Kim SY, Park KS (2010) Molecular characterization of anionic defensin-like peptide in immune response of silkworm, Bombyx mori L. (Lepidoptera). Genes Genome 32:447–453

Srisailam S, Kumar TK, Arunkumar AI, Leung KW, Yu C, Chen HM (2001) Crumpled structure of the custom hydrophobic lytic peptide cecropin B3. Eur J Mol Biol Biochem 268:4278–4284

Steiner H, Hultmark D, Engstrom A, Bennich H, Boman HG (1981) Sequence and specificity of two antibacterial proteins involved in insect immunity. Nature 292:246–248

Sugiyama M, Kuniyoshi H, Kotani E, Taniai K, Kadono-Okuda K, Kato Y et al (1995) Characterization of a Bombyx mori cDNA encoding a novel member of the attacin family of insect antibacterial proteins. Insect Biochem Mol Biol 25:385–392

Suttmann H, Retz M, Paulsen F, Harder J, Zwergel U, Kamradt J, Wullich B, Unteregger G, Stöckle M, Lehmann J (2008) Antimicrobial peptides of the Cecropin-family show potent antitumor activity against bladder cancer cells. BMC Urol 8:5

Tanaka H, Ishibashi J, Fujita K, Nakajima Y, Sagisaka A, Tomimoto K, Suzuki N, Yoshiyama M, Kaneko Y, Iwasaki T, Sunagawa T, Yamaji K, Asaoka A, Mita K, Yamakawa M (2008) A genome-wide analysis of genes and gene families involved in innate immunity of Bombyx mori. Insect Biochem Mol Biol 38:1087–1110. https://doi.org/10.1016/j.ibmb.2008.09.001

Tanaka H, Sagikasa A, Fujita K, Furukuwa S, Ishibashi J, Yamakawa M (2012) BmEts upregulates promoter activity of lebocin in Bombyx mori. Insect Biochem Mol Biol 42:474–481

Ursic-Bedoya R, Buchhop J, Joy JB, Durvasula R, Lowenberger C (2011) Prolixicin: a novel antimicrobial peptide isolated from Rhodnius prolixus with differential activity against bacteria and Trypanosoma cruzi. Insect Mol Biol 20:775–786

Venkatesan P, Perfect JR, Myers SA (2005) Evaluation and management of fungal infections in immune compromised patients. Dermatol Ther 18:44–57

Vonkavaara M, Pavel ST, Holzl K, Nordfelth R, Sjostedt A, Stoven S (2013) Francisella is sensitive to insect antimicrobial peptides. J Innate Immun 5:50–59

Wang D, Qiu X, Wang H, Qiao K, Wang K (2010) Reduced fitness associated with spinosad resistance in Helicoverpa armigera. Phytoparasitica 38:103–110

Wang Q, Guo P, Wang Z, Liu H, Zhang Y, Jiang S, Han W, Xia Q, Zhao P (2018) Antibacterial mechanism of gloverin2 from silkworm, Bombyx mori. Int J Mol Sci 19:2275

Wang M, Zhou Z, Li S, Zhu W, Hu X (2021) Identification and characterization of antimicrobial peptides from butterflies: an integrated bioinformatics and experimental study. Front Microbiol 12:720381

Wei XB, Wu RJ, Si DY, Liao XD, Zhang LL, Zhang RJ (2016) Novel hybrid peptide cecropin A (1–8)-LL37 (17–30) with potential antibacterial activity. Int J Mol Sci 17:983

Wen H, Lan X, Cheng T, He N, Shiomi K, Kajiura Z, Zhou Z, Xia Q, Xiang Z, Nakagaki M (2009) Sequence structure and expression pattern of a novel anionic defensin-like gene from silkworm (Bombyx mori). Mol Biol Rep 36:711–716

Xia L, Zhang F, Liu Z, Ma J, Yang J (2013) Expression and characterization of cecropinXJ, a bioactive antimicrobial peptide from Bombyx mori (Bombycidae, Lepidoptera) in Escherichia coli. Exp Ther Med 5:1745–1751. https://doi.org/10.3892/etm.2013.1056

Xia L, Wu Y, Kang S, Ma J, Yang J, Zhang F (2014) CecropinXJ, a silkworm antimicrobial peptide, induces cytoskeleton disruption in esophageal carcinoma cells. Acta Biochim Biophys Sin 46:867–876

Xia LJ, Wu YL, Ma J, Zhang FC (2018) Therapeutic effects of antimicrobial peptide on malignant ascites in a mouse model. Mol Med Rep 17:6245–6252

Xu XX, Zhong X, Yi HY, Yu XQ (2012) Manduca sexta gloverin binds microbial components and is active against bacteria and fungi. Dev Comp Immunol 38:275–284

Yang W, Cheng T, Ye M, Deng X, Yi H, Huang Y, Tan X, Han D, Wang B, Xiang Z, Cao Y, Xia Q (2011) Functional divergence among silkworm antimicrobial peptide paralogs by the activities of recombinant proteins and the induced expression profiles. PLoS One 6:e18109

Yi HY, Deng XJ, Yang WY, Zhou CZ, Cao Y, Yu XQ (2013) Gloverins of the silkworm Bombyx mori: structural and binding properties and activities. Insect Biochem Mol Biol 43:612–625

Yi HY, Chowdhury M, Huang YD, Yu XQ (2014) Insect antimicrobial peptides and their applications. Appl Microbiol Biotechnol 98:5807–5822

Yun J, Lee DG (2016) Cecropin A-induced apoptosis is regulated by ion balance and glutathione antioxidant system in Candida albicans. IUBMB Life 68:652–662

Zhao WY, Dong BR, Zhou Y (2005) In vitro antimicrobial activity of defensins from rabbit neutrophils against Pseudomonas aeruginosa and its multiple-drug-resistance strains. Sichuan da xue xue bao. Yi xue ban 36:83–85

Chapter 11
Harnessing Antimicrobial Peptides: A Key Defense Against Emerging Respiratory Viral Diseases

Kushal Chaudhary, Sharmistha Chakravarty, Pforeni Jane Sapruna, and Subhasree Dutta

1 Introduction

The global apprehension regarding emerging and reemerging viruses is a pressing issue. Delving into the roots of antimicrobial peptides as plausible tools against human viruses stands as a crucial focus in addressing this challenge. Viral epidemics are becoming increasingly common, with the COVID-19 pandemic alone claiming a minimum of 6.5 million lives worldwide. Viral infections in humans have led to millions of deaths globally, with prevalent diseases, including HIV/AIDS, hepatitis, influenza, the common cold, and more (Evans 2013). Human immunodeficiency virus (HIV) causing AIDS was discovered in 1983 (Gallo and Montagnier 2003). HIV belongs to the retroviridae family containing linear, single-stranded RNA genome (Rambaut et al. 2004). In 2018, according to the World Health Organization (WHO), approximately 770,000 individuals succumbed to HIV-related causes, while around 37.9 million people were living with HIV by the year's end. The discovery of the hepatitis C virus (HCV) in 1989 revealed a linear, positive-sense, single-stranded RNA genome. Belonging to the Flaviviridae family, HCV is categorized into seven major genotypes, with genotypes 1 and 2 being particularly prevalent in human infections, as highlighted by Wilkins et al. in 2010. HCV transmission occurs through various means, including the transfusion of infected blood and blood products, needle sharing, and certain sexual practices, among others. This understanding of the virus's modes of transmission is crucial for implementing effective preventive measures and interventions to curb the spread of hepatitis C.

K. Chaudhary (✉) · P. J. Sapruna · S. Dutta
Department of Zoology, Bodoland University, Kokrajhar, Assam, India

S. Chakravarty
Department of Zoology, Barnagar College, Sorbhog, Assam, India

P. Baindara, S. M. Mandal (eds.), *Evolution of Antimicrobial Peptides*,
https://doi.org/10.1007/978-3-031-67515-7_11

As per the World Health Organization (WHO), approximately 71 million individuals were estimated to be living with chronic hepatitis C virus (HCV) infection, and in the year 2016, nearly 399,000 people succumbed to HCV-related complications globally. Meanwhile, Ebola virus disease (EVD) stands as a lethal ailment marked by sporadic outbreaks that predominantly impact humans and other primates, including monkeys, gorillas, and chimpanzees. The discovery of the Ebola virus traces back to 1976, near the Ebola River in the Democratic Republic of Congo (DRC). This virus is believed to be zoonotic, originating in animals and transmitted to humans primarily through direct contact with the blood, body fluids, and tissues of infected animals. Within the human population, the virus spreads through contact with infected body fluids, gaining entry through broken skin or mucous membranes in the eyes, nose, or mouth. Understanding these modes of transmission is pivotal for devising effective strategies to manage and prevent the spread of Ebola virus disease (EVD). EVD outbreak from 2014 to 2016 in West Africa, documented by Breman et al. in 2016, marked the most extensive and complex episode since the virus's discovery, surpassing all previous outbreaks in both cases and fatalities. Shifting to the realm of herpes viruses, the herpes simplex virus (HSV) belongs to the *Simplexvirus* genus within the Herpesviridae family, characterized by a linear double-stranded DNA genome. HSV is divided into two types: HSV-1 causes prevalent oral herpes infections, while HSV-2, primarily transmitted sexually, leads to genital herpes infections. According to the WHO in 2017, a substantial global burden is evident, with 3.7 billion people under 50 living with HSV-1 and 417 million individuals aged 15–49 grappling with HSV-2 infections. These statistics underscore the significant impact of herpes simplex viruses on global public health.

In all multicellular organisms, antimicrobial peptides (AMPs) form a diverse category of naturally occurring molecules that function as a frontline defense mechanism. These proteins display a broad spectrum of activity, effectively eliminating bacteria, yeasts, fungi, viruses, and even cancer cells. Insects and plants primarily employ AMPs as natural antibiotics to defend against potential pathogenic microbes, whereas microbes produce AMPs to safeguard their environmental habitat. In higher eukaryotic organisms, AMPs, also known as "host defense peptides" (HFPs), "alarmins," and even "defensins," underscore their supplementary immunomodulatory roles. The diverse functions of AMPs are specific to the type of peptide, encompassing various cytokine and growth-factor-like effects that play a pivotal role in preserving the regular balance of the immune system. It is important to understand that when AMPs are expressed inappropriately, they can sometimes cause autoimmune diseases. This emphasizes the significance of comprehending these molecules and their complex activities. Apart from their function in various organisms, AMPs are integral components of the immune systems in plants. In the early 1940s, Stuart and Harris showcased the antimicrobial properties of a crystalline protein isolated from wheat flour against certain human pathogenic bacteria and yeast (Stuart and Harris 1942).

A comprehensive investigation into 156 cationic antiviral peptides (AVPs) unveiled their antiviral activity and modes of action against a diverse array of viruses. Twenty-one different viruses were reported to be influenced by these AVPs. The majority of studies focused on the impact of AVPs against human immunodeficiency virus (HIV), with 73 peptides demonstrating antiviral activity (Masuda et al. 1992; Nakashima et al. 1992). Additionally, 47 peptides exhibited activity against herpes viruses (Zannella et al. 2022; Lehrer et al. 1985), 27 peptides were effective against hepatitis C virus (HCV) (Darke et al. 1999), 11 demonstrated activity against human coronaviruses (Chianese et al. 2022), and 9 displayed antiviral activity against the influenza virus (Zannella et al. 2022).

Other viruses studied included dengue virus (DENV) with eight active peptides (Costin et al. 2010), Zika virus (ZIKV) with eight active peptides (Ji et al. 2019), vaccinia virus with eight active peptides (Howell et al. 2004), human papillomavirus (HPV) with six active peptides (Skeate et al. 2020), Ebola virus with five active peptides (Yu et al. 2020), measles virus with three active peptides (Chianese et al. 2022), Junin virus with two active peptides (Albiol Matanic and Castilla 2004), enterovirus 71 (EV71) with two active peptides (Yu et al. 2021), hepatitis B virus (HBV) with two active peptides (Zhao et al. 2012), human rhinovirus (HRV) with two active peptides (Zhao et al. 2020), Japanese encephalitis virus (JEV) with two active peptides (Jia et al. 2021), BK virus with one active peptide (Dugan et al. 2008), adenovirus with one active peptide (Bastian and Schäfer 2001), Aichi virus with one active peptide (Vilas Boas et al. 2017), human cytomegalovirus (HCMV) with one active peptide (Boffert et al. 2020), and Venezuelan equine encephalitis virus (VEEV) with one active peptide (Ahmed et al. 2019b). This extensive exploration underscores the potential of cationic antimicrobial peptides across a wide spectrum of viral infections.

2 Brief History of AMPs

Alexander Fleming first recognized the presence of a soluble antimicrobial substance produced by humans about 90 years ago. The first human antimicrobial protein, lysozyme, was identified by Alexander Fleming in 1922 from nasal mucus. However, this discovery was somewhat overshadowed in 1928 when Fleming found penicillin, leading to the initiation of the "golden age of antibiotics" in the 1940s. This era shifted attention away from the therapeutic potential of natural AMPs, including lysozyme.

In the 1940s, Fleming, along with others, successfully developed the therapeutic use of penicillin, earning him a share of the 1945 Nobel Prize in Medicine. Unfortunately, this success led to a diminished interest in exploring the therapeutic potential of natural AMPs during that time. In the 1960s, as multidrug-resistant microbial pathogens became apparent, there was a revival of interest in AMPs as host defense molecules. The challenges posed by antibiotic resistance prompted a renewed focus on the probable use of antimicrobial peptides in addressing infectious diseases. A diverse array of life forms has been found to harbor AMPs, with the investigation beginning in plants. Notable discoveries include brombinin in frogs and lactoferrin from milk during the 1960s. A significant breakthrough took place in 1981 when Hans Boman's study showed the induction of powerful AMPs in the cecropia silk moth after bacterial injection, marking the identification of "cecropins" as the first major report of α-helical AMPs. In 1987, Zasloff and his team discovered and studied special AMPs called "magainins" in African clawed frogs. Then, in the mid-1990s, Hoffmann's group made an important finding. They showed that when they removed the ability of fruit flies to make AMPs through genetics, the flies became vulnerable to large fungal infections. This was significant evidence highlighting the essential role of AMPs in protecting insects from harmful invaders.

Moreover, the application of gramicidin, derived from *Bacillus brevis*, not only rapidly disinfected pathogenic bacteria but also effectively suppressed infection in wounds on guinea pig skin. This demonstrated the therapeutic potential of gramicidin for clinical use, as reported by Gause and Brazhnicova in 1944. Likely, the first AMPs for which primary structures were characterized, gramicidins were also the initial antibiotics to be commercially manufactured. Today, gramicidin is accessible as an over-the-counter antibiotic.

Scientific interest surged with the 1994 discovery of AMPs in mammalian skin. Mice lacking a specific type of these peptides underscored their crucial role in mammalian host defense. According to its updated version (October 2021), the antimicrobial peptide database (APD3)1 has reported a total of 3283 AMPs (Wang et al. 2016). The commonalities of various types of AMPs are: that all AMPs are almost cationic (with an average net charge of 3.32) and the number of their residues of amino acids ranges between 10 and 60 (with an average of 33.26; Malkoski et al. 2001).

Nowadays, the antibiotic era has been threatened due to the alarming rate of multidrug resistance (MDR; Barra et al. 2020; Tuttobene et al. 2021). Hence, to address the persistent global issues associated with MDR-related infectious diseases, there is a pressing need to develop new antimicrobial strategies and agents (Ballus et al. 2015; Faya et al. 2018). These peptides can be considered one of the promising strategies for MDR problems since they have broad-spectrum activities, multimodal properties, and minimal resistance generation compared to conventional antibiotics (Deslouches et al. 2020; Tan et al. 2021). Recently, many attempts have been made to develop and identify synthetic AMPs as well (Fjell et al. 2012). Table 11.1 presents a list of antiviral AMPs with associated mechanisms and their associated diseases.

3 Mechanism of Action of Antiviral AMPs

The mechanism of action of antiviral AMPs can be categorized as follows.

Table 11.1 List of antiviral AMPs with associated mechanisms and diseases

AMP	Mechanism	Viral target
Defensins	Interaction with viral envelope (Maleki et al. 2021) Interaction with virus target cells or interference with various stages of the virus life cycle (Ding et al. 2009)	Respiratory syncytial virus (RSV) (Kota et al. 2008) Human immunodeficiency virus (HIV) (Wu et al. 2005) Human papilllomavirus (HPV) Human simplex virus (HSV) (Wang et al. 2013)
Cathelicidins	Disrupting viral capsid Stimulating defensive mechanisms of host cells (Choi and Mookherjee 2012)	Hepatitis C virus (HCV) (Matsumura et al. 2016) Respiratory syncytial virus (RSV) (Harcourt et al. 2016) Influenza A virus (IAV) (Tripathi et al. 2015) Dengue virus (DENV) (Alagarasu et al. 2017)
Transferrins	Inhibiting by interaction with viral structural polypeptide (Pietrantoni et al. 2003)	Hepatitis C virus (HCV) (Tahmoorespur et al. 2020) Dengue virus (DENV) (Chen et al. 2017) Marek's disease virus (MDV) (Giansanti et al. 2005)
Melitttin	Targeting the membrane and causing pore formation and disruption of the membrane (Takahashi et al. 2013)	Influenza A virus (IAV) (Wachinger et al. 1992) Human immunodeficiency virus (HIV) and COVID-19 (Yang et al. 2020)
Hepcidin	Iron regulation in the host cell restricting the availability of iron to invading cells	Influenza A virus (IAV) (Rodriguez et al. 2014) Human immunodeficiency virus 1 (HIV-1) Hepatitis C virus (HCV) Hepatitis C virus (HCV) (Armitage et al. 2014)

3.1 *Disruption of Membrane*

Peptides interact with the viral envelope, destabilizing it and inducing pore formation. Mellitin, for instance, binds to the viral coating and creates pores within the lipid layers (Chen and Lu 2020). Cathelicidins, another peptide, interact with the membrane and undergo assembly to create channels, thereby disrupting the membrane structure (Zeth and Sancho-Vaello 2017). Additionally, Mucropin-M1 has demonstrated a virucidal effect on SARS-CoV, MERS-CoV, and influenza H5N1 viruses by binding to the virus and disrupting its envelope (Li et al. 2011).

3.2 Targeting the Envelope

Certain peptides have been observed to interact with the membrane, thereby inhibiting the fusion activity of the virus. For example, bovine lactoferrin binds to the subunit A2 of influenza hemagglutinin, disrupting fusion (Superti et al. 2019). Similarly, EK1, a synthetic peptide, disrupts viral fusion entry into host cells by interacting with the spike protein on the membrane and inhibiting the HR1 domain (Xia et al. 2020).

3.3 Targeting the Viral Replication Machinery

Protegrin-1 disrupts DENV replication by binding to the viral protease (Rothan et al. 2012). Alpha defensin 1, a peptide consisting of residues 18–45, inhibits influenza A virus replication and protein synthesis (Salvatore et al. 2007). Additionally, Mucropin-M1, a cationic peptide, reduces the expression of replication factors, thus inhibiting HBV replication (Zhao et al. 2012).

3.4 Targeting and Shielding Host Receptor

Certain peptides exert their effects by interacting with receptors, thereby inhibiting their binding with host cells. For instance, the peptide NK-lysin exhibits significant antiviral activity against Spring Viremia of Carp Virus (SVCV) by preventing viral particle binding to host cells (Falco et al. 2019). Peptides can also directly bind to host receptors, shielding them from viral recognition. An example is lectin-like human α-defensin 5 (HD5), which competitively blocks angiotensin-converting enzyme 2 (ACE2) receptors in host cells (Wang et al. 2020). ACE2 receptors serve as primary receptors for coronavirus. Figure 11.1 illustrates the mechanism of action of antiviral peptides (AVPs).

4 AMPs Against SARS-COV-2

In humans, the well-studied classes of antimicrobial peptides (AMPs) include defensins (both α and β) and cathelicidins, notably LL-37 (Doss et al. 2010; Ahmed et al. 2019a, b). Defensins are vital components of the innate immune system, playing a crucial role as part of the initial defense against infections. Their efficacy against various viruses, such as HIV, herpes simplex virus (HSV), influenza, and SARS-CoV, has been extensively established (Kudryashova et al. 2021).

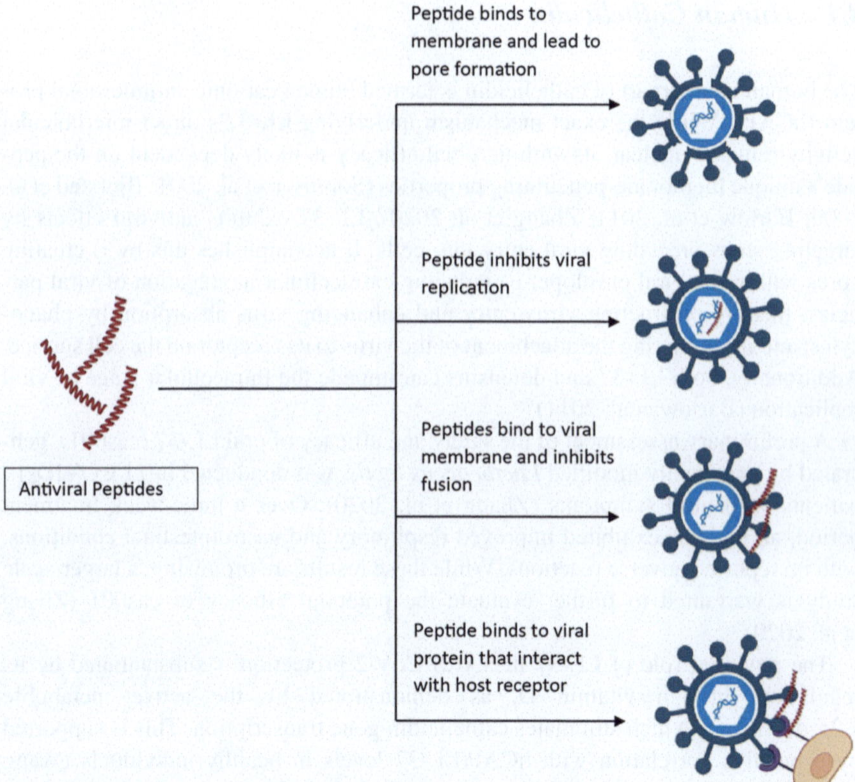

Fig. 11.1 Mechanism of action of antiviral peptides

The connection between SARS-CoV-2 infection and defensins remains unclear. Some researchers have investigated the expression of defensin genes in the oral cavity during COVID-19 infection (Idris et al. 2020). Findings revealed that defensin genes are expressed in the nasopharyngeal/oropharyngeal cavity and are significantly downregulated in SARS-CoV-2-infected patients compared to negative controls. Specifically, defensin beta 4A, 4B, 106B, 107B, and 103A and defensin alpha 1B were notably downregulated. The potential impairment of innate immunity in SARS-CoV-2 infection is indicated by the downregulation of defensin genes, possibly playing a role in disease progression. Hence, the elevation in defensin gene expression and the application of defensin peptides could offer intriguing therapeutic possibilities (Idris et al. 2020; Solanki et al. 2021).

4.1 Human Cathelicidin

The human counterpart of cathelicidin is termed human cationic antimicrobial protein (hCAP). While the exact mechanism underlying LL-37's direct microbicidal activity remains unclear, its antimicrobial efficacy is likely dependent on the peptide's unique membrane-penetrating properties (Sørensen et al. 2001; Björstad et al. 2009; Barlow et al. 2011; Zhang et al. 2020). LL-37 exhibits antiviral effects by targeting steps preceding viral entry into cells. It accomplishes this by i) creating pores within the viral envelope; ii) inducing extracellular aggregation of viral particles, thereby obstructing virus entry and enhancing virus absorption by phagocytes; and iii) hindering the attachment of the virus to its receptor on the cell surface. Additionally, both LL-37 and defensins can impede the intracellular stage of viral replication (Barlow et al. 2011).

A preliminary assessment of the safety and efficacy of oral LL-37 (cas001), generated by genetically modified *Lactococcus lactis*, was conducted in 11 COVID-19 patients with mild symptoms (Zhang et al. 2020). Over a three-week treatment period, all patients exhibited improved respiratory and gastrointestinal conditions, with no reported adverse reactions. While these results are promising, a larger-scale study is warranted to further evaluate the potential efficacy of cas001 (Zhang et al. 2020).

The potential role of LL-37 in SARS-CoV-2 protection is substantiated by its regulation through vitamin D, as demonstrated by the active metabolite 1,25-dihydroxy, which stimulates cathelicidin gene transcription. This is supported by a positive correlation with hCAP/LL-37 levels in healthy individuals (Wang et al. 2004; Dixon et al. 2012). Both vitamin D and LL-37 play crucial roles in lung immunity and the response to respiratory diseases (Tecle et al. 2010). There is evidence suggesting that blood levels and the presence of vitamin D in the body may be a risk factor for SARS-CoV-2 or COVID-19 disease (Kaufman et al. 2020; Arvinte et al. 2020; Maghbooli et al. 2020; Radujkovic et al. 2020). Consequently, a deficiency in vitamin D could lead to reduced levels of LL-37 and a compromised antimicrobial response to SARS-CoV-2 (Crane-Godreau et al. 2020).

Found in diverse mucosal secretions, lactoferrin (LF) is an iron-binding glycoprotein and an AMP that plays a vital role in the innate immune system. Possessing a broad spectrum of antimicrobial activity, LF demonstrates effectiveness against bacteria, fungi, and various viruses, both DNA and RNA based (Vorland 1999). Its antiviral properties extend to inhibiting the replication of diverse human and animal viruses. Notably, LF enhances host immunity against viral infections by preventing virus entry into host cells. This preventive action occurs through mechanisms such as blocking cellular receptors or directly binding to virus particles. Research suggests that a component of breast milk's antiviral properties can be credited to lactoferrin's (LF's) direct antiviral activity, given LF's presence in breast milk (van der Strate et al. 2001). Numerous studies underscore LF's significant role in combating various viruses, including HCV, HSV, HIV, polio, and the rotavirus.

4.2 *Defensins Against Coronavirus*

The application of human defensins as antiviral agents has been substantiated through various animal models, shedding light on their effectiveness and diverse mechanisms against viral pathogens (Memariani and Memariani 2020). Defensins exhibit both direct and indirect actions, and their direct antiviral and immunomodulatory potential has been investigated in the context of coronaviruses such as MERS-CoV and SARS-CoV.

4.2.1 Mouse β Defensins 4–Derived P9

P9, derived from murine β defensin 4, has been identified to bind to the MERS-CoV S2 subunit and maintain co-localization with the virus in host cells (Mustafa et al. 2019). When situated within endosomes, P9's polycationic properties create a basic microenvironment, hindering the acidification of late endosomes. This prevention of endosomal acidification disrupts the pH-dependent activation of viral fusion proteins, consequently impeding the initiation of fusion between the viral and host endosomal membranes. This mechanism of action has shown significant efficacy, extensively inhibiting SARS-CoV, MERS-CoV, as well as a diverse group of influenza viruses, including H1N1, H3N2, H5N1, H7N7, and H7N9.

To assess its prophylactic and therapeutic effects, P9 was tested in a BALB/C mouse model of nonlethal SARS-CoV infection. The administration of a single dose and five doses of the peptide revealed substantial inhibition of viral infection in the treated mouse lungs at day 3 after infection, in comparison to the untreated mouse lungs (Wohlford-Lenane et al. 2009).

A recent demonstration revealed that P9R, another short peptide with a higher net positive charge than its parent peptide P9, exhibits more potent antiviral activity not only against SARS-CoV and MERS-CoV but also against SARS-CoV-2 (Wohlford-Lenane et al. 2009).

4.2.2 Human Intestinal α-Defensins 5

In the course of infection, the SARS-CoV-2 protein S assumes a crucial role in recognizing receptors and facilitating cell membrane fusion (Zhang et al. 2021). Within the S1 subunit, the receptor-binding domain (RBD), identified by Scudiero in 2020, is responsible for recognizing and binding to the host receptor, angiotensin-converting enzyme 2 (ACE2). Investigating the potential to prevent virus-induced mortality, researchers are exploring therapeutic interventions targeting the interactions between SARS-CoV-2 S and ACE2 (Wohlford-Lenane et al. 2009; Zhao et al. 2020).

Human intestinal α-defensin 5 (HD5), a natural lectin-like peptide secreted by Paneth cells, inhibits the crucial protein–protein interactions between SARS-CoV-2

S1 and ACE2 necessary for viral invasion into host cells (Luan et al. 2022). HD5 binds to ACE2, competitively masking several sites in the ligand-binding domain essential for SARS-CoV S/ACE2 interactions. Consequently, HD5 inhibits the binding of the SARS-CoV-2 S1 subunit and the entry of spike protein pseudovirions into enterocytes (Venkataraman et al. 2009). The high-affinity binding between HD5 and the ligand-binding domain of ACE2, facilitated by multiple hydrogen bonds, serves to protect host cells from viral recognition and infection (Luan et al. 2022).

Elevated levels of DEFα5 were observed by Luan et al. (2022) in cells expressing ACE2, specifically in small intestine enterocyte progenitor cells and small intestine enterocytes. This suggests that defensin alpha 5 (DEFA5) in the human intestine may decrease the entry of SARS-CoV-2, providing a partial explanation for the lower incidence of diarrhea in COVID-19 patients.

4.2.3 Human Beta Defensin 2

Human beta defensin 2 (HBD2) is expressed throughout the respiratory epithelium, spanning from the oral cavity to the lungs, and plays a pivotal role in combating respiratory infections (Lehrer et al. 2012). The dysregulated expression of HBD2 in the respiratory epithelium has been linked to the pathogenesis of several respiratory diseases, encompassing asthma, pulmonary fibrosis, pneumonia, tuberculosis, and rhinitis (Diamond et al. 2008; Lehrer et al. 2012; Ooi et al. 2015). In the context of human respiratory syncytial virus (RSV) infection, HBD2 demonstrates the ability to inhibit viral entry by destabilizing and disintegrating the viral envelope (Rivas-Santiago et al. 2005).

HBD2 may play crucial immunomodulatory roles during coronavirus infections; for instance, in a mouse model, HBD2 conjugated to the MERS receptor-binding domain (RBD) demonstrated the ability to stimulate the production of more targeted protective antibodies to RBD compared to RBD alone. In a study by Zhang et al. (2021), the potential of HBD2 as an inhibitory factor against SARS-CoV-2 was investigated. The researchers observed that HBD2 could bind to the receptor-binding motif (RBM) of the RBD of SARS-CoV-2, which is responsible for linking to the ACE2 receptor. Through biochemical and biophysical analyses, it was confirmed that HBD2 not only binds to the RBD but also prevents it from binding to ACE2. Importantly, the study demonstrated that HBD2 inhibits the SARS-CoV-2 spike from invading ACE2-expressing human cells.

4.2.4 Rhesus θ Defensin 1

Rhesus theta defensin 1 (RTD1), a cyclic AMP, demonstrated significant preventive antiviral efficacy in a severe SARS-CoV lung disease mouse model (Barros et al. 2021). In this study, BALB/C mice infected with a mouse-adapted SARS-CoV strain and treated with two intranasal doses of RTD1 exhibited 100% survival, contrasting with approximately 75% mortality in untreated mice (Barros et al. 2021). RTD1 was suggested to function as an immunomodulatory effector molecule, inducing a straightforward pro-inflammatory cytokine response to eliminate SARS-CoV.

Theta defensin peptides, like RTD1, are naturally produced in old-world monkeys and orangutans. In humans, these peptides are expressed as pseudogenes due to the presence of an early stop codon (Ghorbani et al. 2020). Synthetic theta defensin peptides, known as retrocyclins, with sequences corresponding to those encoded within the human pseudogene, exhibit activity against HIV-1, influenza A virus, and herpes simplex viruses (HSVs) (Spinello et al. 2020). It has been shown that aminoglycoside antibiotics can induce termination codon read-through of retrocyclin cDNA (Rivas-Santiago et al. 2005). Considering the demonstrated efficacy of theta defensin peptides against SARS-CoV lung disease and other viruses, along with the potential of aminoglycosides to stimulate the production of these antimicrobial peptides, could prove beneficial in the context of SARS-CoV-2 (Ghorbani et al. 2020; Spinello et al. 2020).

Recent research has explored novel approaches for treating SARS-CoV-2, with Baindara et al. (2023b) conducting a study on the potential use of biosurfactants in this context (Baindara et al. 2023a). In their investigation, the researchers focused on a biosurfactant derived from a probiotic bacterial strain, *Bacillus clausii* TS, isolating and purifying a surfactin-like lipopeptide. The purified lipopeptide demonstrated notable efficacy in both binding to and inhibiting the SARS-CoV-2 spike (S1) protein. This study highlights a promising avenue for potential therapeutic interventions against SARS-CoV-2, leveraging the unique properties of biosurfactants for their inhibitory effects on viral components.

In 2023, Baindara et al. conducted a comprehensive study on host defense peptides (HDPs), specifically defensins and cathelicidins, focusing on their role in combating respiratory infections (Baindara et al. 2023b). The researchers examined all HDPs secreted in the respiratory tract and investigated the regulation of their expression during respiratory disorders. Additionally, Chowdhury et al. (2021) explored the antiviral activity of Lipopeptide D12 (LPD 12), a lipopeptide, against COVID-19. The study involved the docking of seven different lipopeptides, including tsushimycin, daptomycin, surfactin, bacillomycin, iturin, SrfTE, and LPD 12, individually against the spike (S) glycoprotein of SARS-CoV-2. The findings indicated that LPD 12 exhibited the highest efficacy in protein binding with the S-glycoprotein, significantly disrupting its affinity for binding with angiotensin-converting enzyme 2 (ACE2), the crucial receptor facilitating disease development. Furthermore, in 2020, Souvik et al. investigated antiviral peptides targeting fusion proteins and their mechanism to combat the viral fusion process. The fragment-based drug design (FBDD) method was employed to identify a broad-spectrum agent targeting the conserved region of the fusion protein of SARS-CoV-2. These studies collectively contribute valuable insights into potential therapeutic strategies against respiratory infections, particularly in the context of COVID-19.

5 Antimicrobial Peptides as Antiviral Agents

Antimicrobial peptides (AMPs) exhibit characteristics that render them potent antiviral agents. The reasons why AMPs act as antiviral elements are as follows.

5.1 Interaction with Viral Proteins

Antimicrobial peptides (AMPs) function by interacting with distinct viral proteins, disrupting pivotal stages in the viral life cycle. These peptides can bind to viral envelopes or capsids, leading to structural destabilization and hindering entry into host cells. Additionally, AMPs interfere with fusion proteins, impeding membrane fusion during viral entry. Inside the host cell, they target viral enzymes linked to replication and transcription, impeding the production of viral components. This specific interaction extends to the inactivation of crucial viral enzymes. Moreover, AMPs have the potential to modulate host immune responses, augmenting overall antiviral defenses. The specificity exhibited in their protein interactions underscores their promise as potent antiviral agents, positioning AMPs as compelling candidates for therapeutic development against viral infections.

5.2 Immune Modulation

Antimicrobial peptides (AMPs) demonstrate a notable capacity to modulate the host immune response, primarily by orchestrating the activation of immune cells to enhance the efficient clearance of viral infections. These peptides play a pivotal role in stimulating various components of the immune system, such as macrophages and neutrophils, to bolster their antiviral activities. AMPs not only facilitate the recognition and elimination of virally infected cells but also contribute to heightened immune surveillance. Additionally, their potential to exhibit anti-inflammatory properties further amplifies their role in comprehensive control against viral infections. By mitigating excessive inflammation, AMPs help maintain immune homeostasis, preventing collateral tissue damage associated with uncontrolled inflammatory responses during viral infections. The dual ability of AMPs to activate immune defenses while curbing inflammation underscores their significance in providing a balanced and effective immune response against viral threats.

5.3 *Low Likelihood of Resistance Development*

The probability of resistance development against antimicrobial peptides (AMPs) remains low, primarily attributed to their diverse modes of action. Unlike conventional antimicrobial agents that often target specific pathways, AMPs engage in multifaceted interactions with microbial structures. This diversity includes membrane disruption, interference with cellular processes, and modulation of immune responses. The intricate nature of these interactions poses a challenge for microorganisms, including viruses, to develop resistance since doing so would require simultaneous adaptation to multiple mechanisms. Furthermore, the rapid killing ability of AMPs may limit the time available for microorganisms to undergo adaptive changes. The low likelihood of resistance emergence positions AMPs as promising candidates for sustained effectiveness in combating microbial infections, offering a valuable alternative in the face of escalating concerns about antibiotic resistance.

6 Sources of AMPs and Their Known Application Against SARS-CoV-2

6.1 *Human-Derived Antiviral Peptides*

AMPs are sourced from various mammalian tissues, including neutrophil granules, the intestinal tract, and mucosal membrane secretions. In mammals, AMPs primarily belong to the cathelicidin and defensin classes. Cathelicidins, found in various vertebrates like fish, birds, goats, pigs, rabbits, monkeys, horses, and humans, are commonly present in epithelial cells, macrophages, and neutrophils (Kościuczuk et al. 2012). Notable members include indolicidin and human cathelicidin LL-37 (Ladokhin et al. 1999), with LL-37 widely studied and commonly found in human neutrophil granules (Nizet et al. 2001; Nijnik and Hancock 2009).

Cathelicidins, also known as myeloid AMPs, were initially identified in the myeloid cells of mammalian bone marrow (Gombart et al. 2005). Human cathelicidin LL-37 exhibits effectiveness against various viruses, including influenza A virus (IAV), hepatitis C virus (HCV), respiratory syncytial virus (RSV), dengue virus (DENV), and Venezuelan equine encephalitis virus (VEEV) (Alagarasu et al. 2017; Ahmed et al. 2019a, b). Other human-derived cathelicidins with antiviral activity include GE-17, Bovine Antimicrobial Peptide (BMAP 18), protegrin 1, and Sheep Myeloid Antimicrobial Peptide (SMAP-29) (Sousa et al. 2017; He et al. 2018).

Defensins, small cationic peptides typically comprising 18–45 amino acids, were first discovered in human neutrophils. These peptides, derived from mast cells and tissues involved in host defense, exhibit potent antimicrobial activity, particularly against HIV-1 and human papillomavirus 16 (HPV 16) infections (Furci et al. 2012; Wiens and Smith 2017).

6.2 Amphibian-Derived Antiviral Peptides

AMPs sourced from frog skin exhibit potent antibacterial and antiviral properties. Produced by granular glands, these peptides contain hydrophobic and positively charged amino acids, with a majority belonging to the Pipidae family, including *Hymenochirus*, *Pseudhymenochirus*, *Pipa*, *Silurana*, and *Xenopus* (Nicolas and Mor 1995; Belaid et al. 2002; Rinaldi 2002). Noteworthy examples such as caerin 11, caevin 19, and maculatin 11, derived from amphibians, demonstrated significant inhibition of HIV infection in T cells, even at low concentrations. Additionally, they exhibited antiviral activity against murine leukemia virus (MuLV) and Junin virus (JV) (VanCompernolle et al. 2005).

Melittin, an amphipathic hexapeptide extracted from bee venom, has been identified as a therapeutic agent against various viruses, including HIV, herpes simplex virus (HSV), Junin virus (JV), vesicular stomatitis virus (VSV), and tobacco mosaic virus (TMV) (Memariani et al. 2020).

Two peptides, brevinin-2GHK (BR2GK) and fejerlectin, derived from the skin secretion of *Fejervarya limnocharis*, were investigated for their antiviral properties. Fejerlectin prevented HIV-1 entry into host cells by inhibiting Env-mediated membrane fusion, while BR2GK inhibited the early and middle stages of Zika virus (ZKV) (Xiong et al. 2021a, b). Another recent study highlighted the potential therapeutic use of AR-23 against various DNA and RNA viruses. AR-23 interacts with viral particles, leading to inhibition at early stages, and has demonstrated effectiveness against HSV-1, measles morbillivirus (MeV), human parainfluenza (HPIV-2), HCOV-229E, and SARS-CoV-2 (Chianese et al. 2022).

6.3 Plant-Derived Antiviral Peptides

Plant-derived AMPs are grouped into distinct families based on variations in amino acid sequences, featuring both positive and negative charges (Pelegrini et al. 2011). These AMPs fall into several classes, including lipid transfer proteins, thionins, defensins, chitin-binding proteins, cyclotides, and others (Nawrot et al. 2014). Demonstrating effectiveness in contradiction of a diverse range of viral infections, plant-derived AMPs have been found to combat viruses such as HSV, Japanese encephalitis virus (JEV), Middle East respiratory syndrome coronavirus (MERS-COV), HPV, influenza A virus (IAV), and infectious bronchitis virus (IBV) (Gordts et al. 2015; Derby et al. 2018).

Cyclotides, a sizable family of plant-derived disulfide-rich peptides, exhibit exceptional stability and feature a cyclic cystine knot (CCK) structure. Their stability makes them applicable in drug design, and they have demonstrated effectiveness against HIV and influenza virus (Craik and Du 2017). Griffithsin A, elatior lectin, chitinase (chi) like lectin from tamarind (TCLL), *Nicotiana tabacum* (NICTABA) lectin, cycloviolacin Y5, and VY1 stand out as particularly favorable antiviral lectins derived from plants, according to research (Derby et al. 2018).

Table 11.2 Various antimicrobial peptides (AMPs) exhibit activity against the influenza virus, each employing distinct mechanisms of action (Vallon-Eberhard et al. 2008)

The peptide inhibiting virus attachment and virus–cell membrane fusion	
Peptide	**Influenza serotype**
EB peptide	Broad spectrum
Derived EB peptide	Broad spectrum
Defensins	H1N1, H3N2
The peptide-disrupting viral envelop	
Peptide	**Influenza serotype**
LL-37	H1N1, H3N2
Mucroporin-M1	H5N1, H1N1
The peptides inhibiting viral replication	
Peptide	**Influenza serotype**
HNP-1	H3N2
Peptide 6	H1N1, H3N2

6.4 Insect-Derived Antiviral Peptides

Insects employ various defense mechanisms to combat microbial infections, encompassing both cellular and humoral immune responses. Localized melanization and coagulation, phagocytosis, and the secretion of AMPs into the hemolymph represent key components of insects' immune arsenal to counter infections (Ishii et al. 2010; Eleftherianos and Revenis 2010). Notable AMPs secreted by insects include cecropins, melittins, attacins, lysozymes, defensins, dipteracins, and drosomycin (Copley et al. 2007; Mojsoska et al. 2015).

C-lysozyme, an example of an insect-derived antimicrobial peptide, has demonstrated effectiveness against nucleopolyhedrosis (NPV) infection. Lysozymes, as enzymes, are believed to impact viruses by hydrolyzing the viral structural proteins, leading to the production of defective viruses (Chen et al. 2018). Additionally, gloverin proteins have been investigated for their activity against NPV. Positively charged, these small proteins, comprising 200 amino acids, owe their charge to the presence of arginine and lysine residues (Moreno-Habel et al. 2012). Table 11.2 presents various antimicrobial peptides (AMPs) targeting the influenza virus along with their respective mechanisms of action.

7 AMPs as Adjuvants or Carrier Proteins

Recent research underscores the potential of antimicrobial peptides (AMPs) as adjuvants when combined with vaccines or drugs, aiming to bolster their effectiveness. An ideal adjuvant candidate should be capable of eliciting adaptive immunity and augmenting the immune response. Moreover, there is growing interest in the development of peptide-based delivery systems, owing to their safe, straightforward, and economically viable manufacturing processes (Skwarczynski and Toth

2016). Another benefit of employing peptides is their stability and well-understood composition (Skwarczynski et al. 2020). For nearly a decade, alum was the sole adjuvant utilized in licensed human vaccines; however, it demonstrated limited effectiveness in stimulating Th1 immunity (Oleszycka et al. 2018). Antimicrobial peptides (AMPs) are increasingly recognized as promising therapeutic agents for managing infectious diseases. One notable development is the creation of a two-component adjuvant known as IC31, which comprises a synthetic antimicrobial peptide (KLK) and a DNA oligonucleotide (ODN1a) for vaccine delivery. KLK plays a vital role in transporting ODN1a, delivering antigens to specific antigen-presenting cells (APCs), and providing targeted immune stimulation. ODN1a, in turn, promotes a Th1 immune response (Agger 2016).

Numerous peptide-based vaccines are presently undergoing clinical trials, targeting disease-causing agents such as the influenza virus, hepatitis C virus, polymorphic malaria, and HIV (Malonis et al. 2019). In 2023, Roy and Baindara identified a defensin-like bacteriocin called LS25, which is produced probiotically by *Brevibacillus laterosporus* DSM25, as a potential adjuvant peptide. The utilization of immune simulations on LS25 has demonstrated the elicitation of both humoral and cellular responses. These responses manifest as an increase in the population of B cell and T cell memory cells, along with elevated levels of cytokines. These cytokines exhibit inhibitory effects on viral replication and bolster T cell–mediated immunity.

8 Conclusion and Recommendations

In conclusion, the continuous emergence of new viral strains poses a formidable challenge to the effective treatment of viral diseases, elevating the global risk of viral pandemics. Ongoing research efforts focused on understanding viruses and developing innovative treatments are crucial in addressing this pressing issue. Cationic peptides, characterized by positive charges, present a promising avenue for the development of new antiviral drugs. Their ability to interact with the negatively charged surfaces of viruses, leading to various antiviral effects, positions them as valuable candidates with broad-spectrum activity, low toxicity to host cells, and potential resistance limitation. This review emphasizes the significance of antimicrobial peptides (AMPs) in combating various viral diseases, with a specific focus on addressing SARS-CoV-2. While research continues to determine the specific effectiveness of AMPs against COVID-19, current evidence indicates their potential in managing respiratory diseases, offering hope for future antiviral interventions. The natural origin and broad-spectrum targeting capability of AMPs underscore their potential as valuable tools in the ongoing battle against viral infections.

References

Agger EM (2016) Novel adjuvant formulations for delivery of anti-tuberculosis vaccine candidates. Adv Drug Deliv Rev 102:73–82

Ahmed A, Siman-Tov G, Hall G, Bhalla N, Narayanan A (2019a) Human antimicrobial peptides as therapeutics for viral infections. Viruses 11(8):704

Ahmed A, Siman-Tov G, Keck F, Kortchak S, Bakovic A, Risner K, Narayanan A (2019b) Human cathelicidin peptide LL-37 as a therapeutic antiviral targeting Venezuelan equine encephalitis virus infections. Antivir Res 164:61–69

Alagarasu K, Patil PS, Shil P, Seervi M, Kakade MB, Tillu H, Salunke A (2017) In-vitro effect of human cathelicidin antimicrobial peptide LL-37 on dengue virus type 2. Peptides 92:23–30

Armitage AE, Stacey AR, Giannoulatou E, Marshall E, Sturges P, Chatha K, Smith NM, Huang X, Xu X, Pasricha SR, Li N, Wu H, Webster C, Prentice MA, Pellegrino P, Williams I, Norris JP, Drakesmith H, Borrow P (2014) Distinct patterns of hepcidin and iron regulation during HIV-1, HBV, and HCV infections. Proc Natl Acad Sci USA 111:12187–12192. https://doi.org/10.1073/pnas.1402351111

Arvinte C, Singh M, Marik PE (2020) Serum levels of vitamin C and vitamin D in a cohort of critically ill COVID-19 patients of a North American community hospital intensive care unit in May 2020: a pilot study. Med Drug Discov 8:100064

Baindara P, Chowdhury T, Roy D, Mandal M, Mandal SM (2023a) Surfactin like lipopeptide from Bacillus clausii efficiently bind to spike glycoprotein of SARS-COV 2. J Biomol Struct Dyn 41(23):14152

Baindara P, Ganguli S, Chakraborty R, Mandal SM (2023b) Preventing respiratory viral diseases with antimicrobial peptide master regulators in the lung airway habitat. Clin Pract 13(1):125–147

Ballus J, Lopez-Delgado JC, Sabater-Riera J, Perez-Fernandez XL, Betbese AJ, Roncal JA (2015) Surgical site infection in critically ill patients with secondary and tertiary peritonitis: epidemiology, microbiology and influence in outcomes. BMC Infect Dis 15(1):1–6

Barlow PG, Svoboda P, Mackellar A, Nash AA, York IA, Pohl J, Donis RO (2011) Antiviral activity and increased host defense against influenza infection elicited by the human cathelicidin LL-37. PLoS One 6(10):e25333

Barra ALC, Dantas LDOC, Morão LG, Gutierrez RF, Polikarpov I, Wrenger C, Nascimento AS (2020) Essential metabolic routes as a way to ESKAPE from antibiotic resistance. Front Public Health 8:26

Barros EP, Casalino L, Gaieb Z, Dommer AC, Wang Y, Fallon L, Amaro RE (2021) The flexibility of ACE2 in the context of SARS-CoV-2 infection. Biophys J 120(6):1072–1084

Bastian A, Schäfer H (2001) Human α-defensin 1 (HNP-1) inhibits adenoviral infection in vitro. Regul Pept 101(1–3):157–161

Belaid A, Aouni M, Khelifa R, Trabelsi A, Jemmali M, Hani K (2002) In vitro antiviral activity of dermaseptins against herpes simplex virus type 1. J Med Virol 66(2):229–234

Björstad A, Askarieh G, Brown KL, Christenson K, Forsman H, Onnheim K, Bylund J (2009) The host defense peptide LL-37 selectively permeabilizes apoptotic leukocytes. Antimicrob Agents Chemother 53(3):1027–1038

Boffert R, Businger R, Preiß H, Ehmann D, Truffault V, Simon C, Ruetalo N, Hamprecht K, Müller P, Wehkamp J, Schindler M (2020) The human α-defensin-derived peptide HD5(1-9) inhibits cellular attachment and entry of human cytomegalovirus. Antivir Res 177:104779

Breman JG, Heymann DL, Lloyd GL, McCormick JB, Miatudila M, Murphy FA, Muyembe-Tamfun JJ (2016) Discovery and description of Ebola Zaire virus in 1976 and relevance to the West African epidemic during 2013–2016. J Infect Dis 214(Suppl.3):S93–S101

Chen CH, Lu TK (2020) Development and challenges of antimicrobial peptides for therapeutic applications. Antibiotics 9(1):24

Chen J, Fan Y, Lin J, Chen Y, Hsu W, Chiou S (2017) Bovine lactoferrin inhibits dengue virus infectivity by interacting with heparan sulfate, low-density lipoprotein receptor, and DC-SIGN. Int J Mol Sci 18:1957

Chen TT, Tan LR, Hu N, Dong ZQ, Hu ZG, Jiang YM, Lu C (2018) C-lysozyme contributes to antiviral immunity in Bombyx mori against nucleopolyhedrovirus infection. J Insect Physiol 108:54–60

Chianese A, Zannella C, Monti A, De Filippis A, Doti N, Franci G, Galdiero M (2022) The broad-spectrum antiviral potential of the amphibian peptide AR-23. Int J Mol Sci 23(2):883

Choi KY, Mookherjee N (2012) Multiple immune-modulatory functions of cathelicidin host defense peptides. Front Immunol 3:149

Chowdhury T, Baindara P, Mandal SM (2021) LPD-12: a promising lipoprotein to control Covid 19. Int J Antimicrob Agents 57(1):106218

Copley RR, Totrov M, Linnell J, Field S, Ragoussis J, Udalova IA (2007) Functional conservation of Rel binding sites in drosophilid genomes. Genome Res 17(9):1327–1335

Costin JM, Jenwitheesuk E, Lok SM, Hunsperger E, Conrads KA, Fontaine KA, Rees CR, Rossman MG, Isern S, Samudrala R, Michael SF (2010) Michael, structural optimization and de novo design of dengue virus entry inhibitory peptides. PLoS Negl Trop Dis 4(6):e721

Craik DJ, Du J (2017) Cyclotides as drug design scaffolds. Curr Opin Chem Biol 38:8–16

Crane-Godreau MA, Clem KJ, Payne P, Fiering S (2020) Vitamin D deficiency and air pollution exacerbate COVID-19 through suppression of antiviral peptide LL37. Front Public Health 8:232

Darke PL, Jacobs AR, Waxman L, Kuo LC (1999) Inhibition of hepatitis C virus NS2/ 3 processing by NS4A peptides. Implications for control of viral processing. J Biol Chem 274(49):34511–34514

Derby N, Lal M, Aravantinou M, Kizima L, Barnable P, Rodriguez A, Zydowsky TM (2018) Griffithsin carrageenan fast dissolving inserts prevent SHIV HSV-2 and HPV infections in vivo. Nat Commun 9(1):3881

Deslouches B, Montelaro RC, Urish KL, Di YP (2020) Engineered cationic antimicrobial peptides (eCAPs) to combat multidrug-resistant bacteria. Pharmaceutics 12(6):501

Diamond G, Beckloff N, Ryan L (2008) Host defense peptides in the oral cavity and the lung: similarities and differences. J Dent Res 87(10):915–927

Ding J, Chou YY, Chang TL (2009) Defensins in viral infections. J Innate Immun 1(5):413–420

Dixon BM, Barker T, McKinnon T, Cuomo J, Frei B, Borregaard N, Gombart AF (2012) Positive correlation between circulating cathelicidin antimicrobial peptide (hCAP18/LL-37) and 25-hydroxyvitamin D levels in healthy adults. BMC Res Notes 5:1–5

Doss M, White MR, Tecle T, Hartshorn KL (2010) Human defensins and LL-37 in mucosal immunity. J Leukoc Biol 87(1):79–92

Dugan AS, Maginnis MS, Jordan JA, Gasparovic ML, Manley K, Page R, Williams G, Porter E, O'Hara BA, Atwood WJ (2008) Human α-defensins inhibit BK virus infection by aggregating virions and blocking binding to host cells. J Biol Chem 283(45):31125–31132

Eleftherianos I, Revenis C (2010) Role and importance of phenoloxidase in insect hemostasis. J Innate Immun 3(1):28–33

Evans AS (2013) Viral infections of humans: epidemiology and control, 3rd edn. Springer Science & Business Media

Falco A, Medina-Gali RM, Poveda JA, Bello-Perez M, Novoa B, Encinar JA (2019) Antiviral activity of a Turbot (Scophthalmus maximus) NK-lysin peptide by inhibition of low-pH virus-induced membrane fusion. Mar Drugs 17(2):87

Faya M, Kalhapure RS, Kumalo HM, Waddad AY, Omolo C, Govender T (2018) Conjugates and nano-delivery of antimicrobial peptides for enhancing therapeutic activity. J Drug Del Sci Technol 44:153–171

Fjell CD, Hiss JA, Hancock RE, Schneider G (2012) Designing antimicrobial peptides: form follows function. Nat Rev Drug Discov 11(1):37–51

Furci L, Tolazzi M, Sironi F, Vassena L, Lusso P (2012) Inhibition of HIV-1 infection by human α-defensin-5, a natural antimicrobial peptide expressed in the genital and intestinal mucosae, vol 7, p e45208

Gallo RC, Montagnier L (2003) The discovery of HIV as the cause of AIDS. N Engl J Med 349(24):2283–2285. https://doi.org/10.1056/NEJMp038194

Ghorbani M, Brooks BR, Klauda JB (2020) Critical sequence hotspots for binding of novel coronavirus to angiotensin converter enzyme as evaluated by molecular simulations. J Phys Chem B 124(45):10034–10047

Giansanti F, Massucci MT, Giardi MF, Nozza F, Pulsinelli E, Nicolini C, Botti D, Antonini G (2005) Antiviral activity of ovotransferrin derived peptides. Biochem Biophys Res Commun 331(1):69–73

Gombart AF, Borregaard N, Koeffler HP (2005) Human cathelicidin antimicrobial peptide (CAMP) gene is a direct target of the vitamin D receptor and is strongly up-regulated in myeloid cells by 1, 25-dihydroxyvitamin D3. FASEB J 19(9):1067–1077

Gordts SC, Renders M, Férir G, Huskens D, Van Damme EJ, Peumans W, Schols D (2015) NICTABA and UDA, two GlcNAc-binding lectins with unique antiviral activity profiles. J Antimicrob Chemother 70(6):1674–1685

Harcourt JL, McDonald M, Svoboda P, Pohl J, Tatti K, Haynes LM (2016) Human cathelicidin, LL-37, inhibits respiratory syncytial virus infection in polarized airway epithelial cells. BMC Res Notes 9:1–6

He M, Zhang H, Li Y, Wang G, Tang B, Zhao J, Huang Y, Zheng J (2018) Cathelicidin-derived antimicrobial peptides inhibit Zika virus through direct inactivation and interferon pathway. Front Immunol 9:722

Howell MD, Jones JF, Kisich KO, Streib JE, Gallo RL, Leung DY (2004) Selective killing of vaccinia virus by LL-37: implications for eczema vaccinatum. J Immunol 172(3):1763–1767

Idris MM, Banu S, Siva AB, Nagaraj R (2020) Downregulation of Defensin genes in SARS-CoV-2 infection. MedRxiV. 2020-09

Ishii K, Hamamoto H, Kamimura M, Nakamura Y, Noda H, Imamura K, Sekimizu K (2010) Insect cytokine paralytic peptide (PP) induces cellular and humoral immune responses in the silkworm Bombyx mori. J Biol Chem 285(37):28635–28642

Ji M, Zhu T, Xing M, Luan N, Mwangi J, Yan X, Mo G, Rong M, Li B, Lai R, Jin L (2019) An antiviral peptide from alopecosa nagpag spider targets NS2B-NS3 protease of flaviviruses. Toxins 11(10)

Jia X, Guo J, Yuan W, Sun L, Liu Y, Zhou M, Xiao G, Lu W, Garzino-Demo A, Wang W (2021) Mechanism through which retrocyclin targets flavivirus multiplication. J Virol 95(15):e0056021

Kaufman HW, Niles JK, Kroll MH, Bi C, Holick MF (2020) SARS-CoV-2 positivity rates associated with circulating 25-hydroxyvitamin D levels. PLoS One 15(9):e0239252

Kościuczuk EM, Lisowski P, Jarczak J, Strzałkowska N, Jóźwik A, Horbańczuk J, Krzyżewski J, Zwierzchowski L, Bagnicka E (2012) Cathelicidins: family of antimicrobial peptides. A review. Mol Biol Rep 39:10957–10970

Kota S, Sabbah A, Harnack R, Xiang Y, Meng X, Bose S (2008) Role of human β-defensin-2 during tumor necrosis factor-α/NF-κB-mediated innate antiviral response against human respiratory syncytial virus. J Biol Chem 283(33):22417–22429

Kudryashova E, Zani A, Vilmen G, Sharma A, Lu W, Yount JS, Kudryashov DS (2021) SARS-CoV-2 inactivation by human defensin HNP1 and retrocyclin RC-101. bioRxiv. https://doi.org/10.1101/2021.05.27.445985

Ladokhin AS, Selsted ME, White SH (1999) CD spectra of indolicidin antimicrobial peptides suggest turns, not polyproline helix. Biochemistry 38(38):12313–12319

Lehrer RI, Daher K, Ganz T, Selsted ME (1985) Direct inactivation of viruses by MCP-1 and MCP-2, natural peptide antibiotics from rabbit leukocytes. J Virol 54(2):467–472

Lehrer RI, Cole AM, Selsted ME (2012) θ-Defensins: cyclic peptides with endless potential. J Biol Chem 287(32):27014–27019

Li Q, Zhao Z, Zhou D, Chen Y, Hong W, Cao L, Wu Y, Yan H, Li W (2011) Virucidal activity of a scorpion venom peptide variant mucroporin-M1 against measles, SARS-CoV and influenza H5N1 viruses. Peptides 32(7):1518–1525
Luan J, Ren Y, Gao S, Zhang L (2022) High level of defensin alpha 5 in intestine may explain the low incidence of diarrhoea in COVID-19 patients. Eur J Gastroenterol Hepatol 34:e3
Maghbooli Z, Sahraian MA, Ebrahimi M, Pazoki M, Kafan S, Tabriz HM, Holick MF (2020) Vitamin D sufficiency, a serum 25-hydroxyvitamin D at least 30 ng/mL reduced risk for adverse clinical outcomes in patients with COVID-19 infection. PLoS One 15(9):e0239799
Maleki MSM, Restamian M, Madanchi H (2021) Antimicrobial peptides and other peptide-like therapeutics as promising candidates to combat SARS-CoV-2. Expert Rev Anti-Infect Ther 1
Malkoski M, Dashper SG, O'Brien-Simpson NM, Talbo GH, Macris M, Cross KJ, Reynolds EC (2001) Kappacin, a novel antibacterial peptide from bovine milk. Antimicrob Agents Chemother 45(8):2309–2315
Malonis RJ, Lai JR, Vergnolle O (2019) Peptide-based vaccines: current progress and future challenges. Chem Rev 120(6):3210–3229
Masuda M, Nakashima H, Ueda T, Naba H, Ikoma R, Otaka A, Terakawa Y, Tamamura H, Ibuka T, Murakami T, Koyanagi Y, Waki M, Matsumoto A, Yamamoto M, Funakoshi S, Fujii N (1992) A novel anti-HIV synthetic peptide, T-22 ([Tyr5,12,Lys7]-polyphemusin II). Biochem Biophys Res Commun 189(2):845–850
Matanic VCA, Castilla V (2004) Antiviral activity of antimicrobial cationic peptides against Junin virus and herpes simplex virus. Int J Antimicrob Agents 23(4):382–389
Matsumura T, Sugiyama N, Murayama A, Yamada N, Shiina M, Asabe S, Wakita T, Imawari M, Kato T (2016) Antimicrobial peptide LL-37 attenuates infection of hepatitis C virus. Hepatol Res 46(9):924–932
Memariani H, Memariani M (2020) Therapeutic and prophylactic potential of anti-microbial peptides against coronaviruses. Ir J Med Sci (1971 -) 189(4):1153–1154
Memariani H, Memariani M, Moravvej H, Shahidi-Dadras M (2020) Melittin: a venom-derived peptide with promising anti-viral properties. Eur J Clin Microbiol Infect Dis 39:5–17
Mojsoska B, Zuckermann RN, Jenssen H (2015) Structure-activity relationship study of novel peptoids that mimic the structure of antimicrobial peptides. Antimicrob Agents Chemother 59(7):4112–4120
Moreno-Habel DA, Biglang-awa IM, Dulce A, Luu DD, Garcia P, Weers PM, Haas-Stapleton EJ (2012) Inactivation of the budded virus of Autographa californica M nucleopolyhedrovirus by gloverin. J Invertebr Pathol 110(1):92–101
Mustafa S, Balkhy H, Gabere M (2019) Peptide-protein interaction studies of antimicrobial peptides targeting middle east respiratory syndrome coronavirus spike protein: an in silico approach. Adv Bioinforma 2019:1
Nakashima H, Masuda M, Murakami T, Koyanagi Y, Matsumoto A, Fujii N, Yamamoto N (1992) Anti-human immunodeficiency virus activity of a novel synthetic peptide, T22 ([Tyr-5,12, Lys-7] polyphemusin II): a possible inhibitor of virus-cell fusion. Antimicrob Agents Chemother 36(6):1249–1255
Nawrot R, Barylski J, Nowicki G, Broniarczyk J, Buchwald W, Goździcka-Józefiak A (2014) Plant antimicrobial peptides. Folia Microbiol 59:181–196
Nicolas P, Mor A (1995) Peptides as weapons against microorganisms in the chemical defense system of vertebrates. Ann Rev Microbiol 49(1):277–304
Nijnik A, Hancock RE (2009) The roles of cathelicidin LL-37 in immune defences and novel clinical applications. Curr Opin Hematol 16(1):41–47
Nizet V, Ohtake T, Lauth X, Trowbridge J, Rudisill J, Dorschner RA, Gallo RL (2001) Innate antimicrobial peptide protects the skin from invasive bacterial infection. Nature 414(6862):454–457
Oleszycka E, McCluskey S, Sharp FA, Muñoz Wolf N, Hams E, Gorman AL, Lavelle EC (2018) The vaccine adjuvant alum promotes IL-10 production that suppresses Th1 responses. Eur J Immunol 48(4):705–715

Ooi CY, Pang T, Leach ST, Katz T, Day AS, Jaffe A (2015) Fecal human β-defensin 2 in children with cystic fibrosis: is there a diminished intestinal innate immune response? Dig Dis Sci 60:2946–2952
Pelegrini PB, Del Sarto RP, Silva ON, Franco OL, Grossi-de-Sa MF (2011) Antibacterial peptides from plants: what they are and how they probably work. Biochem Res Int 2011:1
Pietrantoni A, Di Biase AM, Tinari A, Marchetti M, Valenti P, Seganti L, Superti F (2003) Bovine lactoferrin inhibits adenovirus infection by interacting with viral structural polypeptides. Antimicrob Agents Chemother 47(8):2688–2691
Radujkovic A, Hippchen T, Tiwari-Heckler S, Dreher S, Boxberger M, Merle U (2020) Vitamin D deficiency and outcome of COVID-19 patients. Nutrients 12(9):2757
Rambaut A, Posada D, Crandall KA, Holmes EC (2004) The causes and consequences of HIV evolution. Nat Rev Genet 5(1):52–61
Rinaldi AC (2002) Antimicrobial peptides from amphibian skin: an expanding scenario: commentary. Curr Opin Chem Biol 6(6):799–804
Rivas-Santiago B, Schwander SK, Sarabia C, Diamond G, Klein-Patel ME, Hernandez-Pando R, Sada E (2005) Human β-defensin 2 is expressed and associated with Mycobacterium tuberculosis during infection of human alveolar epithelial cells. Infect Immun 73(8):4505–4511
Rodriguez R, Jung CL, Gabayan V, Deng JC, Ganz T, Nemeth E, Bulut Y (2014) Hepcidin induction by pathogens and pathogen-derived molecules is strongly dependent on interleukin-6. Infect Immun 82(2):745–752
Rothan HA, Abdulrahman AY, Sasikumer PG, Othman S, Abd Rahman N, Yusof R (2012) Protegrin-1 inhibits dengue NS2B-NS3 serine protease and viral replication in MK2 cells. Biomed Res Int
Salvatore M, García-Sastre A, Ruchala P, Lehrer RI, Chang T, Klotman ME (2007) β-Defensin inhibits influenza virus replication by cell-mediated mechanism (s). J Infect Dis 196(6):835–843
Skeate JG, Segerink WH, Garcia MD, Fernandez DJ, Prins R, Lühen KP, Voss FO, Da Silva DM, Kast WM (2020) Theta-defensins inhibit high-risk human papilloma virus infection through charge-driven capsid clustering. Front Immunol 11:561843
Skwarczynski M, Toth I (2016) Peptide-based synthetic vaccines. Chem Sci 7(2):842–854
Skwarczynski M, Zhao G, Boer JC, Ozberk V, Azuar A, Cruz JG, Giddam AK, Khalil ZG, Pandey M, Shibu MA, Hussein WM, Nevagi RJ, Batzloff MR, Wells JW, Capon RJ, Plebanski M, Good MF, Toth I (2020) Poly(amino acids) as a potent self-adjuvanting delivery system for peptide-based nanovaccines. Sci Adv 6(5):eaax2285
Solanki SS, Singh P, Kashyap P, Sansi MS, Ali SA (2021) Promising role of defensins peptides as therapeutics to combat against viral infection. Microb Pathog 155:104930
Sørensen OE, Follin P, Johnsen AH, Calafat J, Tjabringa GS, Hiemstra PS, Borregaard N (2001) Human cathelicidin, hCAP-18, is processed to the antimicrobial peptide LL-37 by extracellular cleavage with proteinase 3. Blood 97(12):3951–3959
Sousa FH, Casanova V, Findlay F, Stevens C, Svoboda P, Pohl J, Barlow PG (2017) Cathelicidins display conserved direct antiviral activity towards rhinovirus. Peptides 95:76–83
Spinello A, Saltalamacchia A, Magistrato A (2020) Is the rigidity of SARS-CoV-2 spike receptor-binding motif the hallmark for its enhanced infectivity? Insights from all-atom simulations. J Phys Chem Lett 11(12):4785–4790
Stuart LS, Harris TH (1942) Bactericidal and fungicidal properties of a crystalline protein isolated from unbleached wheat flour. Cereal Chem 19:288–300
Superti F, Agamennone M, Pietrantoni A, Ammendolia MG (2019) Bovine lactoferrin prevents influenza A virus infection by interfering with the fusogenic function of viral hemagglutinin. Viruses 11(1):51
Tahmoorespur M, Azghandi M, Javadmanesh A, Meshkat Z, Sekhavati MH (2020) A novel chimeric anti-HCV peptide derived from camel lactoferrin and molecular level insight on its interaction with E2. Int J Pept Res Ther 26:1593–1605

Takahashi T, Nomura F, Yokoyama Y, Tanaka-Takiguchi Y, Homma M, Takiguchi K (2013) Multiple membrane interactions and versatile vesicle deformations elicited by melittin. Toxins 5(4):637–664

Tan P, Fu H, Ma X (2021) Design, optimization, and nanotechnology of antimicrobial peptides: from exploration to applications. Nano Today 39:101229

Tecle T, Tripathi S, Hartshorn KL (2010) Defensins and cathelicidins in lung immunity. Innate Immun 16(3):151–159

Tripathi S, Wang G, White M, Qi L, Taubenberger J, Hartshorn KL (2015) Antiviral activity of the human cathelicidin, LL-37, and derived peptides on seasonal and pandemic influenza A viruses. PLoS One 10(4):e0124706

Tuttobene MR, Pérez JF, Pavesi ES, Perez Mora B, Biancotti D, Cribb P, Mussi MA (2021) Light modulates important pathogenic determinants and virulence in ESKAPE pathogens Acinetobacter baumannii, Pseudomonas aeruginosa, and Staphylococcus aureus. J Bacteriol 203(5):10–1128

Vallon-Eberhard A, Makovitzki A, Beauvais A, Latgé JP, Jung S, Shai Y (2008) Efficient clearance of Aspergillus fumigatus in murine lungs by an ultrashort antimicrobial lipopeptide, palmitoyl-lys-ala-DAla-lys. Antimicrob Agents Chemother 52(9):3118–3126

van der Strate BW, Beljaars L, Molema G, Harmsen MC, Meijer DK (2001) Antiviral activities of lactoferrin. Antivir Res 52:225–239

VanCompernolle SE, Taylor RJ, Oswald-Richter K, Jiang J, Youree BE, Bowie JH, Unutmaz D (2005) Antimicrobial peptides from amphibian skin potently inhibit human immunodeficiency virus infection and transfer of virus from dendritic cells to T cells. J Virol 79(18):11598–11606

Venkataraman N, Cole AL, Ruchala P, Waring AJ, Lehrer RI, Stuchlik O, Cole AM (2009) Reawakening retrocyclins: ancestral human defensins active against HIV-1. PLoS Biol 7(4):e1000095

Vilas Boas LC, de Lima LM, Migliolo L, Mendes GD, de Jesus MG, Franco OL, Silva PA (2017) Linear antimicrobial peptides with activity against herpes simplex virus 1 and Aichi virus. Biopolymers 108(2)

Vorland LH (1999) Lactoferrin: a multifunctional glycoprotein. APMIS 107(11):971–981

Wachinger M, Saermark T, Erfle V (1992) Influence of amphipathic peptides on the HIV-1 production in persistently infected T lymphoma cells. FEBS Lett 309(3):235–241

Wang TT, Nestel FP, Bourdeau V, Nagai Y, Wang Q, Liao J, White JH (2004) Cutting edge: 1, 25-dihydroxyvitamin D3 is a direct inducer of antimicrobial peptide gene expression. J Immunol 173(5):2909–2912

Wang A, Chen F, Wang Y, Shen M, Xu Y, Hu J, Wang J (2013) Enhancement of antiviral activity of human alpha-defensin 5 against herpes simplex virus 2 by arginine mutagenesis at adaptive evolution sites. J Virol 87(5):2835–2845

Wang G, Li X, Wang Z (2016) APD3: the antimicrobial peptide database as a tool for research and education. Nucleic Acids Res 44(D1):D1087–D1093

Wang C, Wang S, Li D, Zhao X, Han S, Wang T, Wang J (2020) Lectin-like intestinal defensin inhibits 2019-nCoV spike binding to ACE2. BioRxiv. 2020-03

Wiens ME, Smith JG (2017) α-Defensin HD5 inhibits human papillomavirus 16 infection via capsid stabilization and redirection to the lysosome. MBio 8(1):10–1128

Wohlford-Lenane CL, Meyerholz DK, Perlman S, Zhou H, Tran D, Selsted ME, McCray PB Jr (2009) Rhesus theta-defensin prevents death in a mouse model of severe acute respiratory syndrome coronavirus pulmonary disease. J Virol 83(21):11385–11390

Wu Z, Cocchi F, Gentles D, Ericksen B, Lubkowski J, DeVico A, Lehrer R, Lu W (2005) Human neutrophil α-defensin 4 inhibits HIV-1 infection in vitro. FEBS Lett 579(1):162–166

Xia S, Liu M, Wang C, Xu W, Lan Q, Feng S, Qi F, Bao L, Du L, Liu S, Qin C, Sun F, Shi Z, Zhu Y, Jiang S, Lu L (2020) Inhibition of SARS-CoV-2 (previously 2019 nCoV) infection by a highly potent pan-coronavirus fusion inhibitor targeting its spike protein that harbors a high capacity to mediate membrane fusion. Cell Res 30(4):343–355

Xiong W, Li J, Feng Y, Chai J, Wu J, Hu Y, Zou M (2021a) Brevinin-2GHk, a peptide derived from the skin of Fejervaryalimnocharis, inhibits Zika virus infection by disrupting viral integrity. Viruses 13(12):2382

Xiong W, Zhou C, Yin S, Chai J, Zeng B, Wu J, Xu X (2021b) Fejerlectin, a lectin-like peptide from the skin of Fejervaryalimnocharis, inhibits HIV-1 entry by targeting Gp41. ACS Omega 6(9):6414–6423

Yang W, Hu FL, Xu XF (2020) Bee venom and SARS-CoV-2. Toxicon 181:69

Yu Y, Cooper CL, Wang G, Morwitzer MJ, Kota K, Tran JP, Bradfute SB, Liu Y, Shao YJ, Zhang AK, Luo MLG, Reid SP, Hinrichs SH, Su K (2020) Engineered human cathelicidin antimicrobial peptides inhibit ebola virus infection. iScience 23(4):100999

Yu J, Dai Y, Fu Y, Wang K, Yang Y, Li M, Xu W, Wei L (2021) Cathelicidin antimicrobial peptides suppress EV71 infection via regulating antiviral response and inhibiting viral binding. Antivir Res 187:105021

Zannella C, Chianese A, Palomba L, Marcocci ME, Bellavita R, Merlino F, Grieco P, Folliero V, Filippis AD, Mangoni M, Nencioni L, Franci G, Galdiero M (2022) Broad-spectrum antiviral activity of the amphibian antimicrobial peptide temporin L and its analogs. Int J Mol Sci 23(4)

Zeth K, Sancho-Vaello E (2017) The human antimicrobial peptides dermcidin and LL-37 show novel distinct pathways in membrane interactions. Front Chem 5:86

Zhang H, Zhao Y, Jiang X, Zhao Y, Yang L, Chen L, Sun G (2020) Preliminary evaluation of the safety and efficacy of oral human antimicrobial peptide LL-37 in the treatment of patients of COVID-19, a small-scale, single-arm, exploratory safety study. MedRxiv. 2020-05

Zhang L, Ghosh SK, Basavarajappa SC, Muller-Greven J, Penfield J, Brewer A, Weinberg A (2021) Molecular dynamics simulations and functional studies reveal that hBD-2 binds SARS-CoV-2 spike RBD and blocks viral entry into ACE2 expressing cells. BioRxiv

Zhao Z, Hong W, Zeng Z, Wu Y, Hu K, Tian X, Li W, Cao Z (2012) Mucroporin-M1 inhibits hepatitis B virus replication by activating the mitogen-activated protein kinase (MAPK) pathway and down-regulating HNF4α in vitro and in vivo. J Biol Chem 287(36):30181–30190

Zhao H, To, K. K, Sze KH, Yung TTM, Bian M, Lam H, Yuen KY (2020) A broad-spectrum virus- and host-targeting peptide against respiratory viruses including influenza virus and SARS-CoV-2. Nat Commun 11(1):4252

Chapter 12
Antimicrobial Peptides and Their Role Against Head and Neck Cancer

Abhilasha Tiwari and Amit Kumar Chowhan

1 Introduction

Human cancer is a serious cause of mortality and disease all around the world continually. There is a need for better, more targeted, and more effective medications (Gaspar et al. 2013). The seventh most prevalent cancer worldwide is head and neck cancer (HNC), with more than 660,000 new cases and 325,000 deaths annually (Gormley et al. 2022). While consumption of carcinogen-containing products is linked to the maximum prevalence of head and neck squamous cell carcinoma (HNSCC) in regions like Southeast Asia and Australia, rising rates of human papillomavirus (HPV) infection in the oropharynx have been dedicated to the maximum prevalence of HNSCC in Western Europe and the USA. While cancer in the pharynx is increasingly linked to HPV, namely, HPV-16, oral cavity, and larynx, cancers are typically connected with tobacco smoking, alcohol addiction, or both (Mehanna et al. 2012). Improved results for patients with HNSCC were anticipated because of developments in drug discovery, biotechnology, robotic surgery, radiation therapy techniques, and molecular characterization of human cancers, including HNSCC, in the twenty-first century (Johnson et al. 2020).

Antimicrobial peptides (AMPs) have received abundant interest against traditional medicines as a possible replacement (Xuan et al. 2023). A diverse quantity of amino acids contributes to antimicrobial peptides. Typically, this ranges from 10 to 60 amino acids. AMPs are small peptides that come across in nature in large quantities and show an important function in the immune system of many kinds of living beings. They possess a wide variety of inhibitory activities for viruses, fungi, parasites, and bacteria. The discovery of AMPs, which have a promising future in

A. Tiwari · A. K. Chowhan (✉)
Department of Pathology and Lab Medicine, All India Institute of Medical Sciences, Raipur, Chhattisgarh, India
e-mail: chowhanpath@aiimsraipur.edu.in

P. Baindara, S. M. Mandal (eds.), *Evolution of Antimicrobial Peptides*,
https://doi.org/10.1007/978-3-031-67515-7_12

various fields such as food, animal husbandry, agriculture, aquaculture, and medicine, was spurred by the advent of antibiotic-resisting microbes and the growth of usage of antibiotics (Huan et al. 2020). All living things, from bacteria to mammals, generate these amino acid sequences, substituting for the primary barrier of protection against a wide range of foreign bodies. Alexander Fleming in 1922 recognized lysozymes that showed the property to lyse bacterial cells from animal tissues and vegetables. It was the first investigated AMP (Fleming and Allison 1922).

While AMPs have mostly been investigated and developed as viable substitutes for combating viral infections, their application in cancer therapy as anticancer peptides (ACPs) in conjunction with several other traditional medications or as their own has been a promising therapeutic approach. Though the development of selective anticancer peptides has proven difficult, they are supposed to be selective for cancerous cells by not compromising the essential biological functioning of the body (Gaspar et al. 2013). In this chapter, the role of antimicrobial peptides against head and neck cancer is covered along with specific anticancer peptides against each anatomical site of the head and neck.

2 Antimicrobial Peptides

Antimicrobial peptides are found in a variety of species such as prokaryotes and humans, and they are evolutionarily conserved molecules. Alexander Fleming isolated the first human antibacterial protein, lysozyme, from nasal mucus in 1922. In 1928, this insight was overtaken by the discovery of penicillin by Fleming, which led to the development of the drug's medicinal application in the 1940s. In 1945, Fleming shared the Nobel Prize with other scientists for this work. Thus, interest in the medicinal potential of natural AMPs, like lysozyme, declined with the "golden age of antibiotics" in the 1940s. The emergence of microorganisms resistant to drugs in the 1960s sparked research into AMPs as host defense molecules.

All multicellular creatures create antimicrobial peptides (AMPs) and proteins, a broad class of naturally occurring chemicals, as their first line of defense. These proteins possess wide-ranging abilities to eradicate bacteria, yeasts, fungi, viruses, and even cancerous cells directly. Although bacteria also make AMPs to defend their environmental niche, insects and plants are the primary hosts of AMPs, which are used as antibiotics to ward off potentially harmful germs. AMPs can also be referred to as "host defense peptides" in higher eukaryotic organisms, highlighting their additional immunomodulatory functions. These varied, AMP-specific activities involve a range of cytokine and growth-factor-like effects that are important for maintaining proper immunological homeostasis.

The majority of the human body's usual microbe-exposed areas, including the skin, intestinal mucosa, oral mucosa, lung, eye, and reproductive tract, have been shown to contain AMPs. It was discovered that the bulk of these peptides are produced in response to infection, inflammation, or damage. However, other AMPs are constitutively expressed. Understanding how they function depends on their unique sites of expression and the stringent controls placed on AMP expression. This could help explain why, while being produced from ancient genes, AMPs continue to be effective antibiotics while antibiotics obtained from pharmaceuticals can quickly lose their effectiveness due to the emergence of bacterial resistance (Zhang and Gallo 2016).

2.1 Classification of AMPs

Classification of natural antimicrobial peptides is challenging due to the variability. They are categorized according to four factors: (1) source, (2) structure, (3) activity, and (4) species that are rich in amino acids (Huan et al. 2020) (Fig. 12.1).

2.1.1 Source of AMPs

Antimicrobial peptides are available in all life forms, including prokaryotes and eukaryotes (bacteria, plants, and animals). The abundant AMPs are reported from eukaryotes such as plants, fungi, and animals (Fig. 12.2). The following sections describe the AMPs in different life forms.

AMPs in Bacteria

By eliminating other species of bacteria that would fight with them for resources and the same niche, antimicrobial peptides in bacteria help individual species. Lantibiotics and non-lantibiotics are the two groups into which bacterial AMPs, often called bacteriocins, are divided. Those containing the nonnatural amino acid lanthionine are known as lantibiotics (Kumar et al. 2018).

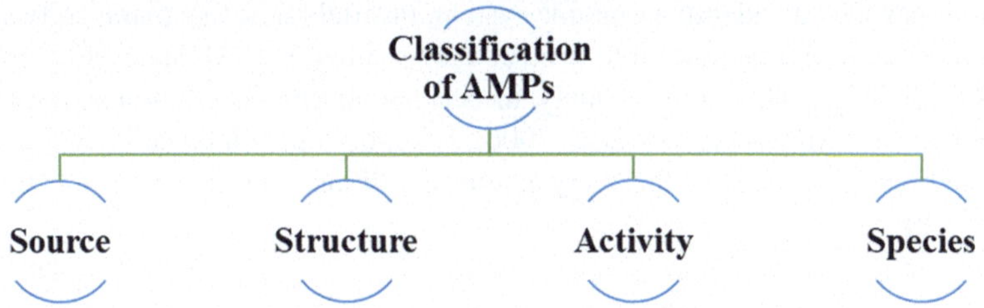

Fig. 12.1 Classification of antimicrobial peptides

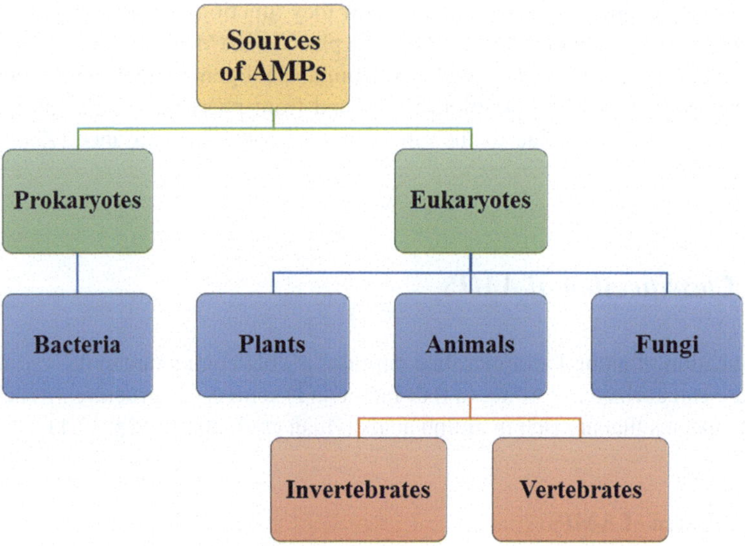

Fig. 12.2 Sources of antimicrobial peptides

AMPs in Plants

AMPs are crucial for the function of innate immunity in eukaryotes. Since plants lack T cell–mediated immunity and B cell–mediated immunity, or adaptive immunity, they are essential to their defense against bacterial and fungal attacks. Many different species of plants contain genes that encode for plant AMPs. Remarkably, all plant AMPs have a high cysteine content and many disulfide linkages. Plant AMPs that have been investigated the most include cyclotides, plant defensins, and thionins. Plant AMPs are present in the tubers, leaves, flowers, and seeds (Tam et al. 2015).

AMPs in Animals (Invertebrates)

Because they lack an adaptive immune system, invertebrates, like plants, rely solely on their innate immune system to protect against infection. All species of invertebrates studied up to this point, namely, insects and marine invertebrates, have been shown to have AMPs (Hancock et al. 2006). Haemolymph, blood cells, white blood cells, and epithelial cells of these organisms are all known to have AMPs (Jenssen et al. 2006).

AMPs in Animals (Vertebrates)

Innate and adaptive immune systems make up vertebrate immunity. Fish, mammals, amphibians, and other animals have all had their AMPs extracted and studied. In addition to granules of white blood cells such as phagocytes, neutrophils, macrophages, natural killer cells, and epithelial tissue found in the mouth, lungs, or skin, body fluids can also be used to identify vertebrate AMPs (Yang et al. 2004). It is interesting to note that AMPs have been abundantly found in amphibian skin glands, with over 500 AMPs identified to date (Jenssen et al. 2006).

2.1.2 Structure of AMPs

Understanding AMPs' structural organization and arrangement is essential to comprehending how they interact with biological targets. To thoroughly examine the biological roles and structure of antimicrobial peptides, a variety of experimental techniques, such as nuclear magnetic resonance (NMR), X-ray crystallography, atomic force microscopy (AFM), and cryo-electron microscopy (cryo-EM), have been combined with computational methods like molecular modeling, docking, and dynamics (Cardoso et al. 2018).

Although there are many other ways to categorize AMPs, the most popular method is to base it on their secondary structure. AMP structures are typically divided into α-helical, β-sheet, mixed (α-helical/β-sheet), and cyclic structures (Fig. 12.3).

The AMP group extensively studied are the α-helix peptides. In α-helical, the two neighboring amino acids in this structure are spaced approximately 0.15 nm apart, and their angle is roughly 100°. The well-known peptides, known as LL-37-derived human cathelicidin, exhibit an amphiphilic α-helix secondary structure in membrane mimetic settings.

The antimicrobial peptides belonging to the second category display β-sheet conformation, wherein at least two β-strands bind together using disulfide bonds. Disulphide bridges are necessary for the structure to stabilize and for peptides to perform their biological purpose. One of the well-known classes of AMPs are Defensins (α and β) with a wide range of antibacterial activities against viruses, fungi, and bacteria. In many different tissues, epithelial cells release β-defensins in significant quantities. The mixed (α-helical/β-sheet), the third class of antimicrobial polypeptides, takes on an α-helix/β-sheet structure, a combination structure that keeps three or four disulfide bridges stable (Büyükkiraz and Kesmen 2022).

2.1.3 Activity of AMPs

According to data, 18 categories are used to categorize antimicrobial activity. Antibacterial, antiviral, antifungal, antiparasitic, anti-human immunodeficiency virus (HIV), and antitumor peptides are a summary of these categories.

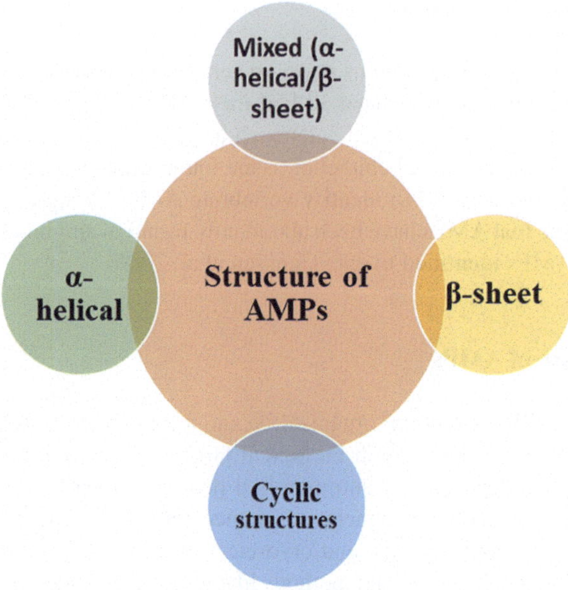

Fig. 12.3 Structure of antimicrobial peptides

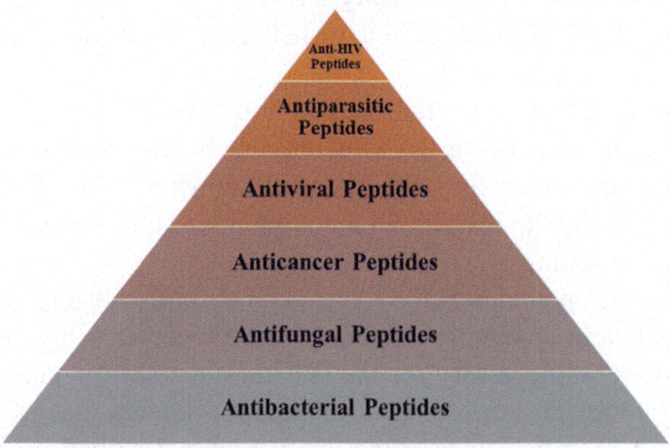

Fig. 12.4 Activity of antimicrobial peptides

The majority, or roughly 60%, are made up of antibacterial peptides. Next, antifungal peptides make up 26%, while the remaining 2–5% are made up of antiviral, antiparasitic, anticancer, and anti-HIV peptides (Fig. 12.4).

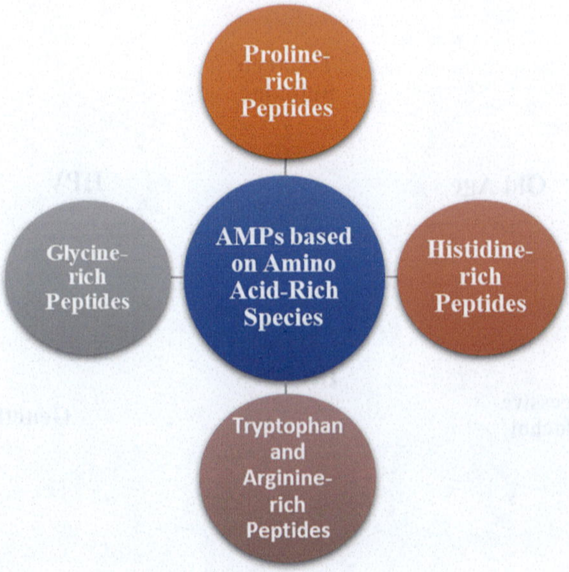

Fig. 12.5 Antimicrobial peptides based on amino-acid-rich species

2.1.4 AMPs Based on Amino-Acid-Rich Species

The AMPs are categorized into four classes as per the amino acid richness: proline-rich peptides (PrAMPs), tryptophan- and arginine-rich antimicrobial peptides, histidine-rich peptides, and glycine-rich antimicrobial peptides (Fig. 12.5).

The pPR-AMP1 demonstrates antimicrobial potential against gram-positive and gram-negative bacteria even though PrAMPs primarily destroy gram-positive bacteria. The two well-known AMPs high in Trp and Arg residues are triptrpticin and indolicidin. HV2 is a histidine-rich AMP observed to suppress the production of tumor necrosis factor α (TNF-α), which has a potent anti-inflammatory action and concentration-dependently prevents bacterial migration. Attacins and diptericins are examples of glycine-rich AMPs that are abundant in nature. Additionally, GG3, a glycine-rich central symmetrical peptide, is a prime candidate for a commercial medication against clinical gram-negative bacteria (Huan et al. 2020).

3 Head and Neck Cancer

Head and neck cancer has many risk factors such as tobacco smoking, excessive alcohol consumption, betel nuts, airborne pollutants, UV exposure, aging and gender, poor diet and nutrition, familial history, and human papillomavirus (HPV) (Fig. 12.6). Head and neck squamous cell carcinoma (HNSCC) therefore can be divided into two groups: HPV positive and HPV negative (Mehanna et al. 2012).

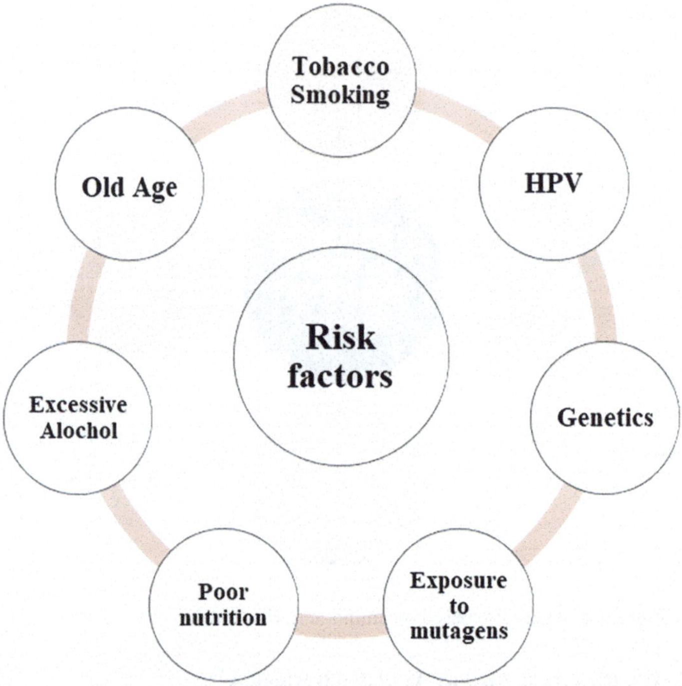

Fig. 12.6 Risk factors of head and neck cancer

3.1 Anatomical Sites of Head and Neck Cancer

Head and neck cancer or head and neck squamous cell carcinoma (HNSCC) occurs in the mucosal epithelium in the anatomical sites of the oral cavity, nasopharynx, oropharynx, hypopharynx, and larynx. The specific sites of the oral cavity are, from posterior to anterior, retromolar trigone, floor of the mouth, anterior tongue, hard palate, buccal mucosa, and lips. The specific sites of the oropharynx are, from posterior to anterior, posterior pharyngeal wall, uvula, soft palate, the base of the tongue, lingual tonsils, and the palatine tonsils. The specific sites of the hypopharynx are the end portion of the throat, starting from the hyoid to the cricoid cartilage (Fig. 12.7).

The tobacco-related head and neck squamous cell carcinoma predominantly originates, from anterior to posterior, in the oral cavity, hypopharynx, and larynx. The HPV-related head and neck squamous cell carcinoma mainly originates from the palatine and lingual tonsils of the oropharynx (Johnson et al. 2020).

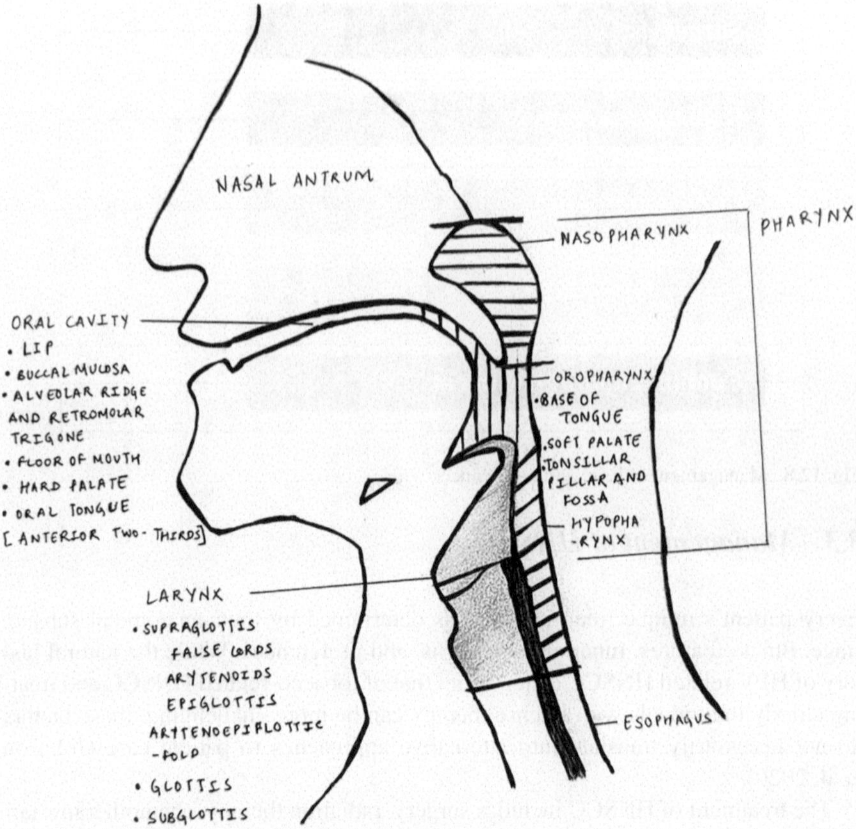

Fig. 12.7 Anatomical sites and subsites of head and neck

3.2 Mechanisms/Pathophysiology of HNSCC

Mucosal epithelial cells that line the oral cavity, pharynx, larynx, and sinonasal tract are the source of HNSCC. Histologically, epithelial cell hyperplasia is the first step toward an organized sequence of events that leads to invasive HNSCC. This sequence includes dysplasia (mild, moderate, and severe), carcinoma in situ, and invasive carcinoma.

Notably, nevertheless, most individuals with HNSCC do not have a prior history of a premalignant lesion. Because HNSCC is heterogeneous, the anatomical location and aetiological agent (virus vs. carcinogen) determine the cell of origin. All these scenarios suggest that normal adult stem or progenitor cells were the original cells, which gave rise to cancer stem cells (CSCs) with pluripotency and self-renewal after undergoing oncogenic transformation. HNSCC CSCs that can grow tumors in transplantation assays make up just a small portion of initial cancer cells (1–3%), but they are appealing targets for new targeted therapies because of their innate resistance to traditional medications (Krishnamurthy et al. 2010).

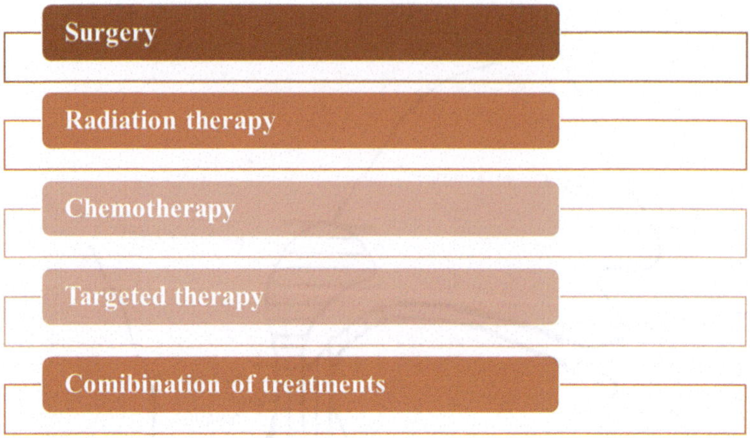

Fig. 12.8 Management of head and neck cancer

3.3 Management of HNSCC

Every patient's unique treatment plan is determined by their anatomical subsite, stage, illness features, functional concerns, and preferences. While the natural history of HPV-related HNSCC differs from that of tobacco-related HNSCC and treating elderly individuals with intense therapy can be more challenging, these factors do not necessarily translate into alternative approaches to patient care (Johnson et al. 2020).

The treatment of HNSCC includes surgery, radiation therapy, chemotherapy, targeted therapy, and a combination of treatments (Fig. 12.8). For locally or locoregionally limited HNSCC, resection, radiation, and systemic therapy are the main therapeutic techniques. Planning a course of treatment should maximize function preservation while aiming for the most highly curative strategy. Single modality intervention (resection or radiation) can yield cure rates of 80% in patients with small primary cancers that involve just one node or no clinical nodal involvement (Lee et al. 2018).

4 Antimicrobial Peptides and Cancer Treatments

The classes of AMPs are categorized as minor and major based on their present anticancer properties that have been studied till now.

4.1 Major Therapeutically Useful AMP Classifications

4.1.1 Defensins

Defensins are produced by eukaryotes, are cationic peptides, and consist of two superfamilies. Natural defensins seem to cause cell cycle arrest and have antiproliferative and proapoptotic impacts on tumor cells (Gerashchenko et al. 2013) that are demonstrated by elevated production of caspase-7 and caspase-9 and other apoptosis markers, decreased activity of transcriptional and cell cycle cyclin-dependent kinases and their catalytic cyclin partners, and increased levels of phosphorylated retinoblastoma protein (Gerashchenko et al. 2014). The aspect of human defensins needs to be addressed: they seem to exhibit a remarkable degree of selectivity for cancer cells but do not seem to have any obvious cytotoxic or cytostatic effects on normal, nonconverted cells. Since human AMPs are endogenously produced, they are not intended to cause overt collateral damage to normal, healthy tissues during an inflammatory response, which could explain this observation. Second, antimicrobial defensins might offer fresh perspectives on unresolved clinical problems like chemotherapeutic resistance (Strzelecka et al. 2017). Defensins also have a strong potential to increase the efficacy of cancer immunotherapy, which is another effective use for them. The conclusion from both in vitro and in vivo experiments showed that immature dendritic cells underwent physiological alterations that are expected to improve adaptive anticancer immunosurveillance (Li et al. 2014).

4.1.2 Lactoferricin B

Lactoferrin is an 80 kDa glycoprotein that sequesters iron and is found in exocrine secretions like milk and polymorphonuclear leukocyte granules. It is one of the AMPs that have been the focus of investigation due to its potential anticancer effects. Because lactoferricin B can selectively and lethally destabilize the cytoplasmic and mitochondrial membranes of cancer cells, it has anticancer properties (Eliassen et al. 2006). Lactoferricin has been demonstrated to direct lymphocytes toward eliminating cancer cells and to have immunomodulatory properties like those of the defensin proteins (Ganz and Lehrer 1995).

4.1.3 Cecropins A and B

Cecropins were originally identified from the silk moth *Hyalophora cecropia*. They are a class of short, basic peptides having a hydrophobic C-terminus and a strongly basic N-terminus, typically ranging between 31- and 39-amino-acid residues. Cecropin A has been demonstrated to cause apoptosis in human hepatocellular carcinoma cells by expressing Fas, Fas-L, caspase-3, and caspase-8, as well as in the promyelocytic cell line HL-60 via a ROS signaling pathway.

4.1.4 Magainin II

An ampicillin-like protein (AMP) called magainin II (MGN-II) was extracted from the skin of the African clawed frog, *Xenopus laevis*, and it has shown strong anticancer properties against a range of solid and hematological cancers. It has been demonstrated that this ionophoric peptide, which has a helical structure, can pierce cancer cell membranes and function as ion channels, cytolyzing malignant cells. The tumoricidal effects of cytotoxic chemotherapy drugs are significantly enhanced by MGN-II. Additionally, MGN-II exhibits tumor selectivity; for example, it has been shown to lyse different hematopoietic tumor and solid tumor cells while having little to no impact on peripheral blood lymphocytes and normal human fibroblasts (Roudi et al. 2017).

4.2 Minor Therapeutic Classes of AMPs (Table 12.1)

5 Antimicrobial Peptides and Their Role Against Head and Neck Cancer

Lehrer and group isolated three short peptides (HNP-1, HNP-2, and HNP-3) from human neutrophils in 1985. These peptides were called "defensins" and were shown to be effective against *Escherichia coli*, *Pseudomonas aeruginosa*, and *Staphylococcus aureus*. Many natural and synthetic AMPs have been identified over the past few decades, and public databases like APD3, CAMP, and DAMP2.0 have been created. These databases provide lists of natural AMPs, the classification of AMP families, and the clinical uses of AMPs, respectively. Additionally, databases of peptides have been created, such as the antiviral AVPdb, the antiparasitic ParaPep, the anticancer CancerPPD, and the defensins knowledgebase (Tornesello et al. 2020).

5.1 Human β-Defensins and Head and Neck Cancer

Human β-defensins (hBDs) are expressed in the oral cavity, gingiva, dental pulp, tongue, salivary glands, and mucosa, among other epithelial tissues in the human body. These are some of the main AMPs that are retrieved in the oral cavity. Three disulfide bonds and a β-sheet make up the easy structure of these tiny, cationic peptides. Because of their cationic properties, hBDs' antimicrobial actions exhibit broad-spectrum antibacterial, antifungal, and antiviral effects via depolarizing and permeabilizing microbial cell membranes. The α-defensins and β-defensins are the

Table 12.1 Minor therapeutic classes of antimicrobial peptides

Sr. no.	Class	Feature
1.	LTX-302	It is derived from its parental peptide LfcinB, 9-mer peptide, and α-helical secondary structure and has stronger anticancer effects (Berge et al. 2010).
2.	Tilapia hepcidins (THs) 1–5	These represent three hepcidin-like AMPs that were isolated from tilapia (*Oreochromis mossambicus*): THs 1–5, TH 2–2, and THs 2–3, discovered that in a variety of human cancer cell lines, THs 1–5 reduced colony formation and accelerated cell death (Chang et al. 2011).
3.	Buforin II	This is a 21-residue α-helical AMP that shares sequence homology with histone H2A's N-terminal region. An artificially created analog of buforin II, buforin IIb, is altered to have a proline residue sandwiched between two α-helices (Lee et al. 2008).
4.	LL-37	It belongs to the class of α-helical AMPs generated from human cathelicidin derivatives. Its precursor, hCAP18, is a signaling molecule and peptide antibiotic that is present in bodily fluids (Dürr et al. 2006).
5.	Chrysophsin-1	This is an α-helical AMP that is unique to red sea bream gill cells because of its hemolytic and antibacterial activities (Mason et al. 2007).
6.	Dermaseptins B1–B6	The skin secretions of the South American frog *Phyllomedusa bicolor* are constitutively expressed with six similar peptides ranging in length from 24 to 33 residues (Van Zoggel et al. 2012).
7.	Gaegurins	These are a family of six AMPs that were identified from *Rana rugosa*. Based on their length and sequence, they can be roughly categorized into two subfamilies. It has strong anticancer effects in the family II peptide GGN6 and its derivative PTP7 (Kim et al. 2003).
8.	Polybia-MPI	This is a cationic peptide isolated from *Polybia paulista*, a social wasp. Despite being largely nontoxic to human red blood cells and normal fibroblasts, polybia-MPI has strong bactericidal (against both gram-positive and gram-negative bacteria), fungicidal, and tumoricidal biochemical characteristics (Wang et al. 2008).
9.	Temporin-1CEa	This is a skin-secreted amphipathic AMP from the Chinese brown frog *Rana chensinensis*. Research has indicated that the application of temporin-1CEa to many human cancer cell lines causes tumor cell death quickly by damaging the cells' mitochondria and cell membranes (Che et al. 2015).
10.	D-K6L9	Phosphatidylserine binds to this peptide and can cause tumor necrosis. Combinations containing D-K6L9 and BP1 peptide, interleukin 12 (IL-12), and glycyrrhizin (an inhibitor of HMGB1 protein) show antitumor efficaciousness (Cichoń et al. 2014).

two categories of human oral defensins. The kind of defensin is determined by differences in the patterns connecting the three disulfide connections and the distance between cysteines (Kompuinen et al. 2023).

These antimicrobial peptides are produced and subsequently found in nasal secretions, saliva, gingival crevicular fluid, and gingival tissues (Öztürk et al. 2020). It has also been shown that several cell types, including monocytes, macrophages, odontoblasts, keratinocytes, and monocyte-derived dendritic cells, can express hBDs in addition to epithelium (Kompuinen et al. 2023). Only four (hBDs 1–4) of the 28 distinct hBDs found in the human body have been found in the gingival epithelium, according to genomic research (Schutte et al. 2002). Up to now, several studies have assessed how head and neck cancer patients' hBD levels have changed clinically (Table 12.2).

Table 12.2 Human beta defensins in head and neck cancers in different studies

Sr. no.	Cancer type	Study population	Findings
1.	Oral squamous cell carcinoma (OSCC)	Case (Oral squamous cell carcinoma) = Thirty Control = Fifteen Leukoplakia = Seventeen	When case (oral squamous cell carcinoma) is compared to leukoplakia, human β-defensin 1 was lower and is reduced in oral squamous cell carcinoma with metastases of lymph nodes (Han et al. 2014).
		Case (Oral squamous cell carcinoma) = Sixty Control = Fifteen Leukoplakia = Thirty	In oral squamous cell carcinoma, human β-defensins 1, 2, and 3 were decreased. Human β-defensins 2 and 3 were associated with oral squamous cell carcinoma lymph node metastases (Wang et al. 2014).
		Case (Oral squamous cell carcinoma) = Seventy-nine Control = Seventy-nine	In oral squamous cell carcinoma, human β-defensin-3 protein and mRNA expressions are increased (Yu et al. 2011).
		Case (Oral squamous cell carcinoma) = Forty-five Control = Sixteen	In oral squamous cell carcinoma, human β-defensin-3 mRNA was increased (Kesting et al. 2009).
		Case (Oral squamous cell carcinoma) = Five Control = Five Irritation fibroma = Five Leukoplakia = Five	Decreased genetic expression of human β-defensin-1, β-defensin-2, and β-defensin-3 gene expression was higher in the oral squamous cell carcinoma (Wenghoefer et al. 2008a).
		Case (Oral squamous cell carcinoma) = Twenty-nine Control = Twenty-three	In contrast to stages 1 and 4, human β-defensin 3 was raised in oral squamous cell carcinoma stages 2 and 3 (Hussaini et al. 2006).
		Case (Oral squamous cell carcinoma) = Four	Downregulation of human β-defensin in squamous cell carcinoma (Abiko et al. 1999).
2.	Pleomorphic adenoma (benign tumor of the salivary gland)	Case (Pleomorphic adenoma) = Five Control = Five Adjacent normal tissue with adenoma = Five Chronic sialadenitis = Five	Lower levels of human β-defensin-1 expression were seen in the case in contrast to controls (Pantelis et al. 2009).

(continued)

Table 12.2 (continued)

Sr. no.	Cancer type	Study population	Findings
3.	Esophageal squamous cell carcinoma	Case (Squamous cell carcinoma) = Fifty-eight Control = Fifty	Tumors had greater human β-defensin-2 immunohistochemistry (IHCI) scores than normal tissues (Shi et al. 2014).
4.	Tumor of salivary gland	Case (Salivary gland neoplasm malign) = Seven (Salivary gland neoplasm benign) = Ten Control = Forty-one	Salivary samples have lower levels of human β-defensin-1, β-defensin-2, and β-defensin-3 expression tissue biopsies from gland tumors (Kesting et al. 2012).
5.	Carcinoma of Tonsil	Carcinoma of Tonsil = Seventeen Nasal polyposis = Fifty Tonsilitis (chronic) = Eleven	Secretions of human β-defensins 1, 2, and 3 were reduced during malignant transformation (Pácová et al. 2009).
6.	Respiratory papillomatosis (recurrent)	Case (RPP) = Fifteen Control = Ten	The levels of human β-defensin-1, β-defensin-2, and β-defensin-3 mRNAs in RPP were greater than healthy tissues of the mouth mucosa (Chong et al. 2006).
7.	Tumors of salivary gland	Case (Salivary gland tumor malign) = Seven (Salivary gland tumor benign) = Seven Control = Seven	The nucleus of salivary gland tumors (malignant) was where human β-defensin 1 migrated (Wenghoefer et al. 2008b).
8.	Squamous cell carcinoma of tongue	Case (Squamous cell carcinoma of the tongue with no regional metastasis) = Twenty-three (Squamous cell carcinoma of the tongue with positive regional metastasis) = Twelve	Decreased expression of genes linked to defensin is linked to localized metastases of early-stage tongue carcinoma to the neck (Lee et al. 2022).
9.	Tumors of the head and neck	Case (Tumors of head and neck) = Fifteen Control (Healthy gingiva) = Fifteen	The expression of human β-defensins 1, 2, and 3 is elevated in oral squamous cell carcinoma (Hoppe et al. 2016).

5.2 Leucine-Leucine-37 and Head and Neck Cancer

The vertebrate host defense peptides known as cathelicidins are a class of proteins; only human cationic antimicrobial protein 18 (hCAP18) has been found; this protein undergoes proteolytic processing to produce active leucine-leucine-37 (LL-37). Well-expressed in immunological and epithelial cells, hCAP18/LL-37 has strong antibacterial properties against a variety of pathogens and plays crucial roles in immune system regulation and inflammation. Furthermore, overexpression of human host defense cationic antimicrobial peptide-18/antimicrobial peptide also plays a complex role in the formation of cancer. It has been demonstrated to inhibit the growth of tumors in the stomach, oral, and hematological cancers.

Normal human tongue and oral mucosa contain hCAP18/LL-37 at transcriptional and protein levels. It was recently shown that in comparison to normal oral mucosa, poorly differentiated oral squamous cell carcinomas (OSCC) expressed weakly hCAP18/LL-37. Tumor development and lymph node metastasis were also linked to decreased expression of hCAP18/LL-37. However, it is yet unknown how hCAP18/LL-37 functions in OTSCC (Vierthaler et al. 2020).

6 FDA-Approved Anticancer AMPs

While many AMPs exhibit toxicity to the intended condition, a significant number also exhibit side effects by impacting other organs. The FDA has only licensed a very small number of AMPs for clinical use. It is advised to only use intravenously or topically because of their extremely short half-lives.

The benefit of AMP development progress can be summed up as each AMP's powerful, targeted, and secure manner of action. Additionally, synthetic analogs and derivatives seemed to be more effective in treating illnesses brought on by germs resistant to drugs. Table 12.3 contains a few examples of FDA-approved Aaticancer AMPs in different cancers (Divyashree et al. 2020).

7 Conclusion

The ACP research has been driven by the pressing need to obtain novel agents for cancer prevention and treatment. Given this, ACPs are thought to be attractive agents for developing novel antitumor medications and immunizations. Clinical

Table 12.3 FDA-approved anticancer AMPs in different cancers

Name	Use
Multiferon/natural interferon-alpha	Used for the treatment of metastatic renal cell carcinoma
Pliditepsin (Aplidin)	Cancer
PM060184	Cancer
Marizomib	Cancer
Human defensin 6	Colon cancer
Lysozyme	Colon cancer
Cathelicidin LL-37	Colon cancer
Human defensin 5	Colon cancer
Human neutrophil peptides 1	Colon cancer
Hepcidin	Colon cancer

trials assessing cancer treatment are covered in this review since some ACPs are proapoptotic and antiproliferative in a variety of cancer cell types, both in vitro and in vivo. The challenges affect every known AMP-based trend that is currently in existence. In this sense, the design of novel ACP agents will be greatly impacted by the ongoing advancement of computational techniques that have contributed to a better understanding of the ACP mechanism of action, including the identification of novel molecular targets.

Pharmaceutical firms continue to find natural and synthesized AMPs to be appealing sources despite the numerous obstacles that must be overcome before clinical applications can be implemented. Genomic and proteomic technologies in conjunction with virtual screening, ACP prediction, and data mining can help advance the commercial development of anticancer peptides. When combined, ACPs may signal the beginning of a fresh approach to creating vaccinations or anticancer medications that will lower the number of new cases and death rates from cancer (Jafari et al. 2022).

References

Abiko Y, Mitamura J, Nishimura M, Muramatsu T, Inoue T, Shimono M, Kaku T (1999) Pattern of expression of beta-defensins in oral squamous cell carcinoma. Cancer Lett 143(1):37–43. https://doi.org/10.1016/s0304-3835(99)00171-8

Berge GM, Eliassen LT, Camilio KA, Bartnes K, Sveinbjørnsson B, Rekdal Ø (2010) Therapeutic vaccination against a murine lymphoma by intratumoral injection of a cationic anticancer peptide. Cancer Immunol Immunother 59(8):1285–1294. https://doi.org/10.1007/s00262-010-0857-6

Büyükkiraz ME, Kesmen Z (2022) Antimicrobial peptides (AMPs): a promising class of antimicrobial compounds. J Appl Microbiol 132(3):1573–1596. https://doi.org/10.1111/jam.15314

Cardoso MH, Oshiro KGN, Rezende SB, De Souza Cândido E, Franco OL (2018) The structure/function relationship in antimicrobial peptides: what can we obtain from structural data? Adv Protein Chem Struct Biol 112:359–384. https://doi.org/10.1016/bs.apcsb.2018.01.008

Chang W, Pan C, Rajanbabu V, Cheng C, Chen J (2011) Tilapia (Oreochromis mossambicus) antimicrobial peptide, hepcidin 1–5, shows antitumor activity in cancer cells. Peptides 32(2):342–352. https://doi.org/10.1016/j.peptides.2010.11.003

Che W, Chen Y, Zhang L, Gong X, Zhou Y, Shang D (2015) Melanoma cell surface-expressed phosphatidylserine as a therapeutic target for cationic anticancer peptide, temporin-1CEa. J Drug Target 24(6):548–556. https://doi.org/10.3109/1061186x.2015.1113539

Chong KT, Xiang L, Wang X, Jun EL, Xi L, Schweinfurth JM (2006) High level expression of human epithelial beta-defensins (hBD-1, 2 and 3) in papillomavirus induced lesions. Virol J 3(1):75. https://doi.org/10.1186/1743-422x-3-75

Cichoń T, Smolarczyk R, Matuszczak S, Barczyk M, Jarosz M, Szala S (2014) D-K6L9 peptide combination with IL-12 inhibits the recurrence of tumors in mice. Arch Immunol Ther Exp 62(4):341–351. https://doi.org/10.1007/s00005-014-0268-z

Divyashree M, Mani MK, Reddy D, Kumavath R, Ghosh P, Azevedo V, Barh D (2020) Clinical applications of antimicrobial peptides (AMPs): where do we stand now? Protein Pept Lett 27(2):120–134. https://doi.org/10.2174/0929866526666190925152957

Dürr U, Sudheendra US, Ramamoorthy A (2006) LL-37, the only human member of the cathelicidin family of antimicrobial peptides. Biochim Biophys Acta 1758(9):1408–1425. https://doi.org/10.1016/j.bbamem.2006.03.030

Eliassen LT, Berge GM, Leknessund A, Wikman M, Lindin I, Løkke C, Ponthan F, Johnsen JI, Sveinbjørnsson B, Kogner P, Flægstad T, Rekdal Ø (2006) The antimicrobial peptide, lactoferricin B, is cytotoxic to neuroblastoma cells in vitro and inhibits xenograft growth in vivo. Int J Cancer 119(3):493–500. https://doi.org/10.1002/ijc.21886

Fleming A, Allison VD (1922) Observations on a bacteriolytic substance ("Lysozyme") found in secretions and tissues. Br J Exp Pathol 3(5):252–260. http://europepmc.org/articles/PMC2047739

Ganz T, Lehrer RI (1995) Defensins. Pharmacol Ther 66(2):191–205. https://doi.org/10.1016/0163-7258(94)00076-f

Gaspar D, Veiga AS, Castanho MARB (2013) From antimicrobial to anticancer peptides. A review. Front Microbiol 4:294. https://doi.org/10.3389/fmicb.2013.00294

Gerashchenko OL, Zhuravel E, Skachkova O, Khranovska NN, Filonenko V, Pogrebnoy P, Soldatkina M (2013) Biologic activities of recombinant human-beta-defensin-4 toward cultured human cancer cells. Exp Oncol 35(2):76–82. https://pubmed.ncbi.nlm.nih.gov/23828379

Gerashchenko OL, Zhuravel E, Skachkova O, Khranovska NN, Pushkarev V, Pogrebnoy P, Soldatkina M (2014) Involvement of human beta-defensin-2 in regulation of malignant potential of cultured human melanoma cells. Exp Oncol 36(1):17–23. https://pubmed.ncbi.nlm.nih.gov/24691279

Gormley M, Creaney G, Schache A, Ingarfield K, Conway DI (2022) Reviewing the epidemiology of head and neck cancer: definitions, trends and risk factors. Br Dent J 233(9):780–786. https://doi.org/10.1038/s41415-022-5166-x

Han Q, Wang R, Sun C, Jin X, Liu D, Zhao X, Wang L, Ji N, Li J, Zhou Y, Ye L, Liang X, Jiang L, Liao G, Dan H, Zeng X, Chen Q (2014) Human beta-defensin-1 suppresses tumor migration and invasion and is an independent predictor for survival of oral squamous cell carcinoma patients. PLoS One 9(3):e91867. https://doi.org/10.1371/journal.pone.0091867

Hancock REW, Brown KL, Mookherjee N (2006) Host defence peptides from invertebrates – emerging antimicrobial strategies. Immunobiology 211(4):315–322. https://doi.org/10.1016/j.imbio.2005.10.017

Hoppe T, Kraus D, Novak N, Probstmeier R, Frentzen M, Wenghoefer M, Jepsen S, Winter J (2016) Oral pathogens change proliferation properties of oral tumor cells by affecting gene expression of human defensins. Tumour Biol 37(10):13789–13798. https://doi.org/10.1007/s13277-016-5281-x

Huan Y, Kong Q, Mou H, Yi H (2020) Antimicrobial peptides: classification, design, application, and research progress in multiple fields. Front Microbiol 11:582779. https://doi.org/10.3389/fmicb.2020.582779

Hussaini HM, Lim P, Rahman R, Abraham T, Cheong C, Zain RB, Ismail M, Nam A (2006) Expression of human b-defensin-3 in oral squamous cell carcinoma using tissue microarray – a preliminary study. J Oral Pathol Med 35:432–433. https://doi.org/10.5281/zenodo.811215

Jafari A, Babajani A, Sarrami-Forooshani R, Yazdani M, Rezaei-Tavirani M (2022) Clinical applications and anticancer effects of antimicrobial peptides: from bench to bedside. Front Oncol 12:819563. https://doi.org/10.3389/fonc.2022.819563

Jenssen H, Hamill P, Hancock REW (2006) Peptide antimicrobial agents. Clin Microbiol Rev 19(3):491–511. https://doi.org/10.1128/cmr.00056-05

Johnson DE, Burtness B, Leemans CR, Lui VWY, Bauman JE, Grandis JR (2020) Head and neck squamous cell carcinoma. Nat Rev Dis Prim 6(1):92. https://doi.org/10.1038/s41572-020-00224-3

Kesting M, Loeffelbein DJ, Hasler RJ, Wolff K, Rittig A, Schulte M, Hirsch T, Wagenpfeil S, Jacobsen F, Steinstraesser L (2009) Expression profile of human beta-defensin 3 in oral squamous cell carcinoma. Cancer Investig 27(5):575–581. https://doi.org/10.1080/07357900802620851

Kesting M, Stoeckelhuber M, Kuppek A, Hasler RJ, Rohleder NH, Wolff K, Nieberler M (2012) Human B-defensins and psoriasin/S100A7 expression in salivary glands. BioDrugs 26(1):33–42. https://doi.org/10.2165/11597570-000000000-00000

Kim S, Kim SS, Bang YJ, Kim S, Lee BJ (2003) In vitro activities of native and designed peptide antibiotics against drug sensitive and resistant tumor cell lines. Peptides 24(7):945–953. https://doi.org/10.1016/s0196-9781(03)00194-3

Kompuinen J, Keskin M, Yılmaz D, Gürsoy M, Gürsoy UK (2023) Human B-befensins in diagnosis of head and neck cancers. Cells 12(6):830. https://doi.org/10.3390/cells12060830

Krishnamurthy S, Dong Z, Vodopyanov D, Imai A, Helman JI, Prince ME, Wicha MS, Nör JE (2010) Endothelial cell-initiated signaling promotes the survival and self-renewal of cancer stem cells. Cancer Res 70(23):9969–9978. https://doi.org/10.1158/0008-5472.can-10-1712

Kumar P, Kizhakkedathu JN, Straus SK (2018) Antimicrobial peptides: diversity, mechanism of action and strategies to improve the activity and biocompatibility in vivo. Biomolecules 8(1):4. https://doi.org/10.3390/biom8010004

Lee HS, Park CB, Kim JM, Jang S, Park IY, Kim MS, Cho JH, Kim SC (2008) Mechanism of anticancer activity of buforin IIb, a histone H2A-derived peptide. Cancer Lett 271(1):47–55. https://doi.org/10.1016/j.canlet.2008.05.041

Lee NC, Kelly J, Park HS, An Y, Judson BL, Burtness B, Husain Z (2018) Patterns of failure in high-metastatic node number human papillomavirus-positive oropharyngeal carcinoma. Oral Oncol 85:35–39. https://doi.org/10.1016/j.oraloncology.2018.08.001

Lee DY, Hah JH, Jeong WJ, Chung E, Kwon T, Ahn SH, Sung M, Kwon SK (2022) The expression of defensin-associated genes may be correlated with lymph node metastasis of early-stage tongue cancer. Clin Exp Otorhinolaryngol 15(4):372–379. https://doi.org/10.21053/ceo.2022.00150

Li D, Wang W, Shi H, Fu Y, Chen X, Chen X, Liu Y, Kan B, Wang Y (2014) Gene therapy with beta-defensin 2 induces antitumor immunity and enhances local antitumor effects. Hum Gene Ther 25(1):63–72. https://doi.org/10.1089/hum.2013.161

Mason AJ, Bertani P, Moulay G, Marquette A, Perrone B, Drake AF, Kichler A, Bechinger B (2007) Membrane interaction of chrysophsin-1, a histidine-rich antimicrobial peptide from red sea bream. Biochemistry 46(51):15175–15187. https://doi.org/10.1021/bi701344m

Mehanna H, Beech T, Nicholson T, El-Hariry I, McConkey CC, Paleri V, Roberts S (2012) Prevalence of human papillomavirus in oropharyngeal and nonoropharyngeal head and neck cancer—systematic review and meta-analysis of trends by time and region. Head Neck 35(5):747–755. https://doi.org/10.1002/hed.22015

Öztürk A, Bayrakdar SK, Avcı B (2020) Comparison of gingival crevicular fluid and serum human beta-defensin-2 levels between periodontal health and disease. Oral Dis 27(4):993–1000. https://doi.org/10.1111/odi.13597

Pácová H, Astl J, Martínek J (2009) The pathogenesis of chronic inflammation and malignant transformation in the human upper airways: the role of beta-defensins, eNOS, cell proliferation and apoptosis. Histol Histopathol 24(7):815–820. https://doi.org/10.14670/hh-24.815

Pantelis A, Wenghoefer M, Haas S, Merkelbach-Bruse S, Pantelis D, Jepsen S, Bootz F, Winter J (2009) Down regulation and nuclear localization of human β-defensin-1 in pleomorphic adenomas of salivary glands. Oral Oncol 45(6):526–530. https://doi.org/10.1016/j.oraloncology.2008.07.014

Roudi R, Syn N, Roudbary M (2017) Antimicrobial peptides as biologic and immunotherapeutic agents against cancer: a comprehensive overview. Front Immunol 8:1320. https://doi.org/10.3389/fimmu.2017.01320

Schutte BC, Mitros J, Bartlett J, Walters JD, Jia HP, Welsh MJ, Casavant TL, McCray PB (2002) Discovery of five conserved β-defensin gene clusters using a computational search strategy. Proc Natl Acad Sci USA 99(4):2129–2133. https://doi.org/10.1073/pnas.042692699

Shi N, Jin F, Zhang X, Clinton SK, Pan Z, Chen T (2014) Overexpression of human β-defensin 2 promotes growth and invasion during esophageal carcinogenesis. Oncotarget 5(22):11333–11344. https://doi.org/10.18632/oncotarget.2416

Strzelecka P, Czaplińska D, Sądej R, Wardowska A, Pikuła M, Lesner A (2017) Simplified, serine-rich theta-defensin analogues as antitumour peptides. Chem Biol Drug Des 90(1):52–63. https://doi.org/10.1111/cbdd.12927

Tam JP, Wang S, Wong KH, Tan W (2015) Antimicrobial peptides from plants. Pharmaceuticals 8(4):711–757. https://doi.org/10.3390/ph8040711

Tornesello AL, Borrelli A, Buonaguro FM, Tornesello ML (2020) Antimicrobial peptides as anticancer agents: functional properties and biological activities. Molecules 25(12):2850. https://doi.org/10.3390/molecules25122850

Van Zoggel H, Carpentier G, Santos CD, Hamma-Kourbali Y, Courty J, Amiche M, Delbé J (2012) Antitumor and angiostatic activities of the antimicrobial peptide dermaseptin B2. PLoS One 7(9):e44351. https://doi.org/10.1371/journal.pone.0044351

Vierthaler M, Rodrigues PC, Sundquist E, Siponen M, Salo T, Risteli M (2020) Fluctuating role of antimicrobial peptide hCAP18/LL-37 in oral tongue dysplasia and carcinoma. Oncol Rep 44(1):325–338. https://doi.org/10.3892/or.2020.7609

Wang K, Zhang B, Zhang W, Yan J, Li J, Wang R (2008) Antitumor effects, cell selectivity and structure–activity relationship of a novel antimicrobial peptide polybia-MPI. Peptides 29(6):963–968. https://doi.org/10.1016/j.peptides.2008.01.015

Wang X, Jiang W, Duan N, Qian Y, Zhou Q, Ye P, Jiang H, Bai Y, Zhang W, Wang W (2014) NOD1, RIP2 and Caspase12 are potentially novel biomarkers for oral squamous cell carcinoma development and progression. Int J Clin Exp Pathol 7(4):1677–1686. https://pubmed.ncbi.nlm.nih.gov/24817964

Wenghoefer M, Pantelis A, Dommisch H, Reich R, Martini M, Allam J, Novak N, Bergé S, Jepsen S, Winter J (2008a) Decreased gene expression of human β-defensin-1 in the development of squamous cell carcinoma of the oral cavity. Int J Oral Maxillofac Surg 37(7):660–663. https://doi.org/10.1016/j.ijom.2008.02.003

Wenghoefer M, Pantelis A, Dommisch H, Götz W, Reich R, Bergé S, Martini M, Allam J, Jepsen S, Merkelbach-Bruse S, Fischer H, Novak N, Winter J (2008b) Nuclear hBD-1 accumulation in malignant salivary gland tumours. BMC Cancer 8(1):290. https://doi.org/10.1186/1471-2407-8-290

Xuan J, Feng W, Wang J, Wang R, Zhang B, Bo L, Chen Z, Yang H, Sun L (2023) Antimicrobial peptides for combating drug-resistant bacterial infections. Drug Resist Updat 68:100954. https://doi.org/10.1016/j.drup.2023.100954

Yang D, Biragyn A, Hoover DM, Łubkowski J, Oppenheim JJ (2004) Multiple roles of antimicrobial defensins, cathelicidins, and eosinophil-derived neurotoxin in host defense. Annu Rev Immunol 22(1):181–215. https://doi.org/10.1146/annurev.immunol.22.012703.104603

Yu S, Wang F, Tang J, Huang H, Wang H, Ma P, Zheng LW, Zwahlen RA, Yang H (2011) Human beta-defensin-3 (hBD-3) upregulated by LPS via epidermal growth factor receptor (EGFR) signaling pathways to enhance lymphatic invasion of oral squamous cell carcinoma. Oral Surg Oral Med Oral Pathol Oral Radiol Endod 112(5):616–625. https://doi.org/10.1016/j.tripleo.2011.02.053

Zhang L, Gallo RL (2016) Antimicrobial peptides. Curr Biol 26(1):R14–R19. https://doi.org/10.1016/j.cub.2015.11.017

Chapter 13
Antimicrobial Peptides in Reproductive Health and Therapeutic Applications

Pradeep Nag B. Subramanyam and Sai Goutham Reddy Yeddula

1 Introduction

Reproduction is the process by which organisms produce offspring, ensuring the continuation and expansion of their genetic lineage. It involves the generation of male and female gametes, their fusion to form a zygote, their development into a blastocyst, and the implantation of a viable embryo in the uterus culminating in fully functional offspring (Wassarman 1987). During this complex process, the reproductive organs are predisposed to a variety of microorganisms that may cause infections. Additionally, hormonal fluctuations that occur throughout the reproductive cycle make it more susceptible to infections during certain stages of the cycle. The establishment of infection occurs when pathogenic bacteria take the upper hand over the host defense. Fortunately, antibiotics have been instrumental in combating these pathogenic organisms (Liu et al. 2022). However, the emergence of antimicrobial resistance presents a significant challenge in therapeutic management. Considering these challenges, the discovery and exploration of antimicrobial peptides (AMPs) within the body have emerged as a promising avenue for the development of novel therapeutic strategies against infections of the reproductive tract.

AMPs, also recognized as host defense peptides, are naturally occurring molecules ubiquitously distributed in diverse organisms, encompassing bacteria to humans. These peptides (7–100 amino acids) exhibit potent antimicrobial properties, capable of targeting and neutralizing a broad spectrum of pathogens, including bacteria, viruses, fungi, and even parasites (Ageitos et al. 2017). The allure of AMPs lies not only in their robust antimicrobial activity but also in their unique mechanisms of action, which often involve disrupting microbial cell membranes or interfering with essential cellular processes (Bin Hafeez et al. 2021). Moreover, AMPs possess remarkable versatility and plasticity, with the ability to modulate immune

P. N. B. Subramanyam (✉) · S. G. R. Yeddula
Department of Animal Sciences, University of Missouri, Columbia, MO, USA

P. Baindara, S. M. Mandal (eds.), *Evolution of Antimicrobial Peptides*,
https://doi.org/10.1007/978-3-031-67515-7_13

responses, promote tissue repair, and even exert anti-inflammatory effects (Huan et al. 2020).

In the context of the reproductive system, AMPs hold immense promise as novel therapeutic agents for combating infections and preserving reproductive health. These peptides are naturally synthesized and secreted by various reproductive tissues, where they serve as frontline defenders against invading pathogens. In the male reproductive tract, AMPs play crucial roles in safeguarding spermatozoa from microbial assault, maintaining optimal sperm function, and protecting against sexually transmitted infections (Valdez-Miramontes et al. 2021). Similarly, in the female reproductive system, AMPs contribute to the innate immune defenses of the vagina, cervix, uterus, and placenta, helping to prevent ascending infections and safeguarding the developing embryo (Yarbrough et al. 2015). Presently, the Antimicrobial Peptide Database (ADP3) reports a total of 3940 AMPs with actions against bacteria, viruses, fungi, and parasites, and some demonstrate anticancer properties. Out of these peptides, 2463 peptides are secreted from animals (Wang et al. 2016). Some of these peptides have been approved by the Food and Drug Administration (FDA) for use as antibacterial, antiviral, and antitumor effects (Mousavi Maleki et al. 2023). Furthermore, the potential of AMPs extends beyond their antimicrobial properties, with emerging evidence suggesting their involvement in modulating reproductive processes such as sperm maturation, ovulation, and implantation. Additionally, some AMPs exhibit spermicidal activity, raising intriguing possibilities for their use as nonhormonal contraceptives. This chapter deals with the role of AMPs in the male and female reproductive system.

2 AMPs of the Reproductive System

While numerous AMPs have been identified across various biological contexts, the repertoire within the reproductive system remains relatively limited. In this chapter, we delve into the specific AMPs documented within the reproductive system, highlighting their diverse roles and contributions. A brief of the antimicrobial peptides involved in reproduction has been given in Table 13.1. Notable among these are defensins, lactoferrin, lysozyme, cathelicidin, calprotectin, elafin, secretory leukocyte protease inhibitor (SLPI), and bacteriocins, each elucidated in detail.

2.1 *Defensins*

Defensins are cysteine-rich cationic proteins that are found in plants, insects, and animals. In animals, defensins contain 18–45 amino acids and can be divided into three subclasses, including α, β, and θ (Ganz 2004). This classification is based on the spatial distribution of the cysteine residues bonds and organ distribution. Humans possess only alpha and beta subtypes; alpha defensins are relatively smaller

Table 13.1 AMPs involved in reproduction, its regulation, and associated pathologies

AMP name	Location in the reproductive tract	Role and mechanism	Regulation and variability	Associated pathologies
Defensins	Various parts of the male and female reproductive tracts, placental tissues	Antiviral, antibacterial, and immunomodulatory. Disrupts microbial cell membranes.	Hormonally regulated. It varies with infection status.	Linked to reduced risk of ascending infections and imbalance associated with preterm birth (Zhai et al. 2023).
hCAP18/ LL37	Endometrium, vagina, cervix, ovary	Antimicrobial and spermicidal. Recruits immune cells to infection sites.	Expression increases during the menstrual cycle and in response to microbial invaders.	Dysregulation associated with bacterial vaginosis and pelvic inflammatory disease (Frew et al. 2014).
SLPI	Fallopian tube, endometrium, cervix, cervicovaginal fluid	Protease inhibition, antimicrobial, anti-inflammatory, and immunomodulatory.	Levels rise in response to estrogen and decrease in endometriosis and certain infections. Overexpression can occur in cancers.	Lower levels were observed in cases of endometriosis and infertility; excessive expression is associated with increased metastatic potential in epithelial tumors, indicating a role in cancer progression (Nugteren and Samsom 2021).
Elafin	Endometrium	Antiprotease and antimicrobial against gram-positive and gram-negative bacteria and anti-inflammatory.	Expressed during menstruation and modulated by inflammation and microbial presence.	Upregulated in ectopic pregnancies (Zakizadeh et al. 2020); reduced levels in recurrent bacterial vaginosis (Stock et al. 2009); potential marker for cervical cancer risk (Clauss et al. 2010).

(continued)

Table 13.1 (continued)

AMP name	Location in the reproductive tract	Role and mechanism	Regulation and variability	Associated pathologies
Lysozyme	Cervix and vagina in FRT and seminal fluid	Lyses bacterial cell walls, providing broad-spectrum antimicrobial protection.	It varies with microbial presence and hormonal changes.	Lower levels may be associated with increased susceptibility to infections. Imbalances contribute to prostatitis, bacterial vaginosis, and infertility (Mårdh and Colleen 1974).
Human epididymis protein 4 (HE4)	Epididymis, fallopian tubes, uterus, cervix, and Bartholin glands	Involved in sperm maturation. Plays a role in cell adhesion and motility.	Expression levels may vary with hormonal changes and are influenced by pathological conditions, especially in ovarian cancer.	It is primarily associated with ovarian cancer. Its levels are significantly elevated in patients with epithelial ovarian cancer (Anastasi et al. 2023).
Calprotectin	Cervix and cervicovaginal fluid	Binds zinc and manganese, inhibiting microbial growth, and modulates inflammation.	It varies in conditions like polycystic ovary syndrome, endometriosis, preeclampsia.	Elevated levels are associated with inflammatory conditions of the reproductive tract and polycystic ovary syndrome (Chen et al. 2017).
Psoriasin (S100A7)	Epithelium of reproductive tracts	Antimicrobial, particularly against *E. coli*, and involved in inflammatory response.	Expression induced by inflammatory cytokines.	Linked to the progression of endometrial carcinoma and potentially other inflammatory diseases or cancers within the reproductive tract (Jia et al. 2014).

(continued)

Table 13.1 (continued)

AMP name	Location in the reproductive tract	Role and mechanism	Regulation and variability	Associated pathologies
SP-A	Vagina	Binds pathogens, enhancing their clearance, and modulates the immune response.	SP-A and SP-D crucially prime the myometrium for labor, activating genes and cytokines that transition the uterus from rest to contraction readiness (Sotiriadis et al. 2015).	Deficiency linked to increased risk of vaginal infections.
SP-D	Endometrium, cervix, vagina, fallopian tubes	Anti-inflammatory effects. Opsonizes pathogens for clearance.		Altered levels associated with reproductive tract infections and inflammation.
Lactoferrin	Ovary, fallopian tubes, endometrium	Anti-inflammatory properties. Binds iron, inhibiting bacterial growth.	Influenced by the menstrual cycle and the presence of pathogens.	Implicated in reducing the risk of infections; potential therapeutic in fertility treatments.

molecules (3.5–4 kDa) (Selsted et al. 1985) than beta defensins (4–6 kDa) (Selsted and Ouellette 2005). These small molecules have a complex tertiary structure with three antiparallel beta-sheets that resemble chemokines. Alpha defensins 1–4 are also termed human neutrophil peptides (HNPs) and are primarily expressed in neutrophils but also expressed in natural killer cells, monocytes, and a few subtypes of T lymphocytes (Xu and Lu 2020). Alpha defensins 5 and 6 are expressed in the Paneth cells that are found in the intestinal crypts (Lehrer and Lu 2012). Human beta defensins (HβDs) are higher in number, over 26 transcriptionally active genes of human beta defensins have been reported among them, and beta defensins 1–4 are widely known to be expressed in mucosal surfaces (Gao et al. 2021). The beta defensins 5–6 are specifically expressed in the epididymis, and HβDs 1–4 have been found in the urogenital mucosal lining, while HβDs 5 and 6 are specific to the epididymis. Although few other beta defensins are identified (HβDs 14, 18, 23, and 26) in epididymis, the functions of these beta defensins remain unknown (Batra et al. 2022). In the female reproductive tract, defensins are present in the mucosal linings of the cervix and vagina robust barrier against pathogens through mechanisms such as pore formation in microbial membranes (Yarbrough et al. 2015). Theta defensins are present in neutrophils, weakly in monocytes while absent in lymphocytes and eosinophils (Tang et al. 1999). The higher concentration of theta defensins in inflammatory exudates indicates their role in fighting microbes (Garcia et al. 2008). The defensins act on the bacterial cell membrane resulting in the formation of pores on the membrane destroying bacteria (Pazgier et al. 2012); few of the defensins act by preventing bacterial adhesion and invasion (Chu et al. 2012). Theta defensins release proteolytic enzymes that lyse the bacterial cell wall alongside the membrane destruction (Wilmes et al. 2014). Overall, defensins are secreted from major immune cells, including neutrophils, natural killer cells, and lymphocytes, and aid in microbial invasion by the innate immune system.

2.2 *Lactoferrin*

Lactoferrin is a nonheme cationic monomeric glycoprotein that belongs to the transferrin family and has an affinity toward iron (Baker and Baker 2009). Lactoferrin can bind to two ferric ions. The iron-free form of lactoferrin is called apo-lactoferrin, while the form bound to one ferric ion is called mono-lactoferrin, and the form bound to two ferric ions is called holo-lactoferrin (Baker and Baker 2004). Lactoferrins are produced by neutrophils as secondary granules and are stored as glycoproteins (Rosa et al. 2017). Lactoferrin acts as an acute-phase protein, which increases during infection (Van Snick et al. 1974). Lactoferrin has a role in cell proliferation and differentiation and exerts immunomodulatory and anti-inflammatory effects (Actor et al. 2009). Additionally, lactoferrins are also involved in protecting cells against the development of cancer and metastasis (Tsuda et al. 2006). Lactoferrin exerts a static effect on the microbes by sequestering the iron that is required for the replication of bacteria (Legrand et al. 2005). Further, they can also cause permeabilization of both gram-positive and gram-negative bacteria, resulting in cell lysis and death (Appelmelk et al. 1994). Additionally, lactoferrin is involved in the disaggregation of the biofilm by different mechanisms (DNase and antiprotease activity and a decrease in the synthesis of exopolysaccharides and cellulose) (Zarzosa-Moreno et al. 2020). Furthermore, it is involved in decreasing the production of toxins and reducing the adhesion of the bacteria to the cell (Asadi et al. 2019). Overall, lactoferrins are key molecules involved in the immune system and responsible for microbial defense. Further, the anti-inflammatory and anticancer properties of this AMP make it a diverse and vital component in host defense.

2.3 *Lysozyme*

Lysozyme is a cationic conserved AMP that belongs to the enzyme class known as glycoside hydrolases. Lysozymes are classified into chicken type (c-type), goose type (g-type), and invertebrate type (i-type) based on their amino acid sequence and biochemical properties (Callewaert and Michiels 2010). Lysozyme is present in majority of the bodily secretions including urine, saliva, milk, and tears (Lesnierowski and Stangierski 2018). Lysozymes are also found on the mucosal surface and in phagocytic cells, including neutrophils, macrophages, and dendritic cells (Pipe 1990). Lysozymes possess strong antibacterial effects against gram-positive bacteria by hydrolyzing the β-1,4-glycosidic bond present on the cell wall, while it has not been very effective against gram-negative bacteria due to its presence of a lipopolysaccharide layer, which prevents it from reaching cell wall (Ferraboschi et al. 2021). Further, lysozyme acts as an immune modulator by releasing pathogen-associated molecular patterns (PAMPs) by the bacteria that activate the immune response by stimulating pro-inflammatory cytokine response and activation of inflammasomes (Ragland and Criss 2017). The extensive presence of this AMP in bodily fluids and its capacity to lyse the bacteria and modulate the immune system makes it a fundamental AMP in fighting against gram-positive bacterial infections.

2.4 *Elafin*

Elafin, also known as trappin-2, belongs to the family of serine protein inhibitors (SERPINs) and is specifically active against the elastin protein (Sallenave 2000). This AMP consists of two domains, namely, the transglutaminase substrate domain (cementoin moiety) and the elastase inhibitor domain (Hiemstra et al. 1996). The expression of elafin was found initially in the skin but soon observed in various tissues including the squamous epithelium of various organs such as the epidermis, large intestine, bronchial and alveolar epithelium, and endometrium. It possesses various functions such as antibacterial (Simpson et al. 2001), anti-inflammatory (Cowan et al. 1996), immunomodulatory (Sallenave et al. 2003), tissue remodeling, and augmentation of adaptive immunity against viruses (Lee et al. 2002). Elafin has a broad spectrum of activity and hence could be used as a potential molecule for modification of inflammatory and immune responses in the host defense.

2.5 *Secretory Leukocyte Protease Inhibitor*

Secretory leukocyte protease inhibitor (SLPI) is a cationic non-glycosylated, acid-stable protein containing 107 amino acids (Thompson and Ohlsson 1986). The *SPLI* gene along with the *elafin* gene constitutes the trappin family genes (Seemuller et al. 1986). SPLI constitutes two domains that are responsible for the antiprotease activity (Sallenave et al. 2003). SPLI is produced by a wide variety of cells such as neutrophils, macrophages (Dallegri and Ottonello 1997), and epithelial cells lining mucous membranes of respiratory, alimentary, and renal tubules (Brown et al. 2004). SPLI is found in secretions such as saliva, synovial fluid, milk, tears, cerebral spinal fluid, nasal and bronchial fluid, seminal fluid, and cervical mucus (Franken et al. 1989; McNeely et al. 1997; Shugars 1999; Pillay et al. 2001; Farquhar et al. 2002). SPLI exhibits antimicrobial activity against bacteria, fungi, and viruses (Tomee et al. 1997; Challacombe and Sweet 2002; Fahey and Wira 2002). Additionally, SPLI also exhibits anti-inflammatory and antiprotease activity (Gipson et al. 1999). SPLI along with elafin has antiprotease activity and is responsible for the protection of tissues from the proteases. Further, SPLI is involved in the protection of the tissues from invasion of microbes, including viral infections.

2.6 Calprotectin

Calprotectin, also known as S100A8/S100A9, is a heterodimer of the calcium-binding proteins (Zackular et al. 2015). S100A8/S100A9 contains two domains for the binding of metals: the first site has a high affinity to zinc, and the second site has an affinity to manganese, ferric, nickel, and zinc (Brophy et al. 2012; Rosen and Nolan 2020). Recently, it has been reported to bind with copper ions (Besold et al. 2018). The activity of this AMP is promoted by an increase in intracellular calcium concentration (Stephan and Nolan 2016). The mechanism by which calprotectin exerts antimicrobial activity is by sequestering the trace minerals from microbes, thus inhibiting its activity (Johnstone and Herzberg 2022). This AMP is predominantly secreted by the neutrophils, monocytes, or apoptotic cells (Fitzgerald and Kagan 2020). The presence of calprotectin in the cytoplasm of neutrophils and epithelial cells restricts the growth of intracellular microbes (Zaia et al. 2009). Apart from antimicrobial action, calprotectin is also involved in immunomodulation by interaction with TLR4 (Ma et al. 2017). Calprotectin is an AMP with diverse properties and interactions that could be vital in host immune defense.

2.7 Cathelicidin

Cathelicidin, similar to defensin, is a mammalian host defense peptide and is a precursor that can release antimicrobial peptides through proteolytic cleavage (Sorensen et al. 2001). Cathelicidin is marked by the presence of the cathelin domain and inhibits cathepsin L (Ritonja et al. 1989). Cleavage of cathelicidin by the serine proteases results in the AMP called LL37 (Sorensen et al. 2001). LL37 is the only antimicrobial peptide found in humans and is a cationic, amphipathic, and helical peptide (Durr et al. 2006). LL37 is expressed throughout the respiratory (Lysenko et al. 2000) and alimentary tract (Fabisiak et al. 2016), leukocytes like neutrophils, monocytes, natural killer cells, T lymphocytes, and B lymphocytes. LL37 regulates the inflammatory process and acts as a chemoattractant (Chertov et al. 1996). The spectrum of activity of LL37 includes antibacterial (gram positive and gram negative), antiviral, antifungal, cytotoxic, wound-healing, and angiogenic properties. Further, this AMP exhibits antiapoptotic properties (Durr et al. 2006). LL37, having a broad spectrum of activity, is an important host defense peptide that is involved in antimicrobial activity as well as an immunomodulatory molecule in mammals.

2.8 Surfactant Protein A (SP-A)

Surfactant protein A belongs to the family of calcium-dependent (C-type) lectins that contain two domains, namely, collagen-like and carbohydrate-binding domains. SP-A is found in bodily secretions such as tears, serum, sputum, amniotic fluid, vaginal and seminal fluids, and synovial fluids (Kishore et al. 2006). SP-A is observed in both the superficial and intermediate layers of the vaginal epithelium, and the concentration varies throughout the menstrual cycle (McNeely et al. 1997). Additionally, SP-A has been involved in host defense due to its ability to detect pathogen-associated molecular patterns (Watson et al. 2019).

2.9 Surfactant Protein D (SP-D)

Similar to SP-A, SP-D is also a C-type lectin that is secreted by epithelial cells and is involved in the inactivation of viruses and possesses phagocytic properties, including opsonization and production of proinflammatory cytokines (Leth-Larsen et al. 2004). This AMP is also involved in the differentiation of various monocyte-derived cells, including dendritic cells (Brinker et al. 2001). The mucosal epithelium of the endometrium, cervical gland, and vaginal squamous epithelium secretes SP-D and is hormonally regulated. Apart from the female reproductive tract, the ovaries itself express this AMP, including theca interna cells and corpus luteum. However, this AMP is absent in the granulosa cells of the ovarian follicle (Leth-Larsen et al. 2004).

2.10 Bacteriocins

Bacteriocins are ribosomal synthesized antimicrobial peptides produced by bacteria and are commonly used in the food industry as preservatives and in starter cultures (Chikindas et al. 2018). Nisin is the first bacteriocin produced by *Lactococcus lactis* (Kleerebezem 2004). Bacteriocins have been classified into different classes, namely, class I or lantibiotics (peptides containing β-lantionine, e.g., nisin and subtilin), class II (heterogeneous class of small thermostable peptides, e.g., pediocin, plantaricin, and lactococcin), and class III (large thermolabile peptides, e.g., helveticin) (Bhattacharya et al. 2022). The presence of the antiboltulinic activity of nisin makes it suitable for use in the food industry (Rilla et al. 2004). The bactericidal activity of bacteriocins is due to its action on cell wall synthesis and formation of membrane pores resulting in leakage of essential components, including K ion, amino acids, and ATP (Breukink et al. 1999). The production of nisin is promoted by nisin itself. Bacteriocins have a spectrum of activity against bacteria, that is, it acts as bacteriostatic and bactericidal. Further, bacteriocins act on viruses and inhibit the proliferation of tumor cells. The lantibiotics are used for the treatment of various medical conditions, including blood pressure, inflammation and allergies, skin infection, mastitis, urogenital infection, and hemolytic uremic syndrome (Balciunas et al. 2013). Additionally, bacteriocins also possess spermicidal action (Sutyak et al. 2008). Bacteriocins being an AMP produced by bacteria can prevent the growth of harmful bacteria and hence could be used in the treatment of various infections.

3 Role of AMPs in Innate Immunity of the Reproductive System

The anatomy of the female reproductive tract is functionally divided into the upper and lower reproductive tract. The vagina and ectocervix constitute the lower reproductive tract, and the endocervix, uterus, and salpinx constitute the upper reproductive tract. The lower reproductive tract is lined by keratinized stratified squamous epithelium, while the upper reproductive tract is lined by columnar epithelium (Hafez 1982). The uterine and vaginal epithelial cell contains intercellular junctions, including adherens junction, tight junction, and desmosomes, which form the barrier to mechanical, chemical, and microbial insult (Blaskewicz et al. 2011). Immune cells are differentially distributed across the female reproductive tract, and hormonal changes influence the secretion of various immune cells, which respond to potential pathogens (Yu et al. 2009). Additionally, the cervicovaginal secretions are rich in AMPs and are involved in defense against pathogens and help in maintaining the uterine environment sterile. These are produced from a variety of cells, including immune cells like neutrophils. AMPs, apart from killing the pathogens directly by acting on them, also involved in cytokine production, chemotaxis, and cell proliferation (Pasupuleti et al. 2012).

AMPs are produced in response to various factors such as infection, inflammation, ischemia, trauma, and other stressors that are recognized by pattern recognition receptors (PRRs) which include pathogen-associated molecular patterns (PAMPs) and damage-associated molecular patterns (DAMPs) (Tang et al. 2012). The most common PRRs involved in the reproductive system include toll-like receptors, Nod-like receptors, Retinoic Acid-Inducible Gene-I (RIG)-like receptors, and C-type lectin receptors (Takeuchi and Akira 2010). Activation of PRRs leads to downstream activation of different transcription factors that specifically increase the expression of AMP genes and produce AMPs that are secreted to the system by a variety of cells, including immune cells (Oeckinghaus et al. 2011).

The AMPs of the reproductive tract are mainly secreted by the epithelial cells and myoepithelial cells of the reproductive tract (Ganz 2003). The factors that influence the activity of AMPs include the phase of the menstrual cycle and the biochemical composition of the reproductive tract (Yanaihara et al. 2004). Additionally, the presence of sperm is also reported to modulate the concentration of AMPs (Wira et al. 2011). AMPs actively involved in the innate immune response of the reproductive tract include defensin, elafin, cathelicidin, secretory leukocyte protease inhibitor (SLPI), lysozyme, and lactoferrin (Yarbrough et al. 2015). In the female reproductive tract, α defensins HNPs 1–4 are secreted by various immune cells, including the azurophilic granules of neutrophils, and HNPs 1–3 are expressed by monocytes, natural killer cells, γδT lymphocytes, eosinophils, and B lymphocytes (Lucovnik et al. 2011). Additionally, HD5 is expressed by Female Reproductive Tract (FRT) from the endometrial, cervical, and vaginal epithelium (Quayle et al. 1998). Beta defensins like HBD2 and HBD3 have been produced in response to the microbial and pro-inflammatory cytokines in epithelial and mucosal tissues

(Presicce et al. 2009). Cathelicidin LL37 is the AMP that is present in cervical and vaginal secretions and is produced by neutrophils and epithelial cells (Di Nardo et al. 2003). SLPI and elafin are protease inhibitors produced from epithelial cells and macrophages. The antiprotease action of these AMPs tends to downgrade the tissue damage and unregulated inflammation that is mediated by proteases. SLPI is very effective against the most common bacteria affecting the reproductive tract, that is, *Escherichia coli*, *Pseudomonas aeruginosa*, and *Staphylococcus aureus* (Hiemstra et al. 1996). Elafin cross-links with the extracellular matrix, including collagen, laminin, elastin, and fibronectin, and aids in maintaining the barrier integrity of the mucosa (Muto et al. 2007). Lactoferrins and lysozyme are secreted from neutrophils, and both AMPs are found in vaginal fluid and cervical mucus (Bals et al. 1998). Lactoferrin interferes with sequestrating iron that is required for the bacteria for its metabolism and survival (Zarzosa-Moreno et al. 2020), and the lysozyme enzymatically affects the bacterial cell membrane by its action on peptidoglycan (Leśnierowski and Yang 2021). Overall, the multidimensional role of AMPs in the reproductive tract is responsible for the maintenance of a sterile environment in the female reproductive tract, thereby facilitating various physiological processes like menstruation and pregnancy.

4 AMPs in Male Reproduction

AMPs similar to the female reproductive system are also present in the male reproductive system. The major host defense peptides in the reproductive system are defensins (Zhai et al. 2023). HNP 1–3 mRNA levels are expressed in the testis, caput, corpus, and cauda epididymis. Additionally, HD5 mRNA is expressed in testes (Com et al. 2003). Among the beta defensins, HDB1 protein is present in the testis, semen, spermatozoa, and prostate gland (Zupin et al. 2019). Apart from acting as anti-infectious agents, beta defensins are involved in spermatogenesis, sperm maturation, sperm movement in the female reproductive tract, sperm reservoir at the point of the oviduct, sperm motility, and capacitation interaction with the oocyte (Batra et al. 2022; Solanki et al. 2023). HDB1 interacts with the chemokine receptor type 6 (CCR6) mediated Cation/Calcium Channel of Sperm (CatSper)-dependent calcium influx that regulates the motility and fertile life of the spermatozoa (Diao et al. 2017). Further, the levels of HDB1 in the fertile men's spermatozoa are higher than the infertile men with asthenospermia or leukospermia. Hence, recent studies have proposed beta defensins as the marker of male fertility (Solanki et al. 2023).

Lysozyme-like gene 4 (LYZL4) protein expression has been observed in the seminiferous epithelium of testis and epididymis, and expression of this protein has been postulated to be involved in sperm motility and fertility (Hsia and Cornwall 2004). Additionally, sperm lysozyme-like protein 1 (SLLP1) is involved in sperm–egg binding and bactericidal action (Mandal et al. 2003). Lactoferrin is a sperm-binding protein that has a role in sperm activity. It is used for the modulation of the protection and regulation of sperm activity in vitro (Kobayashi et al. 2007). Lactoferrin supplementation is beneficial for sperm cryopreservation of stallion sperm. Lactoferrin binds to Fe^{3+} and reduces the lipid peroxidation (de Lamirande et al. 1997).

Scygonadin 2 (SYC2) is dominantly expressed in the ejaculatory ducts of male crabs, is transferred to the female crabs during mating, and is involved in the protection of fertilization by enhancing immunity and maintaining sterility in the spermatheca of the female (Qiao et al. 2016). Semen cryopreservation is widely practiced in assisted reproduction and artificial insemination and during pre-radiation and chemotherapy treatments. Cryopreserved semen is used for IVF, ICSI, and other fertility enhancement treatments (Hughes and Martins da Silva 2022). Semen extenders are commonly used for cryopreservation of semen and used for cryopreservation of semen. Antibiotics are used during the cryopreservation of the semen to prevent microbial contamination (Schulze et al. 2017). Recently, there have been reports that microbes have been resistant to the antibiotics commonly used in extenders. Hence, antimicrobial peptides have been used in the extenders, which include synthetic peptide maganin II, human cationic antimicrobial peptide 18 (hCAP18) that is the precursor for LL38, GL13K, and semenogelin-derived cationic AMP (Vickram et al. 2017). These AMPs have multiple applications, including antibacterial and antiviral activities. As the AMPs are natural peptides and have been secreted by male and female reproductive tracts, it is believed to render the antibiotic functions in the semen extenders.

5 AMPs in Female Reproduction

Antimicrobial peptides apart from antimicrobial activity possess additional functions like anti-inflammatory, anti-tumorigenic, antiapoptotic, and proliferative properties. The role of AMPs has been observed in the menstrual cycle, pregnancy, parturition, prevention of microbial infections, and so on. In this chapter, we will summarize the functions of AMPs in different aspects of female reproduction.

6 AMPs in the Female Reproductive Tract During Menstrual Cycle

The menstrual cycle involves the menses, proliferative phase, and secretory phase. The changes that occur during this phase are due to the hormonal changes that occur due to the changes in the hypothalamic–pituitary–gonadal axis (HPG axis) (Kurita 2011). The female reproductive tract is made under the supportive framework of mesenchymal tissue that contains lamina propria (connective tissue) and muscularis. The epithelial cells of the uterine mucosa are dictated by the bidirectional communication between stroma and epithelium. For instance, the proliferation and differentiation of the uterine epithelial cells are governed by the stimulation of the stromal cells by gonadal hormones, that is, estrogen and progesterone (Cunha et al. 2004).

AMPs present in the female reproductive tract during the different phases of the menstrual cycle have been depicted in Fig. 13.1. The secretory phase of the menstrual cycle creates a window of vulnerability to the reproductive tract due to the suppression of the immune system by the hormonal influence, probably allowing the spermatozoa inside the female reproductive tract to facilitate fertilization, and increases the risk of sexually transmitted infections (Wira and Veronese 2014). To counteract the decreased immune response, antimicrobial peptides are secreted throughout the female reproductive tract, and the concentration of AMPs differs between the phases and has been depicted in Fig. 13.1. All the AMPs observed during the menstrual cycle are increased during most of the phases of the cycle, including HBD2, elafin, lactoferrin, HBD4, calprotectin, SLPI, SP-A, HNPs 1–3, HD5, HBD1, and HBD3. The only exception is SP-A in the lower parts of the reproductive tract during the secretory phase of the cycle (Yarbrough et al. 2015). The concentration of the AMPs SPLI and LL37 was reported to be decreased in menopausal women (Shimoya et al. 2006). The increased expression of lactoferrin during the menses is a response to the shedding of iron-laden RBC (Cohen et al. 1987). Elafin during the menses is involved in tissue regeneration and repair, while S100A8/A9 during both secretory and inhibitory phases helps in the defense against microorganisms (Kunimi et al. 2006). Overall, this evidence suggests the importance of AMPs in the regulation of the menstrual cycle not only to fight against microorganisms but also to remodel the reproductive tract in response to the hormonal changes that occur during the menstrual cycle.

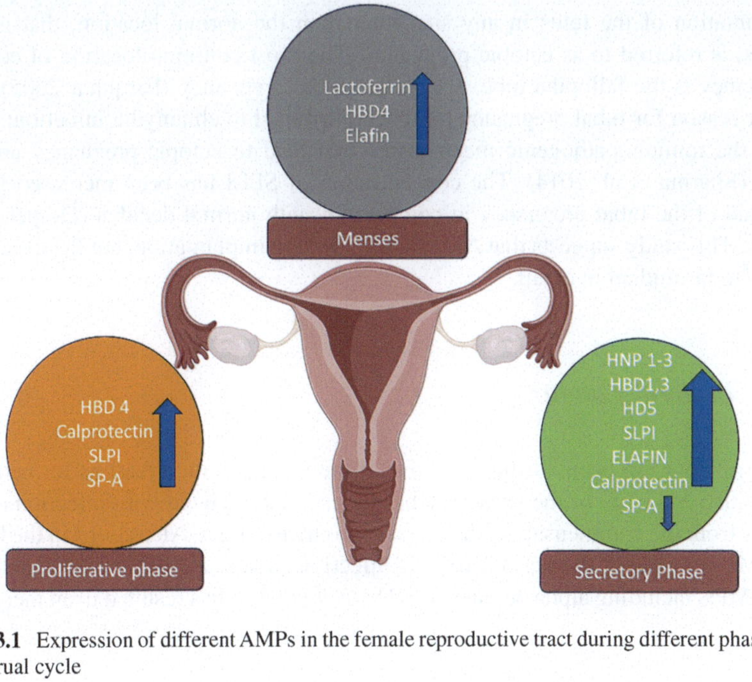

Fig. 13.1 Expression of different AMPs in the female reproductive tract during different phases of menstrual cycle

7 AMPs and Pregnancy

The reproductive tract, especially the upper reproductive tract, should be sterile for the successful establishment of pregnancy. The expression of AMPs during the menses and implantation period suggests their involvement in the maintenance of the sterile female reproductive tract. AMPs are associated with preventing reproductive complications, including ectopic pregnancy and subfertility. Implantation is regarded as a critical process in the establishment of pregnancy. Expression of SPLI during the implantation window reduces the excessive inflammation and facilitates epithelial growth by preventing cleavage of enzyme elastase. Increased expression of alpha defensins during the implantation window is critical for protection against microbial invasion. AMPs are extensively dispersed across the female reproductive tract, fetal tissues, and placenta throughout the pregnancy, indicating they are not only involved in protection against infections but also have diverse roles. These AMPs are involved in protecting the fetus from infection, protecting the skin of the fetus, and remodeling the uterine and placental tissue. Although higher concentrations of AMPs are observed during pregnancy, the role of every AMP in the pregnancy has not been completely elucidated.

8 Pregnancy Complications and AMP

8.1 Ectopic Pregnancy

Implantation of the fetus in any site other than the normal location, that is, the uterus, is referred to as ectopic pregnancy. The most common location of ectopic pregnancy is the fallopian tube, also called tubal pregnancy (Farquhar 2005). The major reason for tubal pregnancy has been attributed to chlamydia infection; however, the immunopathogenic mechanisms that lead to ectopic pregnancy are not clear (Sharma et al. 2014). The concentration of SPLI has been increased in the decidua of the tubal pregnancy in comparison with normal decidua (Draper et al. 1998). This study suggests that AMPs have a role in implantation, but the exact role needs to be studied in detail.

8.2 Preterm Labour

Preterm labor has been attributed as one of the most common reasons for neonatal morbidity. Majority of the preterm labor cases are associated with infection of the uterus from the commensals of the lower reproductive tract. Alterations in the levels of AMPs are one of the reasons for this infection (Son et al. 2021). The increase in the AMPs, including alpha defensins and beta defensins, has resulted in an increased incidence of preterm labor (Para et al. 2020). Similarly, an increased level of elafin was involved in bacterial vaginosis, which is one of the prime reasons for preterm labor. Contrastingly, the decreased concentrations of SPLI and trappin-2 in amniotic fluid are attributed causes of preterm labor (Itaoka et al. 2015). The changes in the proteases during this period may be involved in the remodeling of the collagen in the cervix and result in the production of cytokines and prostaglandins, which cause uterine contraction and lead to preterm labor. Elafin has been observed in higher concentrations in cervical and vaginal secretions of individuals who have short cervix at the 14th week of gestation. The increased concentrations of elafin could be used as a marker for the diagnosis of preterm labor. Overall, these findings suggest the protease inhibitor AMPs, that is, SPLI and elafin, have different roles and mechanisms of action at different locations and are responsible for the maintenance of pregnancy.

9 AMPs Against STDs and Other Infections in the Reproductive System

Sexually transmitted diseases (STDs) refer to those diseases that are transmitted from person to person via sexual contact, which may include sexual intercourse (vaginal or anal) and oral sex. The World Health Organization has reported around one million people around the world are acquiring STDs every day although a majority of them are asymptomatic (Ma et al. 2023). The most common STDs include syphilis, gonorrhea, chlamydia, and trichomoniasis, which are curable, and hepatitis B, herpes simplex virus (HSV), HIV, and human papillomavirus (HPV) are the viral STDs that are currently incurable (Taylor et al. 2022).

Recent studies suggest that AMPs, which belong to the class of antiviral peptides that can inhibit viral replication or can act by various methods to inhibit viral infection, have the potential to cure incurable viral STDs (Madanchi et al. 2020). HIV is a prominent STD that has no therapeutic compounds. A total of 109 HIV AMPs have been reported in the database (http://aps.unmc.edu/AP/database/antiH.php/2019). Advancements in the production of AMP peptide stability and delivery methods elucidate the possibility of these compounds as novel anti-HIV agents and medications (Wang 2012).

Anti-HIV AMPs represent a class of antiviral peptides effective against retroviruses by impeding attachment and fusion, disrupting the viral envelope, and hindering viral replication. Notable peptides with potential roles in inhibiting HIV infection include defensins, LL-37, gramicidin D, caerin1, maximin 3, magainin 2, dermaseptin S1, dermaseptin S4, siamycin I, siamycin II, and RP 71955 (Ogawa et al. 2013; VanCompernolle et al. 2015). LL37, extensively researched for its efficacy against HIV, exhibits diverse mechanisms against the virus. It competes with gp120 for binding to CXCR4 or CXCR5, thereby thwarting viral cell fusion (Vera-Cruz et al. 2021). LL37 also inhibits reverse transcriptase activity, suppressing viral transcription (Madanchi et al. 2020). Moreover, it impedes HIV proteases, disrupting the formation of functional viral proteins. Additionally, LL37 boosts type I interferon production, enhancing antiviral activity (He et al. 2018). Finally, it sequesters anionic lipids in infected cells, leading to the depletion of the viral envelope's lipid content (Zasloff 2002). Enfuvirtide (Fuzeon™) has been commercialized as an anti-HIV drug.

Apart from HIV infection, AMPs have a role in defense against other STDs transmitted by viruses, including herpes simplex virus 2 (HSV-2). AMPs lactoferrin, beta defensin 1, E6-binding peptides, and human alpha defensins 1–3 and 5 are said to be effective against the HPV-causing cervical, cutaneous, and mucosal papilloma viral types (Liu et al. 2004; Mistry et al. 2007). HSV-2 infection of the keratinocytes results in increased production of LL37 in monocyte-derived Langerhans cells (Ogawa et al. 2013). HSV 2 infection has also been attributed as a predisposing factor for HIV infection (Guani-Guerra et al. 2010).

Bacterial STDs include *Neisseria gonorrhoeae*, *Chlamydia psittaci*, *Chlamydia trachomatis*, and syphilis. Among these *Neisseria gonorrhoeae* is common and is a more prevalent infection and fighting against infection is difficult due to the development of multidrug resistance (Bergman et al. 2005). The AMP cell-penetrating peptide (CPPs) and LL37 have been found promising results in vitro, reducing the infection of ME-180 cervical epithelial cells by exhibiting bactericidal activity. Additionally, CPPs reduce the expression of TNF-alpha that was induced in T helper 1 cells by gonococci (Zairi et al. 2007). Chlamydia trachomatis is another organism that has developed antibacterial-resistant strains. AMPs Clavanin-MO and mastoparan-L are found to be infectious bodies of the chlamydia organism. Both AMPs reduce vacuolar degeneration and membrane alterations and are effective against gram-positive and gram-negative organisms (Park et al. 2009). Additionally, Pep-I prevents the growth of both intracellular organisms but not extracellular organisms. Truncated LL37-derived peptide WS22-N-amide is effective against treponema pallidum in vitro. Additionally, this peptide has also been effective against spirochetes (Neshani et al. 2019).

Candidal vaginitis is caused by *Candida albicans* and is one of the most common vaginal infections in the United States. AMPs Aureins 1.2 and 1127, which are derived from AurH1, are effective in the treatment of *C. albicans* infections (Madanchi et al. 2019). AMPs with antifungal activity include aureins, meganin, maximins, brevinine, cecropin, and ranatureins (Sobel 1984).

Trichomoniasis is caused by the organism *Trichomonas vaginalis* and causes cervical infection. It has been reported that the AMP concentrations had been increased during *T. vaginalis* infection (Makinde et al. 2015). The development of metronidazole-resistant *T. vaginalis* strains has a pressing need for alternative therapies. Epinecidin-1 in vitro conditions have been shown to destroy the membrane of *T. vaginalis* a synthetic marine AMP. LL37-derived tryptophan-rich peptide C-amidated tritpticin has also been reported to disrupt the survival and growth of *T. vaginalis* (Huang et al. 2019).

10 Conclusions and Future Perspectives

In conclusion, AMPs represent a compelling frontier in the pursuit of innovative therapies for reproductive system infections. Their innate antimicrobial potency, coupled with their multifaceted biological functions, positions them as promising candidates for addressing the challenges posed by antimicrobial resistance and preserving reproductive health in both men and women. Continued research into the biology and therapeutic potential of AMPs holds the promise of unlocking new strategies for combating reproductive infections and promoting reproductive well-being in the modern era.

References

Actor JK, Hwang SA, Kruzel ML (2009) Lactoferrin as a natural immune modulator. Curr Pharm Des 15:1956–1973. https://doi.org/10.2174/138161209788453202

Ageitos JM, Sanchez-Perez A, Calo-Mata P, Villa TG (2017) Antimicrobial peptides (AMPs): ancient compounds that represent novel weapons in the fight against bacteria. Biochem Pharmacol 133:117–138. https://doi.org/10.1016/j.bcp.2016.09.018

Anastasi E, Farina A, Granato T, Colaiacovo F, Pucci B, Tartaglione S, Angeloni A (2023) Recent insight about HE4 role in ovarian cancer oncogenesis. Int J Mol Sci 24:10479. https://doi.org/10.3390/ijms241310479

Appelmelk BJ, An YQ, Geerts M, Thijs BG, de Boer HA, MacLaren DM, de Graaff J, Nuijens JH (1994) Lactoferrin is a lipid A-binding protein. Infect Immun 62:2628–2632. https://doi.org/10.1128/iai.62.6.2628-2632.1994

Asadi A, Razavi S, Talebi M, Gholami M (2019) A review on anti-adhesion therapies of bacterial diseases. Infection 47:13–23. https://doi.org/10.1007/s15010-018-1222-5

Baker HM, Baker EN (2004) Lactoferrin and iron: structural and dynamic aspects of binding and release. Biometals 17:209–216. https://doi.org/10.1023/b:biom.0000027694.40260.70

Baker EN, Baker HM (2009) A structural framework for understanding the multifunctional character of lactoferrin. Biochimie 91:3–10. https://doi.org/10.1016/j.biochi.2008.05.006

Balciunas EM, Martinez FAC, Todorov SD, de Melo Franco BDG, Converti A, de Souza Oliveira RP (2013) Novel biotechnological applications of bacteriocins: a review. Food Control 32:134–142

Bals R, Wang X, Zasloff M, Wilson JM (1998) The peptide antibiotic LL-37/hCAP-18 is expressed in epithelia of the human lung where it has broad antimicrobial activity at the airway surface. Proc Natl Acad Sci USA 95:9541–9546. https://doi.org/10.1073/pnas.95.16.9541

Batra V, Kumaresan A, Kumar R, Datta T (2022) β-Defensins: antimicrobial peptides at the intersection of immunity and male fertility. In: Current concepts in bovine reproduction. Springer, Singapore, pp 293–317

Bergman P, Johansson L, Asp V, Plant L, Gudmundsson GH, Jonsson AB, Agerberth B (2005) Neisseria gonorrhoeae downregulates expression of the human antimicrobial peptide LL-37. Cell Microbiol 7:1009–1017. https://doi.org/10.1111/j.1462-5822.2005.00530.x

Besold AN, Gilston BA, Radin JN, Ramsoomair C, Culbertson EM, Li CX, Cormack BP, Chazin WJ, Kehl-Fie TE, Culotta VC (2018) Role of calprotectin in withholding zinc and copper from Candida albicans. Infect Immun 86:e00779-17. https://doi.org/10.1128/IAI.00779-17

Bhattacharya D, Nanda PK, Pateiro M, Lorenzo JM, Dhar P, Das AK (2022) Lactic acid bacteria and bacteriocins: novel biotechnological approach for biopreservation of meat and meat products. Microorganisms 10:2058. https://doi.org/10.3390/microorganisms10102058

Bin Hafeez A, Jiang X, Bergen PJ, Zhu Y (2021) Antimicrobial peptides: an update on classifications and databases. Int J Mol Sci 22:11691. https://doi.org/10.3390/ijms222111691

Blaskewicz CD, Pudney J, Anderson DJ (2011) Structure and function of intercellular junctions in human cervical and vaginal mucosal epithelia. Biol Reprod 85:97–104. https://doi.org/10.1095/biolreprod.110.090423

Breukink E, Wiedemann I, van Kraaij C, Kuipers OP, Sahl HG, de Kruijff B (1999) Use of the cell wall precursor lipid II by a pore-forming peptide antibiotic. Science 286:2361–2364. https://doi.org/10.1126/science.286.5448.2361

Brinker KG, Martin E, Borron P, Mostaghel E, Doyle C, Harding CV, Wright JR (2001) Surfactant protein D enhances bacterial antigen presentation by bone marrow-derived dendritic cells. Am J Physiol Lung Cell Mol Physiol 281:L1453–L1463. https://doi.org/10.1152/ajplung.2001.281.6.L1453

Brophy MB, Hayden JA, Nolan EM (2012) Calcium ion gradients modulate the zinc affinity and antibacterial activity of human calprotectin. J Am Chem Soc 134:18089–18100. https://doi.org/10.1021/ja307974e

Brown TI, Mistry R, Collie DD, Tate S, Sallenave JM (2004) Trappin ovine molecule (TOM), the ovine ortholog of elafin, is an acute phase reactant in the lung. Physiol Genomics 19:11–21. https://doi.org/10.1152/physiolgenomics.00113.2004

Callewaert L, Michiels CW (2010) Lysozymes in the animal kingdom. J Biosci 35:127–160. https://doi.org/10.1007/s12038-010-0015-5

Challacombe SJ, Sweet SP (2002) Oral mucosal immunity and HIV infection: current status. Oral Dis 8(Suppl 2):55–62. https://doi.org/10.1034/j.1601-0825.2002.00013.x

Chen S, Jiang M, Ding T, Wang J, Long P (2017) Calprotectin is a potential prognostic marker for polycystic ovary syndrome. Ann Clin Biochem 54:253–257. https://doi.org/10.1177/0004563216653762

Chertov O, Michiel DF, Xu L, Wang JM, Tani K, Murphy WJ, Longo DL, Taub DD, Oppenheim JJ (1996) Identification of defensin-1, defensin-2, and CAP37/azurocidin as T-cell chemoattractant proteins released from interleukin-8-stimulated neutrophils. J Biol Chem 271:2935–2940. https://doi.org/10.1074/jbc.271.6.2935

Chikindas ML, Weeks R, Drider D, Chistyakov VA, Dicks LM (2018) Functions and emerging applications of bacteriocins. Curr Opin Biotechnol 49:23–28. https://doi.org/10.1016/j.copbio.2017.07.011

Chu H, Pazgier M, Jung G, Nuccio SP, Castillo PA, de Jong MF, Winter MG, Winter SE, Wehkamp J, Shen B, Salzman NH, Underwood MA, Tsolis RM, Young GM, Lu W, Lehrer RI, Baumler AJ, Bevins CL (2012) Human alpha-defensin 6 promotes mucosal innate immunity through self-assembled peptide nanonets. Science 337:477–481. https://doi.org/10.1126/science.1218831

Clauss A, Ng V, Liu J, Piao H, Russo M, Vena N, Sheng Q, Hirsch MS, Bonome T, Matulonis U, Ligon AH, Birrer MJ, Drapkin R (2010) Overexpression of elafin in ovarian carcinoma is driven by genomic gains and activation of the nuclear factor kappaB pathway and is associated with poor overall survival. Neoplasia 12:161–172. https://doi.org/10.1593/neo.91542

Cohen MS, Britigan BE, French M, Bean K (1987) Preliminary observations on lactoferrin secretion in human vaginal mucus: variation during the menstrual cycle, evidence of hormonal regulation, and implications for infection with Neisseria gonorrhoeae. Am J Obstet Gynecol 157:1122–1125. https://doi.org/10.1016/s0002-9378(87)80274-0

Com E, Bourgeon F, Evrard B, Ganz T, Colleu D, Jegou B, Pineau C (2003) Expression of antimicrobial defensins in the male reproductive tract of rats, mice, and humans. Biol Reprod 68:95–104. https://doi.org/10.1095/biolreprod.102.005389

Cowan B, Baron O, Crack J, Coulber C, Wilson GJ, Rabinovitch M (1996) Elafin, a serine elastase inhibitor, attenuates post-cardiac transplant coronary arteriopathy and reduces myocardial necrosis in rabbits afer heterotopic cardiac transplantation. J Clin Invest 97:2452–2468. https://doi.org/10.1172/JCI118692

Cunha GR, Cooke PS, Kurita T (2004) Role of stromal-epithelial interactions in hormonal responses. Arch Histol Cytol 67:417–434. https://doi.org/10.1679/aohc.67.417

Dallegri F, Ottonello L (1997) Tissue injury in neutrophilic inflammation. Inflamm Res 46:382–391. https://doi.org/10.1007/s000110050208

de Lamirande E, Jiang H, Zini A, Kodama H, Gagnon C (1997) Reactive oxygen species and sperm physiology. Rev Reprod 2:48–54. https://doi.org/10.1530/ror.0.0020048

Di Nardo A, Vitiello A, Gallo RL (2003) Cutting edge: mast cell antimicrobial activity is mediated by expression of cathelicidin antimicrobial peptide. J Immunol 170:2274–2278. https://doi.org/10.4049/jimmunol.170.5.2274

Diao R, Wang T, Fok KL, Li X, Ruan Y, Yu MK, Cheng Y, Chen Y, Chen H, Mou L, Cai X, Wang Y, Cai Z, Zeng X, Chan HC (2017) CCR6 is required for ligand-induced CatSper activation in human sperm. Oncotarget 8:91445–91458. https://doi.org/10.18632/oncotarget.20651

Draper D, Donohoe W, Mortimer L, Heine RP (1998) Cysteine proteases of Trichomonas vaginalis degrade secretory leukocyte protease inhibitor. J Infect Dis 178:815–819. https://doi.org/10.1086/515366

Durr UH, Sudheendra US, Ramamoorthy A (2006) LL-37, the only human member of the cathelicidin family of antimicrobial peptides. Biochim Biophys Acta 1758:1408–1425. https://doi.org/10.1016/j.bbamem.2006.03.030

Fabisiak A, Murawska N, Fichna J (2016) LL-37: cathelicidin-related antimicrobial peptide with pleiotropic activity. Pharmacol Rep 68:802–808. https://doi.org/10.1016/j.pharep.2016.03.015

Fahey JV, Wira CR (2002) Effect of menstrual status on antibacterial activity and secretory leukocyte protease inhibitor production by human uterine epithelial cells in culture. J Infect Dis 185:1606–1613. https://doi.org/10.1086/340512

Farquhar CM (2005) Ectopic pregnancy. Lancet 366:583–591. https://doi.org/10.1016/S0140-6736(05)67103-6

Farquhar C, VanCott TC, Mbori-Ngacha DA, Horani L, Bosire RK, Kreiss JK, Richardson BA, John-Stewart GC (2002) Salivary secretory leukocyte protease inhibitor is associated with reduced transmission of human immunodeficiency virus type 1 through breast milk. J Infect Dis 186:1173–1176. https://doi.org/10.1086/343805

Ferraboschi P, Ciceri S, Grisenti P (2021) Applications of lysozyme, an innate immune defense factor, as an alternative antibiotic. Antibiotics (Basel) 10:1534. https://doi.org/10.3390/antibiotics10121534

Fitzgerald KA, Kagan JC (2020) Toll-like receptors and the control of immunity. Cell 180:1044–1066. https://doi.org/10.1016/j.cell.2020.02.041

Franken C, Meijer CJ, Dijkman JH (1989) Tissue distribution of antileukoprotease and lysozyme in humans. J Histochem Cytochem 37:493–498. https://doi.org/10.1177/37.4.2926127

Frew L, Makieva S, McKinlay AT, McHugh BJ, Doust A, Norman JE, Davidson DJ, Stock SJ (2014) Human cathelicidin production by the cervix. PLoS One 9:e103434. https://doi.org/10.1371/journal.pone.0103434

Ganz T (2003) Defensins: antimicrobial peptides of innate immunity. Nat Rev Immunol 3:710–720. https://doi.org/10.1038/nri1180

Ganz T (2004) Defensins: antimicrobial peptides of vertebrates. C R Biol 327:539–549. https://doi.org/10.1016/j.crvi.2003.12.007

Gao X, Ding J, Liao C, Xu J, Liu X, Lu W (2021) Defensins: the natural peptide antibiotic. Adv Drug Deliv Rev 179:114008. https://doi.org/10.1016/j.addr.2021.114008

Garcia AE, Osapay G, Tran PA, Yuan J, Selsted ME (2008) Isolation, synthesis, and antimicrobial activities of naturally occurring theta-defensin isoforms from baboon leukocytes. Infect Immun 76:5883–5891. https://doi.org/10.1128/IAI.01100-08

Gipson TS, Bless NM, Shanley TP, Crouch LD, Bleavins MR, Younkin EM, Sarma V, Gibbs DF, Tefera W, McConnell PC, Mueller WT, Johnson KJ, Ward PA (1999) Regulatory effects of endogenous protease inhibitors in acute lung inflammatory injury. J Immunol 162:3653–3662

Guani-Guerra E, Santos-Mendoza T, Lugo-Reyes SO, Teran LM (2010) Antimicrobial peptides: general overview and clinical implications in human health and disease. Clin Immunol 135:1–11. https://doi.org/10.1016/j.clim.2009.12.004

Hafez ES (1982) Structural and ultrastructural parameters of the uterine cervix. Obstet Gynecol Surv 37:507–516. https://doi.org/10.1097/00006254-198208000-00002

He M, Zhang H, Li Y, Wang G, Tang B, Zhao J, Huang Y, Zheng J (2018) Cathelicidin-derived antimicrobial peptides inhibit Zika virus through direct inactivation and interferon pathway. Front Immunol 9:722. https://doi.org/10.3389/fimmu.2018.00722

Hiemstra PS, Maassen RJ, Stolk J, Heinzel-Wieland R, Steffens GJ, Dijkman JH (1996) Antibacterial activity of antileukoprotease. Infect Immun 64:4520–4524. https://doi.org/10.1128/iai.64.11.4520-4524.1996

Hsia N, Cornwall GA (2004) DNA microarray analysis of region-specific gene expression in the mouse epididymis. Biol Reprod 70:448–457. https://doi.org/10.1095/biolreprod.103.021493

Huan Y, Kong Q, Mou H, Yi H (2020) Antimicrobial peptides: classification, design, application and research progress in multiple fields. Front Microbiol 11:582779. https://doi.org/10.3389/fmicb.2020.582779

Huang HN, Chuang CM, Chen JY, Chieh-Yu P (2019) Epinecidin-1: a marine fish antimicrobial peptide with therapeutic potential against Trichomonas vaginalis infection in mice. Peptides 112:139–148. https://doi.org/10.1016/j.peptides.2018.12.004

Hughes G, Martins da Silva S (2022) Sperm cryopreservation for impaired spermatogenesis. Reprod Fertil 4:e220106. https://doi.org/10.1530/RAF-22-0106

Itaoka N, Nagamatsu T, Schust DJ, Ichikawa M, Sayama S, Iwasawa-Kawai Y, Kawana K, Yamashita T, Osuga Y, Fujii T (2015) Cervical expression of elafin and SLPI in pregnancy and their association with preterm labor. Am J Reprod Immunol 73:536–544. https://doi.org/10.1111/aji.12354

Jia J, Duan Q, Guo J, Zheng Y (2014) Psoriasin, a multifunctional player in different diseases. Curr Protein Pept Sci 15:836–842. https://doi.org/10.2174/1389203715081411 28152712

Johnstone KF, Herzberg MC (2022) Antimicrobial peptides: defending the mucosal epithelial barrier. Front Oral Health 3:958480. https://doi.org/10.3389/froh.2022.958480

Kishore U, Greenhough TJ, Waters P, Shrive AK, Ghai R, Kamran MF, Bernal AL, Reid KB, Madan T, Chakraborty T (2006) Surfactant proteins SP-A and SP-D: structure, function and receptors. Mol Immunol 43:1293–1315. https://doi.org/10.1016/j.molimm.2005.08.004

Kleerebezem M (2004) Quorum sensing control of lantibiotic production; nisin and subtilin autoregulate their own biosynthesis. Peptides 25:1405–1414. https://doi.org/10.1016/j.peptides.2003.10.021

Kobayashi J, Suda Y, Takada N, Komine K, Kuroishi T, Kohsaka T, Sasada H, Sato E (2007) 14 motility and fertility of bull spermatozoa frozen in egg yolk extender supplemented with lactoferrin. Reprod Fertil Dev 19:125–126

Kunimi K, Maegawa M, Kamada M, Yamamoto S, Yasui T, Matsuzaki T, Kuwahara A, Furumoto H, Ohmoto Y, Kido H, Irahara M (2006) Myeloid-related protein-8/14 is associated with proinflammatory cytokines in cervical mucus. J Reprod Immunol 71:3–11. https://doi.org/10.1016/j.jri.2005.12.010

Kurita T (2011) Normal and abnormal epithelial differentiation in the female reproductive tract. Differentiation 82:117–126. https://doi.org/10.1016/j.diff.2011.04.008

Lee SK, Lee SS, Hirose S, Park SC, Chi JG, Chung SI, Mori M (2002) Elafin expression in human fetal and adult submandibular glands. Histochem Cell Biol 117:423–430. https://doi.org/10.1007/s00418-002-0405-3

Legrand D, Elass E, Carpentier M, Mazurier J (2005) Lactoferrin: a modulator of immune and inflammatory responses. Cell Mol Life Sci 62:2549–2559. https://doi.org/10.1007/s00018-005-5370-2

Lehrer RI, Lu W (2012) alpha-Defensins in human innate immunity. Immunol Rev 245:84–112. https://doi.org/10.1111/j.1600-065X.2011.01082.x

Lesnierowski G, Stangierski J (2018) What's new in chicken egg research and technology for human health promotion?-a review. Trends Food Sci Technol 71:46–51

Leśnierowski G, Yang T (2021) Lysozyme and its modified forms: a critical appraisal of selected properties and potential. Trends Food Sci Technol 107:333–342

Leth-Larsen R, Floridon C, Nielsen O, Holmskov U (2004) Surfactant protein D in the female genital tract. Mol Hum Reprod 10:149–154. https://doi.org/10.1093/molehr/gah022

Liu Y, Liu Z, Androphy E, Chen J, Baleja JD (2004) Design and characterization of helical peptides that inhibit the E6 protein of papillomavirus. Biochemistry 43:7421–7431. https://doi.org/10.1021/bi049552a

Liu J, Liu ZA, Liu Y, Cheng L, Yan L (2022) Impact of antibiotic treatment for chronic endometritis on pregnancy outcomes in women with reproductive failures (RIF and RPL): a systematic review and meta-analysis. Front Med (Lausanne) 9:980511. https://doi.org/10.3389/fmed.2022.980511

Lucovnik M, Kornhauser-Cerar L, Premru-Srsen T, Gmeiner-Stopar T, Derganc M (2011) Neutrophil defensins but not interleukin-6 in vaginal fluid after preterm premature rupture of membranes predict fetal/neonatal inflammation and infant neurological impairment. Acta Obstet Gynecol Scand 90:908–916. https://doi.org/10.1111/j.1600-0412.2011.01177.x

Lysenko ES, Gould J, Bals R, Wilson JM, Weiser JN (2000) Bacterial phosphorylcholine decreases susceptibility to the antimicrobial peptide LL-37/hCAP18 expressed in the upper respiratory tract. Infect Immun 68:1664–1671. https://doi.org/10.1128/IAI.68.3.1664-1671.2000

Ma L, Sun P, Zhang JC, Zhang Q, Yao SL (2017) Proinflammatory effects of S100A8/A9 via TLR4 and RAGE signaling pathways in BV-2 microglial cells. Int J Mol Med 40:31–38. https://doi.org/10.3892/ijmm.2017.2987

Ma W, Chen Z, Niu S (2023) Advances and challenges in sexually transmitted infections prevention among men who have sex with men in Asia. Curr Opin Infect Dis 36:26–34. https://doi.org/10.1097/QCO.0000000000000892

Madanchi H, Khalaj V, Jang S, Shabani AA, Ebrahimi Kiasari R, Seyed Mousavi SJ, Kazemi Sealani S, Sardari S (2019) AurH1: a new heptapeptide derived from Aurein1.2 antimicrobial peptide with specific and exclusive fungicidal activity. J Pept Sci 25:e3175. https://doi.org/10.1002/psc.3175

Madanchi H, Shoushtari M, Kashani HH, Sardari S (2020) Antimicrobial peptides of the vaginal innate immunity and their role in the fight against sexually transmitted diseases. New Microbes New Infect 34:100627. https://doi.org/10.1016/j.nmni.2019.100627

Makinde H, Novak R, Landay A, Spear G (2015) Increased expression of antimicrobial peptides in the cervix during Trichomonas vaginalis infection (MUC2P.923). J Immunol 194:65.6

Mandal A, Klotz KL, Shetty J, Jayes FL, Wolkowicz MJ, Bolling LC, Coonrod SA, Black MB, Diekman AB, Haystead TA, Flickinger CJ, Herr JC (2003) SLLP1, a unique, intra-acrosomal, non-bacteriolytic, c lysozyme-like protein of human spermatozoa. Biol Reprod 68:1525–1537. https://doi.org/10.1095/biolreprod.102.010108

Mårdh PA, Colleen S (1974) Lysozyme in seminal fluid of healthy males and patients with prostatitis and in tissues of the male uro-genital tract. Scand J Urol Nephrol 8:179–183. https://doi.org/10.3109/00365597409132126

McNeely TB, Shugars DC, Rosendahl M, Tucker C, Eisenberg SP, Wahl SM (1997) Inhibition of human immunodeficiency virus type 1 infectivity by secretory leukocyte protease inhibitor occurs prior to viral reverse transcription. Blood 90:1141–1149

Mistry N, Drobni P, Naslund J, Sunkari VG, Jenssen H, Evander M (2007) The anti-papillomavirus activity of human and bovine lactoferricin. Antivir Res 75:258–265. https://doi.org/10.1016/j.antiviral.2007.03.012

Mousavi Maleki MS, Sardari S, Ghandehari Alavijeh A, Madanchi H (2023) Recent patents and FDA-approved drugs based on antiviral peptides and other peptide-related antivirals. Int J Pept Res Ther 29:5. https://doi.org/10.1007/s10989-022-10477-z

Muto J, Kuroda K, Wachi H, Hirose S, Tajima S (2007) Accumulation of elafin in actinic elastosis of sun-damaged skin: elafin binds to elastin and prevents elastolytic degradation. J Invest Dermatol 127:1358–1366. https://doi.org/10.1038/sj.jid.5700647

Neshani A, Zare H, Akbari Eidgahi MR, Khaledi A, Ghazvini K (2019) Epinecidin-1, a highly potent marine antimicrobial peptide with anticancer and immunomodulatory activities. BMC Pharmacol Toxicol 20:33. https://doi.org/10.1186/s40360-019-0309-7

Nugteren S, Samsom JN (2021) Secretory Leukocyte Protease Inhibitor (SLPI) in mucosal tissues: protects against inflammation, but promotes cancer. Cytokine Growth Factor Rev 59:22–35. https://doi.org/10.1016/j.cytogfr.2021.01.005

Oeckinghaus A, Hayden MS, Ghosh S (2011) Crosstalk in NF-kappaB signaling pathways. Nat Immunol 12:695–708. https://doi.org/10.1038/ni.2065

Ogawa Y, Kawamura T, Matsuzawa T, Aoki R, Gee P, Yamashita A, Moriishi K, Yamasaki K, Koyanagi Y, Blauvelt A, Shimada S (2013) Antimicrobial peptide LL-37 produced by HSV-2-infected keratinocytes enhances HIV infection of Langerhans cells. Cell Host Microbe 13:77–86. https://doi.org/10.1016/j.chom.2012.12.002

Para R, Romero R, Miller D, Panaitescu B, Varrey A, Chaiworapongsa T, Hassan SS, Hsu CD, Gomez-Lopez N (2020) Human beta-defensin-3 participates in intra-amniotic host defense in women with labor at term, spontaneous preterm labor and intact membranes, and preterm prelabor rupture of membranes. J Matern Fetal Neonatal Med 33:4117–4132. https://doi.org/10.1080/14767058.2019.1597047

Park N, Yamanaka K, Tran D, Chandrangsu P, Akers JC, de Leon JC, Morrissette NS, Selsted ME, Tan M (2009) The cell-penetrating peptide, Pep-1, has activity against intracellular chlamydial growth but not extracellular forms of Chlamydia trachomatis. J Antimicrob Chemother 63:115–123. https://doi.org/10.1093/jac/dkn436

Pasupuleti M, Schmidtchen A, Malmsten M (2012) Antimicrobial peptides: key components of the innate immune system. Crit Rev Biotechnol 32:143–171. https://doi.org/10.3109/07388551.2011.594423

Pazgier M, Wei G, Ericksen B, Jung G, Wu Z, de Leeuw E, Yuan W, Szmacinski H, Lu WY, Lubkowski J, Lehrer RI, Lu W (2012) Sometimes it takes two to tango: contributions of dimerization to functions of human alpha-defensin HNP1 peptide. J Biol Chem 287:8944–8953. https://doi.org/10.1074/jbc.M111.332205

Pillay K, Coutsoudis A, Agadzi-Naqvi AK, Kuhn L, Coovadia HM, Janoff EN (2001) Secretory leukocyte protease inhibitor in vaginal fluids and perinatal human immunodeficiency virus type 1 transmission. J Infect Dis 183:653–656. https://doi.org/10.1086/318535

Pipe RK (1990) Hydrolytic enzymes associated with the granular haemocytes of the marine mussel Mytilus edulis. Histochem J 22:595–603. https://doi.org/10.1007/BF01072941

Presicce P, Giannelli S, Taddeo A, Villa ML, Della Bella S (2009) Human defensins activate monocyte-derived dendritic cells, promote the production of proinflammatory cytokines, and up-regulate the surface expression of CD91. J Leukoc Biol 86:941–948. https://doi.org/10.1189/jlb.0708412

Qiao K, Xu WF, Chen HY, Peng H, Zhang YQ, Huang WS, Wang SP, An Z, Shan ZG, Chen FY, Wang KJ (2016) A new antimicrobial peptide SCY2 identified in Scylla paramamosain exerting a potential role of reproductive immunity. Fish Shellfish Immunol 51:251–262. https://doi.org/10.1016/j.fsi.2016.02.022

Quayle AJ, Porter EM, Nussbaum AA, Wang YM, Brabec C, Yip KP, Mok SC (1998) Gene expression, immunolocalization, and secretion of human defensin-5 in human female reproductive tract. Am J Pathol 152:1247–1258

Ragland SA, Criss AK (2017) From bacterial killing to immune modulation: recent insights into the functions of lysozyme. PLoS Pathog 13:e1006512. https://doi.org/10.1371/journal.ppat.1006512

Rilla N, Martinez B, Rodriguez A (2004) Inhibition of a methicillin-resistant Staphylococcus aureus strain in Afuega'l Pitu cheese by the nisin Z-producing strain Lactococcus lactis subsp. lactis IPLA 729. J Food Prot 67:928–933. https://doi.org/10.4315/0362-028x-67.5.928

Ritonja A, Kopitar M, Jerala R, Turk V (1989) Primary structure of a new cysteine proteinase inhibitor from pig leucocytes. FEBS Lett 255:211–214. https://doi.org/10.1016/0014-5793(89)81093-2

Rosa L, Cutone A, Lepanto MS, Paesano R, Valenti P (2017) Lactoferrin: a natural glycoprotein involved in iron and inflammatory homeostasis. Int J Mol Sci 18:1985. https://doi.org/10.3390/ijms18091985

Rosen T, Nolan EM (2020) Metal sequestration and antimicrobial activity of human calprotectin are pH-dependent. Biochemistry 59:2468–2478. https://doi.org/10.1021/acs.biochem.0c00359
Sallenave JM (2000) The role of secretory leukocyte proteinase inhibitor and elafin (elastase-specific inhibitor/skin-derived antileukoprotease) as alarm antiproteinases in inflammatory lung disease. Respir Res 1:87–92. https://doi.org/10.1186/rr18
Sallenave JM, Cunningham GA, James RM, McLachlan G, Haslett C (2003) Regulation of pulmonary and systemic bacterial lipopolysaccharide responses in transgenic mice expressing human elafin. Infect Immun 71:3766–3774. https://doi.org/10.1128/IAI.71.7.3766-3774.2003
Schulze M, Grobbel M, Riesenbeck A, Bruning S, Schaefer J, Jung M, Grossfeld R (2017) Dose rates of antimicrobial substances in boar semen preservation-time to establish new protocols. Reprod Domest Anim 52:397–402. https://doi.org/10.1111/rda.12921
Seemuller U, Arnhold M, Fritz H, Wiedenmann K, Machleidt W, Heinzel R, Appelhans H, Gassen HG, Lottspeich F (1986) The acid-stable proteinase inhibitor of human mucous secretions (HUSI-I, antileukoprotease). Complete amino acid sequence as revealed by protein and cDNA sequencing and structural homology to whey proteins and Red Sea turtle proteinase inhibitor. FEBS Lett 199:43–48. https://doi.org/10.1016/0014-5793(86)81220-0
Selsted ME, Ouellette AJ (2005) Mammalian defensins in the antimicrobial immune response. Nat Immunol 6:551–557. https://doi.org/10.1038/ni1206
Selsted ME, Harwig SS, Ganz T, Schilling JW, Lehrer RI (1985) Primary structures of three human neutrophil defensins. J Clin Invest 76:1436–1439. https://doi.org/10.1172/JCI112121
Sharma H, Tal R, Clark NA, Segars JH (2014) Microbiota and pelvic inflammatory disease. Semin Reprod Med 32:43–49. https://doi.org/10.1055/s-0033-1361822
Shimoya K, Zhang Q, Temma K, Kimura T, Tsujie T, Tsutsui T, Wasada K, Kanzaki T, Koyama M, Murata Y (2006) Secretory leukocyte protease inhibitor levels in cervicovaginal secretion of elderly women. Maturitas 54:141–148. https://doi.org/10.1016/j.maturitas.2004.02.019
Shugars DC (1999) Endogenous mucosal antiviral factors of the oral cavity. J Infect Dis 179(Suppl 3):S431–S435. https://doi.org/10.1086/314799
Simpson AJ, Wallace WA, Marsden ME, Govan JR, Porteous DJ, Haslett C, Sallenave JM (2001) Adenoviral augmentation of elafin protects the lung against acute injury mediated by activated neutrophils and bacterial infection. J Immunol 167:1778–1786. https://doi.org/10.4049/jimmunol.167.3.1778
Sobel JD (1984) Vulvovaginal candidiasis--what we do and do not know. Ann Intern Med 101:390–392. https://doi.org/10.7326/0003-4819-101-3-390
Solanki S, Kumar V, Kashyap P, Kumar R, De S, Datta TK (2023) Beta-defensins as marker for male fertility: a comprehensive reviewdagger. Biol Reprod 108:52–71. https://doi.org/10.1093/biolre/ioac197
Son GH, Lee JJ, Kim Y, Lee KY (2021) The role of antimicrobial peptides in preterm birth. Int J Mol Sci 22:8905. https://doi.org/10.3390/ijms22168905
Sorensen OE, Follin P, Johnsen AH, Calafat J, Tjabringa GS, Hiemstra PS, Borregaard N (2001) Human cathelicidin, hCAP-18, is processed to the antimicrobial peptide LL-37 by extracellular cleavage with proteinase 3. Blood 97:3951–3959. https://doi.org/10.1182/blood.v97.12.3951
Sotiriadis G, Dodagatta-Marri E, Kouser L, Alhamlan FS, Kishore U, Karteris E (2015) Surfactant proteins SP-A and SP-D modulate uterine contractile events in ULTR myometrial cell line. PLoS One 10:e0143379. https://doi.org/10.1371/journal.pone.0143379
Stephan JR, Nolan EM (2016) Calcium-induced tetramerization and zinc chelation shield human calprotectin from degradation by host and bacterial extracellular proteases. Chem Sci 7:1962–1975. https://doi.org/10.1039/C5SC03287C
Stock SJ, Duthie L, Tremaine T, Calder AA, Kelly RW, Riley SC (2009) Elafin (SKALP/Trappin-2/proteinase inhibitor-3) is produced by the cervix in pregnancy and cervicovaginal levels are diminished in bacterial vaginosis. Reprod Sci 16:1125–1134. https://doi.org/10.1177/1933719109341998
Sutyak KE, Anderson RA, Dover SE, Feathergill KA, Aroutcheva AA, Faro S, Chikindas ML (2008) Spermicidal activity of the safe natural antimicrobial peptide subtilosin. Infect Dis Obstet Gynecol 2008:540758. https://doi.org/10.1155/2008/540758

Takeuchi O, Akira S (2010) Pattern recognition receptors and inflammation. Cell 140:805–820. https://doi.org/10.1016/j.cell.2010.01.022

Tang YQ, Yuan J, Osapay G, Osapay K, Tran D, Miller CJ, Ouellette AJ, Selsted ME (1999) A cyclic antimicrobial peptide produced in primate leukocytes by the ligation of two truncated alpha-defensins. Science 286:498–502. https://doi.org/10.1126/science.286.5439.498

Tang D, Kang R, Coyne CB, Zeh HJ, Lotze MT (2012) PAMPs and DAMPs: signal 0s that spur autophagy and immunity. Immunol Rev 249:158–175. https://doi.org/10.1111/j.1600-065X.2012.01146.x

Taylor MM, Wi T, Gerbase A, Thwin SS, Gottlieb S, Babovic MT, Low-Beer D, Alonso M, Mello MB, Ishikawa N, Brink A, Hermez J, Sabry A, Sanni S, Ouedraogo L, Rewari B, Sharma M, Seguy N, Vovc E, Askew I, Doherty M, Broutet N (2022) Assessment of country implementation of the WHO global health sector strategy on sexually transmitted infections (2016-2021). PLoS One 17:e0263550. https://doi.org/10.1371/journal.pone.0263550

Thompson RC, Ohlsson K (1986) Isolation, properties, and complete amino acid sequence of human secretory leukocyte protease inhibitor, a potent inhibitor of leukocyte elastase. Proc Natl Acad Sci USA 83:6692–6696. https://doi.org/10.1073/pnas.83.18.6692

Tomee JF, Hiemstra PS, Heinzel-Wieland R, Kauffman HF (1997) Antileukoprotease: an endogenous protein in the innate mucosal defense against fungi. J Infect Dis 176:740–747. https://doi.org/10.1086/514098

Tsuda H, Fukamachi K, Xu J, Sekine K, Ohkubo S, Takasuka N, Iigo M (2006) Prevention of carcinogenesis and cancer metastasis by bovine lactoferrin. Proc Jpn Acad Ser B Phys Biol Sci 82:208–215. https://doi.org/10.2183/pjab.82.208

Valdez-Miramontes CE, De Haro-Acosta J, Arechiga-Flores CF, Verdiguel-Fernandez L, Rivas-Santiago B (2021) Antimicrobial peptides in domestic animals and their applications in veterinary medicine. Peptides 142:170576. https://doi.org/10.1016/j.peptides.2021.170576

Van Snick JL, Masson PL, Heremans JF (1974) The involvement of lactoferrin in the hyposideremia of acute inflammation. J Exp Med 140:1068–1084. https://doi.org/10.1084/jem.140.4.1068

VanCompernolle S, Smith PB, Bowie JH, Tyler MJ, Unutmaz D, Rollins-Smith LA (2015) Inhibition of HIV infection by caerin 1 antimicrobial peptides. Peptides 71:296–303. https://doi.org/10.1016/j.peptides.2015.05.004

Vera-Cruz A, Tanphaichitr N, Angel JB (2021) Antimicrobial peptide, LL-37, and its potential as an anti-HIV agent. Clin Invest Med 44:E64–E71. https://doi.org/10.25011/cim.v44i3.36657

Vickram A, Kuldeep D, Archana K, Parameswari R, Pathy MR, Iqbal H, Sridharan T (2017) Antimicrobial peptides in semen extenders: a valuable replacement option for antibiotics in cryopreservation-a prospective review. J Exp Biol Agric Sci 5:578–588

Wang G (2012) Natural antimicrobial peptides as promising anti-HIV candidates. Curr Top Pept Protein Res 13:93–110

Wang G, Li X, Wang Z (2016) APD3: the antimicrobial peptide database as a tool for research and education. Nucleic Acids Res 44:D1087–D1093. https://doi.org/10.1093/nar/gkv1278

Wassarman PM (1987) The biology and chemistry of fertilization. Science 235:553–560. https://doi.org/10.1126/science.3027891

Watson A, Phipps MJS, Clark HW, Skylaris CK, Madsen J (2019) Surfactant proteins a and D: trimerized innate immunity proteins with an affinity for viral fusion proteins. J Innate Immun 11:13–28. https://doi.org/10.1159/000492974

Wilmes M, Stockem M, Bierbaum G, Schlag M, Gotz F, Tran DQ, Schaal JB, Ouellette AJ, Selsted ME, Sahl HG (2014) Killing of staphylococci by theta-defensins involves membrane impairment and activation of autolytic enzymes. Antibiotics (Basel) 3:617–631. https://doi.org/10.3390/antibiotics3040617

Wira CR, Veronese F (2014) Hormone regulation of the mucosal environment in the reproductive tract and the prevention of HIV infection. Am J Reprod Immunol 71:487–489. https://doi.org/10.1111/aji.12269

Wira CR, Patel MV, Ghosh M, Mukura L, Fahey JV (2011) Innate immunity in the human female reproductive tract: endocrine regulation of endogenous antimicrobial protection against HIV

and other sexually transmitted infections. Am J Reprod Immunol 65:196–211. https://doi.org/10.1111/j.1600-0897.2011.00970.x
Xu D, Lu W (2020) Defensins: a double-edged sword in host immunity. Front Immunol 11:764. https://doi.org/10.3389/fimmu.2020.00764
Yanaihara A, Otsuka Y, Iwasaki S, Koide K, Aida T, Okai T (2004) Comparison in gene expression of secretory human endometrium using laser microdissection. Reprod Biol Endocrinol 2:66. https://doi.org/10.1186/1477-7827-2-66
Yarbrough VL, Winkle S, Herbst-Kralovetz MM (2015) Antimicrobial peptides in the female reproductive tract: a critical component of the mucosal immune barrier with physiological and clinical implications. Hum Reprod Update 21:353–377. https://doi.org/10.1093/humupd/dmu065
Yu L, Wang L, Chen S (2009) Toll-like receptors, inflammation and tumor in the human female reproductive tract. Am J Reprod Immunol 62:1–8. https://doi.org/10.1111/j.1600-0897.2009.00712.x
Zackular JP, Chazin WJ, Skaar EP (2015) Nutritional immunity: S100 proteins at the host-pathogen interface. J Biol Chem 290:18991–18998. https://doi.org/10.1074/jbc.R115.645085
Zaia AA, Sappington KJ, Nisapakultorn K, Chazin WJ, Dietrich EA, Ross KF, Herzberg MC (2009) Subversion of antimicrobial calprotectin (S100A8/S100A9 complex) in the cytoplasm of TR146 epithelial cells after invasion by Listeria monocytogenes. Mucosal Immunol 2:43–53. https://doi.org/10.1038/mi.2008.63
Zairi A, Tangy F, Ducos-Galand M, Alonso JM, Hani K (2007) Susceptibility of Neisseria gonorrhoeae to antimicrobial peptides from amphibian skin, dermaseptin, and derivatives. Diagn Microbiol Infect Dis 57:319–324. https://doi.org/10.1016/j.diagmicrobio.2006.11.006
Zakizadeh F, Mahmoudzadeh-Sagheb H, Asemi-Rad A, Ghasemi M, Moudi B, Sheibak N, Asadikalameh Z, Heidari Z (2020) Upregulation of elafin expression in the fallopian tube of ectopic tubal pregnancies compared to the normal tubes. J Reprod Immunol 141:103136. https://doi.org/10.1016/j.jri.2020.103136
Zarzosa-Moreno D, Avalos-Gomez C, Ramirez-Texcalco LS, Torres-Lopez E, Ramirez-Mondragon R, Hernandez-Ramirez JO, Serrano-Luna J, de la Garza M (2020) Lactoferrin and its derived peptides: an alternative for combating virulence mechanisms developed by pathogens. Molecules 25:5763. https://doi.org/10.3390/molecules25245763
Zasloff M (2002) Antimicrobial peptides of multicellular organisms. Nature 415:389–395
Zhai YJ, Feng Y, Ma X, Ma F (2023) Defensins: defenders of human reproductive health. Hum Reprod Update 29:126–154. https://doi.org/10.1093/humupd/dmac032
Zupin L, Polesello V, Martinelli M, Luppi S, Giolo E, Zito G, Romano F, Segat L, Crovella S, Ricci G (2019) Human beta-defensin 1 in follicular fluid and semen: impact on fertility. J Assist Reprod Genet 36:787–797. https://doi.org/10.1007/s10815-019-01409-w

Index

P. Baindara, S. M. Mandal (eds.), *Evolution of Antimicrobial Peptides*,
https://doi.org/10.1007/978-3-031-67515-7

SPRINGER NATURE

GPSR Compliance

The European Union's (EU) General Product Safety Regulation (GPSR) is a set of rules that requires consumer products to be safe and our obligations to ensure this.

If you have any concerns about our products, you can contact us on ProductSafety@springernature.com

In case Publisher is established outside the EU, the EU authorized representative is:

Springer Nature Customer Service Center GmbH
Europaplatz 3
69115 Heidelberg, Germany

The manufacturer's authorised representative in the EU is Springer Nature Customer Service Centre GmbH, Europaplatz 3, 69115 Heidelberg, Germany. If you have any concerns regarding our products, please contact ProductSafety@springernature.com

Printed and bound by CPI Group (UK) Ltd, Croydon, CR0 4YY
30/09/2026
02225598-0010